Climate System Modelling

Climate System Modelling

Editor: Bruce Mullan

R CALLISTO REFERENCE

www.callistoreference.com

Callisto Reference,
118-35 Queens Blvd., Suite 400,
Forest Hills, NY 11375, USA

Visit us on the World Wide Web at:
www.callistoreference.com

ISBN: 978-1-63239-852-9 (Hardback)

Cataloging-in-publication Data

Climate system modelling / edited by Bruce Mullan.
   p. cm.
Includes bibliographical references and index.
ISBN 978-1-63239-852-9
1. Bioclimatology. 2. Biodiversity--Climatic factors. 3. Climatic changes. I. Mullan, Bruce.
QH543 .C55 2017
551.6--dc23

Table of Contents

Preface

Climate Models aim to study the varied aspects of climate such as atmosphere, ice, oceans, etc. through quantitative methods to better understand our present climate as well as make future predictions. From theories to research to practical applications, case studies related to all contemporary topics of relevance to this field have been included in this book. This book is a valuable compilation of topics, ranging from the basic to the most complex advancements in the field. The text elaborately discusses the various models that are used for climate modeling. It aims to equip students and experts with the advance topics and upcoming concepts in this area. Through this book, we attempt to further enlighten the readers about the new concepts in this field.

This book has been a concerted effort by a group of academicians, researchers and scientists, who have contributed their research works for the realization of the book. This book has materialized in the wake of emerging advancements and innovations in this field. Therefore, the need of the hour was to compile all the required researches and disseminate the knowledge to a broad spectrum of people comprising of students, researchers and specialists of the field.

At the end of the preface, I would like to thank the authors for their brilliant chapters and the publisher for guiding us all-through the making of the book till its final stage. Also, I would like to thank my family for providing the support and encouragement throughout my academic career and research projects.

Editor

Vulnerability of Breeding Waterbirds to Climate Change in the Prairie Pothole Region, U.S.A.

Valerie Steen[1,2,3]*, Susan K. Skagen[1], Barry R. Noon[2,3]

1 United States Geological Survey, Fort Collins Science Center, Fort Collins, Colorado, United States of America, 2 Department of Fish, Wildlife, and Conservation Biology, Colorado State University, Fort Collins, Colorado, United States of America, 3 Graduate Degree Program in Ecology, Colorado State University, Fort Collins, Colorado, United States of America

Abstract

The Prairie Pothole Region (PPR) of the north-central U.S. and south-central Canada contains millions of small prairie wetlands that provide critical habitat to many migrating and breeding waterbirds. Due to their small size and the relatively dry climate of the region, these wetlands are considered at high risk for negative climate change effects as temperatures increase. To estimate the potential impacts of climate change on breeding waterbirds, we predicted current and future distributions of species common in the PPR using species distribution models (SDMs). We created regional-scale SDMs for the U.S. PPR using Breeding Bird Survey occurrence records for 1971–2011 and wetland, upland, and climate variables. For each species, we predicted current distribution based on climate records for 1981–2000 and projected future distributions to climate scenarios for 2040–2049. Species were projected to, on average, lose almost half their current habitat (-46%). However, individual species projections varied widely, from +8% (Upland Sandpiper) to -100% (Wilson's Snipe). Variable importance ranks indicated that land cover (wetland and upland) variables were generally more important than climate variables in predicting species distributions. However, climate variables were relatively more important during a drought period. Projected distributions of species responses to climate change contracted within current areas of distribution rather than shifting. Given the large variation in species-level impacts, we suggest that climate change mitigation efforts focus on species projected to be the most vulnerable by enacting targeted wetland management, easement acquisition, and restoration efforts.

Editor: R. Mark Brigham, University of Regina, Canada

Funding: We thank the U.S. Geological Survey, National Climate Change and Wildlife Science Center (NCCWSC) and the Plains and Prairie Pothole Landscape Cooperative (PPPLCC) for research funding. The funders had no role in study design, data collection and analysis, decision to publish, or preparation of the manuscript.

Competing Interests: The authors have declared that no competing interests exist.

* E-mail: valerie.steen@gmail.com

Introduction

The Prairie Pothole Region of north-central North America (central Iowa, U.S.A. to central Alberta, Canada; 900,000 km^2) contains one of the largest wetland areas (40,000 km^2) in the world [1]. Historically, most conservation activities have focused on sustaining extensive, high quality duck habitat because of the associated recreational value of duck-hunting across the U.S. [2]. Increasingly, emphasis is being placed on the diversity of ecosystem services offered by prairie pothole wetlands, including carbon sequestration, flood control, groundwater recharge, water quality improvement, and biodiversity [2]. This includes increasing attention to all 115 species of breeding or migrating waterbirds that depend on the region [3].

Successful management of species requires knowledge of habitat preferences. Strategic management of species also requires identifying those species most vulnerable to future threats. Land conversion continues to be a direct threat to waterbird habitat, but climate change will likely exacerbate loss and interact with changes in land cover. Climate models for the Prairie Pothole Region project increasing temperatures and slight or no increases in precipitation, indicating drier conditions affecting hydroperiods, and the extent and quality of wetland habitat [4,5].

Prairie pothole wetlands are susceptible to climatic variation through impacts on wetland hydroperiod, vegetative condition, and water depth in combination with static factors such as basin size [5]. Well-documented causal relations between past variability in wetland condition and extent and waterbird numbers provide insights to future change in waterbird populations under climate change. In dry years, with fewer wet basins, breeding populations of waterbirds are significantly reduced [6,7]. Building on these causal relations, Sorenson et al. [8] projected population changes for waterfowl under future warming scenarios. Their projections indicated that by 2060 duck populations would be half of their current level. Johnson et al. [5] used mechanistic models relating climate to marsh vegetation dynamics, and projected that the Prairie Potholes in North and South Dakota will be too dry to produce suitable wetland vegetative conditions for breeding ducks in the future.

To address how climate change may impact waterbirds in the Prairie Pothole Region, we created empirically-based species distribution models for a focal group of breeding wetland-associated birds. We related bird occurrence (presence/absence) to climate and land cover predictors. As a species' occurrence varies from year to year in response to dynamic wetland conditions, we used multiple years of bird survey data across 41

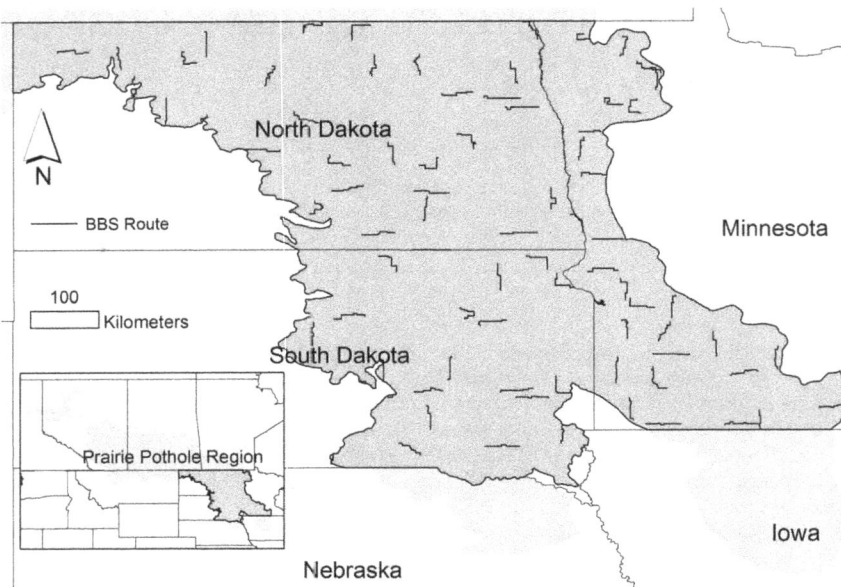

Figure 1. Bird occurrence data were obtained from 77 Breeding Bird Survey (BBS) routes throughout the Prairie Pothole Region (PPR) of North Dakota, South Dakota, and Minnesota. Climate-based projections were also made to the PPR of Iowa.

years, a period that included years of drought and years of heavy precipitation. Although we did not explicitly model wetland condition, we used Random Forests, an ensemble decision tree approach which can capture the interactions between climate variability and the state of wetland basins that drive wetland condition [9]. We projected future waterbird occurrence using species distribution models and future climate projections. To assess how climate change may reduce or expand current suitable habitat, for each species we compared the projections of future distribution to their predicted current distribution, and produced a quantitative estimate of how much habitat would be lost or gained under various climate change scenarios. Additionally, we compared our future projections of waterbird species response to a historic dry period.

Methods

Study Area

The study area (320,000 km^2) was the 45% of the Prairie Pothole Region within four U.S. states (North Dakota, South Dakota, Minnesota, Iowa, Figure 1). The study was restricted to the four states because of available and consistent land cover and downscaled climate data. We excluded Iowa from training the model because too few wetlands remain there to usefully inform the species distribution model, although we did include it in model predictions.

Water-filled glacial depressions termed *potholes* are characteristic of this region and can reach densities greater than 40 per km^2 [10]. Since European settlement, these wetlands have been extensively converted to cropland, with wetland losses greatest in

Table 1. Thirty-one climate and land cover variables used in species distribution models.

Climate		Land Cover	
Temperature	Precipitation	Wetland	Upland
Spring (spr)	Spring	Temporary (temp)	Cropland (crop)
Winter (wint)	Winter	Seasonal (seas)	Grassland (grass)
Fall	Fall	Semipermanent (semi)	Developed (devel)
Summer (sum)	Summer	Lake	Tree
Yearly (1yr)	Yearly	River	
5-year (5yr)	5-year	Shrub	
10-year (10yr)	10-year	Forested (forest)	
5-year std. dev. (5yr_sd)	5-year std. dev.	Total palustrine (pal)	
10-year std. dev. (10yr_sd)	10-year std. dev.	Total	

Temperature calculations were based on averages, while precipitation calculations were based on totals. Land cover variables were based on composition (proportion of total) of that cover type in the landscape. Wetland land cover was apportioned by wetland regime. Total palustrine wetland summed temporary, seasonal, and semipermanent wetlands. Total wetland summed all wetland regimes. Cropland described land planted with crops or fallowed. Grassland included native prairie, conservation reserve program (CRP) land, and hayland.

Vulnerability of Breeding Waterbirds to Climate Change in the Prairie Pothole Region, U.S.A.

3

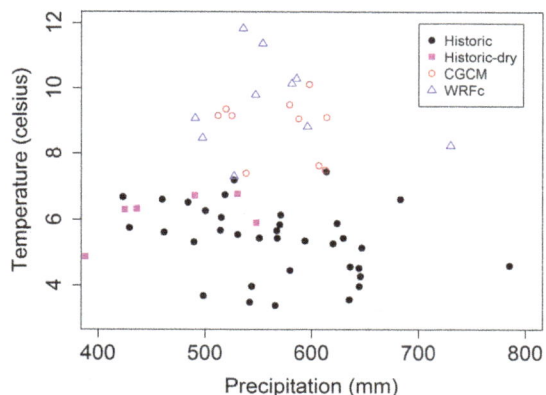

Figure 2. Total precipitation versus average temperature by "bird year" (June of year x-1 to May of year x) for the study area (see Figure 1) for the time periods used to train species distribution models and project future distributions. Historic points showed the years and locations from 1971–2011 used to train the species distribution models with six years withheld. The six years were a dry period from 1987–1992 shown as 'historic-dry'. CGCM and WRFc show two sets of climate projections for the ten year period 2040-49.

the eastern portion of the Prairie Pothole Region: Minnesota (85%), Iowa (95%) and North (49%) and South Dakota (35%) [11,12]. Losses of surrounding upland prairie habitats follow a similar geographic gradient (greatest in the eastern portion of the Prairie Pothole Region) but have been even more severe than wetland losses [3].

Species Occurrence

We obtained species occurrence (presence/absence) data from the North American Breeding Bird Survey (BBS) [13] for waterbirds species with a prevalence of ≥ 0.05. The BBS consists of >3000 routes on secondary roads throughout the continental U.S. and southern Canada. Routes are surveyed once annually during June between 04.45 AM and 10.00 AM. Route locations generally remain the same year after year, although not all routes are surveyed each year and there is variation in the year when a route is initiated. BBS routes are 39.4-km long with 50 stops spaced 0.8 km apart. Three-minute point-count surveys are conducted at each stop. BBS survey data are available for each species and summarized at route totals or 10-stop route segments (https://www.pwrc.usgs.gov/bbs/).

In our study area, BBS surveys took place from late May to early July. This interval extensively overlapped the breeding season (nest-building through brood rearing) for the majority of wetland-dependent species we evaluated. Ten species usually nest during this period and three species occupy brood-rearing habitats. The remainder of the species are engaged in behaviors ranging from incubation to brood-rearing. In addition, seven waterfowl species may be molting body or primary feathers near the end of the survey period.

Even though the breeding cycles of wetland-dependent birds in the Prairie Potholes are not completely synchronous, we believe the BBS survey methods accurately document the presence of all regularly occurring species. Our confidence is based on the overlap between the geographic extent of our survey data, the distribution of our focal species during the breeding season, and the timing of the surveys. The result, we believe, is that the likelihood of correctly documenting the presence of a species was comparable across species, routes, and survey years.

We used data (1971–2011) from high-quality surveys (reported by the BBS as "run type 1") for 77 routes: these were conducted within the correct survey window and not during poor weather. Due to the potential for extensive variation along a route in habitat types, we chose one 10-stop section to model habitat associations rather than use data from the entire route. To accommodate different timing of peak detectability by species, we chose either the first or third section for a species depending which section had

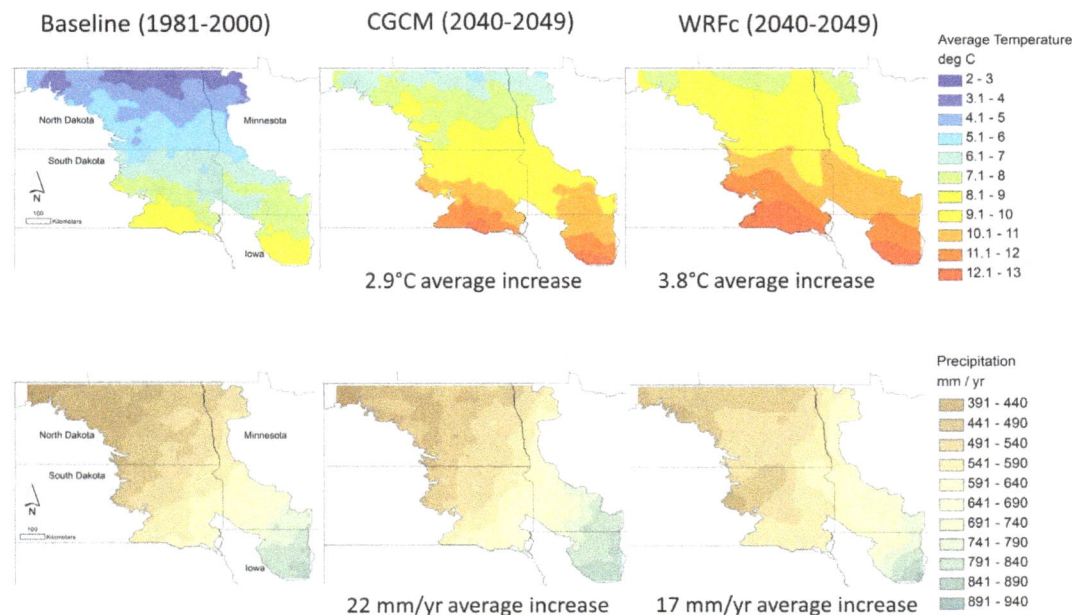

Figure 3. Temperature and precipitation for baseline and two future climate projections for the prairie potholes of North Dakota, South Dakota, Minnesota, and Iowa.

Table 2. Model evaluation showing species prevalence (proportion of data points with species present) for each dataset, and predictive accuracy using the classification matrix and area under the curve (AUC) values.

Common Name	Scientific Name	Prevalence			To withheld data						To withheld dry years	
		Training data (n=974)	Withheld data (n=817)	Dry years data (n=139)	True positive (n)	False positive (n)	True negative (n)	False negative (n)	Overall accuracy (%)	AUC	Overall accuracy (%)	AUC
Canada Goose	Branta canadensis	0.18	0.18	0.09	92	211	457	60	67	0.71	76	0.61
Wood Duck	Aix sponsa	0.06	0.07	0.04	24	187	576	30	73	0.70	84	0.81
Gadwall	Anas strepera	0.11	0.12	0.00	82	183	534	18	75	0.89	74	0.87
American Wigeon	Anas americana	0.05	0.07	0.04	36	129	632	20	82	0.84	71	0.66
Mallard	Anas platyrhynchos	0.60	0.59	0.55	343	101	231	142	70	0.77	57	0.65
Blue-winged Teal	Anas discors	0.36	0.38	0.24	259	139	368	51	77	0.86	63	0.74
Northern Shoveler	Anas clypeata	0.17	0.18	0.09	117	189	481	30	73	0.84	68	0.81
Northern Pintail	Anas acuta	0.21	0.24	0.10	153	161	463	40	75	0.85	69	0.85
Green-winged Teal	Anas crecca	0.05	0.05	0.03	29	180	592	19	77	0.85	72	0.58
Redhead	Aythya americana	0.15	0.12	0.03	87	126	593	11	83	0.91	81	0.95
Ruddy Duck	Oxyura jamaicensis	0.13	0.11	0.06	83	115	612	7	85	0.94	83	0.92
Pied-billed Grebe	Podilymbus podiceps	0.22	0.24	0.11	166	150	474	27	78	0.90	73	0.81
Double-crest. Cormorant	Phalacrocorax auritus	0.09	0.09	0.05	64	201	542	10	74	0.84	68	0.75
American Bittern	Botaurus lentiginosus	0.24	0.27	0.17	190	173	421	33	75	0.84	64	0.76
Great Blue Heron	Ardea herodias	0.07	0.06	0.04	29	180	592	19	76	0.69	88	0.67
Sora	Porzana carolina	0.28	0.27	0.16	178	154	443	42	76	0.86	63	0.71
American Coot	Fulica americana	0.28	0.29	0.15	201	117	467	32	82	0.90	72	0.86
Killdeer	Charadrius vociferus	0.88	0.92	0.86	507	27	41	242	67	0.69	69	0.59
Upland Sandpiper	Bartramia longicauda	0.49	0.48	0.53	308	128	300	81	74	0.82	81	0.84
Willet	Tringa semipalmata	0.16	0.15	0.13	111	163	532	11	79	0.91	70	0.90
Marbled Godwit	Limosa fedoa	0.19	0.18	0.24	121	166	505	28	76	0.88	75	0.90
Wilson's Snipe	Gallinago delicata	0.19	0.20	0.14	140	163	494	23	77	0.90	72	0.80
Wilson's Phalarope	Phalaropus tricolor	0.10	0.12	0.04	79	165	550	23	77	0.86	66	0.84
Franklin's Gull	Leucophaeus pipixcan	0.10	0.11	0.09	64	189	538	26	74	0.81	73	0.86
Ring-billed Gull	Larus delawarensis	0.12	0.16	0.12	103	179	510	25	75	0.84	66	0.78
Black Tern	Chlidonias niger	0.17	0.17	0.07	112	172	505	28	76	0.85	72	0.89
Sedge Wren	Cistothorus platensis	0.27	0.26	0.17	142	168	434	73	71	0.76	81	0.71
Marsh Wren	Cistothorus palustris	0.23	0.24	0.17	166	130	493	28	81	0.89	73	0.78
Common Yellowthroat	Geothlypis trichas	0.83	0.84	0.81	444	41	91	241	65	0.74	60	0.76
Song Sparrow	Melospiza melodia	0.66	0.66	0.57	425	105	172	115	73	0.78	74	0.85

Table 2. Cont.

Common Name	Scientific Name	Prevalence			To withheld data						To withheld dry years	
		Training data (n = 974)	Withheld data (n = 817)	Dry years data (n = 139)	True positive (n)	False positive (n)	True negative (n)	False negative (n)	Overall accuracy (%)	AUC	Overall accuracy (%)	AUC
Yellow-headed Blackbird	Xanthocephalus xanthocephalus	0.54	0.56	0.56	348	86	321	62	82	0.88	64	0.76

The positive and negative rates show the models ability to correctly predict presence and absence data points in the withheld data based on a 0.5 threshold for classification. Overall accuracy was the proportion of true positive and true negative predictions. AUC critical value = 0.70. Dry year predictions were based on models trained without the dry years.

higher detections for that species. Routes were consistently surveyed from stop one, starting around 04.45 AM, to stop 50, ending around 09.00 AM. For all but two species, the first or third section had their highest or second highest number of detections. 'Presence' was defined as ≥ 1 detection at a minimum of one stop along the route segment. We identified focal species based on their prevalence (section-level occurrence rate) with species detected at fewer than 5% of route sections not included.

Land Cover Data

We extracted land cover variables (Table 1) for North and South Dakota from a GIS raster layer created by the U.S. Fish and Wildlife Service (USFWS; USFWS, Region 6 Habitat and Population Evaluation Team, unpublished data); for Minnesota and Iowa from a GIS raster layer created by the USFWS (USFWS, Region 3 Habitat and Population Evaluation Team, unpublished data); and for uplands in the southern portion of the Iowa Prairie Pothole Region from the 1992 National Land Cover Dataset (NLCD). The USFWS data layers were based on Landsat Thematic Mapper Satellite imagery of scenes from 2000–2003, and the NLCD on scenes from the early to mid-1990s. All raster layers were at a 30-m resolution.

Wetland basins in the land cover layers were areas of contiguous wetland extent. The basins were derived from a GIS wetland polygons layer (USFWS National Wetlands Inventory, NWI) where multiple contiguous polygons of differing wetland regimes were dissolved to a single polygon. The USFWS Habitat and Population Evaluation Team followed the procedures of Cowardin et al. [14] and Johnson and Higgins [15] to describe each wetland basin by its most permanent water regime: temporary, seasonal, semipermanent, lake, and river. Generally, temporary wetlands are flooded in spring for a few weeks after snow-melt, seasonal wetlands hold water until summer, and semipermanent wetlands hold water through the growing season; lake and rivers are permanently flooded wetlands [16]. NWI data are based on aerial photographs taken in the late 1970's and early 1980's. Where water pixels extended beyond NWI polygons, they were labeled as water (wetland regimes, see Cowardin et al. [17]). We characterized wetlands into nine classes: temporary, seasonal, semipermanent, lake, river, forested, shrub, total, and total palustrine (Table 1). Total wetland was the combined composition of temporary, seasonal, semipermanent, lake, river, forested, and shrub; total palustrine wetland was temporary, seasonal, and semipermanent.

We described upland habitat using four land cover classes: cropland, grassland, tree, and developed (Table 1). Cropland included areas planted with crops or fallowed. Grassland included native prairie, planted grasses (i.e. previously cropped but now planted with grasses and forbs such as Conservation Reserve Program land), and hayland. Developed land cover included primarily residential areas. Tree habitat included small sections or rows of trees and occasionally areas of forest. Accuracy of the upland land cover data for North and South Dakota, assessed in 2007, was > 90% (M. Estey, personal communication).

To describe habitat associations for our focal waterbird species, we explored composition-based single scale models. In both single-scale and multi-scale models, composition-based predictors, expressed as the amount of a land cover type within a given area, perform better than their distance-based counterparts, expressed as the distance from a sampling location to a land cover type [18]. We used ArcMap 10.0 to calculate land cover composition of the four upland and nine wetland classes at six spatial scales for the BBS route segments. The scales ranged from 335 ha to 32,200 ha and were based on buffering the segments with radii: 0.2-km, 0.4-

Table 3. Values report projected changes in occurrence in the 2040's, relative to 1981–2000 (baseline).

Species	Change in occurrence (%)		
	CGCM	WRFc	Average
Canada Goose	−76	−66	−71
Wood Duck	−70	−37	−54
Gadwall	49	−87	−19
American Wigeon	−58	−71	−65
Mallard	−30	−23	−27
Blue-winged Teal	−9	−4	−7
Northern Shoveler	−51	−62	−57
Northern Pintail	−45	−37	−41
Green-winged Teal	−46	−18	−32
Redhead	−42	−35	−39
Ruddy Duck	−30	−31	−31
Pied-billed Grebe	−40	−18	−29
Double-crested Cormorant	−11	−20	−16
American Bittern	−42	−42	−42
Great-blue Heron	−72	−82	−77
Sora	−94	−98	−96
American Coot	−38	−38	−38
Killdeer	5	0	3
Upland Sandpiper	8	7	8
Willet	−43	−58	−51
Marbled Godwit	−53	−61	−57
Wilson's Snipe	−99	−100	−100
Wilson's Phalarope	−42	−60	−51
Franklin's Gull	−93	−98	−96
Ring-billed Gull	−39	−83	−61
Black Tern	−67	−64	−66
Sedge Wren	−71	−60	−66
Marsh Wren	−40	−42	−41
Common Yellowthroat	−26	−35	−31
Song Sparrow	−38	−41	−40
Yellow-headed Blackbird	−24	−25	−25

Species distribution models projected species occurrence to 4,957 8-km grid points using climate data for the baseline period and two climate projections (CGCM and WRFc). Negative values indicated the proportion of occupied grid cells for each species, projected to be unoccupied in the future. Positive values indicated the proportion by which occupied cells were projected to increase.

km, 1-km, 2-km, 4-km, and 8-km. BBS surveyors record all birds detected within 0.4-km of the survey point. Thus, assuming no decline in detection probability with increasing distance and no landscape effect, we expected 0.4-km to be the appropriate scale to relate land cover to bird occurrence. However, some waterbird species may decline quickly in detection probability with increasing distance from the survey point—therefore, we also explored a 0.2-km scale. Because other species may respond to land cover heterogeneity at broader extents, we also explored a range (1-km to 8-km) of landscape scales. Land cover data were assumed static across current and future years.

Climate Covariates

We used PRISM (PRISM, Parameter-elevation Regressions on Independent Slopes Model) data for historical climate records. These data are available at a 4-km grid scale as monthly

temperature and precipitation and were rescaled to an 8-km grid to match the scale of the projected climate data [19].

Using monthly values of precipitation and temperature, we derived 18 climate variables (Table 1). We calculated mean temperatures for grid points by averaging the minimum and maximum monthly temperatures over different time periods. We delineated seasons as summer (June-August), fall (September-November), winter (December-February), and spring (March-May). We defined year as ending in May to correspond to the June bird surveys. We included seasonal and annual variables because both seasonal and annual climate explain annual variation in the number of prairie pothole wetlands holding water [20]. For semipermanent wetlands and (especially) lakes, wet wetland count is related to long-term climate (at least 3 years) [21]. We included 5-year and 10-year precipitation and mean temperature variables as proxies for long-term climate effects. We also included the

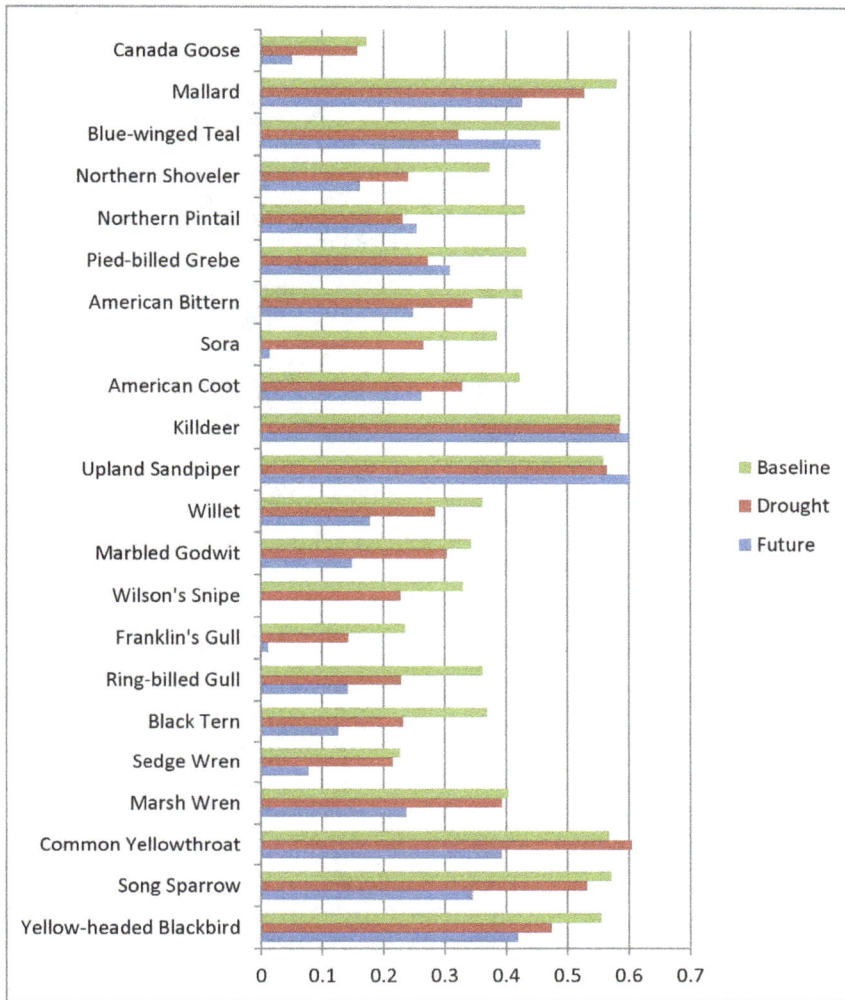

Figure 4. Mean rates of predicted species occurrence at 4,957 8-km grid points. Baseline rate was based on 1981–2000 climate records. Dry years showed predicted occurrence rates for the drought period, 1987–1992. Future rates were based on the average projections of two future climate datasets (CGCM-A2 and WRFc) for 2040–2049.

variances in 5-year and 10-year precipitation and temperature data, because large values of these variables may indicate that wetlands are cycling through wet and dry phases, driving dynamic vegetative conditions [5]. Climate data from 1971–2011 were used to construct the baseline species distribution models. The species distribution models predicted to climate data from 1981–2000 and 2040–2049 to create current and future projections, respectively, of species occurrence.

Future Climate Data

We used statistically downscaled and high resolution climate projections. Statistically downscaled data refine projections from global circulation models (GCMs) using an empirical relationship to local physiography (e.g. topography and water bodies). These projections assume relations will hold into the future and are less computationally intensive than high resolution models. High resolution models nest a dynamical Regional Climate Model within the GCM, re-running the GCM based on mesoscale (a few to a few hundred kilometers) physical relationships with topographical features and surface characteristics [22]. The high resolution projections circumvent the problem associated with lack

of "stationarity" when the relationships between GCM output and the fine-scale climate change over time.

The statistically downscaled projections were based on data obtained from output of GCM CGCM3.1MR (Canadian Centre for Climate Modeling and Analysis Third Generation Coupled Global Climate Model Version 3.1, Medium Resolution) [23] and downscaled to an 8-km grid. The high resolution models used the Community Climate System Model (CCSM) to set the boundary condition and a mesoscale model, Weather Research and Forecasting model (WRF) to refine the data to a 36-km regional scale (J. Stamm, personal communication) [24]. Given that we expected high spatial correlation for monthly temperature and precipitation, we interpolated the 36-km data to the 8-km grid [25]. Both climate models were run with a mid-high IPCC emissions scenario, A2 [26]. The high resolution projections were available for 2000–2049, and the statistically downscaled projections were available for 2000–2100. We term the statistically downscaled data "CGCM" after the GCM these data are based on, and we term the dynamically downscaled data "WRFc" after the mesoscale model these data are based on.

Table 4. Variable importance for Random Forest species distribution models.

Rank	Canada Goose	Wood Duck	Gadwall	American Wigeon	Mallard	Blue-winged Teal	Northern Shoveler	Northern Pintail	Green-winged Teal	Redhead	Ruddy Duck
1	W-lake(+)	W-templ(-)	T-5yr_sd(-)	W-lake(+)	W-pal(+)	W-total(+)	W-pal(+)	U-tree(-)	U-tree(-)	W-semi(+)	W-semi(+)
2	P-wint(+)	P-wint(+)	P-10yr_sd(-)	U-tree(-)	W-semi(+)	W-pal(+)	W-total(+)	W-pal(+)	P-wint(+)	W-total(+)	W-total(+)
3	W-river(-)	U-grass(-)	U-tree(-)	W-semi(+)	W-river(-)	W-seas(+)	W-semi(+)	P-10yrt(-)	W-lake(+)	W-lake(+)	W-lake(+)
4	T-sum(-)	U-crop(-)	T-spr(-)	W-templ(-)	W-total(+)	W-semi(+)	W-templ(-)	W-seas(+)	W-seas(+)	U-tree(-)	W-pal(+)
5	T-spr(-)	U-tree(+)	U-grass(+)	P-5yr(-)	T-10yr_sd(+)	U-grass(+)	W-seas(+)	W-semi(+)	U-crop(-)	W-templ(-)	U-tree(-)
6	T-10yr_sd(+)	U-devel(-)	P-10yrt(m)	P-spr(-)	W-templ(+)	W-templ(+)	U-tree(-)	P-5yr(-)	P-fall(~)	W-pal(+)	W-seas(+)
7	T-10yrt(m)	W-semi(+)	U-crop(-)	W-pal(+)	U-tree(-)	U-tree(-)	U-crop(-)	U-grass(+)	W-shrub(-)	W-river(-)	W-river(-)
8	T-5yr(m)	T-10yr_sd(+)	T-10yr_sd(+)	W-total(+)	T-spr(-)	W-river(-)	U-grass(+)	W-river(-)	W-templ(+)	W-seas(+)	W-templ(-)
9	T-1yrt(-)	T-sum(~)	W-total(+)	P-10yrt(-)	U-grass(+)	U-crop(-)	T-spr(+)	T-spr(-)	P-5yr(-)	U-crop(-)	U-crop(-)
10	W-total(+)	W-total(+)	W-river(-)	P-1yr(-)	T-5yr_sd(+)	P-10yrt(-)	P-10yrt(-)	W-total(-)	T-5yr(-)	U-grass(+)	U-grass(+)

Rank	Pied-billed Grebe	Double-crested Cormorant	American Bittern	Great-blue Heron	Sora	American Coot	Killdeer	Upland Sandpiper	Willet	Marbled Godwit
1	W-semi(+)	W-semi(+)	U-crop(-)	T-spring(-)	P-10yrt(-)	W-semi(+)	W-semi(+)	W-seas(+)	U-tree(-)	U-tree(-)
2	W-total(+)	U-tree(+)	W-total(+)	P-10yrt(+)	T-sum(-)	W-total(+)	W-pal(+)	U-grass(+)	P-10yrt(-)	U-crop(-)
3	W-pal(+)	W-templ(-)	U-grass(+)	U-tree(+)	T-5yr(-)	W-pal(+)	W-river(-)	U-tree(-)	P-5yr(-)	W-lake(+)
4	W-seas(+)	T-10yr_sd(+)	W-semi(+)	P-5yr(+)	W-seas(+)	W-seas(+)	U-tree(-)	P-10yrt(-)	U-grass(+)	W-total(+)
5	U-crop(-)	W-lake(+)	W-pal(+)	P-spr(+)	T-1yr(-)	W-river(-)	W-total(+)	W-pal(+)	W-pal(+)	P-10yrt(+)
6	W-lake(+)	T-5yr(~)	W-seas(+)	P-1yr(+)	W-river(-)	P-10yrt(-)	W-seas(+)	U-devel(-)	W-semi(+)	U-grass(+)
7	T-10yr_sd(+)	P-5yr(+)	P-10yrt(-)	T-10yrt(-)	W-total(+)	W-lake(+)	U-devel(-)	W-river(-)	W-total(+)	T-10yrt(-)
8	U-grass(+)	P-spr(-)	T-5yr(-)	W-shrub(+)	W-pal(+)	U-crop(-)	P-wint(~)	U-crop(-)	U-crop(-)	P-5yr(-)
9	T-sum(-)	P-fall(~)	T-sum(-)	P-sum(+)	P-5yr(-)	U-grass(+)	W-templ(+)	W-semi(+)	T-10yrt(-)	W-semi(+)
10	U-tree(-)	T-5yr_sd(-)	T-1yr(-)	T-1yr(-)	U-grass(+)	T-5yr(-)	U-grass(+)	P-5yr(-)	W-river(-)	W-river(+)

Rank	Wilson's Snipe	Wilson's Phalarope	Franklin's Gull	Ring-billed Gull	Black Tern	Sedge Wren	Marsh Wren	Common Yellow-throat	Song Sparrow	Yellow-headed Blackbird
1	T-5yr(-)	U-tree(-)	U-tree(+)	W-semi(+)	W-seas(+)	P-10yr_sd(-)	W-semi(+)	W-total(+)	U-grass(-)	U-grass(-)
2	T-10yrt(-)	U-grass(+)	W-pal(+)	W-lake(+)	W-pal(+)	U-tree(+)	W-total(+)	W-templ(+)	U-crop(+)	U-crop(+)
3	W-total(+)	U-crop(-)	T-10yrt(-)	W-pal(+)	W-total(+)	U-crop(-)	W-pal(+)	W-seas(+)	U-tree(+)	U-tree(+)
4	U-crop(-)	P-10yrt(-)	W-total(+)	W-total(+)	W-semi(+)	P-10yrt(+)	U-crop(-)	W-pal(+)	P-10yrt(+)	P-10yrt(+)
5	W-pal(+)	P-5yr(-)	W-seas(m)	U-tree(-)	W-templ(+)	T-sum(-)	W-seas(+)	U-tree(+)	W-semi(-)	W-semi(-)
6	W-lake(-)	W-lake(m)	U-grass(+)	U-grass(+)	U-crop(-)	P-5yr(-)	U-grass(+)	W-semi(-)	P-sum(+)	P-sum(-)
7	T-sum(-)	W-semi(+)	W-templ(+)	P-5yr(-)	P-10yrt(-)	T-10yr_sd(+)	W-lake(+)	P-sum(-)	T-10yr_sd(+)	T-10yr_sd(+)
8	U-tree(~)	W-seas(+)	W-lake(+)	W-seas(+)	T-5yr(-)	W-pal(+)	T-sum(-)	U-devel(-)	T-sum(-)	T-sum(-)
9	W-river(-)	P-spr(~)	U-crop(-)	W-river(-)	P-wint(+)	U-grass(+)	W-river(-)	W-lake(+)	T-10yrt(-)	U-tree(-)
10	W-seas(~)	W-total(+)	P-sum(~)	T-5yr(-)	P-5yr(m)	P-1yr(+)	T-10yrt(-)	P-5yr_sd(-)	P-5yr(+)	P-5yr(-)

Top ten variables are shown in descending order of rank. Variable categories were denoted by W (wetland), U (upland), P (precipitation), and T (temperature). Signs indicated the relationship between the predictor and the species response: + (positive), - (negative), m (unimodel), and ~ (equivocal).

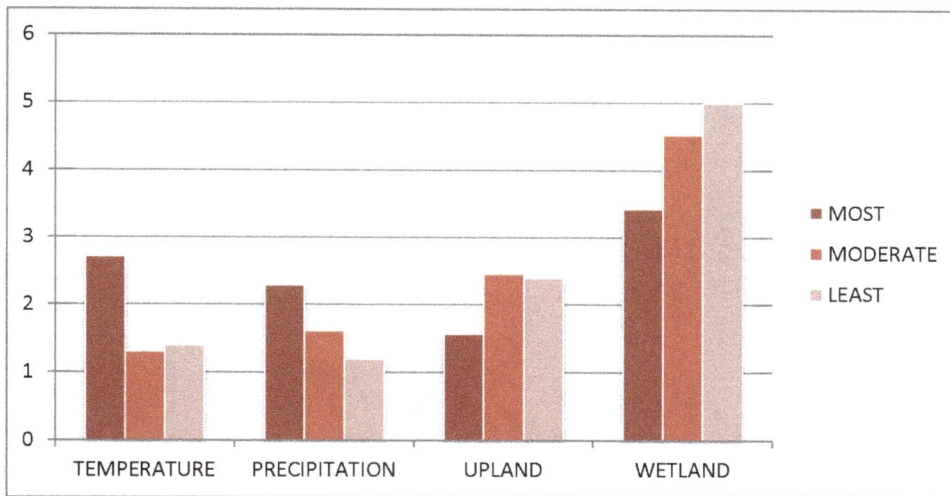

Figure 5. The frequency (y-axis) of variable type (x-axis) in top ten variables for waterbird species distribution models, grouped by species sensitivity to climate change. The most sensitive species were projected to lose ≥66% of their current habitat; moderately sensitive species 33–65%; and least sensitive <33%.

Species Distribution Models

We estimated a species distribution model (SDM) for each waterbird species, relating BBS occurrence records (1971–2011) to climate, and wetland and upland land cover (hereafter grouped as "land cover") predictor variables. We used climate for the same year as the occurrence record from the climate grid point nearest the BBS route segment and land cover surrounding the route segment. We defined occurrence as one or more detection per 10-stop segment by year. The spatial scale used in the final models for land cover calculations was chosen separately for each species based on model performance. We ran six models for each species based on the six different spatial scales of land cover and chose the model with the highest classification accuracy. We used a non-parametric machine learning approach, Random Forests, to create the SDMs [9]. We chose Random Forests because of its high predictive power, ability to model unspecified variable interactions and correlated variables, its ranking of variable importance, and its demonstrated use for bioclimatic species distribution models [27,28]. Random Forests uses an ensemble of classification (categorical response variable) or regression (continuous response variable) trees, each built with a subset of the data, to model the pattern between predictor variables and the response variable. We used permutation procedures to assess variable importance, a method based on reduction in predictive accuracy to internally withheld data when values of a given predictor variable are randomly shuffled. We report the top ten variables for each model. Although the choice of the number of variables to report is arbitrary, we expect the top ten will provide an adequate basis for comparing models.

We used the RandomForests package in R to create our models [29]. We specified 3000 trees which is a sufficiently large number of trees to capture any patterns in the data. Each tree was constructed with a bootstrapped subsample with replacement of the data records (BBS routes). Because the ratio of presence to absence was often skewed, particularly for either very abundant or rare species, we balanced the data by setting Random Forests to randomly use, for each tree, 25 records where the species was present and another 25 where the species was absent [30]. A subsample of five predictor variables was evaluated at each binary split in the tree algorithm.

We partitioned the BBS data in a number of ways to strengthen model evaluation and inference. First, we only excluded consecutive years of surveys to reduce the influence of temporal autocorrelation and maximize information content: the "main training set". The excluded data were used to validate the models created with these data: the "main test set". Second, we separated out six years of data covering a drought period from 1987 through 1992 [21]. We created species distribution models with the drought data to look at variable importance in dry years compared to variable importance for the whole study period (main training set). To assess model transferability, we predicted to the drought data subset using species distribution models created with the remaining wetter years [31,32].

Model Evaluation

To evaluate each model's ability to forecast to the same range of predictor variables, we predicted to the main test set. To evaluate each model's transferability – that is, to project to a new location or time period where predictor variables may be outside the range of the variables used to build the model – we projected to the drought period with models trained with data from the wet years. The transferability assessment should more realistically evaluate how the models extrapolate to a dry future [31].

To assess a model's performance, we report patterns of correct classification in a confusion matrix and the area under the receiver operating characteristic curve (AUC) [33]. From the confusion matrix, we report the counts of true positives, false positives, true negatives, false negatives and overall classification accuracy based on a 0.5 probability of occurrence threshold for concluding presence. Because we set sample sizes of presence and absence points to be equally subsampled in the Random Forests model, we selected a threshold of 0.5 [34]. Overall classification accuracy was calculated by dividing the number of correctly predicted presences and absences by total predictions. AUC is a threshold free assessment of model performance. AUC values range from zero to one and give the probability that a known presence observation has a higher predicted value of presence than an absence observation for a randomly selected pair of presence-absence observations [33]. Models with AUC values of at least 0.7 are

Table 5. Variable importance from 22 Random Forest species distribution models for the particularly dry period, 1987–1992.

Rank	Canada Goose	Mallard	Blue-winged Teal	Northern Shoveler	Northern Pintail	Pied-billed Grebe	American Bittern	Sora	American Coot	Killdeer	Upland Sandpiper
1	P-5yr_sd(+)	W-semi(+)	W-total(+)	W-semi(+)	U-tree(-)	W-lake(+)	U-grass(+)	T-10yr_sd(+)	W-semi(+)	P-10yr_sd(-)	U-grass(+)
2	T-sum(+)	W-total(+)	W-semi(+)	T-1yr(m)	W-semi(+)	U-crop(-)	W-semi(+)	T-sum(-)	W-lake(+)	T-5yr_sd(-)	U-tree(-)
3	T-sprt(-)	W-pal(+)	U-crop(-)	T-wint(m)	T-sprt(-)	U-grass(+)	W-total(+)	W-river(-)	W-total(+)	U-grass(+)	W-seas(+)
4	T-10yr(-)	W-river(-)	W-lake(+)	W-total(+)	P-10yr(-)	W-semi(+)	U-crop(-)	W-total(+)	W-pal(-)	U-devel(+)	U-crop(-)
5	P-10yr_sd(-)	U-crop(-)	P-fall(+)	W-lake(+)	T-1yr(-)	W-seas(m)	W-lake(+)	P-5yr_sd(-)	U-crop(-)	T-fall(-)	P-10yr(+)
6	T-1yr(-)	P 10yr(-)	W-pal(+)	P-10yr_sd(-)	P-fall(+)	W-total(+)	W-pal(+)	W-pal(-)	T-5yr_sd(+)	W-semi(+)	W-total(+)
7	T-wint(-)	W-templ(+)	U-grass(+)	T-5yr(-)	W-total(+)	T-wint(+)	T-sum(-)	W-seas(+)	T-10yr_sd(+)	T-sum(+)	W-pal(-)
8	T-5yr(-)	T-wint(+)	T-wint(+)	U-crop(-)	T-5yr(-)	P-5yr_sd(-)	U-devel(-)	U-tree(-)	P-sprt(-)	U-tree(-)	W-semi(+)
9	W-seas(-)	T-sprt(~)	P-sum(+)	T-10yr(+)	T-wint(-)	W-pal(+)	P-fall(+)	W-temp(+)	W-temp(+)	T-5yr(-)	W-tempt(+)
10	P-10yr(-)	W-shrub(-)	T-sum(-)	W-seas(-)	W-pal(+)	U-devel(-)	W-seas(+)	W-lake(+)	W-seas(m)	P-1yr(-)	T-wint(+)

Rank	Willet	Marbled Godwit	Wilson's Snipe	Franklin's Gull	Ring-billed Gull	Black Tern	Sedge Wren	Marsh Wren	Common Yellowthroat	Song Sparrow	Yellow-headed Blackbird
1	P-10yr(-)	W-lake(+)	W-lake(+)	W-total(+)	W-lake(+)	U-crop(-)	T-sum(-)	W-semi(+)	W-semi(+)	U-grass(-)	W-semi(+)
2	P-5yr(-)	U-tree(-)	W-total(+)	W-lake(+)	U-crop(-)	W-pal(+)	W-total(+)	U-crop(-)	U-crop(-)	U-tree(+)	P-10yr(-)
3	W-total(+)	W-total(+)	T-10yr(-)	T-10yr(-)	W-lake(-)	W-lake(+)	T-1yr(-)	W-temp(+)	W-semi(-)	W-semi(-)	W-pal(+)
4	U-tree(-)	P-10yr(-)	T-5yr(-)	P-fall(~)	P-10yr(-)	W-temp(+)	T-10yr_sd(-)	T-1yr(-)	W-pal(+)	U-crop(+)	W-total(+)
5	W-semi(+)	U-crop(-)	P-5yr_sd(+)	P-5yr_sd(+)	P-sprt(-)	W-semi(+)	U-grass(+)	W-lake(+)	W-total(+)	P-sum(+)	P-5yr(-)
6	W-pal(+)	W-river(-)	W-tempr(+)	T-10yr_sd(-)	U-grass(+)	W-total(+)	T-10yr(-)	W-lake(+)	U-tree(+)	P-10yr(+)	W-seas(+)
7	W-seas(+)	T-10yr(-)	T-1yr(-)	T-sprt(-)	P-5yr(-)	W-total(+)	U-crop(-)	W-semi(-)	P-5yr_sd(-)	P-1yr(+)	U-crop(-)
8	P-10yr_sd(-)	P-5yr(-)	U-tree(-)	T-5yr(-)	T-fall(-)	P-10yr(+)	P-10yr(+)	W-lake(+)	W-lake(+)	U-devel(+)	U-grass(+)
9	T-sum(m)	T-5yr(-)	T-1yr(-)	T-1yr(-)	P-10yr(-)	P-5yr(+)	T-5yr(+)	T-5yr(-)	T-5yr(-)	T-10yr(-)	P-10yr_sd(-)
10	U-crop(-)	W-pal(+)	T-sprt(-)	U-devel(-)	P-10yr_sd(-)	T-5yr(-)	W-forest(+)	W-seas(-)	W-seas(+)	T-5yr(-)	U-tree(-)

Top ten variables shown in descending order of rank. Variable categories were denoted by W (wetland), U (upland), P (precipitation), and T (temperature). Signs indicated the relationship between the predictor and the species response: + (positive), - (negative), m (unimodel), and ~ (equivocal).

considered acceptable, between 0.8 and 0.9 good, and greater than 0.9 outstanding [35].

Projected Distributional Changes

We created current predictions and future projections of probability of occurrence to each grid cell, for each focal species, by applying the SDMs to the baseline land cover and climate data (the 20-year period for baseline climate data being 1981–2000) and to baseline land cover and future climate data (the 10-year period 2040–2049). Ten to twenty-year time periods were chosen to mitigate the influence of short-term variations in climate.

We created current and future predictive distribution maps for each species in ArcMap 10 based on an assignment of grid point locations as suitable or unsuitable. A grid point was determined suitable if the estimated probability of occurrence (over the time period for the baseline or future data) was greater than 0.5. Three breakpoints within suitable (0.625, 0.75, 0.875) and unsuitable (0.125, 0.25, 0.375) locations showed the degree to which a location was predicted suitable or unsuitable.

We indexed changes between predicted baselines and projected future distributions using change in a species' spatial distribution. To assess change in distribution, we calculated the percent loss (or gain) in the number of grid cells classified as suitable.

Results

Baseline mean temperature (years 1981–2000) was 5.9°C and mean yearly precipitation was 548 mm. By 2040–2049, CGCM projected a 2.9°C increase in mean temperature and a 22 mm (3.9%) increase in annual precipitation while WRFc projected a 3.8°C temperature increase and a 17 mm (3.1%) increase in annual precipitation. Projections of future precipitation fall within the range of historic levels of precipitation, whereas future temperatures projected by both climate models exceed historic temperatures (Figure 2). The climate models differed slightly in the spatial distribution of the precipitation increases, with CGCM projecting greater increase in Iowa and WRFc projecting greater increase in North Dakota than other areas (Figure 3).

The number of data occurrence records in the main training set was 975. The number of years of survey data included in the main training set, for a given route, ranged from one to 21, with a mean of 13. The number of survey routes included for a given year ranged from 15 to 35. Thirty-one waterbird species had prevalence $\geq$ 0.05 and were included in the focal group (Table 2). The number of records in the main test set was 817. The number of data records in the dry years set was 139. We adjusted the prevalence cutoff to 0.07 ($\geq$ 10 detections), at which 22 species qualified.

Model Evaluation

Most models based on known distributional patterns were acceptable to excellent, indicated by AUC values (Table 2). Exceptions were SDMs for the Great Blue Heron and Killdeer. When predicting to dry years only, AUC values indicated the following additional models predicted poorly: Canada Goose, American Wigeon, Mallard, and Green-winged Teal. Overall accuracy of dry year predictions suggested that projected distributional changes for some species should be interpreted with caution, including Blue-winged Teal, Sora, and Common Yellowthroat. For the main datasets, model performance was not related to a species' prevalence according to Spearman's rank correlation (-0.09, p-value 0.62).

Vulnerability

Average projected decline in occurrence rate (spatial distribution) across 31 species under two future climate scenarios was 45%. WRFc models projected slightly more severe distributional changes (-48%) than CGCM (-43%; Table 3). Species expected to experience small to no declines in distribution included Blue-winged Teal, Killdeer, and Upland Sandpiper. Species projected to experience severe declines were Franklin's Gull, Sora, and Wilson's Snipe (Table 3). In general, species maps depicted declines within the baseline range, rather than distributional shifts to new areas (Figures S2-S12).

For most species, future projections of change were consistent with responses of species to historic dry periods (Figure 4). Consistent projections were those that exhibited little to no change between the historic dry period and the future, or those that declined more in the future than in the historic dry period. If a species' habitat was not impacted by dry conditions, the species would be expected to experience little to no impact under future dry conditions. Other species may be impacted by drying conditions, thus responding during the historic dry period, and even more if additional drying occurs in the future. However, inconsistent with expectations, models projected reduced distribution of Blue-winged Teal, Northern Pintail, and Pied-billed Grebe in the dry period relative to future projections. Additionally, several species that remained relatively stable in the historic dry period were projected to decrease in distribution under future scenarios, including Canada Goose, Sedge Wren, Marsh Wren, Common Yellowthroat, and Song Sparrow.

Variable Importance

In general, species distributions were strongly influenced by the distribution of wetland basins and land cover classes and moderately influenced by climate, as evidenced by their influence in the SDMs. Land cover variables, wetland and upland, collectively occurred as 67% of the top ten variables in the SDMs but comprised only 42% of the available predictor variables (Table 4); wetland and upland variables were 1.5 and 1.8 times more likely to appear in lists of top ten predictors than in the list of available predictors, respectively. Species associations with all wetland types, except rivers, were generally positive. All associations with cropland were negative except for the Song Sparrow, whereas associations with grassland were primarily positive except for Wood Duck (Table 4). Climate predictor variables were generally underrepresented in the variables of top importance. Collectively, temperature and precipitation comprised 32% of the top ten variables across the 31 species, although they were 58% of the available predictor variables. Temperature and precipitation variables were similarly influential and were 0.5 and 0.6 times more likely to appear in lists of top ten predictors than in the list of available predictors, respectively. In general, probability of species occurrence was negatively associated with temperature; relationships with temperature variability were often positive (Table 4). Associations with precipitation were often negative, except for Great Blue Heron, Sedge Wren, Song Sparrow, and Wood Duck. Variability in precipitation occurred in the top ten variables for only one species' model (Sedge Wren) and was negatively correlated with probability of occurrence.

Land cover variables were highly influential in observed patterns of species distribution. The importance of these variables is visually apparent when spatial distribution of grasslands and wetlands (Figure S1) and observed climate gradients (Figure 3) were compared to baseline distributions (Figures S2-S12). Many breeding waterbirds have a high probability of occurrence in the western portion of the study area where grasslands and wetlands co-occur.

Temperature and precipitation predictor variables were more often in the top ten variables for the species with the greatest expected declines (Figure 5). Conversely, wetland and upland land cover variables were more often in the top ten variables for the species with smallest expected declines.

Variable Importance: main models versus dry-years models

For dry-years models, climate predictor variables represented 45% of the top ten variables across the 22 species versus 31% for the same 22 species in the main models (Tables 4 and 5). Of land cover predictors, 65% included wetland variables in the top ten variables in the dry years and a similar 67% in the main models. However, representation of different wetland types varied with more seasonal wetlands (positive relationships only) appearing in the non-drought years (30% versus 17%) and more lakes included in the dry years (32% versus 20%).

Discussion

Our projections of large range reductions for waterbirds breeding in the Prairie Pothole Region are not surprising. Globally, freshwater habitats are expected to be particularly vulnerable to climate change [36]. If, as the future climate projections we used indicate, temperatures rise by ∼3.0°C and precipitation rises only by 3% by mid-century in the Prairie Pothole Region, many fewer pothole wetlands will exist on the landscape due to an increased deficit in precipitation relative to evapotranspiration. Similarly, other studies of the Prairie Pothole Region have projected a drier future and concomitant reductions in waterbird habitat [5,8,37].

Past studies in the Prairie Pothole Region that extrapolated from relations between climatic factors and wetlands inferred generalized habitat losses for waterfowl [5,8]. Our species-specific approach indicated large variability in the vulnerabilities of waterbird species to climate change. This is expected as patterns of waterbird habitat selection vary among species for wetland attributes such as size, permanence, and vegetative cover [10,38]. Hydrological studies indicate that temperature and precipitation regimes affect not only the number of wetlands and wetland size, but marsh vegetation dynamics and the vegetative coverage patterns at the landscape scale [5]. While reducing the overall number of wetlands, a drier climate will likely lead to more extensive coverage of wetlands by dense vegetation rather than wetland conditions characterized by a mixture of open water and vegetation [5]. Species are expected to respond differentially to these changes in wetland characteristics. Furthermore, individualistic species' responses appear the norm [39,40,41].

Our projections of future change were not always consistent with documented waterbird responses to a historic dry period which represented one possible expression of a drier climate. The dry historic period, a consequence of reduced precipitation, is not a direct analog of future drying which is expected to be driven by increases in evapotranspiration (Figure 2) [42]. Thus, it is unclear to what extent the historic pattern of drought can be used as a benchmark for future climate change. Therefore, the inconsistencies between the dry historic waterbird response relative to projected future responses may indicate our models are under- or over- estimating waterbird response to climate change for some species. It is also possible that changes in temperature versus precipitation may result in divergent, and unprecedented, future

wetland habitat conditions. In that case, divergent waterbird responses, relative to the past responses, would not be surprising.

The historic range of temperature variability did not overlap future projections and so, our SDMs were projecting beyond known climatic boundary conditions. Model extrapolation to novel conditions is common when projecting species response to future climate [43]. Our single values for yearly averages (Figure 2), indicated almost no overlap in temperature range between historic and projected time intervals. However, because of spatial variation in temperature regimes (i.e., warmer in the south, as shown in Figure 3), there were likely many individual grid cells in which future temperatures overlapped the historic range even if the study area yearly means do not. SDMs based on the Random Forest algorithm are constrained when extrapolating beyond the observed values of the predictor variables. For example, when projected temperatures are outside of the range of the training set the algorithm holds the prediction constant at the last known value of temperature [44]. Therefore, if future wetland habitat conditions selected by the species become less common with increased temperatures, our estimates of habitat losses for many species may be underestimates.

Ranking predictor variables by their importance provides additional insights into how the 31 waterbird species may respond to changing environmental factors. We included predictors related to suitable waterbird habitat quality, including the amount and type of wetland basins, and temporally scaled temperature and precipitation covariates. Species projected to be most sensitive to anticipated climate change (changes in temperature and precipitation, Table 4) consistently reflected the ecology of the species. For example, the two diving ducks, Ruddy Duck and Redhead, primarily selected large wetlands, such as semipermanent basins, and were less susceptible to total drying [5,45]. As a consequence of their habitat associations, no climate covariates ranked in the top ten for these two waterbird species. In contrast, waterbirds that rely on shallow water habitat, such as Sora or Sedge Wren, or dynamic habitat such as Black Tern or Mallard, showed a much greater projected change in distribution to future climatic conditions [5,38].

The variable importance ranks also suggested that waterbirds may shift their habitat preferences with increased drying. More climate covariates and more permanent wetland regimes appeared in the top variables for dry years. In the Prairie Pothole Region, wetland function can rapidly change with significant changes in the climate. In dry years, for example, semipermanent wetlands may function more like seasonal wetlands, and seasonal wetlands more like temporary wetlands. This differential sensitivity to climate change explains why seasonal wetlands were less important and lakes more important in dry years.

Because bioclimatic SDMs are generally exploratory with many collinear climate predictors, there is concern that these models over-fit the data and thus misrepresent species distributions [46]. However, the inclusion of many collinear climate predictors is often warranted when causal links between specific climate predictors and species' distributions are not established, leading to better model fit and projections [47]. We found that when we reduced our 18 climate and 13 land cover variables to 14 uncorrelated climate and 10 uncorrelated land cover variables model projections were similar: 45% average range reduction for the full model and 48% for the reduced model (results not shown).

Conclusions

Our results indicated, on average, large decreases in suitable habitat by the 2040s for 31 waterbird species breeding in the

Prairie Pothole Region of the U.S.A. Importantly, our results were consistent between two contrasting future climate scenarios. However, there was substantial variability in species specific responses to projected climate change. Therefore, strategic efforts to mitigate climate change effects should preferentially direct management actions to those species expected to be most vulnerable. In continuing research, we are exploring in greater detail various sources of uncertainty in our projections including additional model algorithms, alternative covariates, and other sources of species distribution data [48].

Supporting Information

Figure S1 Distribution of grassland and palustrine wetlands on the U.S. Prairie Pothole Region landscape. Darker shades represent greater coverage of grassland (versus cropland) and greater areal coverage of wetlands (log transformed).

Figure S2 Map of species distributions for baseline and two future climate projections. Brown indicates areas where the species is predicted to occur and green represents areas where the species is not predicted to occur.

Figure S3 Map of species distributions for baseline and two future climate projections. Brown indicates areas where the species is predicted to occur and green represents areas where the species is not predicted to occur.

Figure S4 Map of species distributions for baseline and two future climate projections. Brown indicates areas where the species is predicted to occur and green represents areas where the species is not predicted to occur.

Figure S5 Map of species distributions for baseline and two future climate projections. Brown indicates areas where the species is predicted to occur and green represents areas where the species is not predicted to occur.

Figure S6 Map of species distributions for baseline and two future climate projections. Brown indicates areas where the species is predicted to occur and green represents areas where the species is not predicted to occur.

Figure S7 Map of species distributions for baseline and two future climate projections. Brown indicates areas where the species is predicted to occur and green represents areas where the species is not predicted to occur.

Figure S8 Map of species distributions for baseline and two future climate projections. Brown indicates areas where the species is predicted to occur and green represents areas where the species is not predicted to occur.

Figure S9 Map of species distributions for baseline and two future climate projections. Brown indicates areas where the species is predicted to occur and green represents areas where the species is not predicted to occur.

Figure S10 Map of species distributions for baseline and two future climate projections. Brown indicates areas where the species is predicted to occur and green represents areas where the species is not predicted to occur.

Figure S11 Map of species distributions for baseline and two future climate projections. Brown indicates areas where the species is predicted to occur and green represents areas where the species is not predicted to occur.

Figure S12 Map of species distributions for baseline and two future climate projections. Brown indicates areas where the species is predicted to occur and green represents areas where the species is not predicted to occur.

Acknowledgments

Disclaimer: Any use of trade, firm, or product names is for descriptive purposes only and does not imply endorsement by the U.S. Government.

Climate data were provided by Linda Joyce and David Coulson (U.S. Forest Service, statistically downscaled climate models), and John Stamm, Parker Norton, and Gary Clow (U.S. Geological Survey, high resolution models). Lucy Burris and Diane Granfors provided figures 3 and S1, respectively. Curt Flather, Helen Sofaer, and anonymous reviewers provided comments that improved the manuscript.

Author Contributions

Conceived and designed the experiments: VS SS. Performed the experiments: VS. Analyzed the data: VS. Wrote the paper: VS SS BN.

References

1. Keddy PA (2000) Wetland ecology: principles and conservation. Cambridge, UK; New York, NY: Cambridge University Press. 614 p.
2. Gleason RA, Euliss NH, Tangen BA, Laubhan MK, Browne BA (2011) USDA conservation program and practice effects on wetland ecosystem services in the Prairie Pothole Region. Ecological Applications 21: S65–S81.
3. Beyersbergen GW, Niemuth ND, Norton MR (2004) Northern Prairie and Parkland waterbird conservation plan. A plan associated with the Waterbird Conservation for the Americas initiative Prairie Pothole Joint Venture, Denver, CO, USA.
4. Solomon S, Intergovernmental Panel on Climate Change, Intergovernmental Panel on Climate Change. Working Group I. (2007) Climate change 2007: the physical science basis: contribution of Working Group I to the Fourth Assessment Report of the Intergovernmental Panel on Climate Change. Cambridge; New York: Cambridge University Press. 996 p.
5. Johnson WC, Werner B, Guntenspergen GR, Voldseth RA, Millett B, et al. (2010) Prairie Wetland Complexes as Landscape Functional Units in a Changing Climate. Bioscience 60: 128–140.
6. Niemuth ND, Solberg JW (2003) Response of waterbirds to number of wetlands in the Prairie Pothole Region of North Dakota, USA. Waterbirds 26: 233–238.
7. Johnson DH, Grier JW (1988) Determinants of Breeding Distributions of Ducks. Wildlife Monographs: 1–37.
8. Sorenson LG, Goldberg R, Root TL, Anderson MG (1998) Potential effects of global warming on waterfowl populations breeding in the Northern Great Plains. Climatic Change 40: 343–369.
9. Breiman L (2001) Random forests. Machine Learning 45: 5–32.
10. Kantrud HA, Krapu GL, Swanson GA (1989) Prairie basin wetlands of the Dakotas: a community profile. Washington, DC: U.S. Dept. of the Interior, Fish and Wildlife Service, Research and Development. 111 p.
11. Johnson RR, Oslund FT, Hertel DR (2008) The past, present, and future of prairie potholes in the United States. Journal of Soil and Water Conservation 63: 84a–87a.
12. Dahl TE (1990) Wetlands losses in the United States 1780's to 1980's. Washington, D.C.: U.S. Department of the Interior, Fish and Wildlife Service.
13. Sauer JR, Hines JE, Fallon J (2007) The North American Breeding Bird Survey, Results and Analysis 1966 – 2006. Version 10.13.2007. Laurel, MD, USGS Patuxent Wildlife Research Center.
14. Cowardin LM, Shaffer TL, Arnold PM, United States National Biological Service (1995) Evaluations of duck habitat and estimation of duck population

sizes with a remote-sensing-based system. Washington, D.C.: U.S. Dept. of the Interior, National Biological Service. 26 p.

15. Johnson R, Higgins K (1997) Wetland resources of eastern South Dakota. Brookings, SD, USA, South Dakota State University.

16. Stewart RE, Kantrud HA (1971) Classification of natural ponds and lakes in the glaciated prairie region.

17. Cowardin LM, Carter V, Golet FC, LaRoe ET (1979) Classification of wetlands and deepwater habitats of the United States. Washington, D.C.: US Department of the Interior/Fish and Wildlife Service.

18. Martin AE, Fahrig L (2012) Measuring and selecting scales of effect for landscape predictors in species-habitat models. Ecological Applications 22: 2277–2292.

19. Coulson DP, Joyce LA (2010) Historical climate (1940–2006) for the conterminous United States at the 5 arc minute grid spatial scale based on PRISM climatology. Fort Collins, Colorado: U.S. Department of Agriculture, Forest Service, Rocky Mountain Research Station.

20. Larson DL (1995) Effects of Climate on Numbers of Northern Prairie Wetlands. Climatic Change 30: 169–180.

21. Winter TC, Rosenberry DO (1998) Hydrology of prairie pothole wetlands during drought and deluge: a 17-year study of the Cottonwood Lake wetland complex in North Dakota in the perspective of longer term measured and proxy hydrological records. Climatic Change 40: 189–209.

22. Giorgi F, Mearns LO (1991) Approaches to the simulation of regional climate change: A review. Reviews of Geophysics 29: 191–216.

23. Coulson DP, Joyce LA, Price DT, McKenney DW, Siltanen RM, et al. (2009) Climate Scenarios for the conterminous United States at the 5 arc minute grid spatial scale using SRES scenarios A1B and A2 and PRISM climatology. Fort Collins, Colorado: U.S. Department of Agriculture, Forest Service, Rocky Mountain Research Station.

24. Skamarock W, Klemp J, Dudhia J, Gill D, Barker D, et al. (2008) A description of the advanced research WRF version 3, NCAR technical note. NCAR/TN–475+ STR.

25. National Center Atmospheric Research Staff (2014) The Climate Data Guide: Regridding Overview.

26. Nakicenovic N, Intergovernmental Panel on Climate Change. Working Group III (2000) Special report on emissions scenarios: a special report of Working Group III of the Intergovernmental Panel on Climate Change. Cambridge; New York: Cambridge University Press. 599 p.

27. Prasad AM, Iverson LR, Liaw A (2006) Newer classification and regression tree techniques: Bagging and random forests for ecological prediction. Ecosystems 9: 181–199.

28. Lawler JJ, White D, Neilson RP, Blaustein AR (2006) Predicting climate-induced range shifts: model differences and model reliability. Global Change Biology 12: 1568–1584.

29. R Development Core Team (2012) R: A language and environment for statistical computing. Vienna, Austria: R Foundation for Statistical Computer.

30. Chen C, Liaw A, Breiman L (2004) Using Random Forest to Learn Imbalanced Data. Berkeley, University of California.

31. Schröder B, Richter O (2000) Are habitat models transferable in space and time? Zeitschrift für Ökologie und Naturschutz 8: 195–205.

32. Guisan A, Thuiller W (2005) Predicting species distribution: offering more than simple habitat models. Ecology Letters 8: 993–1009.

33. Hastie T, Tibshirani R, Friedman JH (2009) The elements of statistical learning: data mining, inference, and prediction.

34. Liu CR, Berry PM, Dawson TP, Pearson RG (2005) Selecting thresholds of occurrence in the prediction of species distributions. Ecography 28: 385–393.

35. Hosmer DW, Lemeshow S (2000) Applied logistic regression: Wiley-Interscience.

36. Kundzewicz ZW, Mata LJ, Arnell N, Doll P, Kabat P, et al. (2007) Freshwater resources and their management.

37. Poiani KA, Johnson WC (1991) Global Warming and Prairie Wetlands - Potential Consequences for Waterfowl Habitat. Bioscience 41: 611–618.

38. Weller MW, Spatcher CS (1965) Role of habitat in the distribution and abundance of marsh birds. Iowa, Ames.

39. Matthews SN, Iverson LR, Prasad AM, Peters MP (2011) Changes in potential habitat of 147 North American breeding bird species in response to redistribution of trees and climate following predicted climate change. Ecography 34: 933–945.

40. Peterson AT (2003) Projected climate change effects on Rocky Mountain and Great Plains birds: generalities of biodiversity consequences. Global Change Biology 9: 647–655.

41. Tingley MW, Koo MS, Moritz C, Rush AC, Beissinger SR (2012) The push and pull of climate change causes heterogeneous shifts in avian elevational ranges. Global Change Biology 18: 3279–3290.

42. Cook B, Smerdon J, Seager R, Coats S (2014) Global warming and 21st century drying. Climate Dynamics: 1–21.

43. Elith J, Leathwick JR (2009) Species Distribution Models: Ecological Explanation and Prediction Across Space and Time. Annual Review of Ecology Evolution and Systematics 40: 677–697.

44. Elith J, Graham CH (2009) Do they? How do they? WHY do they differ? On finding reasons for differing performances of species distribution models. Ecography 32: 66–77.

45. Kantrud HA, Stewart RE (1977) Use of Natural Basin Wetlands by Breeding Waterfowl in North Dakota. The Journal of Wildlife Management 41: 243–253.

46. Beaumont LJ, Hughes L, Poulsen M (2005) Predicting species distributions: use of climatic parameters in BIOCLIM and its impact on predictions of species' current and future distributions. Ecological Modelling 186: 250–269.

47. Braunisch V, Coppes J, Arlettaz R, Suchant R, Schmid H, et al. (2013) Selecting from correlated climate variables: a major source of uncertainty for predicting species distributions under climate change. Ecography 36: 971–983.

48. Beale CM, Lennon JJ (2012) Incorporating uncertainty in predictive species distribution modelling. Philosophical Transactions of the Royal Society B-Biological Sciences 367: 247–258.

2

Present, Future, and Novel Bioclimates of the San Francisco, California Region

Alicia Torregrosa[1]*, Maxwell D. Taylor[2], Lorraine E. Flint[3], Alan L. Flint[3]

1 Western Geographic Science Center, United States Geological Survey, Menlo Park, California, United States of America, **2** Contractor, Western Geographic Science Center, United States Geological Survey, Menlo Park, California, United States of America, **3** California Water Science Center, United States Geological Survey, Sacramento, California, United States of America

Abstract

Bioclimates are syntheses of climatic variables into biologically relevant categories that facilitate comparative studies of biotic responses to climate conditions. Isobioclimates, unique combinations of bioclimatic indices (continentality, ombrotype, and thermotype), were constructed for northern California coastal ranges based on the Rivas-Martinez worldwide bioclimatic classification system for the end of the 20th century climatology (1971–2000) and end of the 21st century climatology (2070–2099) using two models, Geophysical Fluid Dynamics Laboratory (GFDL) model and the Parallel Climate Model (PCM), under the medium-high A2 emission scenario. The digitally mapped results were used to 1) assess the relative redistribution of isobioclimates and their magnitude of change, 2) quantify the loss of isobioclimates into the future, 3) identify and locate novel isobioclimates projected to appear, and 4) explore compositional change in vegetation types among analog isobioclimate patches. This study used downscaled climate variables to map the isobioclimates at a fine spatial resolution -270 m grid cells. Common to both models of future climate was a large change in thermotype. Changes in ombrotype differed among the two models. The end of 20th century climatology has 83 isobioclimates covering the 63,000 km² study area. In both future projections 51 of those isobioclimates disappear over 40,000 km². The ordination of vegetation-bioclimate relationships shows very strong correlation of Rivas-Martinez indices with vegetation distribution and composition. Comparisons of vegetation composition among analog patches suggest that vegetation change will be a local rearrangement of species already in place rather than one requiring long distance dispersal. The digitally mapped results facilitate comparison with other Mediterranean regions. Major remaining challenges include predicting vegetation composition of novel isobioclimates and developing metrics to compare differences in climate space.

Editor: Kimberly Patraw Van Niel, University of Western Australia, Australia

Funding: This work was supported by the US Geological Survey Land Change Science Program. The funders had no role in study design, data collection and analysis, decision to publish, or preparation of the manuscript.

Competing Interests: The authors have declared that no competing interests exist.

* E-mail: atorregrosa@usgs.gov

Introduction

Natural resource managers need tools to assess potential impacts of climate change across their local area of influence. Several approaches have been taken to use the outputs from global circulation models (GCM) to infer potential future change in vegetation, sea levels, and frequency of natural hazards such as wildfires and droughts. Selecting metrics from GCM outputs that facilitate quantitative comparison of biologically relevant changes is a major challenge for those seeking to understand the future effects of climate change on biological systems.

GCM climate variables are often analyzed as individual elements. For example, Cayan and colleagues [1] plot annual temperature anomalies from GCMs using the "A2" medium-high Intergovernmental Panel on Climate Change (IPCC) emission scenario to show marked shifts toward hotter annual temperatures in all future climate projections for California. Managers can use warming trend model convergence to support policy analysis based on a warming world. While raw climatic variables indicate potential magnitudes of change in one climate dimension, more integrated and biologically relevant combinations are better suited for exploring potential ecosystem response to changing climates

and map out the salient differences among future climate projections.

Thompson and colleagues [2] introduced the use of orthogonal axes of temperature and precipitation at continental scales to analyze the climate space of various tree species in the present and thereby forecast potential distributions of these species under future climate projections. Many species-environment models have been developed using various modeling algorithms [3], [4], [5] and some, such as the climate envelope model BIOCLIM use as many as 35 different climate parameters in the form of independent, continuous variables along a gradient [6]. Forecasting regional vegetation change based on species-specific climate based models has been criticized. Most biotic communities undergoing change are comprised of large numbers of species across multiple taxa each with varying amounts of genetic amplitude that determine species response and adaptation to climate change. Modeling more than a few species is time consuming and non-climatic factors can be significant determinants of species distributions such as competition, predation, and dispersal. These are lacking from most models [7], [8] rendering their results incomplete. Forecasting onto future landscapes with

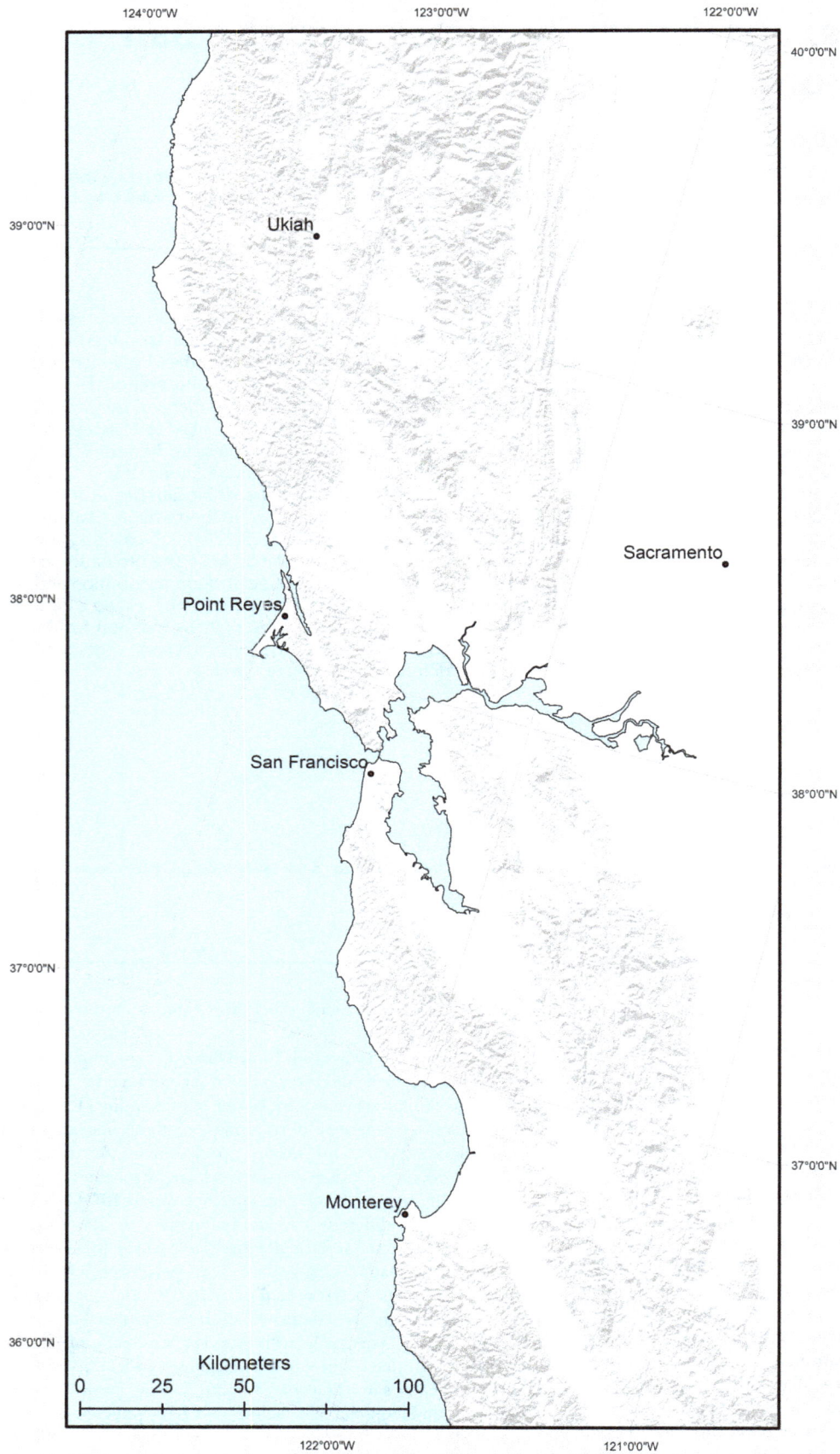

Figure 1. Topographic shaded relief of study area mapped in an Albers equal area projection.

climate envelopes that have no analogs to current conditions further complicates the task [9], [10].

An alternative approach is to work directly with bioclimates as categorical units and then populate these units with the species, communities, or functional traits distributed within these units. For example, several regions across the globe have a long term weather pattern of relatively mild year round temperatures, dry summers, and wet winters –a Mediterranean climate regime. Within this climate regime, an ombrotypic (wet/dry gradient) threshold distinguishes equivalent vegetation types. Those in more xeric conditions form a drought deciduous vegetation: coastal sage in California, garrigue in France, and renosterbos in South Africa; and in less xeric conditions a sclerophyllous evergreen shrub vegetation: chaparral in California, matorral in Chile, macchia in Italy, maquis in France and Israel, and fynbos in South Africa [11]. Focusing on bioclimate units has advanced research into the convergent physiognomy and ecophysiology of the regime [12], [13] as well as provided a better understanding of factors beyond climate that influence species response [14] and biodiversity [15], [16]. Likewise using a bioclimate framework at local scales, may help us think of innovative approaches to prepare for the unprecedented change that natural resource managers will be faced with.

The strong correspondence between vegetation and climate has long been used to map climate zones [17] and conversely to map vegetation [18], [19]. The landmark study of biologically relevant categorical breaks of climate variables by Rivas-Martinez [20], [21], [22] generated a hierarchical classification based on hundreds of thousands of relevés sampled along boreal - tropical latitudinal and elevational gradients. The Rivas-Martinez Worldwide Classification System (RMWBS) and other similar systems [23] provide numerically based methods whose spatial resolution is limited primarily by the resolution of the climate data. We use the RMWBS for our study area because it scales hierarchically and it captures with high sensitivity the precipitation and temperature patterns that differentiate plant communities in the Mediterranean climate [24], [25], [26] of the study area.

Downscaling GCM output to the local scale is particularly important for land management decisions that are implemented at the regional level such as land acquisition strategies to accommodate the dispersal of species of concern into more suitable habitat, restoration prioritization of one habitat patch over another, or managed translocation. Recognizing the challenge and the need for regional level climate change analysis and adaptation planning, the California Energy Commission (CEC) convened expert science panels to provide guidance, tools, and data to assist these efforts [27], [28]. An important guiding assumption for the CEC science panels was that even if future projections are uncertain, the use of the same projections would allow improved collaboration across adaptation management sectors such as energy, agriculture, water

Table 1. Climate indices used to derive the continentality, ombrotype, and thermotype indices, in order of appearance in text.

Index	Description	Calculation	Units
Ic	Continentality	$Ic = Tmax - Tmin$	degrees Celsius
Io	Ombrotype	$Io = Pp/Tp$	ratio
Ios^2	Ombrothermic index of the warmest bimonth of the summer quarter	$[Ios_2 = (Pps_2/Tps_2)\,10]$	scaled ratio
Ios^4	Ombrothermic index of the summer quarter	$[Ios_2 = (Pps_4/Tps_4)\,10]$	scaled ratio
It	Thermicity	$It = (T+m+M)\,10$	scaled (degrees Celsius)
Itc	Compensated Thermicity Index	if $(18.0 > Ic)$ then no compensation; if $(18.0 < Ic < = 21.0)$ then $Itc = It+5$; if $(Ic > 21.0)$ then $Itc = It+((Ic\,-21)+15)$.	scaled (degrees Celsius)
m	Average Minimum temperature of the coldest month	Thirty year average of minimum temperatures for January	degrees Celsius
M	Average Maximum temperature of the coldest month	Thirty year average of maximum temperatures for January	degrees Celsius
Pp	Yearly Positive Precipitation	Total average precipitation of those months whose average temperature is higher than 0°C	mm
Pps_2	Total precipitation of the warmest bimonth of the summer quarter	Thirty year average of the cumulative precipitation for July plus August	mm
Pps_4	Total precipitation of the summer quarter	Thirty year average of the cumulative precipitation for May, June, July, and September	mm
T	Yearly Average Temperature	Thirty year average of the average annual temperature	degrees Celsius
Tmax	Average temperature of warmest month of the year	Thirty year average of July monthly average temperatures	degrees Celsius
Tmin	Average temperature of coldest month of the year	Thirty year average of January monthly average temperatures	degrees Celsius
Tp	Yearly Positive Temperature		degrees Celsius
Tp	Positive Temperature Index	In tenths of degrees Celsius, sum of the monthly average temperature of those months whose average temperature is higher than 0°C	scaled (degrees Celsius)
Tps2	Total temperature of the warmest bimonth of the summer quarter	Temperature (July+Sept)	degrees Celsius

Some indices listed below, for example m –the average minimum temperature of the coldest month, are not specified in the text but are needed to calculate the indices that are described in the text.

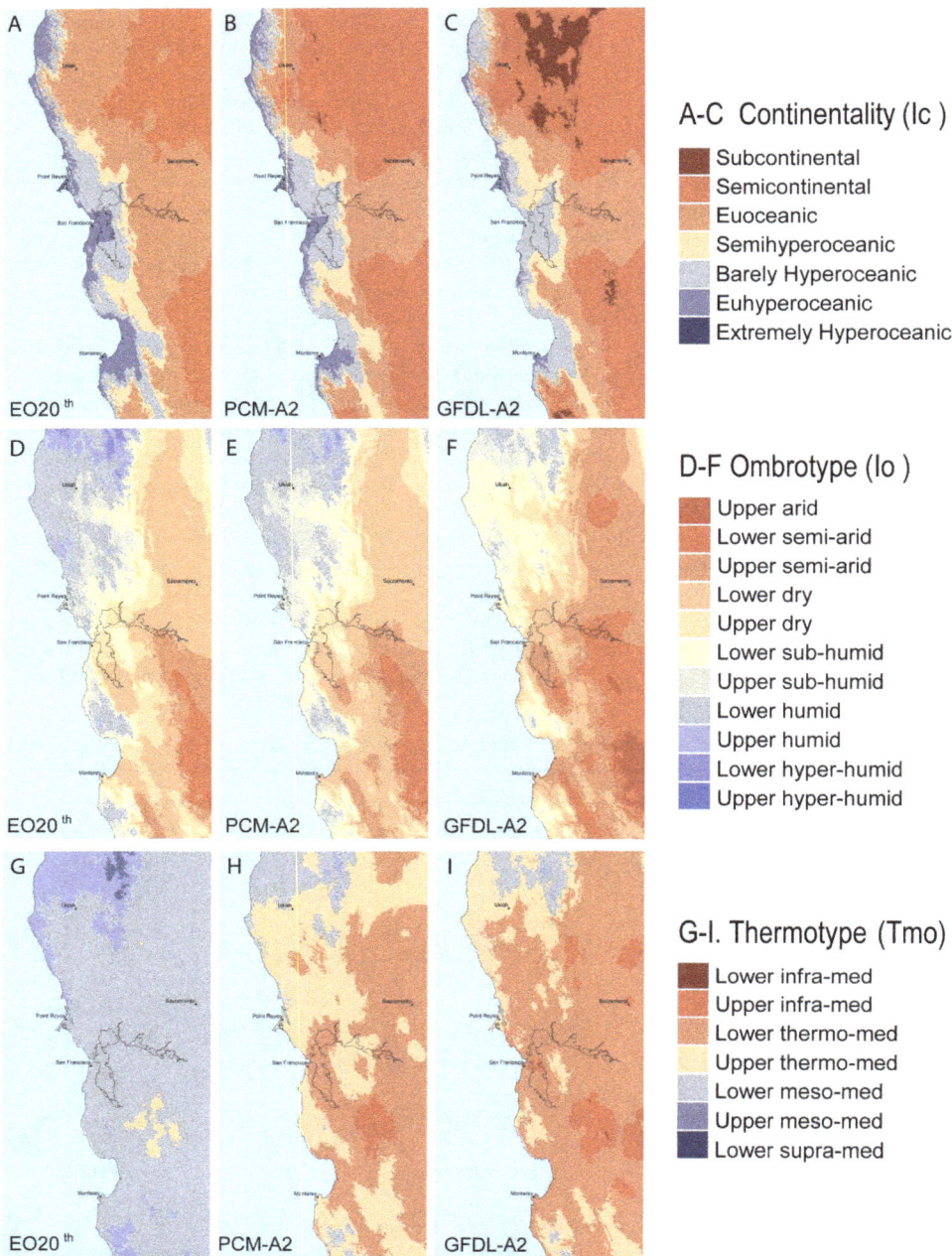

Figure 2. Distribution of three bioclimatic indices across the study area under three climatologies. *A* continentality during 1971–2000. *B* continentality projected for 2070–2099 under PCM-A2. *C* continentality projected for 2070–2099 under GFDL-A2. *D* ombrotype during 1971–2000. *E* ombrotype projected for 2070–2099 under PCM-A2. *F* ombrotype projected for 2070–2099 under GFDL-A2. *G* thermotype during 1971–2000. *H* thermotype projected for 2070–2099 under PCM-A2. *I* thermotype projected for 2070–2099 under GFDL-A2. Legend for thermotype classes abbreviates Mediterranean to "med."

resources, and wildlife. The CEC white paper on climate scenarios for California [1] describes the GCM models and emission scenarios selected to investigate climate change in California. The CEC white paper on downscaling predictor variables [29] describes the two GCMs (PCM and GFDL) and two emission scenarios (A2, medium-high and B1, low) identified for cross-sector use. It also details the derivation of climate variables to drive regional and local scale models. The selection of GCM models,

emission scenario and downscaling methods for this study are based on the results from the CEC science panel.

Traditionally the term macrobioclimate has been used for the five global bioclimate zones (tropical, Mediterranean, temperate, boreal, and polar). The term bioclimate has been used for the 5–7 categories within each macroclimate zone, and isobioclimate for the third nested combination of bioclimatic variants, thermotypes, and ombrotypes [see globalbioclimatics.org for more details]. We follow in this tradition for the products developed and discussed in this paper: 1) high resolution RMWBS isobioclimates for the end of the 20th century, 2) RMWBS isobioclimates for the end of the 21st century projected from two future climate models, PCM and GFDL; 3) change maps and transition matrices created by comparing 1 and 2; and 4) an alternative statistical approach for using bioclimate units to explore future plant community distribution.

The area of high resolution isobioclimates produced for this research coincides with the working boundary delineated by the Terrestrial Biodiversity and Climate Change Collaborative (TBC3), a group of university, government, and non-profit scientists conducting research and developing data sets for climate adaptation efforts in the greater San Francisco Bay Area [30]. This region is the largest biodiversity hotspot in the United States [31] based on the rarity-weighted richness index of rare and imperiled species of the United States [32]. Other studies currently underway by the authors use the same methodology and will expand the area of high resolution isobioclimate coverage to the entire State of California and beyond.

Methods

The study area grid was defined as a gridded rectangle in a modified Albers equal area projection that included all 10 San Francisco Bay Area counties with an additional 100 km buffer north and south and a 30 km buffer to the east for connectivity studies. The rectangle represents six million hectares (62,304 square kilometers – 15 million acres) along the northern coast of California bounded by latitudes 39.55 and 36.29 north and longitudes 123.79 and 120.54 west (Fig. 1) with a grid cell resolution of 270×270 meters (854,651 cells).

Rivas-Martinez Worldwide Bioclimatic Classification

The RMWBS integrates up to 26 climate parameters to derive and then segment into biologically relevant categories 3 primary indices, continentality, a measure of oceanic influence and temperature fluctuations; ombrotype, a measure of aridity; and thermotype, a synthetic measure of temperature regime. All calculations and mapping were done in the ARCGIS (geographic information system) [33]. The bay and delta areas were included in the analysis because bioclimate change also affects estuarine process of interest to resource managers in the region.

Each grid cell was categorized into isobioclimate types using the combined Rivas-Martinez bioclimatic indices of continentality [Ic], ombrotype [Io], and thermotype [Tmo] for two 30-year climatological periods, end of 20th century (EO20th) and end of 21st century (EO21st). Each of the three bioclimate indices is derived using the RMWBS climate parameter definitions, listed as equations in table 1, and the following RMWBS hierarchical classification procedures.

Table 2. R-Squared values for downscaled climate parameter and PRISM data compared to weather station observations.

Climate Parameter	R^2	
	4-km	270-m
Precipitation mm/month	0.6496	0.6497
Temperature (min) degrees C	0.866	0.8729
Temperature (max) degrees C	0.9147	0.9191

(Adapted from Figure 4. **Flint and Flint** *Ecological Processes* **2012** 1:2).

Table 3. Continentality index categories and range of values.

Codes		Continentality type	Value (It, Itc); CO*10
Alpha	Numeric		
Exho	1	Extremely Hyperoceanic	0–4
Euho	2	Euhyperoceanic	4.01–8
Bhoc	3	Barely Hyperoceanic	8.01–11
Seho	4	Semihyperoceanic	11.01–13
Euoc	5	Euoceanic	13.01–17
Seco	6	Semicontinental	17.01–21
Suco	7	Subcontinental	21.01–28

Continentality (Fig. 2A–C) is calculated as the range between the average temperatures of the warmest (Table 1, Tmax) and coldest months (Table 1, Tmin) of the year expressed in degrees Celsius. In the study area, the consistently warmest and coldest months the 100 years of the 20th century are July and January respectively.

Ombrotype (Fig. 2D–F) is calculated as the ratio between the yearly positive precipitation in millimeters (Table 1, Pp) and the yearly positive temperature in degrees Celsius (Table 1, Tp). The yearly positive precipitation index is defined as the total average precipitation of those months whose average temperature is higher than 0°C. Yearly positive temperature is the sum of the monthly average temperature of those months whose average temperature is higher than 0°C. In some Mediterranean regions where the warmest months of the year are closer to the autumnal equinox rather than the summer solstice there is a need to use summer compensated ombrothermic indices to discriminate between isobioclimates at the edges of Mediterranean and Temperate macrobioclimates. None of the cells in the study area require this compensation.

Thermotypes (Fig. 2G-i) were assigned based on thresholds for the thermicity index (Table 1, It), compensated thermicity index (Table 1, Itc), and positive temperature index (Table 1, Tp). The thermicity index is calculated as a sum of the yearly average temperature, the average minimum temperature of the coldest

Table 4. Ombrotype index categories and range of values.

Codes		Ombrotype	Value (Io) mm/C
Alpha	Numeric		
UARI	5	Upper arid	0.6–1
LSAR	6	Lower semiarid	1.01–1.5
USAR	7	Upper semiarid	1.51–2
LDRY	8	Lower dry	2.01–2.8
UDRY	9	Upper dry	2.81–3.6
LSHU	10	Lower subhumid	3.61–4.8
USHU	11	Upper subhumid	4.81–6
LHUM	12	Lower humid	6.01–9
UHUM	13	Upper humid	9.01–12
LHHU	14	Lower hyperhumid	12.01–18
UHHU	15	Upper hyperhumid	18.01–24

Table 5. Thermotype index categories and range of values.

Codes		Thermotype	Value (It, Itc); CO*10	Value (Tp) used if Ic >= 21 or It, Itc<120
Alpha	Numeric			
Lsme	1	Lower supramediterranean	145–210	1200–1500
Umme	2	Upper mesomediterranean	211–280	1501–1825
Lmme	3	Lower mesomediterranean	281–350	1826–2150
Utme	4	Upper thermomediterranean	351–400	2151–2300
Ltme	5	Lower thermomediterranean	401–450	2301–2450
Uime	6	Upper inframediterranean	451–515	2451–2650
Lime	7	Lower inframediterranean	516–580	>2650

month of the year, and the average maximum temperature of the coldest month of the year. In this study area, January is consistently the coldest month of the year. A compensated thermicity index is used when the continentality index value is above 18°C. When continentality is moderate ($18.0 < Ic <= 21.0$) the Itc compensation value of 5 is added and when high ($Ic > 21.0$) the compensation value is calculated as the sum of $((Ic - 21) + 15)$. The positive temperature index is included only when $Ic > 21$ or $It/Itc < 120$. For these cases the positive temperature index is derived as the sum of the monthly average temperature of those months whose average temperature is higher than 0°C. In the study area all months of the year have a monthly average temperature above 0°C.

monthly precipitation (mm), and minimum and maximum monthly air temperature (°C) products [34] that were spatially downscaled to 270-m grid cells using a modification [35] of the gradient plus inverse distance squared (GIDS) interpolation approach [36]. This interpolation scheme generates multiple regression equations for each month for each 4-km cell relative to each target 270-m grid cell using the relation of each climate variable to elevation and location. Assessments [35] comparing measured climate data from weather stations with the two interpolated products, the 4-km PRISM and downscaled 270-m climate grids, calculated higher $R2$ for the 270-m downscaled climate than PRISM for all climate values tested (Table 2). This reflects the capacity of higher precision grid cells to more closely reflect point data, all other things being equal. One advantage of this interpolation technique over kriging is that it does not require the assumption of stationarity of data and it incorporates spatial and temporal changes in adiabatic lapse rate and the influence on local climate.

Bioclimate indices were calculated for the two EO21st projections that were available from Thorne et al. [29]. These 270-m downscaled projections were developed on the basis of 12-km national maps of downscaled GCM output available at http://tenaya.ucsd.edu/wawona-m/downscaled/. Projected values for monthly precipitation (mm), and minimum and maximum monthly temperature (oC) were further downscaled to 270 meters in 3 steps [35]. The first step downscales the 12-km grids to 4-km using the GIDS approach to enable bias-correction to the PRISM grids. The second step applies the bias correction to adjust the mean and standard deviation to match those of the 1950–2000 record developed in PRISM. Once corrected, the 4-km grids are downscaled to 270-m grid using the GIDS approach [35].

To simplify the narrative describing the metrics, analyses, and results from this research only one emission scenario is reported.

We selected the results from the A2-medium-high emission scenario because the CO_2 emissions of both A-2 and the B1-low emission had already been well exceeded by 2012.

Changes in the spatial distribution of bioclimate index categories between the EO20[th] and EO21[st] periods were generated in using cell by cell raster ARCGIS processing. Change was quantified as the number of categories that differed between periods. For example, if a cell in the EO20[th] period has the continentality category (Table 3) of extremely hyperoceanic (EXHO) and in the future period the cell was barely hyperoceanic (BHOC), it was assigned a change index of 2, if the cell had a continentality category of BHOC in both the EO20[th] and future periods it was assigned a change of 0.

Isobioclimates were identified as unique categorical combinations of the three indices, continentality with 7 categories (Table 3), ombrotype with 12 categories (Table 4), and thermotype with 7 categories (Table 5). Novel isobioclimates are defined as those unique isobioclimates that are not present in the study area in the EO20[th] period.

Associations between vegetation and bioclimate indices were developed with canonical correspondence analysis (CCA) implemented using CANOCO 4.5 [37]. Vegetation type abundance and distribution data from the statewide California Department of Forestry and Fire Protection, Fire and Resource Assessment Program Multi-Source Vegetation data layer [38] were clipped to the study area. Of the 77 statewide vegetation types 42 are found in the study area. Nineteen vegetation types were removed using a set of 3 criteria to generate a subgroup of 23 vegetation-types for the CCA ordination. Removal criteria included: 1) anthropogenic types not expected to change due primarily to climate such as urban and agriculture, 2) vegetation types dependent on local hydrology such as riparian and wetland, and 3) types classified as unknown shrubs and conifer. Abundance was defined as the percent cover of each vegetation type found in each isobioclimate with each isobioclimate treated as a plot. Each bioclimate index was treated as an environmental variable in the analysis: continentality, ombrotype, and thermotype.

Results

All the cells in the study area in the EO20[th] period and both projections of the EO21[st] period had a summer ombrothermic index (Ios4) of less than 2 placing the entire study area in the Mediterranean macrobioclimate. Only the driest of the 7 major Mediterranean bioclimates, hyperdesertic Mediterranean, is not represented in the study in either the EO20[th] or EO21[st] periods.

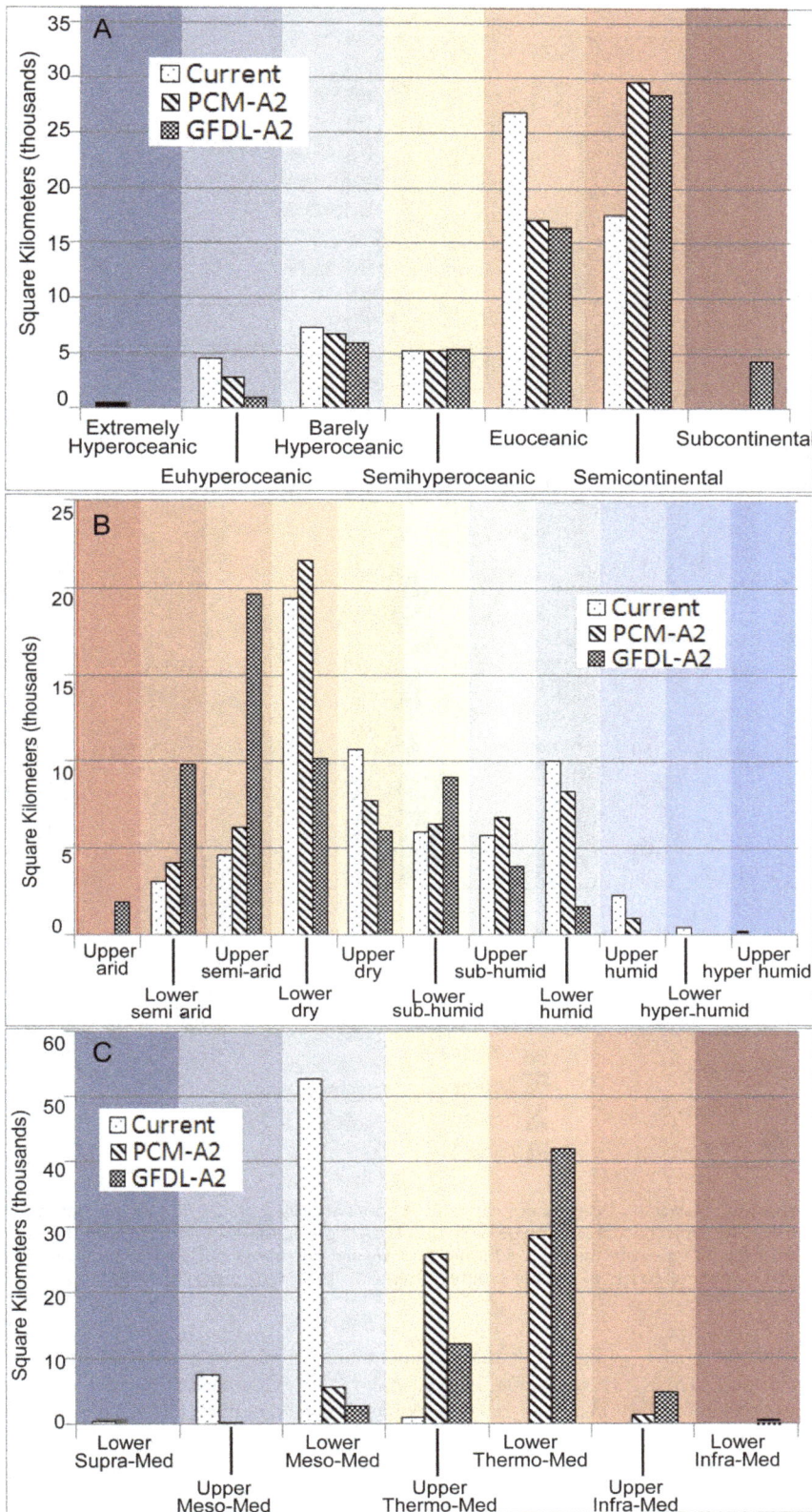

Figure 3. Areal comparison of three bioclimatic indices during the end of the 20th century and two projections of end of the 21st century. *A* continentality. *B* ombrotype. *C* thermotype. Thermotype classes use the abbreviation "med" for Mediterranean.

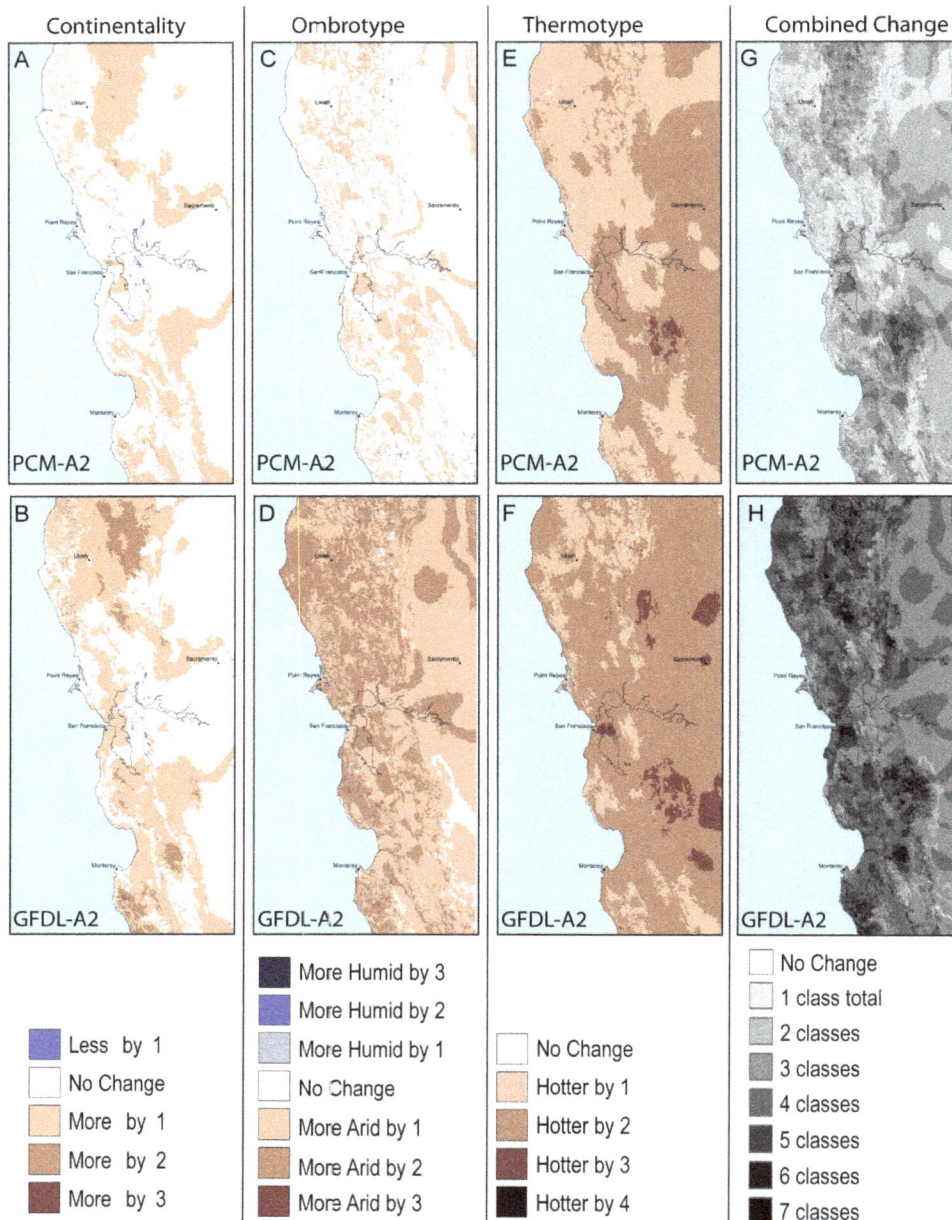

Figure 4. One hundred years of isobioclimate change. Numbers of classes of change in *A* continentality with the PCM-A2 projection. *B* continentality with the GFDL-A2. *C* ombrotype with the PCM-A2 projection. *D* ombrotype with the GFDL-A2 projection. E ombrotype with the PCM-A2 projection. *E* thermotype with the PCM-A2 projection. *F* thermotype with the GFDL-A2 projection. *G* Total number of classes of change summed from each bioclimate index in the PCM-A2 future. *H* Total number of classes of change summed from each bioclimate index in the PCM-A2 future.

The values for continentality (Ic) range from extremely hyperoceanic (EXHO) to semicontinental (SECO) in the EO20[th] period (Table 3). In both modeled future projections an additional category of subcontinental appears (SUCO). In the EO20[th] period the majority of the study area, about 27,000 km^2, is in the euoceanic category (Fig. 2A and Fig. 3A) with an Ic value of 13°–17°C. At the end of the century under both the PCM-A2 and GFDL-A2 projections a majority, 30,000 km^2 and 28,000 km^2 respectively, transitions into the SECO range of 18–21°C (Fig. 2B, C, and Fig. 3A). Under the GFDL projection more area changes Ic values with 4400 km^2 transitioning into a SUCO category predominantly in the north eastern portion of the study area

(Fig. 2C) where topographic complexity and elevation is the highest (Fig. 1). Under the PCM-A2 projection the Point Reyes Peninsula remains extremely hyperoceanic. The marked coastal influence associated with the east-west connection of the bay-delta to the ocean is seen strongly in the EO20[th] period and maintained in both future projections although much less in the GFDL-A2 projection (Fig. 4A and B). Continentality increases in areas of higher elevation and substantially so with GFDL-A2.

The lower dry (LDRY) ombrotype (Io) category covers the most area in both the EO21[st] period and the PCM-A2 future at 19,000 km^2 and 22,000 km^2 respectively (Fig. 2E, f and Fig. 3B). The landscape of the GFDL-A2 future becomes much more arid

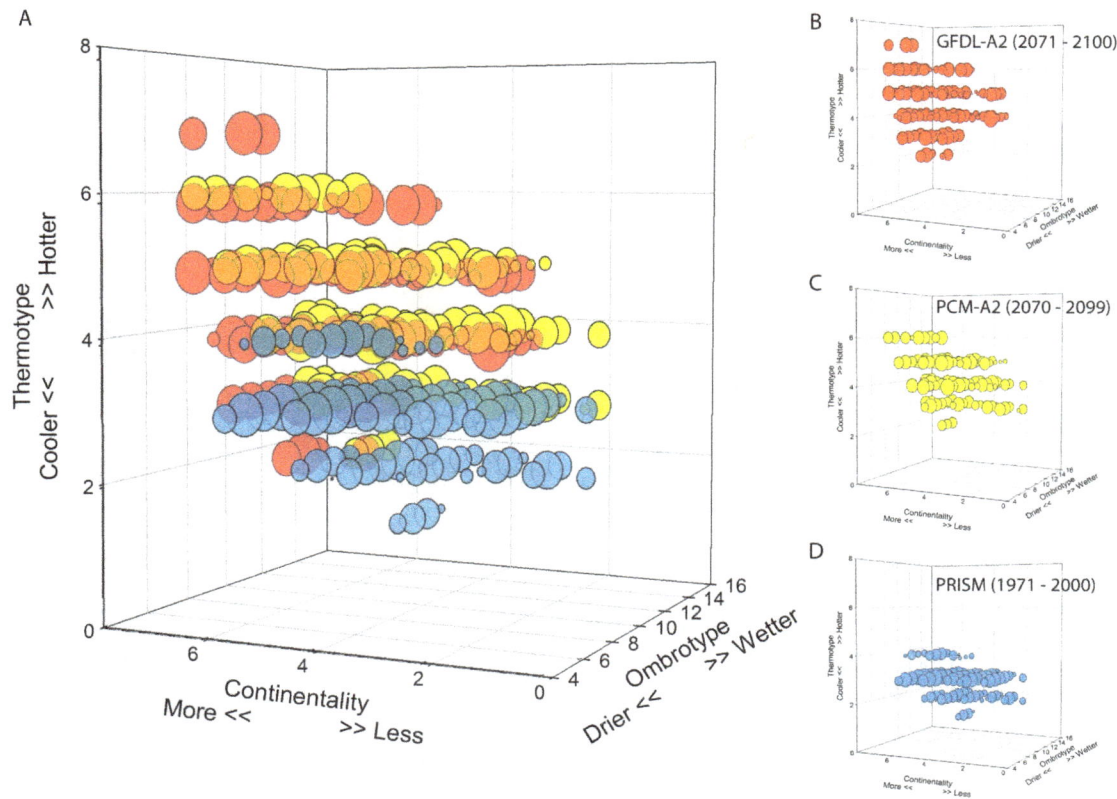

Figure 5. Unique Isobioclimates each located within a 3-dimensional cube of climate space. *A* The entire set of 195 isoclimate types representing climate conditions at the end of the 20th century and two projections of the future, GFDL-A2 and PCM-A2. The size of the semi-transparent circles represents, on a logarithmic scale, the relative area for that unique isobioclimate across the landscape. Each temporal period/climate data source is represented by a different color, red = 2071–2100/GFDL-A2, yellow = 2070/PCM-A2 , and blue = 1971–2000/PRISM. Circles that appear orange are a result of red and yellow circles overlapping and represent isobioclimates that are common to both future projections. *B* 115 isobioclimate types of GFDL-A2 climate projections . *C* 108 isobioclimate types of PCM-A2 climate projections. *D* 83 isobioclimates of the end of 20th century climate conditions.

with the majority of the landscape (20,000 km^2) becoming upper semiarid (USAR). Most of the area in the GFDL-A2 projection shifts one or two categories drier with only a small area of no change in the southeastern portion of the landscape (Fig. 4D). This area of no change is in the driest of categories, upper arid (UARI), however had it become drier it would have brought in the regionally novel Io category, hyperarid. The distribution of Io change in both PCM-A2 and GFDL-A2 futures is closely associated with elevation with the lower elevations experiencing relatively less change than the higher elevations, except in the coastal area northwest of Ukiah under the GFDL-A2 future.

The dominant thermotype (Tmo) category in the study area, lower mesomediterranean (LMME), covers 53,000 km2 under EO20[th] climate (Fig. 2G and 3C). Along the coast and in the northern coastal mountains the thermotypes are the cooler, upper mesomediterranean (UMME) and lower supramediterranean (LSME). The warmest Tmo, upper thermomediterranean (UTME), occurs under EO20[th] conditions as a small patch in the interior, southeast of San Francisco. In both projections more than 99.7% of the landscape becomes warmer by 1–2 categories (Fig. 2H, 2I, 4E, and 4F) while in the GFDL-A2 scenario 84% of the landscape becomes warmer by 2 or more Tmo categories (Fig. 4F). The Point Reyes peninsula is one of the small areas that do not undergo a Tmo change but only under the future PCM-A2 projection (Fig. 4E).

The three bioclimate indices (Io, Ic, Tmo) combined produce 83 unique isobioclimates in the EO20[th] century, 108 in the PCM-A2 projection, and 115 in the GFDL-A2, for a total of 195 unique combinations (Table 6). When graphed in 3-dimensions (3-D) each unique isobioclimate can be represented by an x-y-z coordinate on a 3-D grid. Each climatological period has a distinct climate space that it occupies within the climate cube (Fig. 5). Isobioclimates found in the EO20[th] and both future climate spaces show as overlapping points and represent isobioclimates extant into both projections of future conditions (Fig. 6A, B, and C). Some are present in the EO20[th] but disappear in the future (Fig. 6A), and those that have no overlap are the regionally novel isobioclimates of the future (Fig. 6B and C). Isobioclimates unique to the PCM-A2 projections have a combination of the lowest values for continentality and middle thermotype values while those unique to GDFL-A2 are in the highest thermotype areas of the climate cube (Fig. 5).

The EO20[th] regional landscape has 51 isobioclimates covering close to 40,000 km2 that disappear in both future projections (Table 6, Fig. 5 and Fig. 6A). Eighteen novel isobioclimates are found exclusively in the PCM-A2 projection (Fig. 5) but occupy a small portion of the landscape 327 km^2 (Fig. 6B). The 34 novel isobioclimates exclusive to the GFDL-A2 future occupy a larger area, 4,494 km2 and are concentrated mainly in the highest elevations northeast of Ukiah (Fig. 6C).

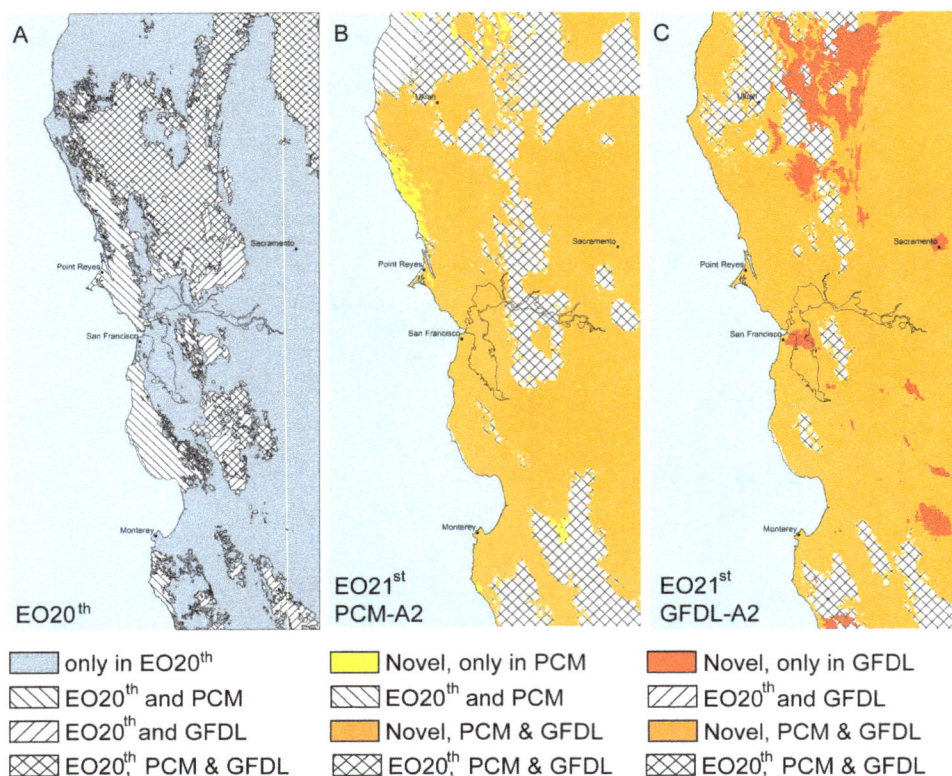

Figure 6. The study area seaprated into 4 categories based on the continuity of isobioclimates. *A* based on end of the 20th century climate conditions, areas where isobioclimates dissapear in the future are coded blue, areas where the isobioclimates dissapear only under the GFDL projections (e.g. continue under PCM projection) are hatched to the right, areas where the isobioclimates dissapear only under the PCM projections (e.g. continue under GFDL projection) are hatched to the left, and areas where the isobioclimate continue under both projections are cross hatched. *B* novel and extant isobioclimates under the PCM-A2 climate projection. *C* novel and extant isobioclimates under the GFDL-A2 climate projection.

Each of the three indices represents a biologically relevant factor as demonstrated by the eigenvector lengths of the three environmental variables (Fig. 7) and the high correlation coefficients of the first three axes of the canonical correspondence analysis (CCA) of vegetation types as a function of the bioclimate index (Table 7). CCA is a constrained ordination, a matrix algebra based eigenanalysis [39] that measures the strength of the relationship between abundance (in this case dominant vegetation type) and environmental factors (in this case three bioclimate indices). The correlation coefficients and eigenvalues of the three indices and axes suggest that ombrotype is the strongest

determinant of vegetation distribution followed by continentality and thermotype. The plot of the first two CCA axes (Fig. 7) shows that Axis 1 is dominated by ombrotype; for example redwood and Douglas-fir are to the left (wetter), montane hardwoods and mixed chaparral occupy the center, and blue-oak, valley oak, and coast live oak woodlands are to the right (drier), with juniper and desert scrub occupying the driest ombrotypes at the far right. The second axis is dominated by continentality - coastal scrub and perennial grasslands occupy the most oceanic areas at the top. Redwood tends toward more oceanic climates than Douglas-fir, while they occupy similar ombrotypes. Similarly, coast live oak and blue-oak

Table 6. Number of isobioclimates common or unique in the EO21st climatology and two modeled projections of the EO22nd climatology.

Category	# of unique isobioclimates	Current	PCM-A2	GFDL-A2
EO20th and found in both future projections	19	13,216	16,754	8,124
EO20th and in PCM-A2	11	6,177	2,002	-
EO20th and in GFDL-A2	2	3,176	-	5
EO20th but not in either future projection	51	39,735	-	-
Novel found in PCM-A2 only	18	-	327	-
Novel found in GFDL-A2 only	34	-	-	4,494
Total	195	62,304	62,304	62,304

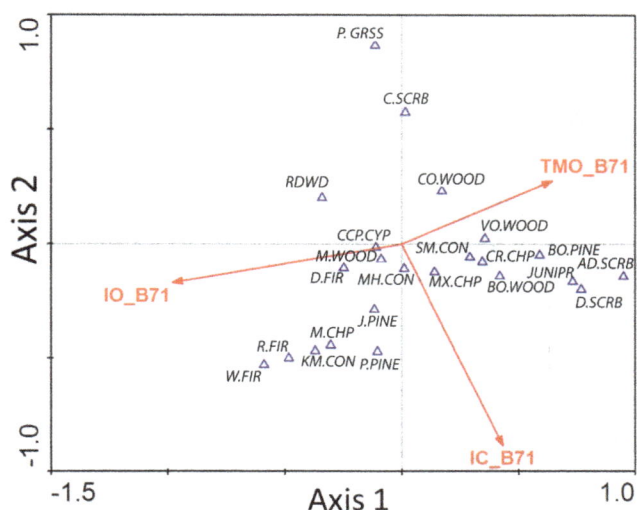

Figure 7. Canonical correspondence analysis (CCA) axes 1 and 2 biplots. The vegetation is ordinated with three 1971–2000 climate indices, continentality (IC_B71), ombrotype (IC_B71), and thermotype (TMO_B71). The eigenvalue vectors for the indices are superimposed on the CCA-biplot ordination to indicate relative influence of each index on species abundance and distribution. Vegetation codes: alkali desert scrub (AD.SCRB), blue oak-foothill pine (BO.PINE), blue oak woodland (BO.WOOD), coastal scrub (C.SCRB), closed-cone pine-cypress (CCP.CYP), coastal oak woodland (CO.WOOD), chamise-redshank chaparral (CR.CHP), Douglas-fir (D.FIR), desert scrub (D.SCRB), Jeffrey pine (J.PINE), juniper (JUNIPR), Klamath mixed conifer (KM.CON), montane chaparral (M.CHP), montane hardwood (M.WOOD), montane hardwood-conifer (MH.CON), mixed chaparral (MX.CHP), perennial grassland (P.GRSS), ponderosa pine (P.PINE), red fir (R.FIR), redwood (RDWD), sierran mixed conifer (SM.CON), valley oak woodland (VO.WOOD), white fir (W.FIR).

Table 7. Regression/canonical coefficients for ombrotype (IO_c71), continentality (IC_C71), and thermotype (TMO_C71) under EO21st climatology for 3 ordination axes.

CODE	EIGENVALUE INDEX	AXIS 1	AXIS 2	AXIS 3
IO_B1	OMBROTYPE	−0.8803	−0.2700	−0.9009
IC_B1	CONTINENTALITY	0.1932	−0.9954	−0.2105
TMO_B1	THERMOTYPE	0.0824	0.2589	−1.2281

woodlands are differentiated by continentality. Although montane conifer types (including Jeffery pine, ponderosa pine, red fir and white fir) are relatively rare in the region (small areas in the northern mountains) they extend into the lower left quadrant (wet ombrotypes, more continental) in a reasonable sequence. These relationships are well known botanically [40], [41] and give added confidence to the use of the bioclimate indices.

Discussion

The Rivas-Martinez classification successfully captured the high spatial climatic variability in the California coast range and allowed for delineation of isobioclimates at a fine spatial resolution. These isobioclimates are correlated with the proportions of vegetation-types in well-known patterns and provide a template for considering potential shifts in vegetation.

Seasonal and diurnal temperature fluctuations that are strongly influenced by ocean-land-atmosphere processes, such as attenuation of temperature extremes in coastal areas by fog, are a measure of the maritime coastal influence. This coastal effect, reducing the difference between maximum and minimum seasonal and daily temperatures is captured by the continentality index. In the study area the strong bay-delta signature, noticeable at the coastal opening of the Golden Gate, extends east. Marine stratocumulus and fog enter into the study area lowering the maximum temperatures during the day, increasing the minimum temperatures at night, and muting seasonal temperature swings. In both future projections, especially in GFDL-A2 there is a shift toward more continentality. The vegetation ordination shows perennial

grassland, coastal scrub, redwood, and coastal oak woodland distributed toward the lower end of Ic (Fig. 7) suggesting these communities types would contract under GFDL-A2 conditions. Improving modeling of ocean-atmosphere-land dynamics would help to better predict potential future changes in fog and therefore improve forecasts of continentality and the impact on vegetation types associated with foggier coastal areas.

The thermotype landscape shows a greater patchiness between types in future projections and has the greater deviation from EO20th conditions as measured by the number of thermotype class difference (Fig. 2G–I and 3E–F). Many future isobioclimate types consist of existing combinations of continentality and ombrotype but become regionally novel isobioclimate types when the thermotype index is added to the combination. The thermotype index has the weakest correlation coefficient in the CCA ordination. Further analysis could clarify if this is due to insufficient categories for successfully quantifying the thermal impact on vegetation or greater thermal tolerance in species adapted to Mediterranean climate regimes.

Unlike the widespread view that vegetation envelopes will move upslope and north into higher latitudes, in this region the cooler refugia are west toward the coast. These results are consistent with Loarie et al. 2008 that showed coastal areas having the highest potential for maintaining cooler mesic habitats. Indeed the west coast of North America has been a biogeographic refuge over geologic time because of the moderating influence of the Pacific Ocean [42], [43], [44], [45]. Global climate-driven changes in offshore currents and coastal upwelling, such as have been documented in the paleo-record [46], [47], as well as more recently [48], highlight the influence of ocean conditions on terrestrial climate.

The high correlation between biota and isobioclimates can be used to simplify complex distributional patterns. The categorical breaks along each of the bioclimate indices represent thresholds that have been defined because they are useful for distinguishing vegetation patterns on the landscape. Climate shifts that jump across several bioclimate categories represent multiple quanta of change across the landscape. Simple metrics of composite change summed up from each dimension suggest a landscape with sufficient complexity to harbor potential ecosystem resilience and could be used to further test hypotheses of species persistence [49]. Identifying areas of greatest expected isobioclimate change helps to identify areas of greatest vulnerability especially when different models project the same locations to change the most (Fig. 4). This geolocational intelligence can assist land managers to identify specific refugia, prioritize adaptive management efforts, and target lands for acquisition.

Future work on isobioclimates such as developing additional metrics to simplify maps of total change (Fig. 4G and 4H) could help to sharpen our understanding of vulnerability as it relates to

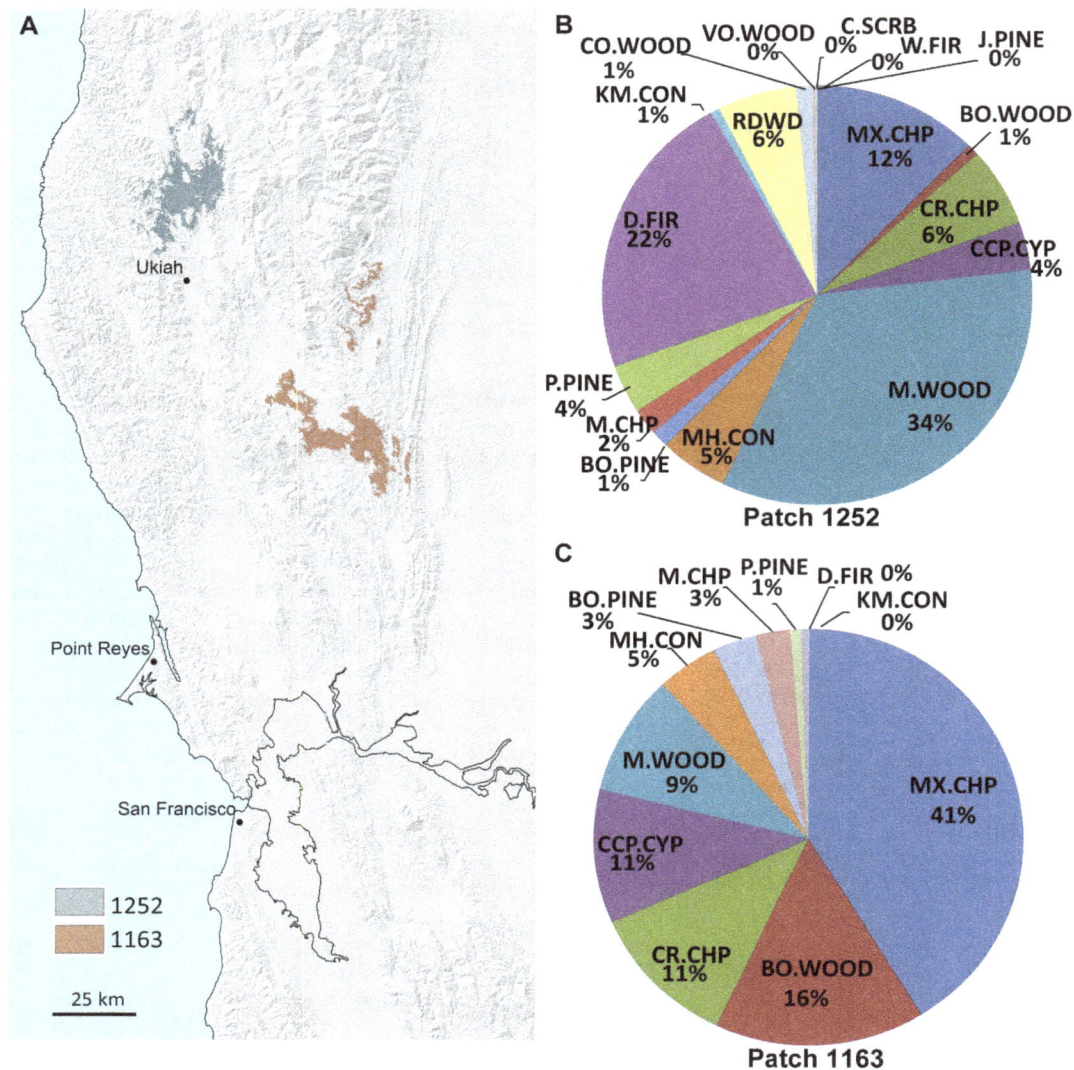

Figure 8. Isobioclimate patch analogs. *A* Map of two locations with different isobioclimates, 1252 and 1163 during the end of 20th century climate condition. Under the climate conditions of the PCM-A2 projection, patch 1252 (the blue area in the north) will have the same isobioclimate as patch 1163 (the red area to the southeast) does at the end of the 20th century. *B* dominant shrub and tree abundance in patch 1252 based on end of 20th century California Fire and Resource Assessment vegetation maps. *C* dominant shrub and tree abundance in patch 1163 from same vegetation map.

different dimensions of climate change. Improved metrics to compare change of isobioclimate across climate space could be used to test hypotheses of relative resilience or determine climate dimensions of greatest impact. For example, some areas of greatest regional novelty such as "lower hyperhumid – hyperoceanic – upper thermomed" conditions in the PCM-A2 projection (small isolated yellow circle in fig. 5 with an isobioclimate climate space location of $X = 11$, $Y = 1$, $Z = 4$) cover relatively small areas (Fig. 6B). Yet their impact could be ecologically quite large if this particular isobioclimate provides a harbor for pathogens that allows them to get a regional foothold. Isobioclimates represent climate conditions averaged over at least 30 years which means that some years will be more wet or dry than others. The expansion and contraction of pathogenic, or invasive, populations during these years of extreme conditions is enhanced if refugia remain somewhere on the landscape. Understanding the geoloca-

tional identity of these foci can help to better understand the spread of propagules of interest.

Current bioclimate analogs can also be used to identify areas in the current landscape that contain conditions similar to those that might be expected in the future. For example in Fig. 8A the patch identified as isobioclimate 1252 currently has lower humid – euoceanic – upper mesomediterranean climate conditions. In the PCM-A2 future it is projected to become isobioclimate 1163 with upper subhumid – semicontinental – lower mesomediterranean conditions. The current, e.g. EO20[th], isobioclimate, 1252, supports primarily montane hardwood, Douglas fir, mixed chaparral, redwood, chamise chaparral, and small percentages of other vegetation types (Fig. 8B). Under the PCM-A2 scenario, isobioclimate patch 1252 will become patch analog 1163, which is currently found to the southeast and is a mix of mixed chaparral, blue oak woodland, closed-cone pine cypress, chamise chaparral, montane hardwood, and small percentages of 6 other vegetation types (Fig. 8C). Note that every vegetation type in patch 1163 is already present in patch 1252 suggesting that long distance dispersal, at least at the dominant species level, is not necessary. Therefore, vegetation change will primarily be a local rearrangement of species already in place. However, several dominant species, in particular redwood, are not currently found in patch 1163, and these vegetation types would be expected to be lost when this location transitions into isobioclimate 1252 if future projections hold.

Adaptive management strategies implemented by natural resource managers could include identifying and protecting small local patches of the more arid vegetation/species that provide foci for spread. For example, the 1% blue oak woodland found in patch 1252 serves as the nucleus for expansion to the relative abundance of 16% currently found in patch 1163. Another strategy, albeit more controversial, is managed translocation of species from analog patches to enhance habitat value. Very fine scale topoclimate variability becomes an important factor because short distance dispersal, such as from the other side of a canyon, may provide propagules for community recombination. This suggests differentiating the landscape into areas where the velocity of climate change is more or less important depending on the dispersal abilities of species [50], [51], [52] and taking advantage of higher resolution elevation data to better incorporate topoclimatic variation. Developing adaptive management strategies will be further challenged by ecosystem processes that are stimulated or exacerbated by increasing temperatures or changing precipitation patterns such as wildfires [53] and the increased photosynthetic efficiency in response to increasing levels of CO_2 [54], [55].

The geography of climate change is complex and multi-scaled [56]. The relationship between climate and vegetation mosaics in California mountains is a fine scale process driven by topoclimatic effects (solar radiation, cold air pooling) and below the scale of this analysis. The next step but beyond the scope of this study, is to investigate analog patterns of bioclimatic -vegetation relationships using higher resolution isobioclimate and vegetation data across larger areas using CCA or other statistical techniques [57], [58].

Extending high resolution isobioclimate mapping efforts to larger areas, will affect what is labeled a novel isobioclimate. Isobioclimate novelty is related to the spatial scale of analysis. Isobioclimates that are novel on a regional scale may lose their novelty at the state-wide scale if analogs exist for them at the state-wide scale. The concept of regionally novel isobioclimates is none-the-less important for ecological conservation purposes. Conceptually it is similar to the distinction made between local and global rarity of plants species. These distinctions are important for conservation and protection of rare species and arise in part from research into the process of extinction and geographic fragmentation.

The analyses in this paper characterize the landscape in ways that can be used for land management decisions. The biologically relevant categories that define individual isobioclimate units facilitate their use as analytic units to explore change across the landscape. Isobioclimate analogs provide a framework to generate hypotheses and forecasts of shifts in vegetation community structure in a response to climate change. Implementing a worldwide bioclimatic classification system at regional to local scales provides a multi-scale framework for investigating the response of biotic systems to climate change.

Acknowledgments

We thank Roger Sayre, Mara Tongue, and two anonymous reviewers for excellent feedback and suggestions and Stuart Weiss for insightful discussions.

References to non-USGS products and services are provided for information only and do not constitute endorsement or warranty, express or implied, by the USGS, USDOI, or U.S. Government

Author Contributions

Conceived and designed the experiments: AT. Performed the experiments: AT MT LF AF. Analyzed the data: AT MT LF AF. Contributed reagents/materials/analysis tools: AT LF AF. Wrote the paper: AT MT.

References

1. Cayan D, Maurer E, Dettinger M, Tyree M, Hayhoe K (2008) Climate change scenarios for the California region. Climatic Change 87: 21–42.
2. Thompson RS, Hostetler SW, Bartlein PJ, Anderson KH (1998) A strategy for assessing potential future change in climate, hydrology, and vegetation in the western United States. Washington, D.C. , USA.: U.S. Geological Survey.
3. Theurillat J-P, Guisan A (2001) Potential Impact of Climate Change on Vegetation in the European Alps: A Review. Climatic Change 50: 77–109.
4. Elith J, Leathwick JR (2009) Species Distribution Models: Ecological Explanation and Prediction Across Space and Time. Annual Review of Ecology, Evolution, and Systematics 40: 677–697.
5. Notaro M, Mauss A, Williams JW (2012) Projected vegetation changes for the American Southwest: combined dynamic modeling and bioclimatic-envelope approach. Ecological Applications 22: 1365–1388.
6. Beaumont LJ, Hughes L, Poulsen M (2005) Predicting species distributions: use of climatic parameters in BIOCLIM and its impact on predictions of species' current and future distributions. Ecological Modelling 186: 251–270.
7. Pearson R, Dawson T (2003) Predicting the impacts of climate change on the distribution of species: Are bioclimate envelope models useful? Global Ecol Biogeog 12: 361.
8. Wisz MS, Pottier J, Kissling WD, Pellissier L, Lenoir J, et al. (2012) The role of biotic interactions in shaping distributions and realised assemblages of species: implications for species distribution modelling. Biological reviews of the Cambridge Philosophical Society.
9. Stralberg D, Jongsomjit D, Howell CA, Snyder MA, Alexander JD, et al. (2009) Re-Shuffling of Species with Climate Disruption: A No- Analog Future for California Birds? PLoS ONE 4.
10. Williams JW, Jackson ST (2007) Novel climates, no-analog communities, and ecological surprises. Front Ecol Environ 5: 475–482.
11. Cody ML, Mooney HA (1978) Convergence Versus Nonconvergence in Mediterranean-Climate Ecosystems. Annual Review of Ecology and Systematics 9: 265–321.
12. Castri Fd, Mooney HA (1973) Mediterranean type ecosystems : origin and structure. Berlin: Springer.
13. Mooney HA, editor(1977) Convergent evolution in Chile and California: Mediterranean climate ecosystems. Stroudsburg, Pa, USA: Dowden, Hutchinson and Ross Inc. 238 p.
14. Rundel PW (1982) Nitrogen utilization efficiencies in mediterranean-climate shrubs of California and Chile. Oecologia 55: 409–413.
15. Sala OE, Chapin III SF, Armesto JJ, Berlow E, Bloomfield J, et al. (2000) Global Biodiversity Scenarios for the Year 2100. Science 287: 1770–1774.
16. Klausmeyer KR, Shaw MR (2009) Climate Change, Habitat Loss, Protected Areas and the Climate Adaptation Potential of Species in Mediterranean Ecosystems Worldwide. PLoS ONE 4: e6392.
17. Köppen W (1900) Versuch einer Klassifikation der Klimat, Vorsuchsweize nach ihren Beziehungen zur Pflanzenwelt. Geographische Zeitschrift 6: 593–611, 657–612
18. Holdridge LR (1947) Determination of world plant formations from simple climatic data. Science 105: 367–368.
19. Sanderson M (1999) The Classification of Climates from Pythagoras to Koeppen. Bulletin of the American Meteorological Society 80.
20. Rivas-Martinez S (1997) Syntaxonomical synopsis of the potential natural plant communities of North America. Itinera Geobotánica 10: 5–148.
21. Rivas-Martínez S (1995) Clasificación bioclimática de la Tierra (Bioclimatic Classification System of the World). Folia Bot Matritensis 16: 1–25.
22. Rivas-Martínez S, Penas A, Luengo MA, Rivas-Sáenz S (2003) Worldwide Bioclimatic Classification System. In: H.. Lieth, editor. CD-Series II: Climate and Biosphere. Spain: Phytosociological Research Center.
23. Metzger MJ, Bunce RGH, Jongman RHG, Sayre R, Trabucco A, et al. (2012) A high-resolution bioclimate map of the world: a unifying framework for global biodiversity research and monitoring. Global Ecology and Biogeography, DOI: 10.1111/geb.12022.
24. Rivas-Martínez S (1996) Geobotánica y Climatología. Discurso investidura "honoris causa". Granada: Universidad de Granada

25. Cress JJ, Sayre R, Comer P, Warner H (2009) Terrestrial ecosystems – Isobioclimates of the conterminous United States. Scientific investigations Map 3084. 1.0. ed. [Reston, Va.]: U.S. Dept. of the Interior, U.S. Geological Survey. pp. 1 online resource (1 map).

26. Sayre R, Comer P, Warner H, Cress J (2009) A new map of standardized terrestrial ecosystems of the conterminous United States. pp. 17.

27. Franco G, Wilkinson R, Sanstad AH, Wilson M, Vine E (2003) Climate Change Research, Development, and Demonstration Plan. Sacramento, CA: California Energy Commission.

28. Franco G (2009) California Energy Commission Climate Change Research Program. Sacramento, CA: California Energy Commission.

29. Thorne J, Boynton R, Flint L, Flint A, Le T-Ng, et al. (2012) Development and Application of Downscaled Hydroclimatic Predictor Variables for Use in Climate Vulnerability and Assessment Studies. Publication number: CEC-500-2012-010.

30. TBC3 website. Available: http://tbc3.org/. Accessed: 2013 Feb 13.

31. Stein BA, Kutner LS, Adams JS, editors (2000) Precious heritage: the status of biodiversity in the United States: Oxford University Press, USA. 416 p.

32. Hot Spots of Rarity and Richness website figure. Available http://www.natureserve.org/images/precious/6-9.gif. Accessed: 2013 Feb 13].

33. ESRI: 2011. ArcGIS Desktop: Release 10. Redlands, CA: Environmental Systems Research Institute.

34. PRISM Climate Group, Oregon State University website. Available http://prism.oregonstate.edu. Accessed 2012 July 15

35. Flint LE, Flint AL (2012) Downscaling future climate scenarios to fine scales for hydrologic and ecological modeling and analysis. Ecological Processes 1.

36. Nalder I, Wein R (1998) Spatial interpolation of climatic normals: test of a new method in the Canadian boreal forest. Agric Forest Meteor 92: 211–225.

37. ter Braak CJF, Smilauer P (2002) CANOCO Reference manual and Canodraw for Windopws User's guide: software for Canonical Community Ordination (version 4.5) Ithaca, NY, , USA: Microcomputer Power.

38. Fire and Resource Assessment Program F (2010) The 2010 Forests and Rangelands Assessment. Sacramento.

39. Ter Braak CJF (1996) Unimodal models to relate species to environment. Wageningen, Netherlands: DLO-Agricultural Mathematics Group.

40. Barbour MG, Keeler-Wolf T, Schoenherr AA (2007) Terrestrial vegetation of California. Berkeley, Calif.: University of California Press. xvii, 712 p. p.

41. Sawyer JO, Keeler-Wolf T, Evans J (2008) Manual of California Vegetation. Second Edition. Sacramento, CA.: California Native Plant Society.

42. Vallis GK (2011) Climate and the Oceans. Princeton, NJ: Princeton University Press.

43. Fischer DT, Still CJ, Williams AF (2009) Significance of summer fog and overcast for drought stress and ecological functioning of coastal California endemic plant species. Journal of Biogeography 36: 783–799.

44. Brunsfeld S, Sullivan J, Soltis D, Solts P (2001) Comparative phylogeography of northwestern North America: A synthesis. In: Silvertown J, Antonovics J, editors.

45. Latch EK, Heffelfinger JR, Fike JA, Rhodes Jr OLIN (2009) Species-wide phylogeography of North American mule deer (Odocoileus hemionus): cryptic glacial refugia and postglacial recolonization. Molecular Ecology 18: 1730–1745.

46. Barron JA, Heusser L, Herbert T, Lyle M (2003) High-resolution climatic evolution of coastal northern California during the past 16,000 years. Paleoceanography 18.

47. Briles CE, Whitlock C, Bartlein PJ, Higuera P (2008) Regional and local controls on postglacial vegetation and fire in the Siskiyou Mountains, northern California, USA. Palaeogeography, Palaeoclimatology, Palaeoecology 265: 159–169.

48. Lima FP, Wethey DS (2012) Three decades of high-resolution coastal sea surface temperatures reveal more than warming. Nature Communications, DOI: 10.1038/ncomms1713.

49. Lancaster LT, Kay KM (2013) Origin And Diversification Of The California Flora: Re-Examining Classic Hypotheses With Molecular Phylogenies. , doi:10.1111/evo.12016.

50. Loarie SR, Duffy PB, Hamilton H, Asner GP, Field CB, et al. (2009) The velocity of climate change. Nature 462: 1052–1055.

51. Boeye J, Travis JMJ, Stoks R, Bonte D (2012) More rapid climate change promotes evolutionary rescue through selection for increased dispersal distance. Evolutionary Applications DOI: 10.1111/eva.12004.

52. Eklöf A, Kaneryd L, Münger P (2012) Climate change in metacommunities: dispersal gives double-sided effects on persistence. Phil Trans R Soc B367.

53. Westerling A, Bryant B (2008) Climate change and wildfire in California. Climatic Change 87: 231–249.

54. Drake BG, Gonzàlez-Meler MA, Long SP (1997) More Efficient Plants: A Consequence of Rising Atmospheric CO2? Annu Rev Plant Physiol Plant Mol Biol 48: 609–639.

55. Leakey ADB, Ainsworth EA, Bernacchi CJ, Zhu X, Long SP, et al. (2012) Photosynthesis in a CO2-Rich Atmosphere. In: Eaton-Rye JJ, Tripathy BC, Sharkey TD, editors. Photosynthesis: Plastid Biology, Energy Conversion and Carbon Assimilation, Advances in Photosynthesis and Respiration: sPRINGER. pp. 733–768.

56. Ackerly DD, Loarie SR, Cornwell WK, Weiss SB, Hamilton H, et al. (2010) The geography of climate change: implications for conservation biogeography. Diversity and Distributions 16: 476–487.

57. Torregrosa A (1999) Vegetation dynamics in the Sierra Buttes, CA: Predicting species occurrence by merging CCA and GIS [M.S. thesis]: San Francisco State University. 70 p.

58. Van de Ven CM, Weiss SB, Ernst WG (2007) Plant Species Distributions under Present Conditions and Forecasted for Warmer Climates in an Arid Mountain Range. Earth Interactions 11: 1–33.

3

A Time-Series Analysis of the 20th Century Climate Simulations Produced for the IPCC's Fourth Assessment Report

Francisco Estrada[1,2]*, Pierre Perron[3], Carlos Gay-García[1], Benjamín Martínez-López[1]

1 Department of Atmospheric Sciences, Centro de Ciencias de la Atmósfera, Universidad Nacional Autónoma de México, Mexico City, Mexico, **2** Department of Environmental Economics, Institute for Environmental Studies, Vrije Universiteit Amsterdam, The Netherlands, **3** Department of Economics, Boston University, Boston, Massachusetts, United States of America

Abstract

In this paper evidence of anthropogenic influence over the warming of the 20th century is presented and the debate regarding the time-series properties of global temperatures is addressed in depth. The 20th century global temperature simulations produced for the Intergovernmental Panel on Climate Change's Fourth Assessment Report and a set of the radiative forcing series used to drive them are analyzed using modern econometric techniques. Results show that both temperatures and radiative forcing series share similar time-series properties and a common nonlinear secular movement. This long-term co-movement is characterized by the existence of time-ordered breaks in the slope of their trend functions. The evidence presented in this paper suggests that while natural forcing factors may help explain the warming of the first part of the century, anthropogenic forcing has been its main driver since the 1970's. In terms of Article 2 of the United Nations Framework Convention on Climate Change, significant anthropogenic interference with the climate system has already occurred and the current climate models are capable of accurately simulating the response of the climate system, even if it consists in a rapid or abrupt change, to changes in external forcing factors. This paper presents a new methodological approach for conducting time-series based attribution studies.

Editor: Vanesa Magar, Plymouth University, United Kingdom

Funding: FE acknowledges financial support for this work from the Consejo Nacional de Ciencia y Tecnología (http://www.conacyt.gob.mx) under grant CONACYT-310026. The funders had no role in study design, data collection and analysis, decision to publish, or preparation of the manuscript.

Competing Interests: The authors have declared that no competing interests exist.

* E-mail: feporrua@atmosfera.unam.mx

Introduction

For more than two decades a debate regarding the time-series properties of global and hemispheric temperatures has taken place in the climate change literature (e.g., [1–5]), and it has hardly been settled at the present time [6–10]. The underlying quest behind this discussion is the detection and attribution of climate change, both of them critical issues that have proven to be well beyond pure scientific interest, being highly relevant for example for policy- and decision-making.

This paper analyzes the time-series properties of several General Circulation Models (GCM) runs of the 20th Century Climate Experiment (20c3m) conducted for the Intergovernmental Panel on Climate Change's (IPCC) Fourth Assessment Report (AR4) and a set of the radiative forcing series used to drive the 20c3m simulations to investigate four main issues:

1) Can the nonstationarities in global temperatures be tracked to the anthropogenic radiative forcing? Analyzing the time-series properties of climate models simulations offers the advantage of knowing the experimental design from which they were generated, therefore facilitating the detection and attribution of the nonstationarities present in temperature data.

2) Is the assumption of unit roots in global temperatures consistent with the physics of the climate system? GCM represent the state-of-the-art of climate modeling and the most advanced and complete knowledge of the physics that govern the climate system available to this date. As such, one approach for testing whether or not a unit root representation is a valid assumption for global temperatures in terms of the climate physics is to analyze the time-series properties of GCM simulations.

3) Is the unit root representation adequate for the radiative forcing series? While there has been a long debate regarding the time-series properties of global and hemispheric temperatures, radiative forcing variables have received little attention in this respect, and have usually been assumed to be integrated processes. Here we test the statistical adequacy of this assumption.

4) Are current climate models capable of reproducing important properties of observed temperature series such as structural changes and nonlinear trends? This could be considered as another characteristic to evaluate GCM performance for reproducing current climate and their ability for representing the "climate change forcing signal".

To answer these questions, we present a new methodological approach based on recent advances in econometric methods that provides an alternative to the cointegration approach commonly used for attribution studies. As has been discussed in the literature, the latter approach could lead to incorrect inferences, such as spurious cointegration, since the data generating process of temperature series has been previously misidentified [6], [10]. In addition, the proposed methodological approach is broad enough to have wide applicability in the analysis of trending variables and their long-term relationships in climate research.

The remainder of this paper is organized as follows. The next section describes the data used and briefly discusses some advantages and limitations of the 20c3m experiment that are relevant for the purposes of our study. In the same section, the fundamental aspects of the econometric methodology are described, while a more formal discussion of the methods is offered in the online supporting information. The third section investigates the data generating processes of the simulated global temperatures and radiative forcing series using different standard unit root/ stationarity tests and contrasts these results with those of a new generation unit root test that allows for a one-time break in the trend function. Attribution of climate change is then investigated using a nonlinear nonparametric co-trending test and by the analysis of the regressions residuals of global temperature simulations on radiative forcing series. The last section presents a summary of the main findings of this paper.

Data and Methodology

Data Description and Source

The time-series properties of 15 GCM simulations of the global 2-meter air temperature produced for the IPCC's AR4 20c3m are analyzed. Due to the large number of realizations, using subsets of the 20c3m is a common practice when investigating particular features of this climate modeling experiment. The sample of model runs in this paper was chosen to include the most commonly used general circulation models and was influenced by data availability at the time of writing this paper. The differences in the number of runs per model depend on the modeling groups' decisions on how many simulations they contributed to the 20c3m and on their availability. The uniformity of the results presented in the following sections suggests that similar conclusions may be expected from other 20c3m simulations. Two simulations correspond to the Bergen Climate Model version 2 of the Bjerknes Centre for Climate Research (BCCR_BCM2.0) and to the model of the Canadian Centre for Climate Modelling and Analysis (CCCMA); four to the European Centre-Hamburg model version 5 of the Max Planck Institute for Meteorology (MPI_ECHAM5); three to the Geophysical Fluid Dynamics Laboratory Climate Model version 2 of the National Ocean and Atmosphere Administration (GFDL_CM2.1) and one to the previous version of the same model (GFDL_CM2.0); two to the Hadley Centre Coupled Model version 3 (HADLEY_CM3); two to the Goddard Institute for Space Studies, Atmosphere-Ocean Model (GIS-S_AOM); and one to the Institute Pierre Simon Laplace climate model (IPSL). All simulations were obtained from the Royal Netherlands Meteorological Institute's Climate Explorer (http://climexp.knmi.nl/selectfield_co2.cgi?someone@somewhere).
Figure 1 plots the simulated global temperatures, and as can be seen from visual inspection the GFDL's realizations are the noisiest with large realizations occurring in the 1880 decade. The observed global surface temperature series used in this paper corresponds to the Climate Research Unit HadCRUT3 (available at http://www.metoffice.gov.uk/hadobs/hadcrut3/).

Analyzing the time-series properties of climate models simulations offers the advantages of knowing the experimental design on which they were generated. Unlike the observed data, the 20c3m climate simulations are part of a controlled experiment for which the forcing factors that could impart secular movement to simulated global temperatures are explicitly identified. In this case, the relationship between the exogenous model inputs and endogenous model outputs is unambiguous and therefore the analysis of the radiative forcing variables can provide critical information about the warming trend of the 20th century. In particular, if the attribution of climate change is to be proven by means of currently available statistical models, both temperature and radiative forcing should share similar time-series properties, although the internal variability of climate models may modify some of their particular aspects.

In order to take advantage of the information contained in these temperature simulations, an analysis of the time-series properties of one of the sets of radiative forcing series that were used to run the 20c3m experiments is presented, and the existence of a common secular trend between temperature and forcing variables is investigated. Unfortunately, the 20c3m does not have a unique common set of radiative forcing variables and therefore simulations differ in which forcings are used (different forcing variables and sources) and on how they are incorporated into the different models [11]. The latter is particularly problematic in the case of the radiative forcing of the sulfate aerosols (direct and indirect effects) since they depend not only on the different datasets used for prescribing them [12] but also in the particular implementation of the climate model. As such, even when using the same loading patterns and time variation, the resulting radiative forcing would vary from model to model and most of these time series are not publicly available.

As a consequence, the attribution analysis presented in this paper is based on the well-mixed greenhouse radiative forcing, a variable included in all of the simulations and for which the different datasets are broadly similar. Results including other radiative forcing variables are presented to provide a sensitivity analysis to assess the robustness of our conclusions.

The radiative forcing set selected for this study is the GISS-NASA database [13] (available at http://data.giss.nasa.gov/modelforce/RadF.txt), covering the period 1880–2010 and including the following variables (in W/m^2): well-mixed greenhouse gases (WM_GHG; carbon dioxide, methane, nitrous oxide and chlorofluorocarbons); ozone; stratospheric water vapor; solar irradiance; land use change; snow albedo; stratospheric aerosols; black carbon; reflective tropospheric aerosols; and the indirect effect of aerosols. These time series were used to construct the forcing trends in Figure 1:1) WM_GHG, which is mostly human-induced; 2) solar forcing (SOLAR); 3) TRF, defined as the sum of all forcing variables above with the exception of stratospheric aerosols. Stratospheric aerosols can be considered stationary around a constant and therefore cannot impart the trending behavior in the level of the total radiative forcing, nor on temperature series (the Augmented Dickey-Fuller test statistic value for this series is −4.92, which is significant at the 1% level).

Econometric Methodology

Unit root tests and the identification of the data generating process. Two types of nonstationary stochastic processes have been commonly proposed for modeling global temperature series: trend stationary (TS) and difference stationary (DS). These processes offer contrasting views on how the climate system works and on the importance and effects of changes in anthropogenic forcing over climate, and require different

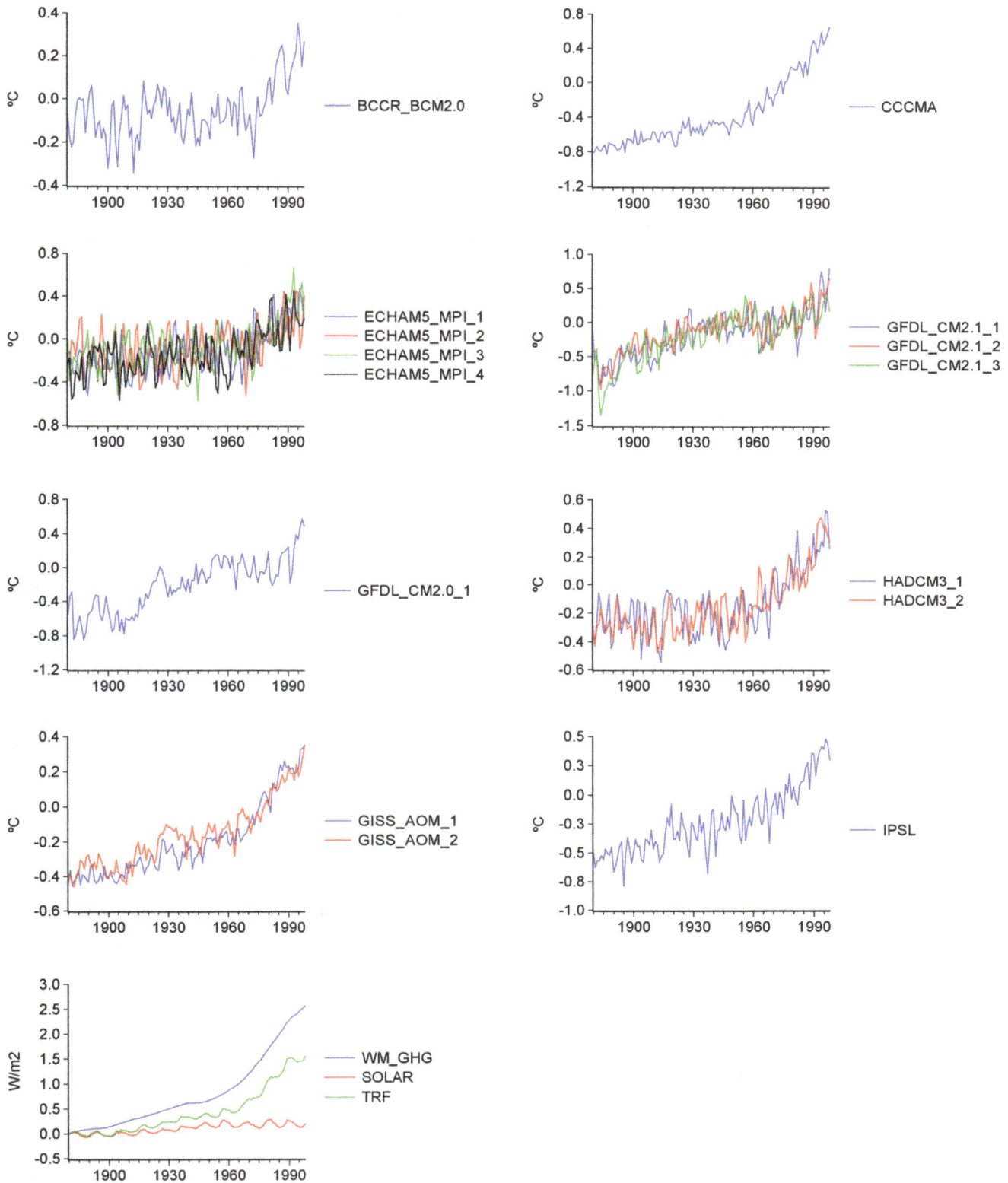

Figure 1. IPCC's AR4 20 cm3 global temperature simulations and radiative forcing variables. WM_GHG includes carbon dioxide, methane, nitrous oxide and chlorofluorocarbons; SOLAR is solar forcing; TRF includes WM_GHG, solar irradiance, reflective tropospheric aerosols, indirect effect of aerosols, ozone stratospheric water vapor, land use change, snow albedo and black carbon. Climate models' simulations are shown as anomalies with respect to their 1961–1990 mean values.

approaches for conducting time-series based attribution studies. If these processes are misidentified, a wide range of statistical models, tests and procedures can produce misleading results and inferences. A brief description of TS, DS and cointegrated processes is provided in the online supporting information (Text S1, section 1.1).

As a first step for investigating the data generating process of the simulated global temperatures and radiative forcing trend series described above, five commonly used unit root and stationarity tests are applied (Text S1, section 1.2). Nevertheless, it is important to consider that these tests can be severely affected when the trend function is subject to changes in level and/or slope. As shown in the literature, the sum of the first order autoregressive coefficients is highly biased towards unity if there is a shift in the trend function [14]. In this case, the unit root null is hardly rejected even if the series is composed of white noise disturbances around a trend. Furthermore, if the break occurs in the slope of the trend function, unit root tests are not consistent, i.e., the null hypothesis of a unit root cannot be rejected even asymptotically [15].

The existence of change points in the trend functions of temperature and radiative forcing series has been documented [16–20], [6], [12] and therefore standard unit root tests may not be adequate for investigating the data generating process of these variables. In consequence, we apply two new generation econometric procedures explicitly designed for addressing this problem: the Perron-Yabu structural change testing procedure [21], [22] and the Kim-Perron unit root test that allows for an unknown one-time structural break in the trend function [23]. These methodologies are briefly described in the Supporting Information (Text S1, sections 1.3 and 1.4). The main ingredient underlying the construction of the estimates and tests is the following specification of the trend function for a given series y_t:

$$y_t = \mu + \beta_1 t + \gamma DT_t^* + u_t \qquad (1)$$

where $DT_t* = t - T_B$ if $t > T_B$ and 0 otherwise. Here T_B is the break date, β_1 is the pre-break slope of the trend, γ is the change in the slope at the time of the break, while $\delta = \beta_1 + \gamma$ is the post-break slope. u_t is a random process whose properties need to be investigated, i.e., stationary or integrated.

Nonlinear Nonparametric Co-trending Test and the Attribution of Climate Change

Cointegration techniques have been commonly applied in attribution studies due to the fact that these techniques offer the possibility, under the DS assumption, of investigating the existence of a common long-term trend between temperatures and radiative forcing variables. However, unit root processes are not the only type of nonstationary processes that can show a common secular movement and cointegration analysis is only one possibility for relating the trends of nonstationary variables. Relationships between nonstationary variables can be established when linear combinations of different time series cancel out some "common features" such as trends and breaks [24].

Once we establish that global temperature simulations and radiative forcing series are better characterized as TS, we apply the nonparametric nonlinear co-trending analysis proposed by Bierens ([25]; Text S1, section 1.5) to investigate the attribution of climate change. Nonlinear co-trending is a special case of common features in which one or more linear combinations (called co-trending vectors) of nonstationary time series are stationary about a linear trend or a constant, indicating that the series share common nonlinear deterministic time trends. With r denoting the

number of co-trending vectors, n series share a common nonlinear trend if one cannot reject the null hypothesis that r = n − 1, while the null hypothesis that r = n can be rejected.

Results and Discussion

Standard Unit Root Tests

The results of applying standard unit root and stationarity tests to the global temperature models simulations and to the radiative forcing trends reveal that for all tests and series, with the possible exception of the GFDL_CM2.1 simulation 2 and the ECHAM5 simulation 4, the unit root hypothesis cannot be rejected (Table S1). Similar findings have been reported for observed global and hemispheric temperatures as well as for radiative forcing series using these tests (e.g., [4–6]). From these results it could be erroneously concluded that both global temperatures and radiative forcing are integrated processes and that cointegration techniques would be adequate for investigating their long-run relationships. Nevertheless, as is shown below, the finding of unit roots in global temperature and radiative forcing series is due to an incorrect specification of the trend function. Unit root tests that allow for a better representation of the trend function in these variables provide contrasting results.

Unit Root Tests Allowing for a One-time Structural Change

As argued in the literature [6], [10], given the time-series properties of temperature series, standard unit root tests can cause to erroneously classify these series as having stochastic trends. This can also be the case for the radiative forcing series.

Visual inspection of temperature series in Figure 1 suggests the existence of structural breaks in the slope of the trend functions similar to the one in observed global temperature series discussed in previous publications (e.g., [6], [16–17]). The existence of changes in the rates of growth of the various greenhouse gases is frequently discussed in the climate policy and mitigation contexts (e.g., [19], [20]) and is also clearly suggested by Figure 1. Therefore, it is important to assess whether the results from standard unit root tests are affected by the presence of structural changes. However, this is a circular problem given that most of the tests for structural breaks require to correctly identify if the data generating process is stationary or integrated. Depending on this outcome, the limit distribution of these tests are different and, if the process is misidentified, the tests will have poor properties. The Perron-Yabu procedure offers a way to break this circular problem allowing to test for structural changes in level and/or slope whether the noise component is stationary or integrated [21], [22].

The results of this procedure are presented in Table 1 column 3. The test statistic values for all temperature simulations are significant at the 5% level, with the exception of GFDL_CM2.1_3 which is significant at the 10% level and of GFDL_CM2.1_2 which is not significant at any conventional levels (not reported in Table 1). In the case of the forcing variables TRF and WM_GHG the test statistic values are significant at the 1% levels, while for SOLAR it is at the 10% level, indicating in all cases the presence of structural changes in their rates of growth.

Consequently, unit root tests that allow for possible structural changes are required for investigating the type of data generating process that best describes temperature and forcing series. For this task, the Kim-Perron unit root test was applied and, as discussed below, once a break in the trend function is allowed the results of standard unit root tests are completely reversed.

The results in Table 1 are quite striking and uniform across all series clearly rejecting the null hypothesis of a unit root at

Table 1. Tests for a unit root with a one-time break in the trend function.

Series	T_b	W	k	$\hat{\beta}_1$	$t_{\hat{\beta}_1}$	$\hat{\gamma}$	$t_{\hat{\gamma}}$	δ	$d(g)$	$\hat{\alpha}$	$t_\alpha\left(\hat{\lambda}_{tr}^{AO}\right)$
OBSERVED	1977	3.59[a]	0	0.0035	**10.84**	0.0142	**7.85**	0.0177	0.00%	0.50	−5.73[a]
ECHAM5_1	1968	8.04[a]	1	0.0011	**2.63**	0.0135	**8.29**	0.0146	−17.84%	0.04	−9.20[a]
ECHAM5_2	1978	4.55[a]	2	0.0015	**3.41**	0.0167	**6.22**	0.0182	2.61%	0.16	−9.77[a]
ECHAM5_3	1973	8.27[a]	1	0.0010	**2.24**	0.0161	**7.45**	0.0171	−3.59%	−0.07	−4.86[a]
ECHAM5_4	1961	3.76[a]	2	0.0013	**2.70**	0.0100	**5.45**	0.0114	−35.78%	0.25	−5.81[a]
BCCR	1974	2.32[b]	0	0.0004	1.57	0.0136	**7.82**	0.0140	−20.93%	0.50	−5.88[a]
CCCMA	1961	5.80[a]	0	0.0042	**19.07**	0.0230	**32.55**	0.0273	53.82%	0.27	−8.35[a]
GFDL_CM2.1_1	1888	1.95[b]	2	−0.0086	**−2.71**	0.0166	**4.69**	0.0079	−55.68%	0.46	−4.51[a]
GFDL_CM2.1_3	1885	1.72[c]	2	−0.0062	−1.47	0.0144	**3.19**	0.0083	−53.18%	0.59	−7.37[a]
GFDL_CM2.0_1	1889	2.53[b]	0	−0.0152	**−6.45**	0.0231	**8.87**	0.0079	−55.22%	0.65	−4.44[a]
HADLEY_CM3_1	1963	9.59[a]	2	0.0007	1.76	0.0161	**10.33**	0.0167	−5.80%	0.30	−5.15[a]
HADLEY_CM3_2	1958	6.59[a]	0	0.0010	**2.84**	0.0127	**10.18**	0.0137	−22.86%	0.36	−7.47[a]
GISS_AOM_1	1966	13.67[a]	0	0.0030	**23.07**	0.0124	**20.70**	0.0154	−13.20%	0.35	−7.37[a]
GISS_AOM_2	1973	5.93[a]	0	0.0035	**22.35**	0.0107	**10.99**	0.0142	−19.84%	0.55	−5.56[a]
IPSL	1969	10.99[a]	0	0.0037	**10.69**	0.0163	**9.25**	0.0200	12.46%	0.17	−8.80[a]
TRF	1960	5.63[a]	2	0.0064	**20.82**	0.0221	**28.98**	0.0285	–	0.84	−4.24[b]
WM_GHG	1960	62.79[a]	7	0.0105	**64.95**	0.0351	**87.76**	0.0456	–	0.90	−3.97[b]
SOLAR	1959	1.80[c]	2	0.0031	**16.33**	−0.0032	**−6.793**	−0.0001	–	0.58	−8.82[a]

The regression model for the unit root tests is defined in equations (4) and (6) in the Supporting Information. The symbols are defined as follows: T_b is the estimated time of the break; W is the Perron-Yabu Exp-Wald statistic with 5% trimming; k is the number of lagged differences added to correct for serial autocorrelation; $\hat{\beta}$, $\hat{\gamma}$ are the regression coefficients of the slope of the trend function and $t_{\hat{\beta}}$, $t_{\hat{\gamma}}$ the corresponding t-statistic values. Bold numbers denote statistical significance at 5% levels. $\delta = (\beta_1 + \gamma)$ is the post-break slope and $d(g)$ is the percent difference with respect to the observed global temperature. $\hat{\alpha}$ is the sum of the first order autoregressive coefficients and $t_\alpha\left(\hat{\lambda}_{tr}^{AO}\right)$ is the Kim-Perron unit root test statistic.

[a], [b], [c]denotes statistical significance at the 1%, 5% and 10%, respectively (for W, for critical values taken from [21], Table 2.b; Kim-Perron unit root test critical values taken from [39], Table 1). Results for Observed taken from [6].

the 1% significance level, for all of the model simulations (see Text S1 section 1.4.1 for a robustness analysis of the unit root test results). In the case of the GFDL_CM2.1_2 simulations the Perron-Yabu test does not reject the null of no break. However, the ADF test with no break rejects the null hypothesis of a unit root for this series (Table S1). Hence, we can conclude that it is TS and no further analysis is needed. As expected from TS series, the estimates of the sum of the autoregressive coefficients of the simulated temperature series are now quite far from unity, ranging from -0.07 (ECHAM5_3) to 0.65 (GFDL_CM2.0_1), with a mean value of 0.34. As in the case of observed global temperature reported previously in the literature, assuming a unit root would have erroneously attributed too much persistence to temperature variability, a fact not supported by the data [6].

While there has been a debate regarding the time-series properties of global and hemispheric temperatures, radiative forcing variables have received little attention in this respect and have usually been assumed to be integrated processes when conducting attribution studies based on observed records and time series analysis [4–5]. The two main arguments for justifying this assumption are: 1) the results of standard unit root tests, which as discussed above are not adequate for this task given the presence of structural breaks, and; 2) the long residence time of greenhouse emissions in the atmosphere produces an accumulation process. However, it should be noticed that cumulative processes are not

necessarily unit root processes, any type of trending process would produce the same effect.

The last column of Table 1 reveals that when allowing for a better representation of the trend function, the conclusions that can be drawn are markedly different from what has been reported previously in the literature. The null of a unit root is strongly rejected in favor of trend stationary processes with a one-time permanent break in the rate of growth of the forcing variables.

Although radiative forcing variables are more persistent than temperatures, the sum of the autoregressive coefficients is far from unity. Shocks in concentrations and radiative forcing do dissipate as opposed to the case of a unit root in which the persistence of shocks is infinite. This finding has important implications for the attribution of climate change since it shows that there are no differences in the order of integration of these variables and that all of them can be better described as trend stationary processes with a change in their rates of growth.

The dates of the break in the slope of the trend of the simulated temperatures vary from 1885 to 1978 (Table 1, column 2). This wide range is mainly due to the GFDL simulations which show large realizations (possible outliers) around the 1880's decade that may affect the estimation of the break date. If these simulations are excluded, the average break date is 1968 which is close to those that have been reported in the literature [16–19], [26–27]. The break dates in the slope of the radiative forcing trends are estimated around 1960, previous to those of observed and simulated global temperatures (Table 1, column 2).

Confidence intervals for the break dates in Table 1 were constructed using the Perron-Zhu procedure [28]. Figure 2 shows that for almost half of the model simulations, the estimated break date is not statistically different from that of the observed series. Excluding the GFDL models, although the confidence intervals do not necessarily overlap with the observed one, they are separated by only a few years and most of them cannot be considered statistically different from each other. Furthermore, with the exception of GFDL_CM2.1, all of the models for which more than one run was considered (ECHAM5, HADC3M, GISS_AOM) provide similar estimates of the break date from run to run.

The break dates of the radiative forcing trend variables are neither statistically different from each other nor from about half of the temperature simulations. However, the break date of observed global temperatures is statistically different from those of TRF and WM_GHG. The apparent delay in the response of the climate system could be related to a change in the Atlantic Multidecadal Oscillation (AMO) to its negative phase around the early 1960s, possibly obscuring the global increase in temperatures due to anthropogenic forcing [29–30]. One possible factor contributing to the differences in the break dates between the observed and simulated series could be associated to the fact that the 20c3m simulations are not constrained to reproduce observed variability. Therefore, natural variability and the models' internal variability do not have to match and neither do the occurrence of changes in the phase of AMO [31–32]. Furthermore, current climate models tend to underestimate inter-annual low-frequency natural climate variability, producing fewer deviations (and of shorter duration) that could mask the warming trend [12].

The fact that runs from different models and models with multiple runs that have similar or identical forcing but different initial conditions give broadly similar estimates of the break date

provides further evidence of its exogenous nature: this common feature of model simulations cannot be interpreted as part of internal variability, but as a result of the changes in radiative forcing.

Figures 3A and 3B show the point estimates and the corresponding 95% confidence intervals of the coefficients of the pre-break slopes and of their changes after the break, respectively. For most of the simulations, a positive and statistically significant pre-break trend is present, nevertheless the coefficients are not statistically different from that of the observed temperature series only for IPSL, GISS_AOM, CCCMA models (Figure 3). When comparing the magnitude of the pre-break slope coefficients of the model simulations with that of the observed one, even if the GFDL models are excluded (for this model the range of the estimates of the pre-break slope coefficient vary from −534.29% to 20% in comparison with the observed estimate), the differences are quite large and the range of values span from −88.57% to 20%. Most of the models underestimate the first warming trend of the 20th century, possibly due to large realizations of observed natural variability [33].

In contrast, the changes in the slope coefficients induced by the structural change are not statistically different from each other for all the simulated and observed temperature series, with the exception of CCCMA (Figure 4). The similitude in these parameter values provides evidence to support the fact that climate models can accurately simulate the response of the climate system to changes in external forcing factors, even if rapid or abrupt, and therefore gives more confidence in their ability to produce credible climate change scenarios at least at the global scale. Note however that, as has been discussed in the literature, this high level of agreement between models occurs despite their large differences in key factors such as climate sensitivity and climate forcing (e.g., [34–35]).

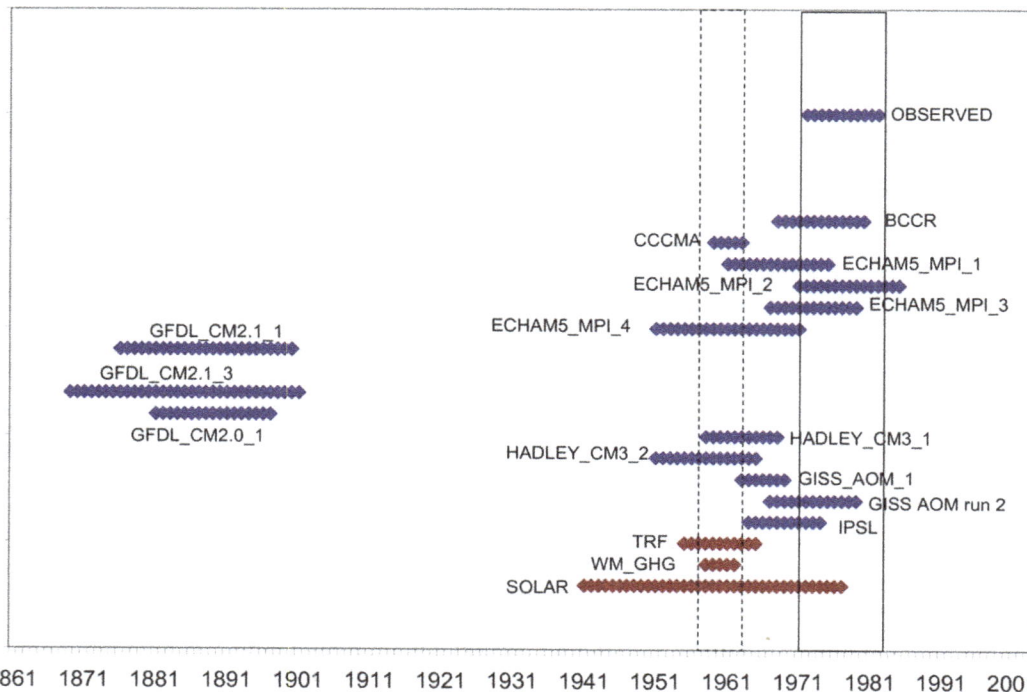

Figure 2. Confidence intervals for the break dates in Table 1. Solid and dashed lines indicate the 95% confidence interval for the break dates for observed global temperatures and WM_GHG, respectively.

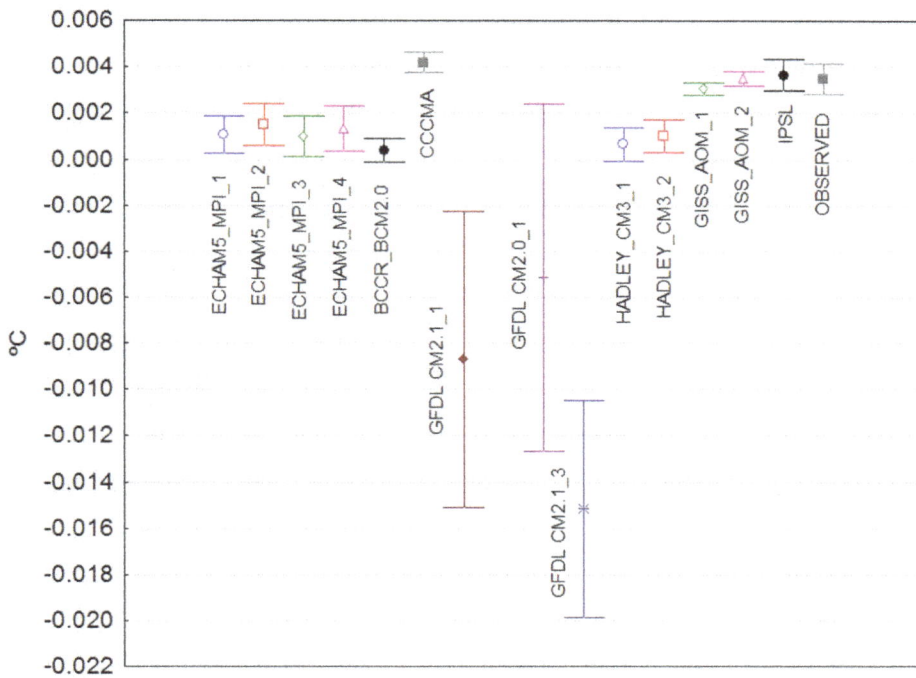

Figure 3. Point estimates and 95% confidence intervals of the pre-break slope coefficients (°C/yr) in Table 1.

Finally, when comparing the post-break slope value (pre-break plus change in slope at the break) to that of the observed global temperature, it becomes apparent that, at least in this sample of models and simulations, climate models included in the IPCC's AR4 tend to underestimate the warming trend that was observed in the second part of the 20th century. As depicted by columns δ and $d(g)$ in Table 1, twelve of the models simulations underestimate the observed trend of the last part of the century (some of them severely, up to 65%). The remaining simulations show from slight overestimations (EC-HAM5_2 and IPSL) to large overestimations (CCCMA, about 50%).

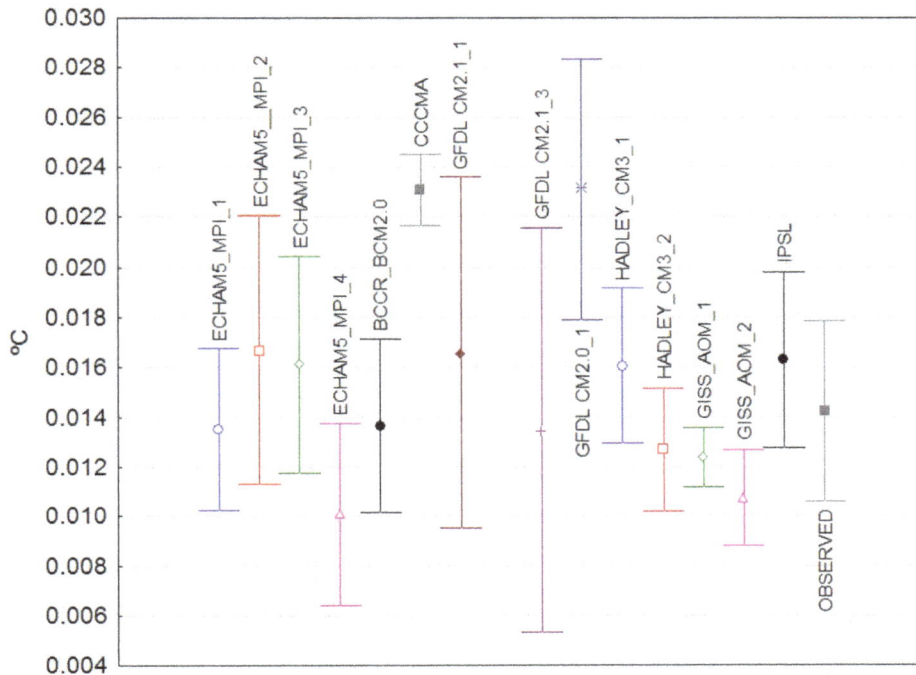

Figure 4. Point estimates and 95% confidence intervals of the post-break change coefficients (°C/yr) in Table 1.

Attribution of the 20th Century Warming Trend

A nonparametric nonlinear co-trending test [25] is applied to investigate if the radiative forcing trends, in particular WM_GHG, and the average of the 20c3m simulations (T^{avg}, depicted in Figure 5) share a common nonlinear trend. An analysis of the residuals of ordinary least squares regressions is also presented to further illustrate the existence of common secular trends. The GFDL's simulations were excluded because of their poor performance in reproducing the observed global trend. However, the results presented below are robust to the inclusion of the GFDL simulations.

The co-trending test results provide strong evidence for the attribution of climate change to the anthropogenic forcing represented by WM_GHG, an input common to all of the 20c3m simulations, and shows that the existence of a common nonlinear trend is robust to the inclusion of other forcing factors. The empirical evidence obtained by this test can be summarized as follows (see Table S2):

1. There is a unique co-trending vector (r = 1) between T^{avg} and WM_GHG, indicating that these variables share a common nonlinear trend.

2. The existence of a unique co-trending vector is robust to the inclusion of all the other forcing factors in TRF.

3. TRF, WM_GHG and T^{avg} share the same nonlinear trend (two co-trending vectors, r = 2).

4. SOLAR and T^{avg} show a distinct long-run secular movement, suggesting that the observed warming can hardly be approximated by the main natural factor (r = 0).

5. There is a unique co-trending vector (r = 1) between T^{avg} and the observed global temperature series.

These results not only support the findings in the previous subsections regarding that temperature and radiative forcing variables are stationary around a common nonlinear trend, but provide strong evidence of attribution of the warming of the 20th century to anthropogenic activities. Results 1, 2 and 3 suggest that, although other forcing factors have had an important effect modulating the net forcing, the nonlinear trend defining the secular movement of TRF and global temperatures during the past century is largely defined by that of WM_GHG. Result 5 shows that, in spite of the ensemble's members differences reported above, the observed and the average of the simulated global

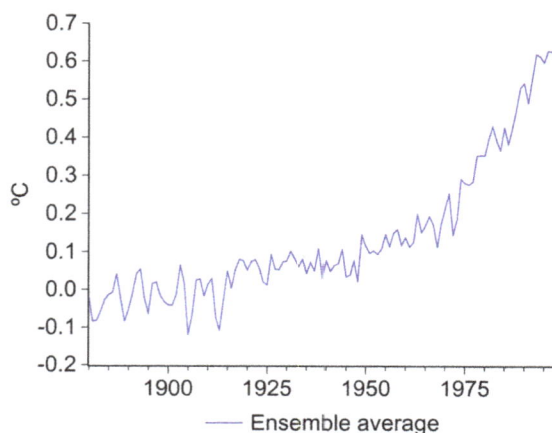

temperatures share the same nonlinear trend, further confirming the ability of current GCM to reproduce the 20th century warming trend.

Given that both radiative forcing and global temperature series have been shown to be TS processes, their long-term relationship can also be investigated using simple OLS regressions involving global temperature and radiative forcing series as the dependent and independent variables, respectively, and analyzing the associated residuals.

The residuals from the regression of T^{avg} on SOLAR reveal that the trend of this variable could only account for part of the warming in the first half of the 20th century (Figure 6). The large positive trend in the residuals since the 1950s confirms that these variables follow different secular trends. However, when using TRF or WM_GHG as the explanatory variable the residuals of the regression are stationary, indicating that these series can indeed reproduce the nonlinear warming trend of the 20th century. As such, visual inspection of the residuals strongly suggest that the main source of the secular movement in both TRF and global temperatures is WM_GHG, although other internal and external forcing factors have modulated them.

The ADF test [36–37] with no deterministic terms (Table S3) confirms that both the residuals from the regressions using TRF and WM_GHG can be considered as stationary variations around the zero line (the test statistics are about 2.5 times the 1% critical value), while the residuals obtained from a regression using SOLAR are clearly nonstationary (the test statistic is not significant at any conventional level).

Overall, the results are in strong agreement with previous attribution studies based on GCM simulations under different combinations of external forcing factors, indicating that the warming of the 20th century cannot be reproduced without the inclusion of the main anthropogenic forcing factors [12], [26–27], [30], [38].

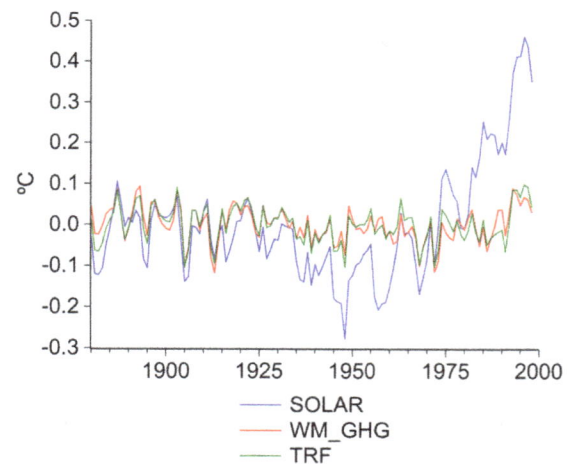

Figure 6. Residuals of the regressions of T^{avg} on SOLAR, TRF and WM_GHG. T^{avg} is the average of models runs; WM_GHG includes carbon dioxide, methane, nitrous oxide and chlorofluorocarbons; SOLAR is solar forcing; TRF includes WM_GHG, solar irradiance, reflective tropospheric aerosols, indirect effect of aerosols, ozone stratospheric water vapor, land use change, snow albedo and black carbon.

Figure 5. Average of models runs (T^{avg}).

Conclusions

This paper presents a new approach for investigating the attribution of climate change based on state-of-the-art econometric techniques that are appropriate for the time-series properties of global temperature and radiative forcing. The results reveal sound statistical evidence underlying the large anthropogenic contribution to the warming of the 20th century. It is shown that WM_GHG, TRF and T^{avg} share a unique common nonlinear trend which is also shown to match the warming trend in observed global temperatures. In contrast, the nonlinear trend describing the secular movement of solar forcing is statistically distinct from that of the observed and simulated temperatures, being particularly unable to explain their evolution during the second part of the century.

By means of new generation unit root and structural change testing procedures, strong evidence is presented suggesting that both global temperatures and radiative forcing series have been misidentified in previous studies as being unit root processes. All these series share similar time-series properties and can be better characterized as stationary processes around nonlinear deterministic trends with time-ordered breaks that are spaced in a way consistent with what could be expected from climate physics. Given the experimental design of the 20c3m and the similitude between different models and runs, this finding provides an unambiguous causal explanation for the increase in the rate of warming during the second part of the 20th century.

The results offer additional evidence regarding the capacity of current climate models to accurately simulate the response of the climate system to changes in external forcing factors, even if rapid or abrupt. This finding contributes to increase confidence in the ability of these models to produce credible climate change scenarios at least for such large spatial scales.

Supporting Information

Table S1　Standard unit root tests applied to global temperature simulations and radiative forcing series.

Table S2　Nonparametric nonlinear co-trending test for TRF, SOLAR, WM_GHG, T^{avg} and T^{cru}.

Table S3　ADF test on the residuals of the regressions of the ensemble average of global temperature simulations on: 1) TRF; 2) WM_GHG and; 3) SOLAR.

Author Contributions

Conceived and designed the experiments: FE PP CG-G BM-L. Performed the experiments: FE PP. Analyzed the data: FE PP BM-L. Contributed reagents/materials/analysis tools: FE PP BM-L CG-C. Wrote the paper: FE PP.

References

1. Galbraith J, Green C (1992) Inference about trends in global temperature data. Clim Change 22: 209–221.
2. Zheng X, Basher RE (1999) Structural time series models and trend detection in global and regional temperature series. J Clim 12: 2347–2358.
3. Woodward WA, Gray HL (1993) Global warming and the problem of testing for trend in time series data. J Clim 6: 953–962.
4. Kaufmann RK, Stern DI (1997) Evidence for human influence on climate from hemispheric temperature relations. Nature 388: 39–44.
5. Kaufmann RK, Kauppi H, Stock JH (2006) Emissions, concentrations, & temperature: a time series analysis. Clim Change 77: 249–278.
6. Gay C, Estrada F, Sanchez A (2009) Global and hemispheric temperature revisited. Clim Change 94: 333–349.
7. Kaufmann RK, Kauppi H, Stock JH (2010) Does temperature contain a stochastic trend? Evaluating conflicting statistical results. Clim Change 101: 395–405.
8. Mills TC (2010) 'Skinning a cat': alternative models of representing temperature trends. An editorial comment. Clim Change 101: 415–426.
9. Mills TC (2010) Is global warming real? Analysis of structural time series models of global and hemispheric temperatures. J Cosmol 8: 1947–1954.
10. Estrada F, Gay C, Sánchez A (2010) Reply to 'Does temperature contain a stochastic trend? Evaluating conflicting results by Kaufmann et al. Clim Change 101: 407–414.
11. Hegerl G, Meehl G, Covey C, Latif M, McAveney B, et al. (2003) 20C3M: CMIP collecting data from 20th century coupled model simulations. Clivar Exchanges 26: S3–S5.
12. IPCC-WGI (2007) Climate Change 2007: The Physical Science Basis. Contribution of Working Group I to the Fourth Assessment Report of the Intergovernmental Panel on Climate Change. Solomon S, Qin D, Manning M, Chen Z, Marquis M, Averyt KB, Tignor M, Miller HL, editors. Cambridge, UK: Cambridge University Press. 1009 p.
13. Hansen J, Sato M, Kharecha P, von Schuckmann K (2011) Earth's energy imbalance and implications. Atmos Chem Phys 11: 13421–13449.
14. Perron P (1989) The great crash, the oil price shock, and the unit root hypothesis. Econometrica 57: 1361–1401.
15. Perron P (2006) Dealing with structural breaks. In: Mills TC, Patterson K, editors. Palgrave Handbook of Econometrics, Vol. 1. New York: Palgrave Macmillan. 278–352.
16. Ruggieri E (2012) A Bayesian approach to detecting change points in climatic records. Int J Climatol In press.
17. Ivanov MA, Evtimov SN (2010) 1963: The break point of the Northern Hemisphere temperature trend during the twentieth century. Int J Climatol 30(11): 1738–1746.
18. Seidel DJ, Lanzante JR (2004) An assessment of three alternatives to linear trends for characterizing global atmospheric temperature changes. J Geophys Res-Atmos 109: L02207.
19. IPCC-WGIII (2007) Climate Change 2007: Mitigation of Climate Change. Contribution of Working Group III to the Fourth Assessment Report of the Intergovernmental Panel on Climate Change. Metz B, Davidson OR, Bosch PR, Dave R, Meyer LA, editors. Cambridge: Cambridge University Press. 862 p.
20. Raupach MR, Canadell JG (2010) Carbon and the Antropocene. Curr Opin Environ Sustainability 2(4): 210–218.
21. Perron P, Yabu T (2009) Testing for shifts in trend with an integrated or stationary noise component. JBES 27: 369–396.
22. Perron P, Yabu T (2009) Estimating deterministic trends with an integrated of stationary noise component. J Econom 151: 56–69.
23. Kim D, Perron P (2009) Unit root tests allowing for a break in the trend function under both the null and the alternative hypotheses. J Econom 148: 1–13.
24. Engle RF, Kozicki S (1993) Testing for common features. JBES 11: 369–395.
25. Bierens HJ (2000) Nonparametric nonlinear cotrending analysis, with an application to interest and inflation in the United States. JBES 18: 323–337.
26. Stott PA, Tett SFB, Jones GS, Allen MR, Mitchell JFB, et al. (2000) External control of 20th century temperature by natural and anthropogenic forcings. Science 290: 2133.
27. Meehl GA, Washington WM, Wigley TML, Arblaster JM, Dai A (2003) Solar and greenhouse gas forcing and climate response in the 20th century. J Climate 16: 426–444.
28. Perron P, Zhu X (2005) Structural breaks with deterministic and stochastic trends. J Econom 129: 65–119.
29. Wang X, Brown PM, Zhang Y, Song L (2011) Imprint of the Atlantic Multidecadal Oscillation on Tree-Ring Widths in Northeastern Asia since 1568. PLoS ONE 6(7): e22740.
30. Wu Z, Huang NE, Wallace JM, Smoliak BV, Chen X (2011) On the time-varying trend in global-mean surface temperature. Clim Dyn 37: 759–773.
31. Zhang R, Delworth TL, Held IM (2007) Can the Atlantic Ocean drive the observed multidecadal variability in Northern Hemisphere mean temperature? Geophys Res Lett 34: L02709.
32. Kravtsov S, Spannagle C (2008) Multidecadal Climate Variability in Observed and Modeled Surface Temperatures. J Climate 21: 1104–1121.
33. Delworth TL, Knutson TR (2000) Simulation of early 20th century global warming. Science 287: 2246.
34. Kiehl JT (2007) Twentieth century climate model response and climate sensitivity. Geophys Res Lett 34: L22710.
35. Kerr RA (2007) Another global warming icon comes under attack. Science 317: 28–29.
36. Dickey DA, Fuller WA (1979) Distribution of the estimators for autoregressive time series with a unit root. J Am Statist Assoc 74: 427–431.

37. Said E, Dickey DA (1984) Testing for unit roots in autoregressive moving average models of unknown order. Biometrika 71: 599–607.

38. Broccoli AJ, Dixon KW, Delworth TL, Knutson TR (2003) Twentieth-century temperature and precipitation trends in ensemble climate simulations including natural and anthropogenic forcing. J Geophys Res 108: 4798.

39. Perron P, Vogelsang T (1993) Erratum: The Great Cash, the Oil Price Shock and the Unit Root Hypothesis. Econometrica 61: 248–249.

4

Combining a Climatic Niche Model of an Invasive Fungus with Its Host Species Distributions to Identify Risks to Natural Assets: *Puccinia psidii* Sensu Lato in Australia

Darren J. Kriticos[1,2]*, Louise Morin[1], Agathe Leriche[1¤], Robert C. Anderson[3], Peter Caley[4]

1 Commonwealth Scientific and Industrial Research Organisation, Ecosystem Sciences, Canberra, Australian Capital Territory, Australia, **2** International Science & Technology Policy & Practice, Department of Applied Economics, University of Minnesota, St. Paul, Minnesota, United States of America, **3** University of Hawai'i, College of Tropical Agriculture and Human Resources, Honolulu, Hawaii, United States of America, **4** Commonwealth Scientific and Industrial Research Organisation, Mathematics Informatics and Statistics, Canberra, Australian Capital Territory, Australia

Abstract

Puccinia psidii sensu lato (s.l.) is an invasive rust fungus threatening a wide range of plant species in the family Myrtaceae. Originating from Central and South America, it has invaded mainland USA and Hawai'i, parts of Asia and Australia. We used CLIMEX to develop a semi-mechanistic global climatic niche model based on new data on the distribution and biology of *P. psidii* s.l. The model was validated using independent distribution data from recently invaded areas in Australia, China and Japan. We combined this model with distribution data of its potential Myrtaceae host plant species present in Australia to identify areas and ecosystems most at risk. Myrtaceaeous species richness, threatened Myrtaceae and eucalypt plantations within the climatically suitable envelope for *P. psidii* s.l in Australia were mapped. Globally the model identifies climatically suitable areas for *P. psidii* s.l. throughout the wet tropics and sub-tropics where moist conditions with moderate temperatures prevail, and also into some cool regions with a mild Mediterranean climate. In Australia, the map of species richness of Myrtaceae within the *P. psidii* s.l. climatic envelope shows areas where epidemics are hypothetically more likely to be frequent and severe. These hotspots for epidemics are along the eastern coast of New South Wales, including the Sydney Basin, in the Brisbane and Cairns areas in Queensland, and in the coastal region from the south of Bunbury to Esperance in Western Australia. This new climatic niche model for *P. psidii* s.l. indicates a higher degree of cold tolerance; and hence a potential range that extends into higher altitudes and latitudes than has been indicated previously. The methods demonstrated here provide some insight into the impacts an invasive species might have within its climatically suited range, and can help inform biosecurity policies regarding the management of its spread and protection of valued threatened assets.

Editor: David R. Andes, University of Wisconsin Medical School, United States of America

Funding: This work was partially funded by the Commonwealth Scientific and Industrial Research Organisation Biosecurity Flagship and Sustainable Agriculture Flagships, and New Zealand's Foundation for Research, Science & Technology through contract CO2X0501, the Better Border Biosecurity (B3) programme (www. b3nz.org). The funders had no role in study design, data collection and analysis, decision to publish, or preparation of the manuscript.

Competing Interests: The authors have declared that no competing interests exist.

* E-mail: Darren.kriticos@csiro.au

¤ Current address: Institut Méditerranéen de Biodiversité et d'Ecologie marine et continentale (IMBE UMR 7263 CNRS/IRD), Aix-Marseille Université, Europole de l'Arbois, Aix-en-Provence, France

Introduction

Bioclimatic niche models for invasive alien species (IAS) have become fundamental tools for pest risk assessment (PRA) [1,2]. They are of most value prior to, or soon after a pest species becomes established in a new region, to delimit the extent of the endangered area within the area of interest of the PRA [3]. This information allows the assets at risk within the PRA area to be identified and quantified, enabling biosecurity managers to decide how best to allocate scarce management resources into pre- and post-border activities to limit the spread of these unwanted organisms, or to initiate eradications. While some attention has been paid recently to methods for quantifying the potential economic impact IAS might have on productive assets such as plantation forests [4], we are unaware of any efforts to identify

systematically assets of particular biological conservation concern within areas threatened by IAS.

Puccinia psidii sensu lato (s.l.) is a plant pathogenic rust fungus native to South and Central America, and possibly the Caribbean, that is commonly known as guava or eucalyptus rust [5,6]. It was first recorded outside its native range in Florida, USA in the late 1970s [7]. More recently it was discovered in 2005 in California [8] and Hawai'i, USA [9], Japan in 2007 [10], China in 2009 [11] and Australia in 2010 [12], where it was originally identified as *Uredo rangelii* (commonly referred to as myrtle rust) [13]. *Puccinia psidii* s.l. is most feared because it has a very wide host range within the family Myrtaceae [14]. A recent study with an accession of *P. psidii* s.l. from Australia has shown that it can infect species across all 15 tribes of the subfamily Myrtoideae in the family Myrtaceae present in Australia [14]. Based on pathogenicity tests, there is evidence that several 'races' ('strains') exist within *P. psidii* s.l., each with a slightly different host range [14,15,16,17], but a comprehensive inventory of the total number of races has never been undertaken. Possible differences in climatic preference between these races have not been investigated. The rust only infects young, actively growing foliage of plants, and as a result has more adverse impacts on young plants [6]. Severe damage has been reported in the field in some years on a range of hosts in both the native and introduced ranges [6,15]. Considering the dominance of myrtaceaeous species in Australia, there is great concern that recurrent epidemics of *P. psidii* s.l. could transform many of the major natural ecosystems and forestry plantations.

There have been at least three previous published attempts to estimate the geographic invasive potential of *P. psidii* s.l. Booth et al. [18] used a homoclime analysis, with a special focus on Australia. This model was subsequently revised and results and methods published in Glen et al. [6] and Booth and Jovanovic [19], respectively. Magarey et al. [20] used NAPPFAST (NCSU APHIS Plant Pest Forecasting System), to estimate the global potential for *P. psidii* s.l. infection based on its known distribution in South America and the Caribbean. Most recently, Elith et al. [21] used the MaxEnt model to explore the effects of taxonomic uncertainty on the potential range of *P. psidii* s.l. using known distribution data, including some from Hawai'i. While these models provided initial indications of the potential distribution of the rust, they were either based on empirical biological knowledge of the organism's response to environmental variables [20], or its known distribution [18,21], but not both, and therefore may be sub-optimal.

There are few methods that are well-equipped for estimating the potential distribution of an organism in a novel environment such as on a new continent or under climate change scenarios [22,23]. CLIMEX [24,25] is one of these methods. It has become a popular climatic niche modelling tool because it is very well suited to modelling the potential distribution of invasive organisms, allowing the modeller to take advantage of knowledge of the biology and phenology of the organism, as well as its geographical distribution. By considering information from various knowledge domains, it is possible to robustly cross-validate parameter estimates, increasing confidence in the resulting models. CLIMEX Compare Locations models have been used successfully to estimate the potential distribution and relative climate suitability of a range of plant pathogens (e.g., [26,27,28,29,30,31]).

The recent invasion of California, Hawai'i, Japan, China and Australia [8,9,10,11,12], and the earlier invasion of Florida [7] by *P. psidii* s.l. present opportunities to study in greater detail the geographic and climatic limitations for the establishment of this pathogen. Distribution data from Hawai'i are particularly useful because the extremely steep climatic gradients cover a wide range of conditions within a very limited area. Further, Hawai'i also has some very susceptible hosts to *P. psidii* s.l. that are widespread. Despite being present in Hawai'i for a relatively short period of time, it is likely that *P. psidii* s.l. has had the opportunity to spread to all suitable climates available within the area. Given that the known distribution of *P. psidii* s.l. is primarily tropical, the highest altitude location records on the islands of Hawai'i are of most interest to biosecurity agencies, as these could indicate the cold tolerance limits for the pathogen. A caution in making use of this information is that Hawai'i has relatively few climate stations sampling very steep climatic gradients. The temperature variables are the most critical here. Fortunately, these are strongly influenced by altitude, so it is possible to use splining techniques to interpolate climatic averages reasonably accurately [32].

In this study, we firstly gathered data on the worldwide distribution of *P. psidii* s.l., including its distribution in regions where recent incursions have occurred – California (2005), Hawai'i (2005), Japan (2007), China (2009) and Australia (2010) [8,9,10,11,12]. A laboratory experiment to gather biological data on an accession of *P. psidii* s.l. from Australia was also performed to add to existing published data on other accessions from Brazil [33,34]. The distribution data, except for Australia, China and Japan, was then combined with all empirical biological information available into the CLIMEX modelling software to develop an improved climatic niche model and generate more robust estimates of the rust's potential distribution worldwide. The fit of the model was tested using the Australian distribution data and the point records in China and Japan. Finally, we used Australia as a case study to demonstrate the benefits of combining the CLIMEX model of *P. psidii* s.l. with the distribution of potential Myrtaceae host plants, including threatened species and forestry plantations, to identify areas that are most at risk from the rust.

Methods

Effect of temperature on *Puccinia psidii* s.l. urediniospore germination on agar disks

Urediniospores of a single-uredium isolate of *P. psidii* s.l. (DAR 81284) were produced on *Syzygium jambos* plants, and harvested as described in Morin et al. [14]. A small amount of spores was mixed with liquid paraffin oil (previously found to have stimulatory effects on germination; [35]) and the suspension was applied with a fine camel hair paint brush onto the surface of 8 mm diameter disks of 2% water agar (Merck) with the aid of a dissecting microscope. The water agar provided the necessary moisture required for spore germination [36]. The spore suspension density was adjusted prior to application by adding more oil to ensure that spores were well separated from each other on the agar once applied. Thirty-two inoculated disks were placed in each of the base of seven 90 mm diameter plastic Petri dishes. A four by eight grid pattern with cells numbered from 1 to 32 printed on paper was fixed on the outside of each dish base to provide a unique identification number to each inoculated disk. Each Petri dish was then wrapped in foil and placed in one of seven different compartments of a uni-directional temperature gradient plate similar to that of Barbour and Racine [37]. The Perspex compartments experienced the following average temperatures (standard deviation): 8.8°C (0.19), 11.2°C (0.20), 15.6°C (0.25), 19.1°C (0.35), 21.8°C (0.32), 26.9°C (0.38) and 29.7°C (0.74). The recorded temperature in each compartment was based on measurements taken over a three week period with a Hobo data logger (Onset Computer Corporation). Four disks (replicates) per plate were removed at random (based on randomly generated numbers) at 3, 6, 9, 12, 18, 24, 30 and 42 h after the commencement of the experiment and each placed on a drop of lacto-glycerol blue stain on a glass slide. The lacto-glycerol blue was absorbed rapidly by the disks, and germination was arrested without disturbance of the disk upper surface. Slides with stained disks were stored in large Petri dishes lined with moist paper towel and placed in a refrigerator until microscopic assessment was completed. For each disk several non-selectively chosen fields of view were examined using a compound microscope until the germination of a total of 100 urediniospores had been assessed. Urediniospores were considered to have germinated when the length of the germ-tube was greater than half the width of the spore. The experiment was performed twice.

Data analysis was performed using the statistical package R (release 2.13.0) [38]. Germination proportion data were analysed using a generalised additive mixed-effects (GAM) model [39,40], assuming a binomial error with a log link function to estimate response curves with estimated temperature and duration on agar disks as covariates. Trial number was included in the model as a random effect. The model included interactions between temperature and duration on agar disks, allowing for two-dimensional smoothing. A scale-invariant tensor product smooth, which uses a lattice of bendy strips with different flexibility in different directions was employed [39]. This smoother was chosen as temperature and duration on agar disks are anistropic (measured in different units), hence it is appropriate for the degree of smoothing to differ between them (c.f. isotropic smoothers).

Geographical distribution of *Puccinia psidii* s.l.

The historical distribution of *P. psidii* s.l. used to build the CLIMEX model was assembled from several sources. Booth et al. [18] reports nine point locations where the rust has been recorded in South America. Unpublished data on counties or locations where the rust has been recorded in Florida, California and Hawai'i were obtained from a range of collaborators (personal communications: Anne Marie La Rosa, United States Forest Service and Cherisa Coles, Janice Uchida and Mee Sook Kim, University of Hawai'i). Additional data on the distribution in Hawai'i were obtained from a roadside survey on the islands of Kaua'i, O'ahu, Maui and Hawai'i [41]. The country-level occurrence records of *P. psidii* s.l. comprised in the CABI Distribution Maps of Plant Diseases [42] and the point distribution records indicated in Magarey et al. [20] represent the centroids of countries or regions, and not true point locations, limiting their value for model calibration. The single point records in Japan and China, reported by Kawanishi et al. [10] and Zhuang and Wei [11], respectively, and the distribution records for Australia, as of April 2012, that had been collected separately by the Queensland, New South Wales (NSW) and Victorian State Governments and kindly provided by various agencies, were not used to build the model but rather used to validate the model fit. The known global distribution of *P. psidii* s.l. based on all these sources is presented graphically in Figure 1.

Base climate data and CLIMEX model

The primary base climatology used to build the model was the CliMond 10′ climate normals centred on 1975 (CM10_1975H_V1_1) [43]. We chose to use this hybrid dataset as it provided improved spatial resolution over the traditional 0.5 degree climate datasets, and improved data quality. We attempted to supplement the available data for Hawai'i through the US National Climatic Data Centre and the United States Geological Survey, because the existing climate data is very scant, and on these islands the climatic gradients are extremely steep for both rainfall and temperature. While it was possible to supplement the data with additional temperature, and to a lesser extent rainfall records, it was impossible to locate any additional source of data variables to indicate atmospheric wetness (e.g., absolute humidity, relative humidity, vapour pressure, or vapour pressure deficit). In an attempt to overcome this problem, a new set of surfaces were generated using NewLocClim [44]. A digital elevation model (DEM) was developed from the Shuttle Radar Topography Mission data [45] by extracting values for a 0.025 degree regular grid of points within the land areas of the State of Hawai'i. This DEM was provided to NewLocClim as a means of adjusting the interpolation of meteorological station data to account for topographic variation. The thin-plate spline was selected as the interpolation method as this has been used successfully in Australia and elsewhere to produce high quality climate surfaces [46,47]. The resulting surfaces for Hawai'i appear to represent the temperature surfaces and the rainfall for the wetter areas adequately; however, the representation of precipitation in the drier areas is not satisfactory when compared with the Rainfall Atlas of Hawai'i [48].

The CLIMEX niche modelling package [24,25] was used to create a model for *P. psidii* s.l. (Table 1). The model is based on previous work [49], in which the model stresses were fitted before location records in Japan, China and Australia had been collected. In the model presented in this paper, only the growth index parameters were adjusted to accommodate the observed germination responses of urediniospores of an Australian accession of *P. psidii* s.l. to a range of temperatures under experimental conditions.

The Compare Locations function in CLIMEX calculates an annual index of climatic suitability, the Ecoclimatic Index (EI), which reflects the combined potential for population growth during favourable periods and survival during stressful periods (Equation 1). The annual Growth Index (GI_A) describes the potential for growth of the modelled organism as a function of average weekly soil moisture (Moisture Index; MI) and temperature (Temperature Index; TI) during favourable conditions only (Equation 2; $GI_W = TI_W \times MI_W$). Stress indices describing cold (CS), wet (WS), hot (HS), and dry (DS) and their interactions with one another can be used to describe the population response to climatically unfavourable conditions. The individual components of stress are combined into a stress index (SI) and a stress interaction index (SX) (Equations 3 and 4; CDX = Cold-Dry Stress, CWX = Cold-Wet Stress, HDX = Hot-Dry Stress and HWX = Hot-Wet Stress) [25].

$$EI = GI_A \times SI \times SX \qquad (1)$$

$$GI_A = 100 \sum_{i=1}^{52} GI_{W_i}/52 \qquad (2)$$

$$SI = (1\text{-}CS/100)(1\text{-}DS/100)(1\text{-}HS/100)(1\text{-}WS/100) \qquad (3)$$

$$SX = \qquad (4)$$
$$(1-CDX/100)(1-CWX/100)(1-HDX/100)(1-HWX/100)$$

The three main sources of information for fitting the Temperature Index function were the ecophysiological observations of Ruiz et al. [33] and Ferreira [34] and those obtained in the laboratory experiment with the Australian accession of *P. psidii* s.l. included in this paper. Values for the Temperature Index parameters suggested by the experiment reported here are somewhat lower than the values indicated by experiments using Brazilian accessions of the rust [33,34]. The parameters adopted in the model span results from all three sets of experiments, taking the lower values from the Australian accession, and the upper values from the Brazilian accessions. The resulting climate suitability maps therefore portray the combined risks from all known sources. Using these parameters it is apparent that the low temperature requirements for germination are not limiting the distribution of *P. psidii* s.l. because the fitted cold stress temperature threshold (TTCS) is considerably different to the experimentally-estimated minimum temperature for development (DV0).

While the Growth Index components are best informed by direct experimental observations [50] or inferred from phenological observations [51], stresses, which indicate negative population

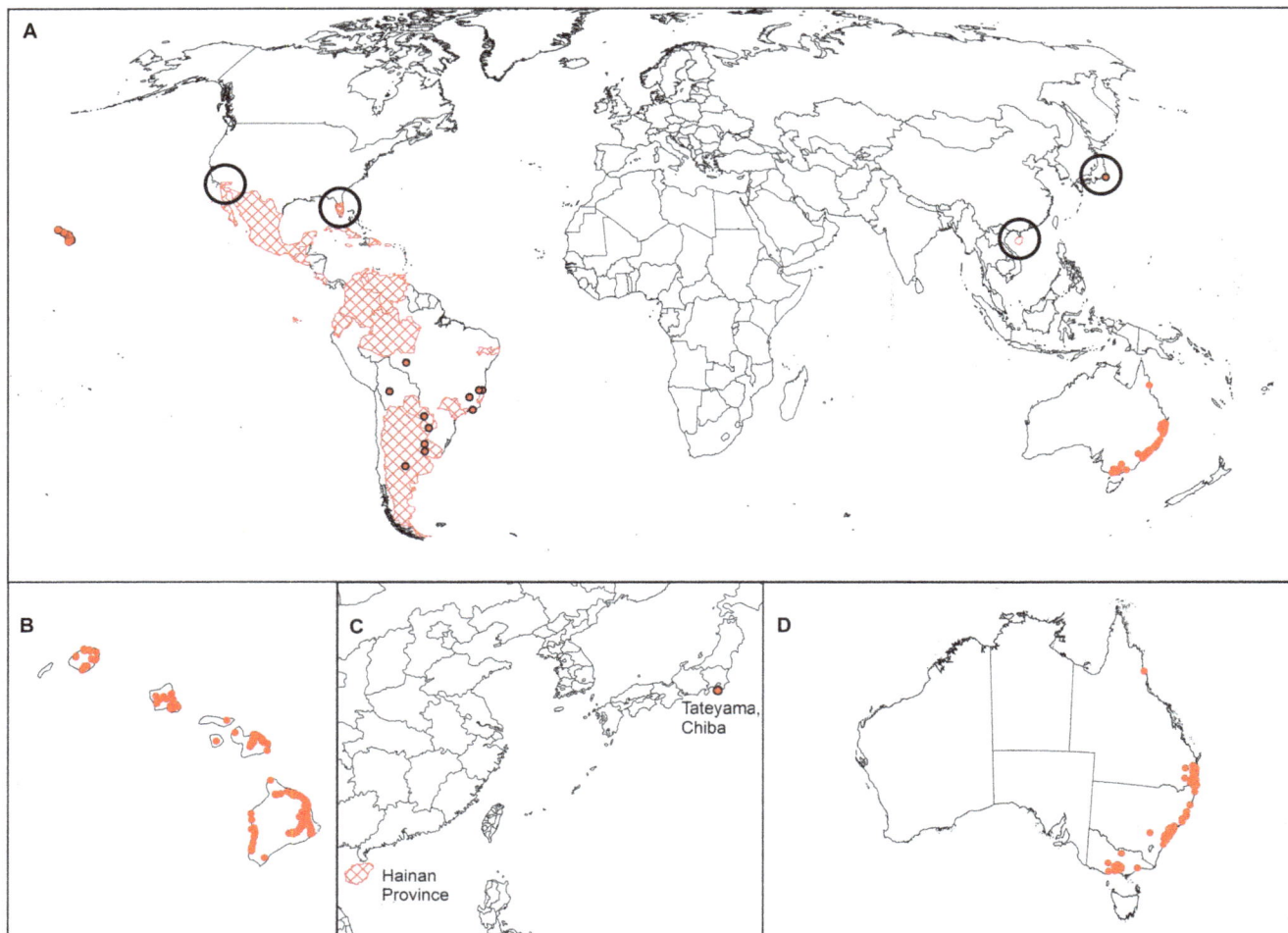

Figure 1. Global distribution of *Puccinia psidii* s.l. based on most recent literature and records from Australia, as of April 2012 (A). Call-outs of distribution in Hawai'i (B), Asia (C) and Australia (D). Dots indicate point location records. Cross-hatched areas indicate administrative regions. Black circles draw attention to small administrative regions where the fungus has been recorded.

growth, are best inferred from distribution data; though critical thresholds can still be informed by experiments and phenological observations. The stress parameters were adjusted until the EI value was positive at all recorded locations. As explained above, the country level data reported in the CABI Distribution Maps of Plant Diseases [42] and elsewhere [20] were not used directly in the model fitting as they were too coarse to guide parameter selection. The role of these records in the modelling process is to provide some level of "fuzzy logic" model verification: where model and data agreement required that somewhere in each of the countries that have reliable presence records is projected to be climatically suitable.

The soil moisture indices were set in consideration of the distribution of the rust in Hawai'i and Argentina. The upper soil moisture values were adjusted to allow *P. psidii* s.l. to thrive in the extremely wet areas to the north of Hilo on the island of Hawai'i, where the rust has been recorded. The lower soil moisture threshold for development (SM0) was adjusted downwards to allow it to persist barely at Misiones in south western Argentina, at the dry end of its distribution (Fig. 1).

The minimum annual heat sum for population reproduction (980 degree days above DV0) was set to allow persistence at the highest elevation location record on the island of Maui in Hawai'i.

Two forms of cold stress were employed in this model to limit further its potential to persist in cold climates. A monthly average daily minimum of 2.5°C is associated with frost events (about one per week). This limit for TTCS is probably associated with the temperature tolerances of the known hosts, rather than a direct physiological impact on *P. psidii* s.l. A degree day cold stress (DTCS) was also employed in order to achieve a satisfactory fit to the known distribution data. The DTCS threshold of 17 degree days per week above DV0 (10°C) suggests that the pathosystem needs to be actively growing by a small amount each day in order to offset respiration losses; extended periods with insufficient heat could lead to population reductions as energy reserves are run down. Cold stress was used to limit the potential range of *P. psidii* s.l. near the single Bolivian record (Fig. 1). Using these inferred parameters, the model also limits the range about 100 km south west of the westernmost location record in Argentina (Misiones). The spatial independence of these extreme range records provides some additional confidence in this limit.

Dry stress is likely to affect the pathogen indirectly, through the plant host; reducing turgor and the availability of photosynthate, and ultimately, the availability of plant hosts. We infer dry stress by virtue of the fact that *P. psidii* s.l. has not been recorded in areas experiencing drought stress, and from the general observation that

Table 1. CLIMEX Compare Locations model parameters for *Puccinia psidii* s.l. (mnemonics are taken from Sutherst et al. [25]).

Index	Parameter	Value[a]
Temperature	DV0 = lower threshold	10°C
	DV1 = lower optimum temperature	14°C
	DV2 = upper optimum temperature	25°C
	DV3 = upper threshold	32°C
Moisture	SM0 = lower soil moisture threshold	0.24
	SM1 = lower optimum soil moisture	1
	SM2 = upper optimum soil moisture	1.5
	SM3 = upper soil moisture threshold	2
Cold stress	TTCS = temperature threshold	2.5°C
	THCS = stress accumulation rate	−0.0045 Week^{-1}
	DTCS = degree day threshold	17°C days
	DHCS = stress accumulation rate	−0.0015 week^{-1}
Hot stress	TTHS = temperature threshold	32°C
	THHS = stress accumulation rate	0.002 Week^{-1}
Dry stress	SMDS = soil moisture threshold	0.2
	HDS = stress accumulation rate	−0.012 Week^{-1}
Annual Heat Sum	PDD = degree-day threshold[b]	980°C Days

[a]Values without units are a dimensionless index of a 100 mm single bucket soil moisture profile.
[b]Minimum annual total number of degree-days above DV0 needed for population persistence.

it requires the presence of plant hosts that are sensitive to drought stress. Accordingly, the dry stress parameters were set to allow marginally suitable persistence at the driest location record on the island of O'ahu, Hawai'i. Using these parameters, there is a considerable amount of dry stress (64%) at the single high elevation location record in Bolivia (Fig. 1). This is consistent with ecophysiological expectations; given the credibility of the record, and the dry winters in this location, modelled dry stress should be present, but non-lethal at this point in Bolivia.

On the island of Hawai'i, *P. psidii* s.l. is present in all areas sampled throughout the area west of Hilo on the eastern side of the island (Fig. 1B), which, according to Giambelluca et al. [48] receives extremely large rainfall totals throughout the year. Accordingly, wet stress parameters were not used to limit the range of *P. psidii* s.l. The effect of hot-wet stress as a limit on the range extent in the Amazon basin was investigated. Using hot-wet stress to limit the distribution to the western most point location in Brazil had the undesirable effect of making Florida appear unsuitable. It also severely limited the potential range in Amazonas province in Brazil, where the rust has been recorded, albeit imprecisely (Fig. 1) [42]. In view of these results, it was decided to remove hot-wet stress from the model, accepting that this aspect must be treated with caution as it is possible that the model overestimates the potential for growth under very warm and wet conditions.

The CLIMEX model was partially validated using records from Japan, China and Australia, by assessing the model sensitivity numerically. Due to the unstable nature of the range of an invasive species, it is not possible to satisfy the data requirements for a formal assessment of model specificity (i.e. the proportion of false

positives), hence the model specificity was only assessed subjectively.

Distribution of potential Myrtaceae hosts within the modelled climatic envelope in Australia

Geographical patterns of species richness. The geographical distribution of potential Myrtaceae hosts of *P. psidii* s.l. in Australia was gauged by querying the Australian Virtual Herbarium [52], and selecting all 197 491 records where Family equalled 'Myrtaceae', and the latitude and longitude coordinates fell within the area where *P. psidii* s.l. was modelled as having an EI>0. To estimate the geographical patterns of species richness of the putative host plants of *P. psidii* s.l. at the scale used for the climate modelling, the number of species records within each cell was summed, ignoring multiple records for the same species within each cell. Because of the fluid and inconsistent nature of the taxonomic classification of records at the sub-specific level, only genus and species epithets were considered when determining taxonomic uniqueness within each cell. To avoid biasing the diversity counts within each cell, where the species epithet was blank (1 926 records) or 'sp' (940 records) in the extracted herbarium dataset, the species field was filled with 'sp.' using the search and replace option in Microsoft Excel. The Myrtaceae species location records were then imported into ArcGIS 10.3 (ESRI, Redlands, Ca., USA) and spatially joined to the cell identifiers for the corresponding 10′ climate data, before being re-exported to Microsoft Excel, where the Remove Duplicates option was employed. A total of 121 377 non-unique combinations of genus, species and cell identifier were found and removed, leaving 59 814 taxonomically and geographically unique records. The unique records were re-imported into ArcGIS and spatially joined to the climate dataset. The species richness for Myrtaceae in each 10′ climate cell was then gauged by dividing the number of taxonomically unique records in each 10′ climate grid cell by the projected area in km^2.

Threatened Myrtaceae species. A map of the Myrtaceae species listed as threatened under the Australian Government Environment Protection and Biodiversity Conservation (EPBC) Act [53] that fall within the area where *P. psidii* s.l. was modelled as having an EI>0 was generated as above.

Hardwood plantations. The major hardwood (Myrtaceae) and mixed hardwood and softwood forest plantations present in Australia were extracted from the Australian National Forest Inventory for 2010 [54]. The dataset was projected using an Albers equal area projection to estimate areas for the forest plantations. These data were spatially intersected with the CLIMEX model results and the Australian State boundaries to estimate the areas of hardwood and mixed hardwood and softwood plantations climatically suitable for supporting the establishment of populations of *P. psidii* s.l. based on the EI. For mapping purposes, the 10′ grid cells were used to indicate the presence of any hardwood forest coups identified in the Australian Land Use dataset V4 [55].

Results

Effect of temperature on *Puccinia psidii* s.l. urediniospore germination on agar disks

The smooth terms (temperature and duration on agar disks, which correlates with exposure to moisture required for germination) in the model were highly significant based on approximate measures of significance ($F_{18,18} = 85.2$, P<0.001). More importantly, a plot of Pearson residuals (the difference between the observed and fitted proportions divided by the standard error of the fitted proportion) versus fitted model values revealed no bias or outliers (Fig. 2); meaning that the GAM model fitted the observed germination data very well over a wide range of germination probabilities. Germination percentages were overall higher in the first trial compared to the second trial, but the response trend to treatments was similar for the two trials. At most temperatures, urediniospores began to germinate within 3 h of application onto the agar surface, although a preferred temperature range was already evident (Fig. 3). After 6 h on agar disks, germination had already reached 35% at temperatures between approximately 15 and 18°C. It continued to increase slightly as the period on agar disks extended until it reached a maximum of 45% germination after more than 30 h on agar disks. Germination occurred over the complete range of temperatures tested (8.8–29.7°C), but was the highest (>35%) between temperatures of approximately 12 and 20°C (Fig. 3). The lower temperature threshold for germination observed here is considerably lower than that indicated by the experiments of Ruiz et al. [33] and Ferreira [34]. Accordingly, DV0 was adjusted downwards to 10°C, leaving allowance for the moderating effect of the averaging process in calculating the long-term climate data compared with the instantaneous measurements in the laboratory studies.

CLIMEX model

The CLIMEX EI world map of *P. psidii* s.l. shows its preference for moist climates with moderate temperatures throughout the wet tropics and sub-tropics, and even extending into some cool regions with a mild Mediterranean climate (Fig. 4a).

In the native range, the regions of highest climatic suitability fall in Brazil and Paraguay in south-eastern South America, and also in a narrow band in the north-west of South America. The Caribbean and surrounding moist regions of Central America and the invaded areas of the USA are also highly climatically suitable. Its range appears limited by cold and dry stress in the south and western parts of Argentina, dry stress near the Rio Grande River in New Mexico and Texas in the USA, and by cold stress elsewhere in the USA (data not shown).

In other regions where *P. psidii* s.l. has invaded, there are several, highly climatically favourable areas throughout the State of Hawai'i (except for the saddle on the main island of Hawai'i, Fig. 4b) and most of South-East Asia (Fig. 4a). In Eastern and Southern Asia, it is only limited by cold stress in northern China, and hot and dry stress in India, respectively. In Australia (Fig. 4c), favourable climates are restricted mostly to near-coastal southern and eastern regions where it is limited by cold stress (southern parts of the Great Diving Range), and by hot and dry stress in the interior (data not shown).

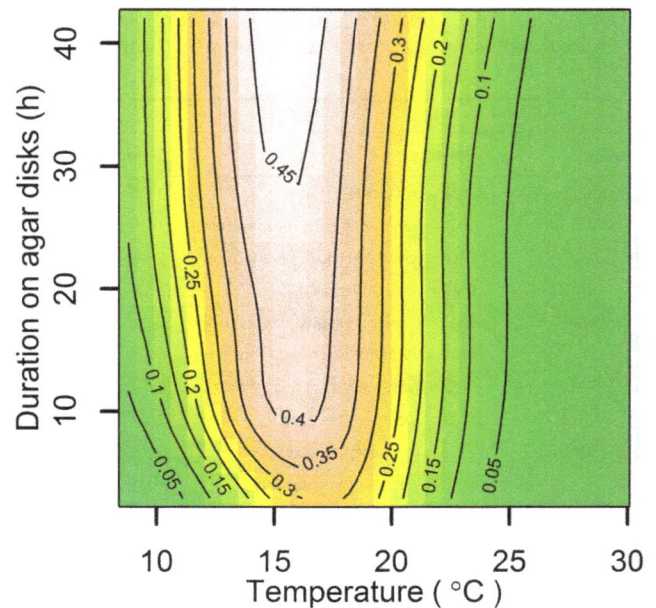

Figure 3. Effect of temperature on germination of urediniospores of *Puccinia psidii* s.l. (Australian accession; DAR 81284) on agar disks, which provided the necessary moisture for germination. Proportion of spores that germinated at each treatment combination (seven temperatures × eight durations on agar disks, and four replicates per treatment combination in each trial) is represented by the isolines (proportion of germinated urediniospores increases from dark green to pale orange). Results are pooled from two trials.

Suitable climates for *P. psidii* s.l. can also be found in some of the regions that have not yet been invaded by the rust. Highly favourable climates exist in coastal parts of the Mediterranean (restricted by hot and dry stress), high elevation areas throughout Central Africa, and eastern coastal regions of South Africa and Madagascar (Fig. 4a). In Australasia, the island chains of the Solomon Islands, New Caledonia and Vanuatu are also highly suitable climatically (Fig. 4c). In New Zealand, much of the North Island and a very small area in the north of the South Island are climatically suitable.

The potential for *P. psidii* s.l. to infect and grow under favourable conditions only, without taking into account potential survival during stressful periods, is indicated by the GI_A (Fig. 5). A positive GI_A occurs in many areas depicted in the CLIMEX EI map (Fig. 4) as being unsuitable for persistence. This situation indicates that the amount of population growth during the favourable season(s) is insufficient to offset the population declines during the stressful season(s). The ongoing presence of *P. psidii* s.l. in these areas would depend on seasonal reinvasion from nearby source areas.

Model Validation

The CLIMEX EI world map shows that all known recorded point locations for *P. psidii* s.l. are modelled as being climatically suitable, and all region records include at least some locations that are modelled as being climatically suitable (Figs 1, 4a), indicating perfect model sensitivity (i.e. sensitivity = 1). This includes all records in Australia, China and Japan reserved from model fitting, providing strong validation of the model. The cold and dry stress constraints appear to be well supported with independent point locations in North and South America and Hawai'i lying on

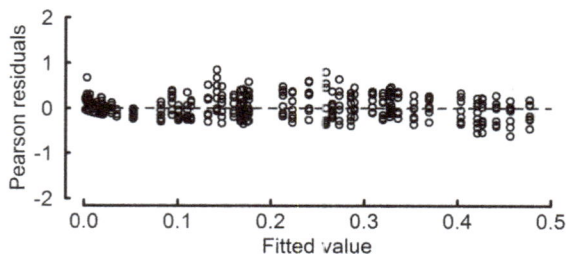

Figure 2. Pearson residuals (the difference between the observed and fitted proportions divided by the standard error of the fitted proportion) versus fitted model values for the GAM model of the effect of temperature and duration on agar disks on germination of *Puccinia psidii* s.l. urediniospores.

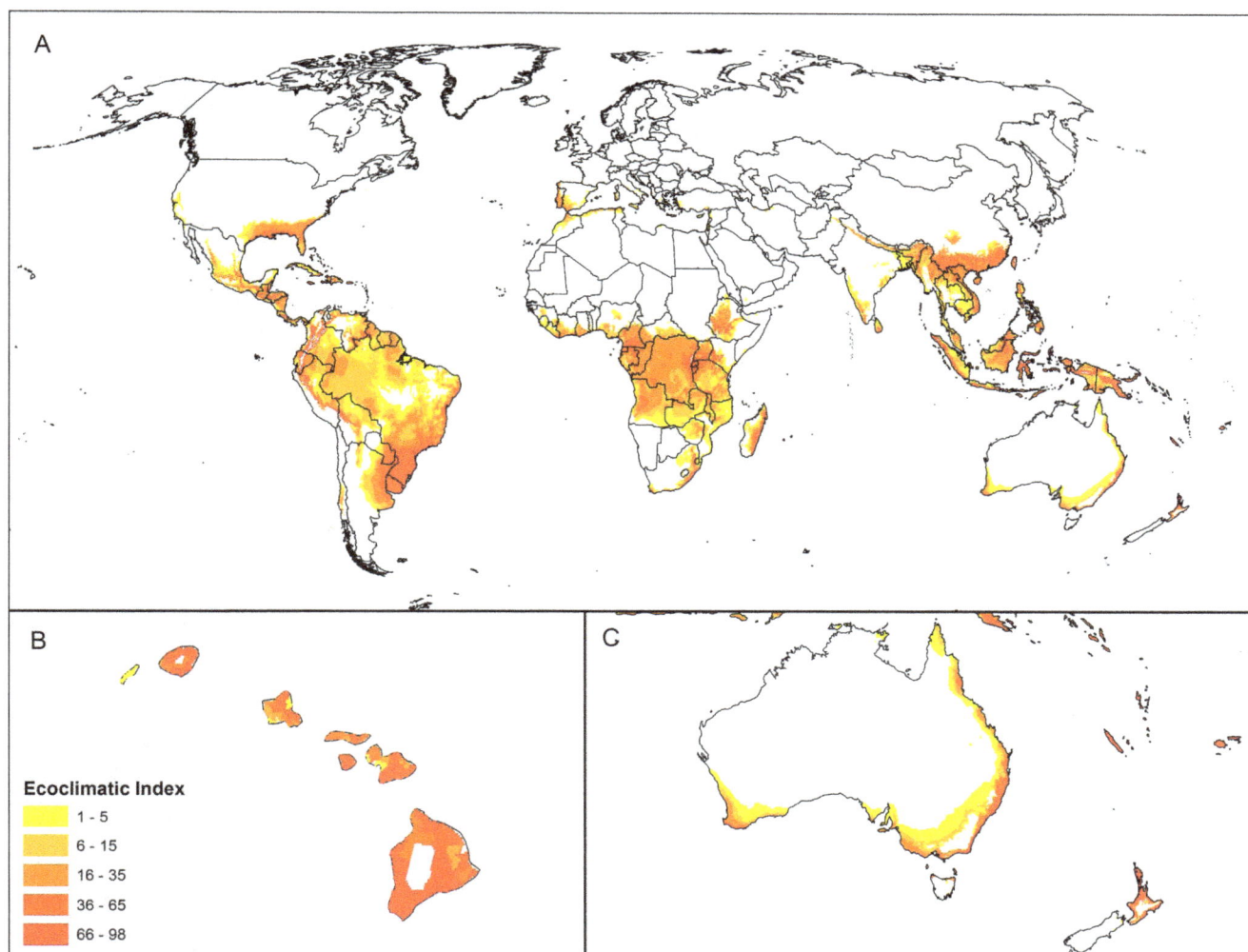

Figure 4. Relative climate suitability for *Puccinia psidii* s.l. as indicated by the CLIMEX Ecoclimatic Index (EI) modelled using the CliMond CM10 1975H V1_1 climate dataset [43]. A) World, B) Hawai'i, and C) Australasia. The EI reflects the combined potential for population growth during favourable periods and survival during stressful periods. Values of zero indicate no potential for establishment, while values of 100 indicate optimal conditions for growth and survival year round.

similar cool limits. The model specificity is satisfactory. The areas indicated as climatically suitable, but where there are no records (e.g., central Africa), appear ecologically plausible, sharing a similar climate to areas that are presently occupied by *P. psidii* s.l. During the initial model-fitting, there was evidence that *P. psidii* s.l. was found in slightly cooler locations in its introduced range (e.g., in southern California), than was known from within its presumed native range in South America and the Caribbean. However, in the absence of more extensive surveys such as gradsect trapping [56], the potential exists for *P. psidii* s.l. to be present, but undetected in cooler locations in the native range.

Distribution of potential Myrtaceae hosts within the modelled climatic envelop in Australia

In Australia, pockets of high densities of myrtaceous species (Fig. 6b) are scattered throughout the area climatically suitable for persistence of *P. psidii* s.l. (EI>0) (Figs 4c or 6a). The highest species richness of myrtaceous hosts is found in four main areas: in descending order, the south west of Western Australia (WA), the coastal wet tropics and south-eastern region of Queensland and the coastal hinterland of central and northern NSW (Fig. 6b). Within the modelled climatic envelop, the greatest density of threatened Myrtaceae species listed under the EPBC Act falls in 1) the Sydney Basin and the eastern border region of NSW and Queensland, 2) a small western coastal region to the south of Geraldton in WA, and 3) a small south coastal region of WA between Albany and Esperance (Fig. 6c).

Most of the hardwood forestry operations in Australia fall within the climatic potential range for *P. psidii* s.l. to persist (EI>0) (Table 2). The greatest concentrations of forests at risk are in southern and south-western parts of WA, south-eastern South Australia, throughout the mesic midlands of Victoria and coastal northern NSW and south-eastern Queensland (Fig. 6d, Table 2). Small pockets also exist in the northern extremes of Tasmania and the wet tropics of northern Queensland (Fig. 6d).

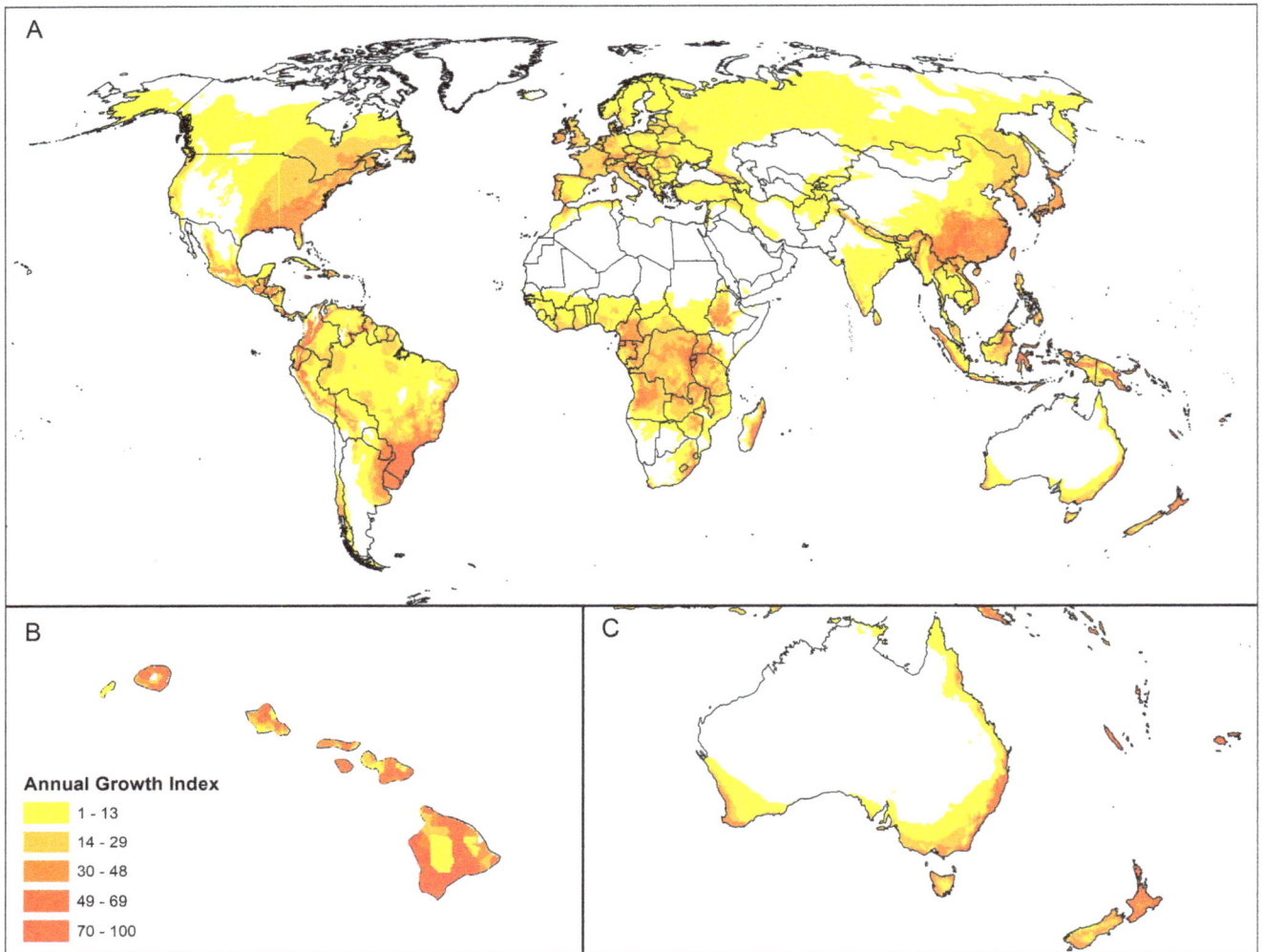

Figure 5. Relative growth potential for *Puccinia psidii* s.l. as indicated by the CLIMEX annual Growth Index (GI$_A$) modelled using the CliMond CM10 1975H V1_1 climate dataset [43]. A) World, B) Hawai'i, and C) Australasia. The GI$_A$ summarises the potential for population growth across the year, ignoring the effects of stresses and minimum annual heat sum requirements to complete a generation. Outside of the area of suitable climate indicated by the Ecoclimatic Index (Fig. 4), the GI$_A$ estimates the relative potential for infection and population growth.

Discussion

Potential Host Risks

The family Myrtaceae comprises a large number of species that are widespread across the Australian continent, often occurring as major components of natural ecosystems [57,58]. *Eucalyptus* spp. are among the most widely planted species in the family Myrtaceae and are popular in many parts of the world as forestry resources with more than 20 million ha planted (e.g., Australia, Brazil, China, India, South Africa and Thailand) [59]. They are also valuable water-use-efficient amenity trees grown in many areas of the world (e.g., California) [60]. In some places eucalypts have even taken on an iconic character (e.g., on the Argentine pampas they have frequently been planted along roadsides) [61]. Based on our model, these regions all appear climatically suited to *P. psidii* s.l., and depending on host suitability, the eucalyptus assets of regions that have not yet been invaded by the rust may be at risk if dispersal barriers are overcome.

Given the rapid spread of *P. psidii* s.l. to Japan, China and Australia in the last few years, and the large spatial gaps involved,

it would seem inevitable that the identified uninvaded risk areas could become invaded eventually. Biosecurity efforts should therefore be aimed at slowing its global spread with, for example, the use of relatively low-cost phytosanitary precautions, and the development and deployment of resistant host varieties, rather than attempting to prevent its spread through quarantine barriers. In regions where Myrtaceae are non-native, the numbers of valuable myrtaceaeous species are limited. This makes it potentially feasible to undertake programmes to develop and deploy plant varieties that are resistant to *P. psidii* s.l. In Brazil, for example, before resistant clones were deployed, *P. psidii* s.l. was reported to cause major damage in eucalyptus plantations, particularly in the first two years of planting [62]. The existence of different pathotypes of *P. psidii* s.l. (e.g. [15,16,17]) however, means that any attempt to manage the invasion risks using resistant hosts should, if possible, confirm that resistance is effective across the known pathotypes.

In Australia, the key concerns associated with the recent invasion by *P. psidii* s.l. include the additional risks to native species and ecosystems important for conservation, as well as forestry

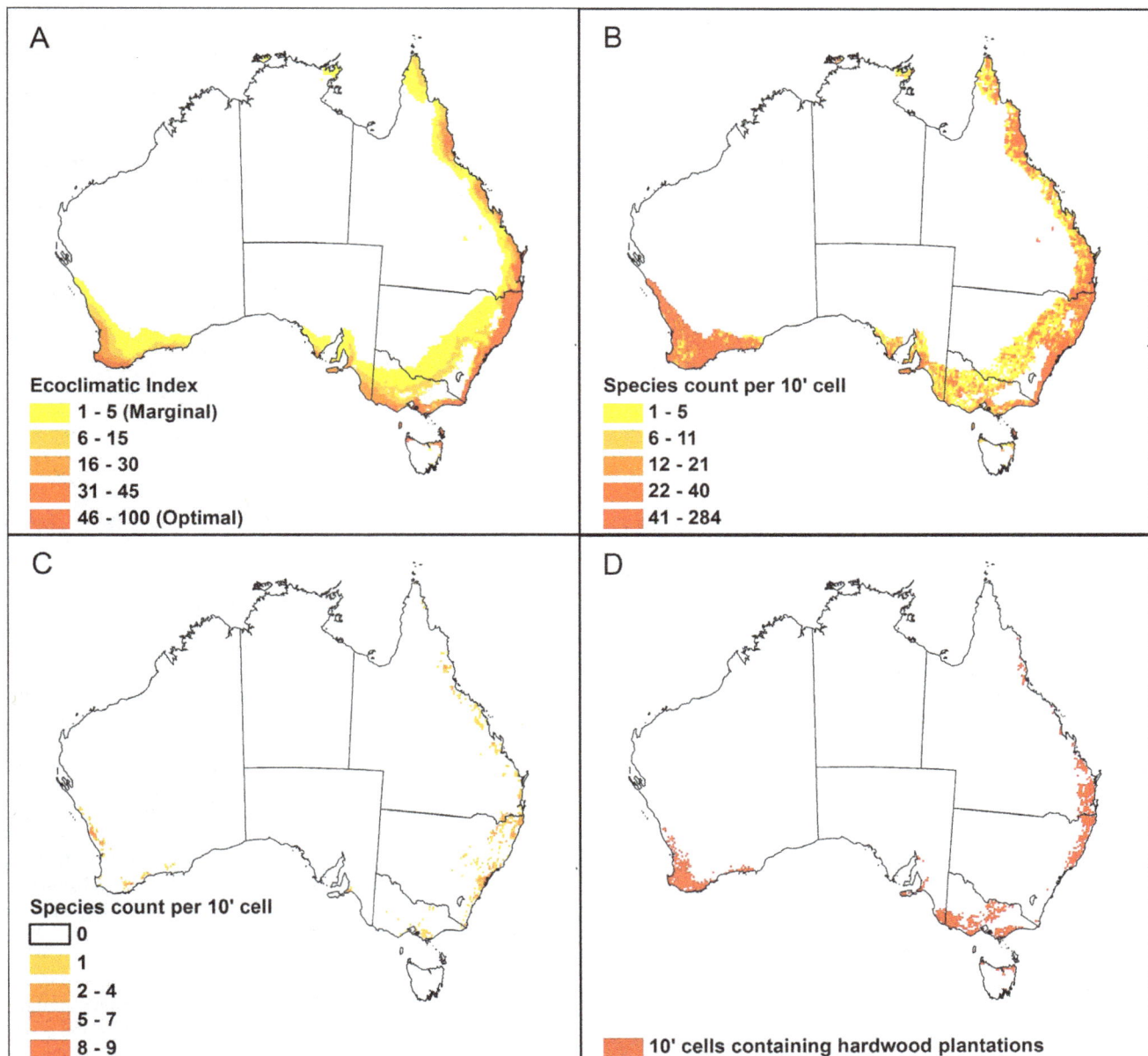

Figure 6. Australian natural assets at risk from *Puccina psidii* s.l. Comparison of the climate suitability map for *Puccinia psidii* s.l. in Australia as indicated by the CLIMEX Ecoclimatic Index (A) with the geographical distribution of potential Myrtaceae hosts within the climate suitability envelop. B) Species richness of Myrtaceae (species per 10′ cell), C) density of threatened species {unique records of species listed under the Environmental Protection and Biodiversity Conservation Act 1999 (Australia) [53]}, and D) 10′ cells containing hardwood and mixed hardwood and softwood forest plantations.

plantations based on *Eucalyptus* and *Corymbia* spp. In contrast to other regions where there are limited numbers of Myrtaceae plant hosts to manage, in Australia there is a multiplicity of species belonging to that family, and consequently more limited options with which to respond to this biological invasion. The large number of plant species that could be impacted, the putative large variation in susceptibility, both within and between species, and the significant differences between field and laboratory responses of pathosystems [14] make it extremely difficult to approach this problem with systematic foresight.

Modelled risk

The CLIMEX model presented in this paper, which combined updated distribution as well as biological data for *P. psidii* s.l. gleaned from our experimentation, extends the estimated global area at risk from this invasive fungus into cooler climates than some previously published modelling attempts [6,18,19]. The global climatically suitable envelope for *P. psidii* s.l., based on the modelled EI values, spans the wet tropical and subtropical regions of the world with moderate temperatures, and includes some cooler regions with a mild Mediterranean climate. In contrast, the GI_A for *P. psidii* s.l., which only takes into account the potential for growth in favourable conditions and not the potential survival

Table 2. Areas of hardwood and mixed hardwood and softwood plantations in Australia by State and Territory that fall within the climatic potential range for *Puccinia psidii* s.l. (Ecoclimatic Index>0) (Areas calculated from Australia's plantations inventory data [54]).

State	Area at risk of establishment (ha)	Area where establishment is unlikely (ha)	Total plantation area (ha)[b]	Proportion of total plantation area at risk (%)
Northern Territory	0	762	762	0
Tasmania	40 607	191 648	232 254	17.5
New South Wales	83 453	2 172	85 625	97.5
Queensland	26 569	469	27 038	98.3
South Australia	58 385	4	58 393	100.0
Victoria	187 781	19 478	207 259	90.6
Western Australia	289 295	1 435	290 730	99.5
Total	686 090	215 972	902 062	76.1

NB. Total areas do not equal those quoted in [54] due to the exclusion in this analysis of data supplied to the National Forest Inventory in tabular form at the State or Territory level. For the same reason, data are not available for the Australian Capital Territory.

during stressful periods, indicates that there is an opportunity for seasonal development of the rust to occur over a much larger area of the world than that indicated by positive EI values. These additional areas identified by the GI$_A$, however, are unlikely to be suitable under current climates for persistent rust populations to become established. Accordingly, the potential impacts in these climatically marginally suitable areas are likely to be limited. Where the requirement for a minimum annual heat sum will not be met, it is likely that in these areas any infection will not develop to sporulation unless the host plants are located in a warm microsite.

Model reliability

The potential distribution model for *P. psidii* s.l. accords with all known locations where field infections have been noted. The location records in Japan, China and Australia were collected after the model stresses had been fitted [49], thus providing strong geographically independent support for the model. Given the rate at which biological invasions tend to proceed, it is unsurprising that there are large areas that are projected to be climatically suitable for establishment by *P. psidii* s.l., but for which there are no distribution records. For example, there are large areas in central Africa and the Mediterranean that appear climatically suitable for *P. psidii* s.l., and where at least some suitable myrtaceous hosts are present [14], but for which there are no records of *P. psidii* s.l. These areas are climatically similar to areas occupied by *P. psidii* s.l., as indicated by the Köppen-Geiger zonations [43]. There are numerous possible explanations for the lack of records in these apparently climatically suitable areas, including a lack of introduced propagules, environmental and demographic stochasticity, poorly-suited hosts, and a lack of appropriate search effort.

The significant uncertainty surrounding the wet and hot-wet tolerance limits for *P. psidii* s.l. should be kept in mind when considering the modelled risks in the tropics. Should this aspect of the model become critical to some important decisions, then transect surveys using susceptible trap plants could be undertaken along critical climatic gradients [56]. The modelled poleward limits of *P. psidii* s.l. should be considered indicative, rather than prescriptive. For example, where *P. psidii* s.l. climatic suitability appears to diminish in the Australian Alps and in Tasmania, the climatic gradients are extremely steep in relation to the scale of the climate data used in this exercise, and due to the precision of the climate data used in this exercise, and due to the precision of the climatic dataset [63] there may be favourable microhabitats within the areas modelled as unsuitable.

Comparison with previous modelling efforts

The modification of the original Kriticos and Leriche model [49] to encompass empirical results on the germination of urediniospores of an Australian accession of *P. psidii* s.l. under different temperatures had a trivial impact on the extent of the modelled potential geographical range. Including the new experimental data in the model improved the apparent relative climatic favourability of warm temperate climates *within* the modelled potential geographical range.

Compared to the models of Booth et al. [18] and Booth and Jovanovic [19], the potential distribution model presented here draws on a broader range of distribution data, and benefits from the inclusion of information drawn from experimental observations of the germination and growth of the fungus [33,34]. Our model indicates a significantly greater ability of the organism to tolerate cold stress than these other models.

The model of Magarey et al. [20] is broadly analogous to the GI$_A$ in CLIMEX, and differs from the EI in ignoring those climatic stress factors that may reduce the potential for long-term persistence at a location. The net result of our study is a model that displays a generally more conservative risk picture than that portrayed by Magarey et al. [20] in regions experiencing seasonally hot dry conditions and continental climates, e.g., north-eastern USA and northern China. Curiously, the model of Magarey et al. [20] indicates that south-eastern Australia (south of Sydney) is unsuitable for *P. psidii* s.l. and yet more extreme continental climates (north-eastern USA and northern China) are suitable.

Elith et al. [21] presents results of several MaxEnt models, exploring the taxonomic uncertainty within the *P. psidii* s.l. complex. The "Puccinia_94" results in Fig. 2 of Elith et al. [21], were generated using data for a grouping loosely referred to as *P. psidii* s.l. and are taxonomically the most comparable to the CLIMEX model presented here, and generally accord reasonably well with our results. However, there are some notable contrasts in the MaxEnt and CLIMEX models. The CLIMEX results for current climate indicate that southern Tasmania and the South Island of New Zealand are unsuitable for establishment, and indicate a larger area at risk in WA than indicated by the MaxEnt Puccina_94 model [21]. The latter indicates that warm moist tropical environments such as Irian Jaya are climatically unsuitable, which appears at odds with experimental data [33,34]. Illogically, the two Uredo models in Elith et al. [21], which, by definition are based on a smaller environmental envelope than Puccinia_94, result in a significantly larger area modelled as suitable in Australia. By changing the background with the input locations, Elith et al. [21] may have confounded modelling artefacts with the taxonomic treatment effects in their results. The resulting MaxEnt Uredo maps indicating that xeric regions of Australia may be marginally suitable for invasion and that Alpine regions are highly suitable for invasion are clearly nonsensical. The reason for these results, which we consider implausible, clearly requires further investigation.

Species richness hypothesis

The map of species richness of Myrtaceae within the suitable climatic envelop for *P. psidii* s.l. in Australia presented in this paper shows the areas where epidemics may be more likely to be frequent and severe due to large numbers of different Myrtaceae species present. We hypothesise that regions rich in Myrtaceaeous species are more likely to contain one or more hosts susceptible to *P. psidii* s.l. Complementary development rates (e.g., [64]) and phenologies [65] amongst these hosts would also provide *P. psidii* s.l. with suitable foliage for infection throughout several periods of the year. All else being equal, higher rust inoculum loads might develop in these areas, and thus pose a greater threat to susceptible hosts than in areas of lower Myrtaceae richness. By comparing the climate suitability map for *P. psidii* s.l. with the map of Myrtaceae species richness, it is possible to identify hypothetical hotspots for epidemics where the EI is projected to be optimal and the highest diversity of Myrtaceae is found. Those hotspots are located in a narrow strip along the eastern coast of NSW, including the Sydney Basin, in the Brisbane and Cairns areas in Queensland and in the coastal region from the south of Bunbury to Esperance in WA. Sites within these areas would be ideal to implement long-term monitoring experiments to quantify the impacts of *P. psidii* s.l. [66] to guide future management responses.

Additional threats to vulnerable species

Puccinia psidii s.l. poses a new threat to plant species of the family Myrtaceae that are already recognised as threatened by a range of other factors under the Federal EPBC Act and/or various state Acts in Australia [53]. The map of the threatened Myrtaceae species in combination with the climatic suitability and Myrtaceae species richness maps presented in this paper could facilitate the identification of areas where threatened species are most at risk from *P. psidii* s.l., and thus deserving surveillance and management attention. It remains to be established whether *P. psidii* s.l. poses an extinction risk to already threatened species, or whether background levels of resistance are sufficient for plant communities to adapt to the presence of this invasive rust.

Threats to forest production

Puccinia psidii s.l. has so far only been reported on a few *Eucalyptus* spp. in native forests in Australia [67,68], and has been found only once severely affecting seedlings in a eucalypt plantation (A. Carnegie, personal communication). Nonetheless, there is potential for disease development in forestry plantations in Australia because nearly 80% are located within the climatic suitability envelop for *P. psidii* s.l. Under current climate, the most northerly portion of Tasmania, an area dense with eucalypt plantations, appears suitable for *P. psidii* s.l. The only plantations that have no apparent risk of supporting persistent rust populations are those located in the Northern Territory.

Spread in Australia

Considering the rapid spread of *P. psidii* s.l. via the nursery industry and wind-borne spores since its introduction in Australia [12,69], it would seem inevitable that the risk areas identified by our CLIMEX model will eventually become invaded. At present, the spread of the fungus has been limited to eastern Australia, which is isolated by deserts from the areas of WA that contain significant Myrtaceae diversity and eucalypt plantations. These deserts and the prevailing western wind patterns may slow the natural spread of *P. psidii* s.l. into Western Australia, supporting the efforts of biosecurity managers there to prevent its spread via human movement of infected plant material. At present, only a small area of Tasmania appears climatically suited for persistent populations of *P. psidii* s.l., although most of Tasmania is suitable to support some population growth. Under a warming climate we would expect this area of suitability to increase. The separation of Tasmania from mainland Australia via the Bass Straight may provide a hindrance to the spread of *P. psidii* s.l. into Tasmania, though the presence of islands in Bass Strait may provide the opportunity for a stepping-stone invasion pathway. Tasmanian biosecurity managers may wish to conduct routine surveillance of these islands with a view to eradicating isolated infections as a means of slowing the spread of *P. psidii* s.l. into Tasmania.

General applicability of the methods

The modelling and analytical methods demonstrated in this paper can help biosecurity and conservation managers to identify areas at heightened risk from IAS. These methods add to pest risk mapping methods available to assist these managers to gauge the size or value of the threats and to target interventions to manage them [1,2,70]. The pre-border pest risks from *P. psidii* s.l. to Africa, most of Asia, south-western Europe and New Zealand were identified using well-established niche modelling methods. The post-border example for Australia presented in this paper demonstrates how these broad scale bioclimatic risk patterns can be downscaled to identify and quantify relative risks to assets within the climatically suitable range. By combining bioclimatic niche modelling with non-climatic factors such as host species richness, we are able to move beyond simply identifying invasion risks, to provide some insight into the potential impacts that invasive species might have if they expand their range into climatically suitable areas, and distinguishing between invasiveness and impact [71]. In so doing we are able to inform biosecurity policies regarding the management of spread of IAS, and protection of valued assets.

Acknowledgments

Annemarie La Rosa (USFS) and Janice Uchida (University of Hawai'i at Manoa) provided valuable advice on the Hawai'ian survey. Alison Vaughan and David Cantrill (Victorian Herbarium) provided distribution records for Australian Myrtaceae from the Australian Virtual Herbarium, which is now part of the Atlas of Living Australia. Suzy Perry, Fiona Giblin, Geoff Pegg and Heather Taylor (Department of Agriculture, Fisheries and Forestry Queensland) provided distribution data for *P. psidii* s.l. in Queensland and Paul Mahon (Office of Environment and Heritage NSW) and David Smith (Victoria Department of Primary Industries) provided distribution information for NSW and Victoria, respectively. Adam Gerrand, Australian Bureau of Resource Sciences and Mijo Gavran of Australian Bureau of Agricultural and Resource Economics and Sciences kindly supplied plantation inventory data. We also thank Helen Murphy and John K. Scott (CSIRO Ecosystem Sciences) for their useful comments on a draft of this paper.

Author Contributions

Conceived and designed the experiments: DJK LM AL. Performed the experiments: DJK LM. Analyzed the data: DJK LM PC. Contributed reagents/materials/analysis tools: AL RCA PC. Wrote the paper: DJK LM.

References

1. Eyre D, Baker RHA, Brunel S, Dupin M, Jarošik V, et al. (2012) Rating and mapping the suitability of the climate for pest risk analysis. EPPO Bulletin 42: 48–55.
2. Baker RHA, Benninga J, Bremmer J, Brunel S, Dupin M, et al. (2012) A decision support scheme for mapping endangered areas in pest risk analysis. EPPO Bulletin 42: 65–73.
3. FAO (2006) International standards for phytosanitary measures: 1 to 24. Rome: Secretariat of the International Plant Protection Convention.
4. Kriticos DJ, Leriche A, Palmer D, Cook DC, Brockerhoff EG, et al. (2012) Linking climate suitability, spread rates and host-impact when estimating the potential costs of invasive pests. PLoS One 8: e54861.
5. Coutinho TA, Wingfield MJ, Alfenas AC, Crous PW (1998) Eucalyptus rust: A disease with the potential for serious international implications. Plant Disease 82: 819–825.
6. Glen M, Alfenas AC, Zauza EAV, Wingfield MJ, Mohammed C (2007) *Puccinia psidii*: a threat to the Australian environment and economy - a review. Australasian Plant Pathology 36: 1–16.
7. Marlatt RB, Kimbrough JW (1979) *Puccinia psidii* on *Pimenta dioica* in south Florida. Plant Disease Reporter 63: 510–512.
8. Mellano V (2006) Rust on myrtle found in San Diego County. Healthy Garden-Healthy Home, University of California Cooperative Extension Retail Nursery Newsletter 1: 3.
9. Uchida J, Zhong S, Killgore E (2006) First report of a rust disease on Ohi'a caused by *Puccinia psidii* in Hawai'i. Pant Disease 90: 524.
10. Kawanishi T, Uematsu S, Kakishima M, Kagiwada S, Hamamoto H, et al. (2009) First report of rust disease on Ohi'a and the causal fungus, *Puccinia psidii*, in Japan. Journal of General Plant Pathology 75: 428–431.
11. Zhuang J-Y, Wei S-X (2011) Additional materials for the rust flora of Hainan Province, China. Mycostema 30: 853–860.
12. Carnegie AJ, Lidbetter JR, Walker J, Horwood MA, Tesoriero L, et al. (2010) *Uredo rangelii*, a taxon in the guava rust complex, newly recorded on Myrtaceae in Australia. Australasian Plant Pathology 39: 463–466.
13. Simpson JA, Thomas K, Grgurinovic CA (2006) Uredinales species pathogenic on species of Myrtaceae. Australasian Plant Pathology 35: 549–562.
14. Morin L, Aveyard R, Lidbetter JR, Wilson PG (2012) Investigating the host-range of the rust fungus *Puccinia psidii* sensu lato across tribes of the family Myrtaceae present in Australia. PLoS One 7: e35434.
15. MacLachlan JD (1938) A rust of the pimento tree in Jamaica, B.W.I. Phytopathology 28: 157–170.
16. Ferreira FA (1983) Ferrugem do eucalipto [Eucalyptus rust]. Revista Arvore 7: 91–109.
17. Coelho L, Alfenas AC, Ferreira FA (2001) Variabilidade fisiologica de Puccinia psidii - ferrugem do eucalipto. [Physiological variability of *Puccinia psidii* - the rust of eucalyptus]. Summa Phytopathologica 27: 295–300.
18. Booth TH, Old KM, Jovanovic T (2000) A preliminary assessment of high risk areas for *Puccinia psidii* (Eucalyptus Rust) in the Neotropics and Australia. Agriculture Ecosystems and Environment 82: 295–301.
19. Booth TH, Jovanovic T (2012) Assessing vulnerable areas for *Puccinia psidii* (eucalyptus rust) in Australia. Australasian Plant Pathology 41: 425–429.
20. Magarey RD, Fowler GA, Borchert DM, Sutton TB, Colunga-Garcia M, et al. (2007) NAPPFAST: An internet system for the weather-based mapping of plant pathogens. Plant Disease 91: 336–345.
21. Elith J, Simpson J, Hirsch M, Burgman MA (2013) Taxonomic uncertainty and decision making for biosecurity: spatial models for myrtle/guava rust. Australasian Plant Pathology 42: 43–51.
22. Kriticos DJ, Randall RP (2001) A comparison of systems to analyse potential weed distributions. In: Groves RH, Panetta FD, Virtue JG, editors. Weed Risk Assessment. Melbourne, Australia: CSIRO Publishing. pp. 61–79.
23. Webber BL, Yates CJ, Le Maitre DC, Scott JK, Kriticos DJ, et al. (2011) Modelling horses for novel climate courses: insights from projecting potential distributions of native and alien Australian acacias with correlative and mechanistic models. Diversity and Distributions 17: 978–1000.
24. Sutherst RW, Maywald GF (1985) A computerised system for matching climates in ecology. Agriculture, Ecosystems and Environment 13: 281–299.
25. Sutherst RW, Maywald GF, Kriticos DJ (2007) CLIMEX Version 3: User's Guide: Hearne Scientific Software Pty Ltd.
26. Brasier CM, Scott JK (1994) European oak declines and global warming: a theoretical assessment with special reference to the activity of *Phytophthora cinnamomi*. EPPO Bulletin 24: 221–232.
27. Lanoiselet V, Cother EJ, Ash GJ (2002) Climex and Dymex simulations of the potential occurrence of rice blast disease in south-eastern Australia. Australasian Plant Pathology 31: 1–7.
28. Hoddle MS (2004) The potential adventive geographic range of glassy-winged sharpshooter, *Homalodisca coagulata* and the grape pathogen *Xylella fastidiosa*: Implications for California and other grape growing regions of the World. Crop Protection 23: 691–699.
29. Yonow T, Kriticos DJ, Medd RW (2004) The potential geographic range of *Pyrenophora semeniperda*. Phytopathology 94: 805–812.
30. Watt MS, Kriticos DJ, Alcaraz S, Brown AV, Leriche A (2009) The hosts and potential geographic range of Dothistroma needle blight. Forest Ecology and Management 257: 1505–1519.
31. Pardey PG, Beddow JM, Hurley TM, Kriticos DJ, Park RF, et al. (2013) Right-sizing stem rust research. Science 340: 147–148.
32. Hutchinson MF, Gessler PE (1994) Splines - more than just a smooth interpolator. Geoderma 62: 45–67.
33. Ruiz RAR, Alfenas AC, Ferreira FA, Vale FXR (1989) Influência da tempuratura, do tempo de molhamento foliar fotoperiodo e da intensidade de luz sobre a infecção de *Puccinia psidii* em eucalipto [Influence of temperature, leaf wetness period, photoperiod and light intensity on the infection of *Puccinia psidii* in Eucalyptus]. Fitopatologia Brasiliensis 14: 55–61.
34. Ferreira FA (1981) Ferrugem do Eucalipto - ocorrências, temperatura para germinação de uredosporos, produção de teliosporos, hospedeiro alternativo e resistência [Eucalyptus rust - occurrence, temperature for uredospore germination, production of teliospores, alternative host and resistance]. Fitopatologia Brasileira 6: 603–604.
35. Tessmann DJ, Dianese JC (2002) Hentriacontane: a leaf hydrocarbon from *Syzygium jambos* with stimulatory effects on the germination of urediniospores of *Puccinia psidii*. Fitopatologia Brasileira 27: 538–542.
36. Bonde MR, Berner DK, Nester SE, Frederick RD (2007) Effects of temperature on urediniospore germination, germ tube growth, and initiation of infection in soybean by Phakopsora isolates. Phytopathology 97: 997–1003.
37. Barbour MG, Racine CH (1967) Construction and performance of a temperature-gradient bar and chamber. Ecology 48: 861–863.
38. R Development Core Team (2011) R: A language and environment for statistical computing. Vienna, Austria: R Foundation for Statistical Computing.
39. Wood S (2006) Generalized Additive Models: An Introduction with R. Boca Raton, FL: Chapman & Hall.
40. Hastie TJ, Tibshirani RJ (1990) Generalized Additive Models. Boca Raton, FL.: Chapman & Hall.
41. Anderson RC (2012) A baseline analysis of the distribution, host range, and severity of the rust *Puccinia psidii* in the Hawai'ian Islands 2005–2010: Hawai'i Cooperative Studies Unit. University of Hawai'i at Hilo. 39 p.
42. CABI (2005) Distribution Maps of Plant Diseases, *Puccinia psidii* Available: http://www.cabi.org/dmpd/?loadmodule = review&page = 4050&reviewid = 87711&site = 165.Accessed 2012 November 12.
43. Kriticos DJ, Webber BL, Leriche A, Ota N, Bathols J, et al. (2012) CliMond: global high resolution historical and future scenario climate surfaces for bioclimatic modelling. Methods in Ecology and Evolution 3: 53–64.
44. Gommes R, Grieser J, Bernardi M (2004) FAO agroclimatic databases and mapping tools. European Society for Agronomy Newsletter 26.
45. Farr TG, Rosen PA, Caro E, Crippen R, Duren R, et al. (2007) The Shuttle Radar Topography Mission. Review of Geophysics 45: RG2004.
46. Hutchinson M, Xu T, Houlder D, Nix H, McMahon J (2009) ANUCLIM 6.0 User's Guide. Canberra: Australian National University, Fenner School of Environment and Society.
47. Hutchinson MF (1995) Interpolating mean rainfall using thin plate smoothing splines. International Journal of Geographical Information Systems 9: 385–403.
48. Giambelluca TW, Nullet MA, Schroeder TA (1986) Rainfall Atlas of Hawai'i. Honolulu, Hawai'i: Water Resources Research Centre, University of Hawai'i at Manoa. R76. 267 p.
49. Kriticos DJ, Leriche A (2008) The current and future potential distribution of guava rust, *Puccinia psidii* in New Zealand. Rotorua, New Zealand: Scion. MAF Biosecurity New Zealand Technical Paper No: 2009/28.
50. Kriticos DJ, Sutherst RW, Brown JR, Adkins SA, Maywald GF (2003) Climate change and the potential distribution of an invasive alien plant: *Acacia nilotica* ssp. *indica* in Australia. Journal of Applied Ecology 40: 111–124.
51. de Villiers M, Hattingh V, Kriticos DJ (2012) Combining field phenological observations with distribution data to model the potential range distribution of the fruit fly *Ceratitis rosa* Karsch (Diptera: Tephritidae). Bulletin of Entomological Research 103: 60–73.
52. The Council of Heads of Australasian Herbaria (2012) Australia's Virtual Herbarium. Available: http://avh.chah.org.au. Accessed 2011 December.
53. Department of Sustainability Environment Water Population and Communities (2009) EPBC Act list of threatened flora. Available: http://www.environment.gov.au/cgi-bin/sprat/public/publicthreatenedlist.pl?wanted = flora. Accessed 2012 June 8.
54. Gavran M, Parsons M (2010) Australia's Plantations 2010 Inventory Update. Canberra: National Forest Inventory, Bureau of Rural Sciences.
55. ABARES (2010) Land Use of Australia, Version 4, 2005/2006 (September 2010 release). Available: http://adl.brs.gov.au/anrdl/metadata_files/pa_luav4g9ab l07811a00.xml. Accessed 2012 November 5.

56. Kriticos DJ, Potter KJ, Alexander N, Gibb AR, Suckling DM (2007) Using a pheromone lure survey to establish the native and potential distribution of an invasive lepidopteran, *Uraba lugens*. Journal of Applied Ecology 44: 853–863.

57. Ladiges PY, Udovicic F, Nelson G (2003) Australian biogeographical connections and the phylogeny of large genera in the plant family Myrtaceae. Journal of Biogeography 30: 989–998.

58. Myerscough PJ (1998) Ecology of Myrtaceae with special reference to the Sydney region. Cunninghamia 5: 787–807.

59. GIT Forestry Consulting (2008) Global Eucalyptus Map. Available: www.git-forestry.com/download_git_eucalyptus_map.htm. Accessed 2013 February 7.

60. McCarthy HR, Pataki DE, Jenerette GD (2011) Plant water-use efficiency as a metric of urban ecosystem services. Ecological Applications 21: 3115–3127.

61. Chiani RG (1959) Aplicaciones de la madera de eucalipto. El eucalipto 'cara familiar' del paisaje pampeano, resulta un importante productor de maderas para diversos usos [Uses of Eucalypt timber. Eucalypts, a familiar sight in the Pampas territory, constitute an important source of timber for different purposes]. Boletin Argentino Forestal 17: 10–11.

62. Xavier AA, da Silva RL (2010) Evolução da silvicultura clonal de Eucalyptus no Brasil. Agronomía Costarricense 34: 93–98.

63. Kriticos DJ, Leriche A (2009) The effects of spatial data precision on fitting and projecting species niche models. Ecography 33: 115–127.

64. Woodall GS, Dodd IC, Stewart GR (1998) Contrasting leaf development within the genus Syzygium. Journal of Experimental Botany 49: 79–87.

65. Keatley MR, Fletcher TD, Hudson IL, Ades PK (2002) Phenological studies in Australia: Potential application in historical and future climate analysis. International Journal of Climatology 22: 1769–1780.

66. Underwood AJ (1994) On beyond BACI - sampling designs that might reliably detect environmental disturbances. Ecological Applications 4: 3–15.

67. NSW Department of Primary Industries (2012) Field hosts of myrtle rust recorded in NSW. Available: http://www.dpi.nsw.gov.au/biosecurity/plant/myrtle-rust/hosts. Accessed 2012 June 8.

68. Queensland Department of Agriculture Fisheries and Forestry (2012) List of host plants affected by myrtle rust. Available: http://www.daff.qld.gov.au/4790_19789.htm. Accessed 2012 June 8.

69. Carnegie AJ, Cooper K (2012) Emergency response to the incursion of an exotic myrtaceous rust in Australia. Australasian Plant Pathology 40: 346–359.

70. Venette RC, Kriticos DJ, Magarey R, Koch F, Baker RHA, et al. (2010) Pest risk maps for invasive alien species: a roadmap for improvement. Bioscience 80: 349–362.

71. Ricciardi A, Cohen J (2007) The invasiveness of an introduced species does not predict its impact. Biological Invasions 9: 309–315.

5

Evaluating Habitat Suitability for the Establishment of *Monochamus spp.* through Climate-Based Niche Modeling

Sergio A. Estay[1]*, Fabio A. Labra[2], Roger D. Sepulveda[1], Leonardo D. Bacigalupe[1]

1 Instituto de Ciencias Ambientales y Evolutivas, Facultad de Ciencias, Universidad Austral de Chile, Valdivia, Chile, 2 Centro de Investigación e Innovación para el Cambio Climático, Facultad de Ciencias, Universidad Santo Tomas, Santiago, Chile

Abstract

Pine sawyer beetle species of the genus *Monochamus* are vectors of the nematode pest *Bursaphelenchus xylophilus*. The introduction of these species into new habitats is a constant threat for those regions where the forestry industry depends on conifers, and especially on species of *Pinus*. To obtain information about the potential risk of establishment of these insects in Chile, we performed climate-based niche modeling using data for five North American and four Eurasian *Monochamus* species using a Maxent approach. The most important variables that account for current distribution of these species are total annual precipitation and annual and seasonal average temperatures, with some differences between North American and Eurasian species. Projections of potential geographic distribution in Chile show that all species could occupy at least 37% of the area between 30° and 53°S, where industrial plantations of *P. radiata* are concentrated. Our results indicated that Chile seems more suitable for Eurasian than for North American species.

Editor: Daniel Doucet, Natural Resources Canada, Canada

Funding: Funded by CONICYT grant N° 79100021. The funders had no role in study design, data collection and analysis, decision to publish, or preparation of the manuscript.

Competing Interests: The authors have declared that no competing interests exist.

* Email: sergio.estay@uach.cl

Introduction

Currently, there are hundreds to thousands of exotic species established outside their native ecosystems [1]. Probably these numbers will increase in the future as a result of the steady growth in international trade which produces human-aided long-distance dispersal of organisms [2].

Forests in Asia, Europe and North America have experienced the introduction of insect pests which have caused ecological, social and economic damage to natural forest, industrial plantations and urban trees. Given that eradication of established invasive species often implies large economic costs concurrent with a low probability of success, the logical recommendation for governments is to place the highest priority on preventing introduction of such species [3–4]. In this regard, pest risk assessment (PRA) is a key procedure that encompasses several methodologies that aim to evaluate the likelihood of an exotic species being introduced to a region and causing damage to agriculture [5]. Thus, PRA uses biological and economic information to determine whether some species should be regulated and the strength of the sanitary measures to be taken against it [6].

One of the necessary steps of a PRA is the assessment of the suitability of the new habitat for the establishment of the exotic organism [6–8]. Over the last decade, ecologists have developed several tools with solid bases in mathematics, statistics and information theory that facilitate these analyzes [4,9–12]. Among these, climate-based ecological niche modeling is commonly used

in risk assessment [13–14]. Climate-based ecological niche models may be considered as a subset of the more general species distribution models, which are numerical tools that combine observations of species (either presences or presences and absences) in a set of locations with environmental variables to obtain ecological and evolutionary insights and to predict distributions across landscapes [11,15]. In recent years, niche models have been used to predict potential geographic distribution of several forest pests such as the Asian longhorn beetle [16], pine shoot beetle [17], European woodwasp [18], redbay ambrosia beetle [19] and emerald ash borer [14].

One of the most serious threats to pine forests in the world is pine wilt disease, caused by the pinewood nematode, *Bursaphelenchus xylophilus*. This disease is native to North America where it is a secondary pathogen of native pines, but is the cause of pine wilt disease in non-native pines [20]. In countries where the pinewood nematode has been introduced, such as Japan, pine wilt is an important non-native disease [21–22], with estimated losses of 46 million m^3 of wood in the last 50 years [23]. Although this nematode may be carried by several xylophagus insects, successful transmission to conifers has only been demonstrated for the pine sawyer beetles of the genus *Monochamus* [24–25].

There are no native species of *Monochamus* in South America and they are included in the list of insects recommended for regulation as quarantine pests of the COSAVE (Regional plant protection organization of the Southern Cone of South America). The potential introduction of these species to a continent where *Pinus* plantations are a key component in the forest industry [26] could have serious economic consequences. In the case of Chile, commercial plantations of *P. radiata* are the basis for the forestry industry. Currently, Chile has 1.5 million ha of *P. radiata* plantations established across several site types and climate conditions that vary from 30° to 43°S latitude [27]. In addition, urban trees of this species as well as of other *Monochamus* hosts (*Picea*, *Abies*, *Cedrus* and *Pseudotsuga*) may be found in most Chilean cities all over the country. In this study we used ecological niche modeling methods to obtain insights on the role of climate in shaping the current distribution of nine species of *Monochamus* vectors of *B. xylophilus* and the relative importance of each variable analyzed in determining native geographic ranges for each species. We then use these models to generate a map of the potential distribution of each of these species in Chile, which may be used as a proxy of the suitability of the new habitat in a PRA.

Materials and Methods

Species occurrence

Records of confirmed presences (i.e. confirmed establishment) of *Monochamus* species were obtained from multiple primary sources. The primary sources used were the open databases Invasive Species Compendium [28] and the EPPO Plant Quarantine Data Retrieval System (PQR, [29]). Both datasets are considered within the PRATIQUE initiative of EPPO [30]. To complement this information, we also used information from Dillon and Dillon [31] and Cherepanov [32] for North American and Eurasian *Monochamus*, respectively. When no geo-referenced localities (just locality names) were provided, geographic coordinates were obtained from official gazetteers (GeoNet, [33]; TGN, [34]). We restricted our study to species with at least 20 confirmed records. These procedures allowed us to obtain datasets for five North American species, and four Eurasian species. The species considered and the respective number of data points were as follows. In North America: *M. carolinensis* (34), *M. marmorator* (25), *M. notatus* (36), *M. scutellatus* (47), *M. titillator* (39). For Eurasia: *M. alternatus* (32), *M. galloprovincialis* (49), *M. saltuarius* (24) and *M. sutor* (47) (Table S1–S2). All these species are either known to be vectors of *B. xylophilus* or are considered potential vectors [24–25]. All confirmed records were used, making no difference between native and exotic distributions [35–40].

Climatic variables

Current global climatic conditions grids with a spatial resolution of 2.5 arc-minutes were obtained from the WorldClim database [41]. These grids contain variables compiled from monthly data collected from 1950 to 2000. Based on the biological knowledge about these species [24–25,42–43], we selected six ecologically relevant bioclimatic variables: annual mean temperature, mean temperature of the coldest quarter, mean temperature of the warmest quarter, annual accumulated degree days (base 5°C), mean relative humidity and total annual precipitation. The "coldest" and "warmest" quarter are defined according to the Worlclim database: the mean temperature of the three-months period with the lowest and highest average temperature, respectively. We also incorporate altitude as a descriptor of topography to obtain seven explanatory variables in our modeling procedure (Table 1).

Modeling methods

Because of our datasets were based on presence-only localities, we used a maximum entropy modeling approach to estimate climate-based niche models for all 9 species. Analysis was performed with the Maxent 3.3.3 k software [44–48]. Comparison

Table 1. Ranges of the environmental variables observed into the 95% geographic kernel defined for each species.

Region	Species	Ann T (°C)	T°Col (°C)	T° War (°C)	ADD	% RH	PP (mm)	Altitude (masl)
North America	M. carolinensis	−5.7–25.5	−22.8–20.8	7.6–30.5	0–7190	0–81.5	192–1970	−6–3625
	M. marmorator	−5.7–19.2	−24.2–11.0	9.7–27.0	0–4891	0–81.5	393–1970	−6–1294
	M. notatus	−7.1–20.9	−26.0–14.0	4.0–28.8	0–5605	0–86.4	192–3098	−6–3625
	M. scutellatus	−16.1–25.1	−32.5–21.7	−12.0–33.4	0–7051	0–86.4	51–3573	−88–3748
	M. titillator	−2.5–26.2	−19.3–23.8	7.5–30.4	0–7562	0–80.7	205–1970	−6–3625
Eurasia	M. alternatus	−11.5–28.1	−25.5–26.5	−4.3–30.5	0–8448	0–83.3	0–5576	−2–6512
	M. galloprovincialis	−23.2–28.0	−49.4–18.0	−2.5–37.6	0–7707	0–90.3	0–2718	−416–3355
	M. saltuarius	−23.2–21.9	−49.4–15.7	−2.5–29.1	0–5240	0–90.3	0–2953	−51–5909
	M. sutor	−23.2–19.6	−49.4–12.5	−2.5–31.3	0–5031	0–90.3	0–2838	−41–6098
Chile	Pinus Plantations	−5.0–17.4	−9.4–12.8	−0.6–22.4	0–3346	0–86.8	0–3073	0–4339

Ann T° = mean annual temperature, T°Col = mean temperature of the coldest season, T° War = mean temperature of the warmest season, ADD = annual accumulated degree-days, % RH = annual mean relative humidity and PP = total annual precipitation. Pinus plantations refers to the area of Chile covered with *Pinus radiata* plantations (see methods for details).

Table 2. Jacknife statistics of model performance and relative importance of each variable.

Region	Species	AUC	Gain	Environmental variables						
				Ann T°	T°Col	T° War	ADD	% RH	PP	Altitude
North America	M. carolinensis	0.73	0.283	0.141–0.036	0.121–0.043	0.154–0.000	0.151–0.044	0.141–0.070	0.102*–0.123†	0.141–0.070
	M. marmorator	0.70	0.362	0.275–0.103	0.281–0.146	0.232–0.070	0.268–0.098	0.276–0.016	0.227*–0.201†	0.276–0.016
	M. notatus	0.65	0.326	0.134–0.063	0.129–0.056	0.111–0.058	0.138–0.050	0.134–0.000	0.072*–0.168†	0.133–0.000
	M. scutellatus	0.74	0.740	0.433–0.140	0.456–0.147	0.414–0.117	0.420–0.123	0.433–0.134	0.307*–0.404†	0.434–0.134
	M. titillator	0.64	0.249	0.071–0.045	0.096–0.038	0.052*–0.081	0.058–0.037	0.060–0.080	0.106–0.148†	0.059–0.080
	M. alternatus	0.72	0.414	0.207*–0.305†	0.305–0.259	0.268–0.118	0.290–0.119	0.271––0.046	0.236–0.246	0.271––0.046
Eurasia	M. galloprovincialis	0.66	0.406	0.059–0.117	0.140–0.002	0.087–0.041	0.129––0.057	0.055––0.026	−0.026*–0.195†	0.055––0.026
	M. saltuarius	0.77	0.671	0.329–0.148	0.460–0.064	0.447–0.039	0.442––0.012	0.396––0.034	−0.040*–0.461†	0.391–0.034
	M. sutor	0.72	0.368	0.033*–0.088	0.138–0.043	0.137––0.023	0.083––0.025	0.125––0.073	0.035–0.215†	0.126––0.073

For each species, the table shows the area under the curve (AUC) and regularized training gain (Gain). For each variable first value correspond to the gain of a model fitted using all variables except the focal one. The more important variable according to this criterion is marked with *. The second value corresponds to the gain of a model fitted using just the focal variable. The more important variable according to this criterion is marked with †, (see methods for details). Abbreviations as in table 1.

of the prediction accuracy across several niche modelling methods showed Maxent to be among the best modeling approaches for presence-only data [48]. Briefly, Maxent is a machine-learning algorithm that works by minimizing the relative entropy of the probability densities calculated from the presence records versus those probability densities were calculated from random sampling over the study region [44,46–47]. It is important to note that Maxent is a density estimation method, and not a regression method, and as such it has properties that make it robust to limited amounts of training data (small samples) [11,45]. Also, its results are less affected by variable autocorrelation and it allows flexible modeling of different types of functions between environmental variables and the probability of species occurrence [44].

We examined the output of the fitted model in logistic format, to indicate the suitability of the habitat of each species in the landscape. The study area to fit the model was restricted to the 95% spatial kernel for North America and Eurasia according the current registered presence of each species. Models were then evaluated using area under the curve (AUC) of the Receiver operating characteristic (ROC) curve and regularized training gain. The ROC curve corresponds to the plot between 1-specificity (proportion of false positives) versus sensitivity (proportion of true positives, [45]). The AUC index measures the ability (probability) of the maxent model to discriminate between presence sites versus background sites [44,49–50]. To complement the model evaluation by AUC values, we also used regularized training gain (hereafter gain), which corresponds to the logarithm of the average ratio between the likelihood assigned to an observed presence site and the likelihood assigned to a background site. The observed value of gain was also used to estimate the relative importance of each variable by using a jackknife method. Briefly, the decrease in gain by fitting a model using all variables except the focal one was compared with the gain of the previously full model (including all variables). Next, we fit a model using only the focal variable and compared the gain in relation to the full model. This procedure yielded an estimate of the relative importance of each variable in the model. Modeling results to a 20-fold cross-validation scheme considering the usual highly correlation between climatic variables [50–51]. This cross-validation scheme divides the dataset into 20 subsets. In each step the model is fitted using 19 subsets and using the last one (independent) to test (validate) the fitting. This procedure is repeated 20 times, and the AUC and jackknife values reported correspond to the average value of the 20 testing procedures.

Fitted models of each species were later projected over the continental Chilean territory using the same environmental variables described previously. Given the logistic scale used, these maps may be interpreted as a measure of the suitability of the habitat (0 = unsuitable, 1 = highly suitable) and are a proxy of how favorable the habitat is for the establishment of these pests. To estimate the extent of Chilean territory these species could occupy, original logistic maps were converted to binary maps (0 = absence, 1 = presence) applying a threshold that maximizes test sensitivity and specificity [52]. These binary maps were projected on Chilean territory and on the proportion of territory covered by *Pinus* plantations. The percentage of all territory and *Pinus* plantations potentially covered for each species was calculated. Area of *Pinus* plantations was obtained using the VII national agricultural, livestock and forestry census [53]. This map corresponds to agricultural districts that contain at least one commercial *Pinus* plantation.

Manipulation of environmental layers was performed in R environment [54], Quantum GIS 1.8.0 [55] and GRASS 6.4.2 [56].

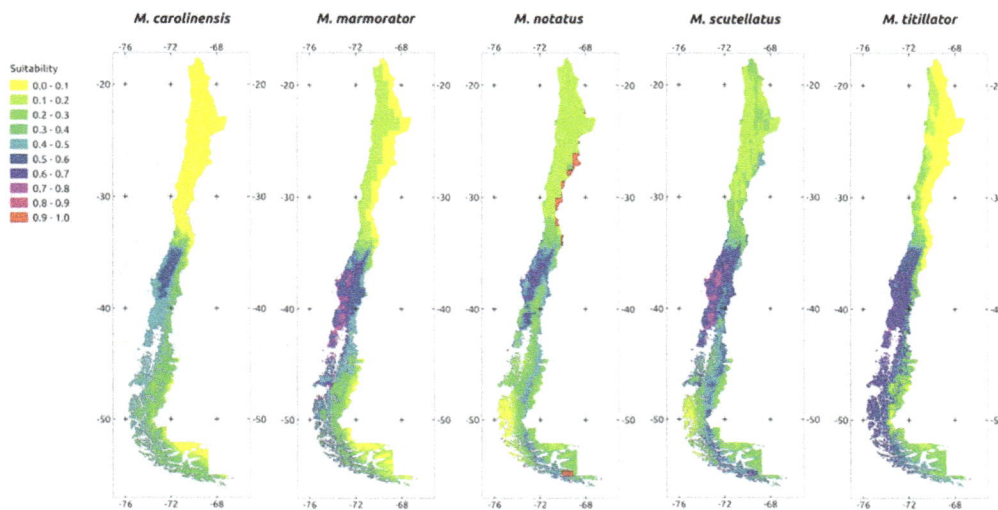

Figure 1. Projections of the Maxent model fitted for each North American species into Chile. Colors represent the probability of each pixel being a suitable habitat for the corresponding species.

Results

All fitted models showed high values of AUC, which makes us confident of a high discriminative ability. The lowest AUC (0.64) was obtained for the North American *M. titillator*, while the highest (0.77) was obtained for the Eurasian *M. saltuarius* (Table 2).

In general, models fitted using all variables except the focal one, showed that the exclusion of total annual precipitation and mean temperature of the warmest season caused the highest reduction in gain (Table 2). The analysis of models including just one variable showed that models fitted using total annual precipitation, mean temperature of the coldest season and mean annual temperature reached the highest gain (Table 2).

When we separate North American and Eurasian species, some differences appear. Models excluding the focal variable showed that for North American species (Fig. S1–S5) the exclusion of total annual precipitation caused the highest reduction in gain, but for Eurasian species (Fig. S6–S9) the highest reduction is caused by total annual precipitation and annual mean temperature (Table 2). On the other hand, using one variable, North American and Eurasian models showed that the variable with the highest gain was total annual precipitation in almost all species (Table 2).

Projections of the models into the Chilean territory showed that climate in this region is moderately to highly suitable for most species (Fig. 1, 2). Specifically, the central and southern regions

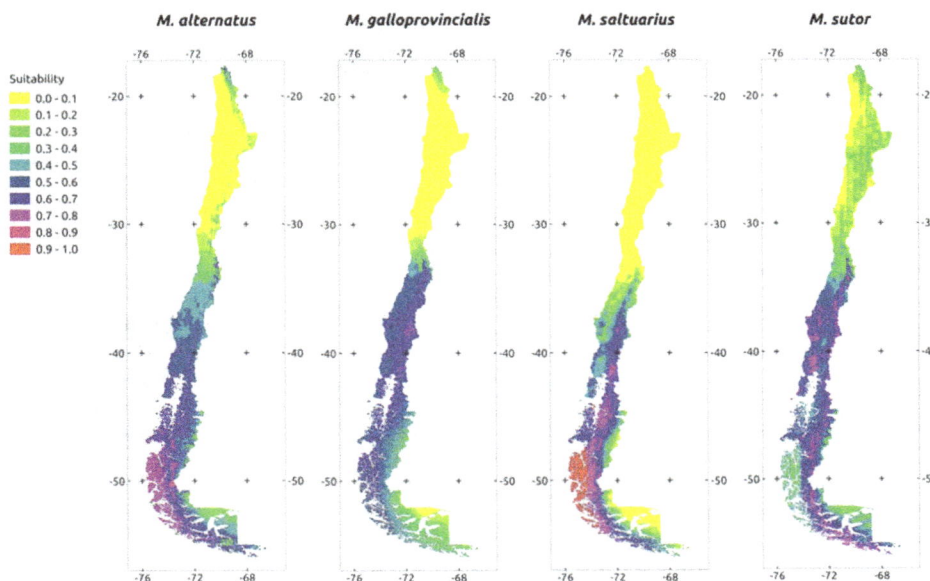

Figure 2. Projections of the Maxent model fitted for each Eurasian species into Chile. Colors represent the probability of each pixel being a suitable habitat for the corresponding species.

Table 3. Percentage of potential area covered for each species.

Region	Species	Threshold	% All	% *Pinus*
	M. carolinensis	0.442	16.6	61.0
	M.marmorator	0.514	21.7	64.7
North America	M.notatus	0.510	13.9	37.1
	M. scutellatus	0.522	22.4	72.4
	M. titillator	0.494	32.2	71.9
	M. alternatus	0.472	43.9	72.5
Eurasia	M. galloprovincialis	0.430	45.9	95.5
	M. saltuarius	0.446	36.0	54.1
	M. sutor	0.473	46.0	92.6

Threshold is the logistic threshold applied to obtain a binary map. This threshold correspond to the value that maximize test sensitivity plus specificity. % All is the percentage of Chilean territory that could be potentially covered by the species. % Pinus is the percentage of territory covered by *Pinus* plantations that could be potentially covered by each *Monochcmus* species.

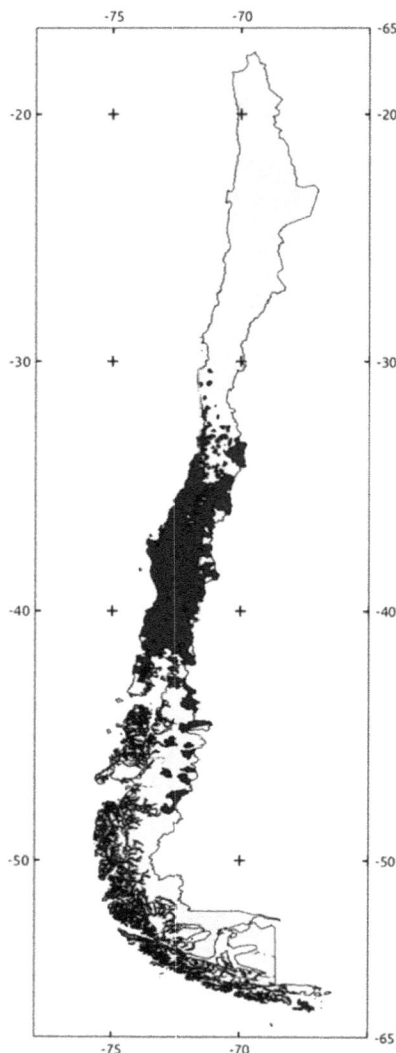

Figure 3. Agricultural districts of Chile that contains at least one commercial plantation of *P. radiata* **(dark areas, www. odepa.cl).**

(35°–55°S) of Chile seem more suitable for the establishment of *Monochamus* species than the northern region (18°–35°S). The proportion of territory corresponding to suitable and unsuitable habitat showed a clear distinction between species. For North American species the main proportion of suitable habitat is between 35° to 44°S, but for Eurasian species it occurs from 35° to 56°S (Table 3).

Discussion

In this study, we performed climate-based niche modeling for five North American and four Eurasian *Monochamus* species. Interestingly, most models showed an acceptable discriminatory power (>0.7, [49]). However, average values of AUC for North American and Eurasian species were very similar (Table 2), suggesting that model quality was equivalent between regions.

The relative importance of each variable for the fitted models showed that total annual precipitation is commonly the most important variable for species of North America and Eurasia. The decrease in gain by excluding this variable represents the amount of information provided by the excluded variable that is not present in other variables and is lost in the model by excluding it. The same situation appears in the results of models fitted using just one variable. For both regions total annual precipitation is again the most important variable. Hence, this variable could be considered as providing the highest amount of information, independently if this information is contained or not in other variables.

To the best of our knowledge, the greater importance of precipitation over temperature in conditioning the distribution of *Monochamus* species is an unexpected result. In a recent study, Chen et al. [57] pointed out that precipitation is important for the population dynamics of *M. alternatus*, but only as a secondary variable and less important than temperature. One potential explanation could be related to the link between precipitation and the distribution of host trees [58] or the influence of water content of the soil on the incidence of the symbiont nematode *B. xylophilus* [59].

Mean temperature of the warmest season, mean temperature of the coldest season and mean annual temperature are all indicators of the thermal restrictions that an organism experiences in the field. Thermal restrictions for completing development and lower thermal developmental threshold have been described for North

American (*M. carolinensis* [60]) and Eurasian *Monochamus* (*M. alternatus* [42]; *M. saltuarius* [61] and *M. galloprovincialis* [62]). Ma et al. (2006) [43] even propose the $-10°C$ January mean temperature isotherm as the northern limit of *M. alternatus* potential distribution in China. Therefore, the inclusion of these variables in our models is not surprising. However, considering the importance of thermal requirements of ectotherms we expected a higher importance of accumulated degree days, but this variable had little influence in most species.

When models were projected into Chilean territory two important results arise. First, there are important differences in the potential suitable habitat between species. On average climate in Chile seems to be more suitable for Eurasian species than for North American species, especially in the area covered for *Pinus* plantations. The reasons behind these differences may be related to the range in climatic conditions experiences by each species in its native range. In general, North American species show a more restricted distribution than the Eurasian species analyzed [28–29,31–32].

Pinus plantations in Chile are primarily *P. radiata*, a species with controversial evidence about susceptibility to the pine wilt disease. In its native distribution, a survey performed in 1988 found no evidence of infection [63]. However, Furuno et al [64] reported approximately 80% mortality of *P. radiata* due to pine wilt disease in Japan in a 30-year experiment. Due to the contradictory evidence, EPPO classify *P. radiata* as a moderately susceptible species to *B. xylophilus* [24]. Our results show that areas with the highest probability of being suitable for *Monochamus* species are located in Central and Southern Chile mainly between 30° and 53°S. However, commercial plantations in Chile are restricted to 30°–43°S (Fig. 3). The area between 45°S and 53°S is composed mainly of conservation areas with native forest (national parks), and therefore, this region could be considered at low risk of *Monochamus* establishment. However, the region between 30° and 43°S could be considered to be at moderate to high risk of establishment of *Monochamus* (Fig. 3), if enough individuals arrives. Also, this region is the more populated part of the country and contains a high number of terrestrial, aerial and maritime ports where several interceptions of *Monochamus* have occurred in the past [65]. The combination of several potential points of introductions due to ports (high propagule pressure) and highly suitable habitat (high probability of introduction) suggest that efforts for early detection of these species should be concentrated in this region. However, it is necessary to note that low suitability habitat or low probability of establishment does not mean zero risk, and reasonable monitoring levels as well as preventive activities should be carried out even outside the 30°S–43° region.

Climate-based niche modeling has proved to be useful in forecasting the potential distribution of pest species, especially in the initial phase of a risk assessment. However, the addition of complementary distributional information (e.g. real absences) and variables other than climatic ones will reduce uncertainty in long-term risk assessment. Difficulties in the interpretation of correlative models (such as MaxEnt) have been previously highlighted [66–67]. Correlative models seems to be sensitive to the training data set and the addition of new information (new presences from new habitats) could caused increases in the sensitivity of the model (detection of true positives) jointly with increases of estimated prevalence [67]. In our case, the absence of independent data sets impedes the quantification of the estimated prevalence and sensitivity. This situation is common in the risk analysis of potential forest pests where information is poor and in some cases even the native distribution of the organism is not clearly defined. To overcome this problem the use of mechanistic models, that link physiological characteristics with habitat occupation provide an alternative approach [67–68].

The incorporation of these approaches in plant health management will help planning and design of activities aimed at preventing establishment of pest species and improving phytosanitary status of forestry and agriculture in developing countries.

Supporting Information

Figure S1 Projections of the fitted models into the 95% geographic kernel defined for *M. carolinensis*. Colors represent habitat suitability (0 = unsuitable, 1 = highly suitable). Red points correspond to the presence points used in the study.

Figure S2 Projections of the fitted models into the 95% geographic kernel defined for *M. marmorator*. Colors represent habitat suitability (0 = unsuitable, 1 = highly suitable). Red points correspond to the presence points used in the study.

Figure S3 Projections of the fitted models into the 95% geographic kernel defined for *M. notatus*. Colors represent habitat suitability (0 = unsuitable, 1 = highly suitable). Red points correspond to the presence points used in the study.

Figure S4 Projections of the fitted models into the 95% geographic kernel defined for *M. scutellatus*. Colors represent habitat suitability (0 = unsuitable, 1 = highly suitable). Red points correspond to the presence points used in the study.

Figure S5 Projections of the fitted models into the 95% geographic kernel defined for *M. titillator*. Colors represent habitat suitability (0 = unsuitable, 1 = highly suitable). Red points correspond to the presence points used in the study.

Figure S6 Projections of the fitted models into the 95% geographic kernel defined for *M. alternatus*. Colors represent habitat suitability (0 = unsuitable, 1 = highly suitable). Red points correspond to the presence points used in the study.

Figure S7 Projections of the fitted models into the 95% geographic kernel defined for *M. galloprovincialis*. Colors represent habitat suitability (0 = unsuitable, 1 = highly suitable). Red points correspond to the presence points used in the study.

Figure S8 Projections of the fitted models into the 95% geographic kernel defined for *M. saltuarius*. Colors represent habitat suitability (0 = unsuitable, 1 = highly suitable). Red points correspond to the presence points used in the study.

Figure S9 Projections of the fitted models into the 95% geographic kernel defined for *M. sutor*. Colors represent habitat suitability (0 = unsuitable, 1 = highly suitable). Red points correspond to the presence points used in the study.

Table S1 Geographic coordinates of the presence points used for each North American species.

Table S2 Geographic coordinates of the presence points used for each Eurasian species.

Author Contributions

Conceived and designed the experiments: SAE FAL. Performed the experiments: SAE FAL. Analyzed the data: SAE FAL RDS LDB. Wrote the paper: SAE FAL RDS LDB.

References

1. Mack RN (2003) Global plant dispersal, naturalization, and invasion: pathways, modes and circumstances. In: Invasive species: vectors and management strategies. Ed. by Ruiz G, Carlton JT. Island Press, Washington. 3–30.
2. Pimentel D (2002) Biological invasions: economic and environmental costs of alien plant, animal, and microbe species. CRC, Florida.
3. Campbell FT (2001) The science of risk assessment for phytosanitary regulation and the impact of changing trade regulations. BioScience 51: 148–153.
4. Jiménez-Valverde A, Peterson AT, Soberón J, Overton JM, Aragón P, et al. (2011) Use of niche models in invasive species risk assessments. Biol. Invasions 13: 2785–2797.
5. Gray GM, Allen JC, Burmaster DE, Gage SH, Hammitt JK, et al. (1998) Principles for Conduct of Pest Risk Analyses: Report of an Expert Workshop. Risk Anal. 18: 773–780.
6. IPPC (2012) Pest risk analysis for quarantine pests including analysis of environmental risks. International Standards for Phytosanitary Measures. N°11. FAO. Rome.
7. Stohlgren TJ, Jarnevich CS (2009) Risk assessment of invasive species. In: Invasive species management: a handbook of principles and techniques. Ed. by Clout MN, Williams PA. Oxford University Press, New York. 19–35.
8. Leung B, Roura-Pascual N, Bacher S, Heikkilä J, Brotons L, et al. (2012) TEASIng apart alien species risk assessments: a framework for best practices. Ecol. Lett. 15: 1475–1493.
9. Peterson AT, Vieglais DA (2001) Predicting species invasions using ecological niche modeling: new approaches from bioinformatics attack a pressing problem. Bioscience 51: 363–371.
10. Peterson AT (2003) Predictability of the geography of species' invasions via ecological niche modeling. Q. Rev. Biol. 78: 419–433.
11. Franklin J (2009) Mapping Species Distributions: Spatial Inference and Prediction. Cambridge University Press, Cambridge.
12. Kulhanek SA, Leung B, Ricciardi A (2011) Using ecological niche models to predict the abundance and impact of invasive species: application to the common carp. Ecol. Appl. 21: 203–213.
13. Herborg LM, Drake JM, Rothlisberger JD, Bossenbroek JM (2009) Identifying suitable habitat for invasive species using ecological niche models and the policy implications of range forecasts. In: Bioeconomics of Invasive Species: Integrating Ecology, Economics, Policy and Management. Ed. yy Keller RP, Lodge DM, Lewis MA, Shogren JF. Oxford University Press. 63–82.
14. Sobek-Swant S, Kluza DA, Cuddington K, Lyons DB (2012) Potential distribution of emerald ash borer: What can we learn from ecological niche models using Maxent and GARP?. Forest Ecol. Manag. 281: 23–31.
15. Elith J, Leathwick JR (2009) Species distribution models: ecological explanation and prediction across space and time. Ann. Rev. Ecol. Evol. S. 40: 677–697.
16. MacLeod A, Evans HF, Baker RHA (2002) An analysis of pest risk from an Asian longhorn beetle (*Anoplophora glabripennis*) to hardwood trees in the European community. Crop Prot. 21: 635–645.
17. McKenney D, Hopkin AA, Campbell KL, Mackey BG, Foottit R (2003) Opportunities for improved risk assessments of exotic species in Canada using bioclimatic modeling. Environ. Monit. Assess. 88: 445–461.
18. Carnegie AJ, Matsuki M, Haugen DA, Hurley BP, Ahumada R, et al. (2006) Predicting the potential distribution of *Sirex noctilio* (Hymenoptera: Siricidae), a significant exotic pest of *Pinus* plantations. Ann. For. Sci. 63: 119–128.
19. Koch FH, Smith WD (2008) Spatio-temporal analysis of *Xyleborus glabratus* (Coleoptera: Circulionidae: Scolytinae) invasion in eastern US forests. Environ. Entomol. 37: 442–452.
20. Dwinell LD, Nickle WR (1989) An overview of the pine wood nematode ban in North America. General Technical Report SE-55. USDA Forest Service.
21. Liebhold AM, Macdonald WL, Bergdahl D, Mastro VC (1995) Invasion by exotic forest pests: a threat to forest ecosystems. Forest Science Monographs N°30.
22. Futai K (2008) Pine wilt in Japan: From first incidence to the present. In Zhao BG, Futai K, Sutherland JR, Takeuch Y (eds) Pine wilt disease. Springer. p. 5–12.
23. Zhao BG, Futai K, Sutherland JR, Takeuchi Y (2008) Pine wilt disease. Springer. Tokyo.
24. Evans HF, McNamara DG, Braasch H, Chadoeuf J, Magnusson C (1996) Pest risk analysis (PRA) for the territories of the European Union on *Bursaphelenchus xylophilus* and its vectors in the genus *Monochamus*. EPPO Bulletin 26: 199–249.
25. Akbulut S, Stamps WT (2011) Insect vectors of the pinewood nematode: a review of the biology and ecology of *Monochamus* species. Forest Pathol. 42: 89–99.
26. FAO (2010) Global Forest Resources Assessment 2010: Main Report. Food and Agriculture Organization of the United Nations. Forestry Department. Rome.
27. INFOR (2010) Estadísticas forestales chilenas 2008. Instituto Forestal (INFOR). Available: http://www.infor.cl. (In Spanish). Accessed 18 February 2014.
28. CABI (2013) Invasive Species Compendium. CAB International. Available: http://www.cabi.org/isc. Accessed 18 February 2014.
29. EPPO (2013) PQR - EPPO database on quarantine pests. Paris, France. Available: http://www.eppo.int. Accessed 18 February 2014.
30. Baker RHA, Battisti A, Bremmer J, Kenis M, Mumford J, et al. (2009) PRATIQUE: a research project to enhance pest risk analysis techniques in the European Union. EPPO Bulletin 39: 87–93.
31. Dillon LS, Dillon ES (1941) The tribe Monochamini in the western hemisphere (Coleoptera: Cerambycidae). Reading Public Museum and Art Gallery Scientific Publications N°1.
32. Cherepanov AI (1983) Cerambycidae of Northern Asia. Vol. 3. Part I. Academy of Sciences of the USSR. Siberian Division.
33. Geospatial-Intelligence Agency (2013) GEOnet Names Server. Available: http://geonames.nga.mil. Accessed 18 February 2014.
34. Getty Information Institute (2013) TGN - Thesaurus of Geographic Names. Available: http://www.getty.edu. Accessed 18 February 2014.
35. Broennimann O, Treier UA, Müller-Schärer H, Thuiller W, Peterson AT, et al. (2007) Evidence of climatic niche shift during biological invasion. Ecol. Lett. 10: 701–709.
36. Broennimann O, Guisan A (2008) Predicting current and future biological invasions: both native and invaded ranges matter. Biol. Letters 4: 585–589.
37. Pearman PB, Guisan A, Broennimann O, Randin CF (2008) Niche dynamics in space and time. Trends Ecol. Evol. 23: 149–158.
38. Medley KA (2010) Niche shifts during the global invasion of the Asian tiger mosquito, *Aedes albopictus* Skuse (Culicidae), revealed by reciprocal distribution models. Global Ecol. Biogeogr. 19: 122–133.
39. Strange JP, Koch JB, Gonzalez VH, Nemelka L, Griswold T (2011) Global invasion by *Anthidium manicatum* (Linnaeus) (Hymenoptera: Megachilidae): assessing potential distribution in North America and beyond. Biol. Invasions 13: 2115–2133.
40. Bidinger K, Lötters S, Rödder D, Veith M (2012) Species distribution models for the alien invasive Asian Harlequin ladybird (*Harmonia axyridis*). J. Appl. Entomol. 136: 109–123.
41. Hijmans RJ, Cameron SE, Parra JL, Jones PG, Jarvis A (2005) Very high resolution interpolated climate surfaces for global land areas. Int. J. Climatol. 25: 1965–1978.
42. Park NC, Moon YS, Lee SM, Park JD, Kim KS (1992) Effects of temperature on the development of *Monochamus alternatus* hope (Coleoptera: Cerambycidae). The Research Reports of the Forestry Research Institute 44: 151–156. (In Korean).
43. Ma RY, Hao SG, Tian J, Sun JH, Kang L (2006) Seasonal variation in cold-hardiness of the Japanese pine sawyer *Monochamus alternatus* (Coleoptera: Cerambycidae). Environ. Entomol. 35: 881–886.
44. Phillips SJ, Anderson RP, Schapire RE (2006) Maximum entropy modeling of species geographic distributions. Ecol. Model. 190: 231–259.
45. Phillips SJ, Dudík M (2008) Modeling of species distributions with Maxent: new extensions and a comprehensive evaluation. Ecography 31: 161–175.
46. Elith J, Phillips SJ, Hastie T, Dudík M, Chee YE, et al. (2011) A statistical explanation of MaxEnt for ecologists. Divers. Distrib. 11: 43–57.
47. Dudik M, Phillips S, Schapire R (2004) Performance guarantees for regularized maximum entropy density estimation. Proceedings of the 17th Annual Conference on Computational Learning Theory 472–486.
48. Elith J, Graham CH, Anderson RP, Dudik M, Ferrier S, et al. (2006) Novel methods improve prediction of species' distributions from occurrence data. Ecography 29: 129–51.
49. Hosmer DW, Lemeshow S (1989) Applied Logistic Regression. Wiley, Sons, New York.
50. Fielding AH, Bell JF (1997) A review of methods for the assessment of prediction errors in conservation presence/absence models. Environ. Conserv. 24: 38–49.
51. Hijmans RJ (2012) Cross-validation of species distribution models: removing spatial sorting bias and calibration with a null model. Ecology 93: 679–688.
52. Liu C, White M, Newell G (2013) Selecting thresholds for the prediction of species occurrence with presence-only data. J. Biogeogr. 40: 778–789.
53. ODEPA (2007) VII Censo nacional agropecuario y forestal. INE/ODEPA. Chile. (In Spanish).
54. R Core Team (2013) R: A language and environment for statistical computing. R Foundation for Statistical Computing, Vienna, Austria. Available: http://www.R-project.org. Accessed 18 February 2014.
55. Quantum GIS Development Team (2012) Quantum GIS Geographic Information System. Open Source Geospatial Foundation. Available: http://qgis.osgeo.org. Accessed 18 February 2014.
56. GRASS Development Team (2012) Geographic Resources Analysis Support System (GRASS) Software, Version 6.4.2. Open Source Geospatial Foundation. Available: http://grass.osgeo.org. Accessed 18 February 2014.
57. Chen SL, Du RQ, Gao WL, Wu H, Yu PW, et al. (2010) Analysis of the factors influencing population dynamics of *Monochamus alternatus* Hope (Coleoptera:

Cerambycidae) in Wuyishan Scenic Spot. Acta Entomol. Sinica 53: 183–191. (In Chinese).

58. Rutherford TA, Webster JM (1987) Distribution of pine wilt disease with respect to temperature in North America, Japan, and Europe. Can. J. Forest Res. 17: 1050–1059.

59. Mamiya Y (1983) Pathology of the pine wilt disease caused by *Bursaphelenchus xylophilus*. Annu. Rev. Phytopathol. 21: 201–220.

60. Pershing JC, Linit MJ (1986) Development and seasonal occurrence of *Monochamus carolinensis* (Coleoptera: Cerambycidae) in Missouri. Environ. Entomol. 15: 251–253.

61. Jikumaru S, Togashi K (1996) Effect of temperature on the post-diapause development of *Monochamus saltuarius* (Gebler)(Coleoptera: Cerambycidae). Appl. Entomol. Zool. 31: 145–148.

62. Naves P, de Sousa E (2009) Threshold temperatures and degree-day estimates for development of post-dormancy larvae of *Monochamus galloprovincialis* (Coleoptera: Cerambycidae). J. Pest Sci. 82: 1–6.

63. Bain J, Hosking G (1988) Are NZ *Pinus radiata* plantations threatened by pine wilt nematode Bursaphelenchus xylophilus? New Zealand Forestry 32: 19–21.

64. Furuno T, Nakai I, Uenaka K, Haya K (1993) The pine wilt upon the exotic pine species introduced in Kamigamo and Shirahama Experiment Station of Kyoto University- Various resistances among genus pinus to pinewood nematode, *Bursaphelenchus xylophilus*. Report of the Kyoto University Forests 25: 20–34.

65. Ferrada R, Canales RA, Ide S, Valenzuela J (2007) Intercepciones de insectos vivos realizadas en embalajes de madera de internación en el período: 1995–2005. Ministerio de Agricultura, Chile. (In Spanish).

66. Elith J, Graham C (2009) Do they? How do they? Why do they differ? On finding reasons for differing performances of species distribution models. Ecography 32: 66–77.

67. Webber BL, Yates CJ, Le Maitre DC, Scott JK, Kriticos DJ, et al. (2011) Modelling horses for novel climate courses: insights from projecting potential distributions of native and alien Australian acacias with correlative and mechanistic models. Divers. Distrib. 17: 978–1000.

68. Kearney M, Porter W (2009) Mechanistic niche modelling: combining physiological and spatial data to predict species' ranges. Ecol. Lett. 12: 334–350.

6

Predicting Potential Global Distributions of Two *Miscanthus* Grasses: Implications for Horticulture, Biofuel Production, and Biological Invasions

Heather A. Hager[1]*, Sarah E. Sinasac[1], Ze'ev Gedalof[2], Jonathan A. Newman[1]

1 School of Environmental Sciences, University of Guelph, Guelph, Ontario, Canada, **2** Department of Geography, University of Guelph, Guelph, Ontario, Canada

Abstract

In many regions, large proportions of the naturalized and invasive non-native floras were originally introduced deliberately by humans. Pest risk assessments are now used in many jurisdictions to regulate the importation of species and usually include an estimation of the potential distribution in the import area. Two species of Asian grass (*Miscanthus sacchariflorus* and *M. sinensis*) that were originally introduced to North America as ornamental plants have since escaped cultivation. These species and their hybrid offspring are now receiving attention for large-scale production as biofuel crops in North America and elsewhere. We evaluated their potential global climate suitability for cultivation and potential invasion using the niche model CLIMEX and evaluated the models' sensitivity to the parameter values. We then compared the sensitivity of projections of future climatically suitable area under two climate models and two emissions scenarios. The models indicate that the species have been introduced to most of the potential global climatically suitable areas in the northern but not the southern hemisphere. The more narrowly distributed species (*M. sacchariflorus*) is more sensitive to changes in model parameters, which could have implications for modelling species of conservation concern. Climate projections indicate likely contractions in potential range in the south, but expansions in the north, particularly in introduced areas where biomass production trials are under way. Climate sensitivity analysis shows that projections differ more between the selected climate change models than between the selected emissions scenarios. Local-scale assessments are required to overlay suitable habitat with climate projections to estimate areas of cultivation potential and invasion risk.

Editor: Jose L. Gonzalez-Andujar, Instituto de Agricultura Sostenible (CSIC), Spain

Funding: This research was funded by the Ontario Ministry of Agriculture and Food and Ministry of Rural Affairs (project 027167, www.omafra.gov.on.ca), a Mitacs Elevate post-doctoral fellowship (www.mitacs.ca), and the Natural Sciences and Engineering Research Council of Canada (http://www.nserc-crsng.gc.ca). The funders had no role in the study design, data collection and analysis, decision to publish, or preparation of the manuscript.

Competing Interests: The authors have declared that no competing interests exist.

* E-mail: hhager@uoguelph.ca

Introduction

Plant species are often introduced to new regions through human intervention. Plants that were introduced historically for medicinal, agricultural, or horticultural uses compose a large proportion (>60%) of the currently naturalized angiosperms in the United States and elsewhere [1]. Once established, these species have the potential to become invasive, with subsequent negative ecological and economic effects [2]. Many jurisdictions have introduced weed risk assessment methods to evaluate the risk that deliberately introduced plant species will become invasive in the future (e.g., [3], [4], [5], [6]). However, for species that were introduced prior to widespread use of such methods, their proliferation through increasing horticultural sales or industrial cultivation could increase their risk of escape. Assessing the risks and potential invasive outcomes of such species is therefore important in developing best management practices [7], [8], [9].

Several species of *Miscanthus* have been introduced to novel ranges in North America, Europe, and Scandinavia for both horticultural and agricultural purposes. These are tall, perennial, rhizomatous, C4 grasses native to temperate, humid subtropical, and tropical savannah climates of Asia [10]. *Miscanthus sinensis* Andersson was first introduced to North America as an

ornamental plant in the 1890 s; it has since escaped cultivation in the northeastern United States and is considered invasive in some states [11]. Less is known about the first introduction of *M. sacchariflorus* (Maxim.) Franch. as an ornamental plant in North America, but escaped specimens were noted in the mid-western United States by 1950 [12], and the first escaped specimens in Ontario, Canada, were collected in 1952 [13]. These two species produce a sterile hybrid *M. x giganteus* J.M. Greef & Deuter ex Hodkinson & Renvoize that has been encountered infrequently in the wild [14]. *M. sinensis* has become a very popular ornamental plant in areas of the United States and Canada [15], and all three species are of interest as potential biofuel crops (e.g., [16], [17], [18]). Breeding programs for horticultural and agricultural improvement could enhance the potential of these species to be invasive [19], [11].

The role of weed risk assessment is to evaluate the risk of escape from cultivation and the extent of possible economic and environmental damage [4]. Thus, many assessments include estimating the potential climate suitability of the risk assessment area for candidate species for import (e.g., [3], [4], [5], [6]) because climate is a good predictor of plant distributions [20], [21]. Climate suitability provides a coarse-scale estimate of a species' distribution while ignoring factors such as substrate geology, biotic interactions, and infrastructure, which affect regional habitat suitability.

Various methods have been used to estimate potential species distributions, for example, plant hardiness zones [22], [6], climate regions occupied [23], and a wide range of bioclimatic and niche models (e.g., [24], [25], [9]). The latter use detailed information on temperature and moisture in the species' native range to determine areas with potentially suitable climate for population persistence. Estimating species preferences for certain growing conditions can be difficult, leading to uncertainty in model parameterization. However, model sensitivity to the choice of parameter values is seldom evaluated (but see [26], [27]), but is important in understanding which parameters have the greatest effect on model output and therefore should be estimated most carefully or where results should be treated most cautiously. Such information can be used in interpreting model output and in directing future research to improve model reliability.

Estimating species' potential ranges under projected climate change is also a priority. This work is of particular concern for species of agronomic and horticultural importance, as well as for invasive and potential pest species [28]. Shifts in distributions of agronomic and horticultural species are likely to keep pace with climate change if people continue to plant them where conditions are suitable [29]. Distribution shifts of pests and invaders could also be favoured under climate change because these species tend to have wide physiological tolerances and traits that allow them to take advantage of long-distance (human-assisted) dispersal vectors [30]. Previous assessments of climatic suitability indicate that the modelled distributions differ depending on the choice of climate change model and scenario (e.g., [31], [32]). Comparing multiple models and climate change scenarios therefore allows assessment of the sensitivity of results to these choices.

The purpose of this analysis was threefold. First, we estimated the current potential global distributions of *M. sacchariflorus* and *M. sinensis* using a commonly employed niche model. Second, we evaluated the model sensitivity to estimate which parameters are most critical and whether this differs between the species. Third, we examined how the potentially suitable area is projected to change under future climates, as well as how sensitive these results are to the selection of climate model and emissions scenario. The resulting potential distributions indicate areas that might be suitable for horticultural or agricultural cultivation of these species, but also susceptible to their invasion, should they escape.

Methods

We estimated the current global native and introduced distributions of *M. sacchariflorus* and *M. sinensis* using species occurrence data from the Global Biodiversity Information Facility (GBIF; www.gbif.org), botanical garden records, and a literature search of agronomic trials. All data were sorted to remove duplicate records and were separated based on whether geolocation information (latitude and longitude) was provided or whether we could geocode the record using information such as street name, city, county, or region ("inferred" location). Records that did not provide location information below the country level were omitted.

We used the native distribution data to model the potential climatic range of each species using the CLIMEX Compare Locations model [33], [34]. We used 0.5° world grid climate data provided with the software (from the Climate Research Unit at Norwich, UK [35]). CLIMEX assumes that the geographical distribution of the species is limited by climate; it does not generally account for biotic interactions [36] or substrate type. It calculates an annual growth index based on the species' fitted temperature and moisture response functions, as well as four stress indices (hot, cold, dry, wet, and their combinations) to calculate an ecoclimatic index (EI). The EI is an estimate of a location's climatic suitability to support a persistent population of the species being modelled [34]. EI ranges from 0 to 100, with 0 meaning the area is not suitable for species persistence, and 100 meaning the climate is optimal for the species year-round. In practice, EI of 100 is only attained for species in stable and ideal climate [34].

For each species, we began with parameter values from the default temperate template and adjusted them iteratively based on the species' biology until the modelled distributions approximated the native distributions [34]; these included humid subtropical (*M. sacchariflorus*, *M. sinensis*) and tropical savannah (*M. sinensis*) climates of Asia. The model was then validated by comparison with the observed distribution records in the species' introduced ranges in Europe and North America. Given the satisfactory fit, no further adjustments were required at this stage for either model.

Parameter Fitting

Temperature and cold/hot stress. Based on their native distributions, both *M. sacchariflorus* and *M. sinensis* are well suited for cold-temperate regions (e.g., [37], [38]). Their percentage shoot emergence at experimental, low temperatures ranges from 10 to 100% at 7 to 15°C [39]. Their *M. x giganteus* hybrid shows only minor reduction in leaf photosynthetic capacity when grown at 10°C or 14°C compared to 25°C [40], [41]. Therefore, the limiting low temperature (DV0) and lower optimal temperature (DV1) were set at 5°C and 15°C, respectively, for both species (Table 1). This and the cold stress (below) accounted for the northernmost occurrence of both species in far northeastern China and the southeastern Primorsky Krai region of Russia.

The upper optimal temperature (DV2) and limiting high temperature (DV3) were adjusted based on maximum temperatures occurring in the native region [42] and observations that the experimental optimal temperature for photosynthesis of their hybrid is between 30°C and 35°C [43]. The native distribution of *M. sinensis* extends much further south into tropical regions than does that of *M. sacchariflorus* [37]. Therefore, both DV2 and DV3 for *M. sinensis* were greater than those for *M. sacchariflorus* (Table 1).

For the cold stress index, we used parameters related to overwinter survival (lethal temperatures: cold stress temperature threshold, TTCS, and cold stress temperature rate, THCS) and cold stress affecting metabolism (cold stress degree day threshold, DTCS, and cold stress degree day rate, DHCS). Both species have strong cold tolerance and overwinter underground as rhizomes. The lethal temperature at which 50% of rhizomes were killed in a freezing experiment ranged from −3.4°C to −6.3°C for both species [44]. Therefore, TTCS was set to −5°C and THCS to a very low accumulation rate below this threshold air temperature because of the expectation that rhizomes would be insulated by plant litter and snow pack in colder areas. DTCS was adjusted downward slightly from the temperate template (15°C) to 12°C for *M. sacchariflorus* and 14°C for *M. sinensis*, with a very low accumulation rate (DHCS; Table 1) because both species have high cold tolerance. DTCS was lower for *M. sacchariflorus* than for *M. sinensis* because native distribution records show *M. sacchariflorus* persisting at slightly higher latitudes.

There has been little investigation of heat stress effects on the growth of *Miscanthus*. Therefore, heat stress parameters (heat stress threshold, TTHS, and heat stress rate, THHS) were adjusted based on native distribution records for the species. TTHS was set to begin accumulating at or above the upper optimal growth temperature for *M. sacchariflorus* and *M. sinensis*, respectively, with a

Table 1. Fitted parameter values used to generate bioclimatic envelope models of *Miscanthus sacchariflorus* and *M. sinensis* distributions using CLIMEX.

Parameter description	Parameter	*Miscanthus sacchariflorus*	*Miscanthus sinensis*
Moisture			
Limiting low moisture[a]	SM0	0.25	0.25
Lower optimal moisture[a]	SM1	0.8	0.8
Upper optimal moisture[a]	SM2	1.2	1
Limiting high moisture[a]	SM3	1.8	2.5
Temperature			
Limiting low temperature[b]	DV0	5	5
Lower optimal temperature[b]	DV1	15	15
Upper optimal temperature[b]	DV2	28	30
Limiting high temperature[b]	DV3	32	35
Cold Stress			
Cold stress temperature threshold[b]	TTCS	−5	−5
Cold stress temperature rate[c]	THCS	−0.0002	−0.0002
Cold stress degree-day temperature threshold[d]	DTCS	12	14
Cold stress degree-day rate[c]	DHCS	−0.00005	−0.00005
Heat Stress			
Heat stress temperature threshold[b]	TTHS	32	36
Heat stress temperature rate[c]	THHS	0.06	0.05
Dry Stress			
Dry stress moisture threshold[a]	SMDS	0.1	0.1
Dry stress rate[c]	HDS	−0.02	−0.02
Wet Stress			
Wet stress moisture threshold[a]	SMWS	1.8	3
Wet stress rate[c]	HWS	0.02	0.05
Hot-wet Stress			
Hot-wet temperature threshold[b]	TTHW	31	-
Hot-wet moisture threshold[a]	MTHW	1	-
Hot-wet stress rate[c]	PHW	0.01	-
Degree-days			
Minimum degree days above DV0 to complete one generation[d]	PDD	600	600

[a]Proportion soil capacity.
[b]°C.
[c]Week^{-1}.
[d]°D.

high accumulation rate (THHS). TTHS was higher and THHS was slightly lower for *M. sinensis* than for *M. sacchariflorus* to account for the former species' more tropical native distribution.

Soil moisture and dry/wet stress: Limiting low soil moisture (SM0) and lower optimal soil moisture (SM1) were set according to the temperate template. Upper optimal soil moisture (SM2) and limiting high soil moisture (SM3) were set according to the wet-tropical template for *M. sinensis*. In comparison, SM2 and SM3 were reduced slightly for *M. sacchariflorus* in conjunction with the wet stress parameters to limit its potential distribution from occurring widely in wet tropical areas.

The dry stress threshold (SMDS) and dry stress rate (HDS) were set to accommodate the high drought tolerance of both species [45]. Both species were assigned an SMDS of 0.1, which is close to the minimum moisture content at which plants can extract water from the soil [46], [47]. HDS was set at a moderate rate, given that

the hybrid can recover from short-term (30 days) but not long-term (60 days) drought [48].

Both *M. sacchariflorus* and *M. sinensis* grow well in saturated areas such as drainage ditches (HAH, personal observation), but not when submerged such as in streams [49], [50]. Therefore, the wet stress threshold (SMWS) and wet stress rate (HWS) were set so that the species would grow in wet areas but would experience a high rate of stress accumulation. Parameters for *M. sacchariflorus* were adjusted to restrict its southerly distribution.

Hot-wet stress: A hot-wet stress was added to the model for *M. sacchariflorus* to exclude its distribution from tropical equatorial areas of southeastern Asia-Pacific [37]. No hot-wet stress was used for *M. sinensis*.

Sensitivity to Model Parameters

Once the models were validated, we performed a sensitivity analysis for each species to determine the response of the modelled distributions to small changes in parameter values. Each parameter was adjusted upward or downward while holding all other parameters constant, and the resulting EI values for the global distribution were generated in CLIMEX. Rate and soil moisture variables were adjusted by 10%; temperature variables were adjusted by 1°C (D. Kriticos, *personal communication*). EI was divided into five classes of suitability for species growth for further analysis: 0, unsuitable; 1–10, marginal; 10–20, suitable; 20–30, favourable; and >30, highly favourable [51], [34].

EI values from each iteration of the sensitivity analysis were compared with those of the original model to determine the change in area for each EI class. To do this, EI data generated by CLIMEX were reprojected to a cylindrical equal area projection in ArcGIS 10.1, masked to land area, and a nearest neighbour, inverse distance weighting interpolation was performed so that each grid cell represented an equal amount of area (50×50 km). The resulting grid cells were then reclassified corresponding to the five EI classes and the number of cells counted to determine the total area in each class. We then computed the proportional change in area from the original model for each EI class. Sensitivity was evaluated as a greater proportional change in area than in parameter value.

Climate Change Scenarios

To compare the species' potential distributions under future climate change predictions, and to assess the sensitivity of these potential distributions to model selection and emissions assumptions, we used two general circulation models with two emissions scenarios (A2 and B1). We used the Bergen Climate Model 2.0 (BCM) from the Bjerknes Center for Climate Research, and the Coupled Global Climate Model 3 (CGCM3-T63) from the Canadian Center for Climate Modeling and Analysis [52]. The selected emissions scenarios represent comparatively low (B1) and high (A2) future greenhouse gas concentrations, thus spanning the likely range of probable future conditions [53].

Climate projection data for each model–scenario combination were input into CLIMEX as the 30-yr mean for three standard time periods: baseline (1971–2000), 2050s (2041–2070), and 2080s (2071–2100). First, monthly means of maximum temperature, minimum temperature, and relative humidity, and daily means of precipitation for the world (excluding Greenland, Antarctica, and the Arctic) were downloaded from the Canadian Climate Change Scenarios Network (CCCSN: http://www.cccsn.ec.gc.ca/?page = dd-gcm) for each model–scenario combination. The 30-yr means were then calculated to obtain a single value for each climate variable for each $2.75° \times 2.75°$ grid cell. Because CLIMEX requires monthly climate data, the average daily precipitation data were multiplied by the number of days in the month to produce monthly means. CLIMEX also requires two relative humidity values, RH% at 0900 h and at 1500 h. Although only mean monthly values are available from the CCCSN, there is a strong diurnal cycle in RH%, with maximum values at night and minimum values in mid-afternoon. Daily mean values are therefore reasonably close to observed values at 0900 h and were considered RH% at 0900 h. RH% at 1500 h was estimated using the CLIMEX method of multiplying the 0900 h value by 0.85 [34], [31].

Using these model–scenario data, the CLIMEX bioclimatic model was run for each species. EI values were obtained for a total of 20 model runs ([circulation model baseline + two scenarios × two future time periods] × two models × two species). The EI data were reprojected into a cylindrical equal area projection, masked to land area, and interpolated using inverse distance weighting of the three nearest points in ArcGIS 9.3. The areas of suitable climate under baseline and future conditions were calculated by reclassifying the raw EI data into the five EI classes. Projected changes in the potential distributions of each species were determined by overlaying future and baseline projections.

For each species and time, areas of agreement between models or scenarios were determined by overlaying projected distributions based on the two models for a given scenario or the two scenarios for a given model, respectively. To quantify the similarity between pairs of projected potential distributions, we calculated a simple index of agreement by dividing the area where the two projections agree about the potential presence of *Miscanthus* (all area classified as suitable, favourable, and highly favourable, or EI >10) by the area where the two projections disagree. The higher the value, the more similar the two projections are. Values >1.0 indicate that the models agree more than they differ; values <1.0 indicate more disagreement than agreement.

Results

Occurrence records indicate that both the native and introduced distributions of *M. sacchariflorus* are less widespread than those of *M. sinensis* (Fig. 1). In the native range, both species occur throughout Japan, Korea, and south-central and eastern China, and in parts of northern and northeastern China. However, *M. sinensis* extends further south and east into Taiwan, the Philippines, and Vanuatu [37] (Fig. 1). In the introduced range, *M. sacchariflorus* occurs mainly in northwestern Europe, Denmark, Sweden, northeastern United States, and southeastern Canada. *M. sinensis* has been introduced to these areas as well as southeastern and parts of western United States, Mexico, Puerto Rico, Colombia, Chile, Argentina, Uruguay, southern Australia, Tasmania, and New Zealand.

For the *M. sacchariflorus* model, 100% of given native, 90.1% of inferred native, 89.1% of given introduced and 97.7% of inferred introduced occurrence records are in areas deemed suitable, favourable, or highly favourable (EI>10). Zero and 9.4% of given and inferred native, and 10.9% and zero of given and inferred introduced occurrences are in marginal areas. One inferred native record (0.6%) and one inferred introduced record (2.3%) are in areas deemed unsuitable. For the *M. sinensis* model, 100% of given native, 94.2% of inferred native, 99.7% of given introduced, and 97.5% of inferred introduced occurrences are in areas deemed suitable, favourable, or highly favourable. Zero and 5.8% of given and inferred native, and 2.0% and 2.5% of given and inferred introduced occurrences are in marginal areas. No observed occurrences are in areas deemed unsuitable. The models indicate that *M. sinensis* has a wider potential global distribution than does *M. sacchariflorus* (Fig. 1).

Sensitivity to Model Parameters

The modelled distributions differed in their sensitivity to a ±10% or 1°C change in parameter values (Table 2). Eight of 22 (36%) parameters for *M. sacchariflorus* and 5 of 19 (26%) parameters for *M. sinensis* showed sensitivity in at least one EI class. Models for both species were very sensitive to changes in heat stress temperature threshold (TTHS) and upper optimal and limiting temperatures (DV2, DV3) and were mildly sensitive to changes in upper optimal and limiting moistures (SM2, SM3). In addition, the *M. sacchariflorus* model was highly sensitive to an increase in dry stress rate (HDS) and a decrease in hot-wet temperature threshold (TTHW), and mildly sensitive to a decrease in wet stress moisture threshold (SMWS).

Sensitive temperature parameters all related to upper temperatures and heat tolerance thresholds (Table 2). A decrease in the

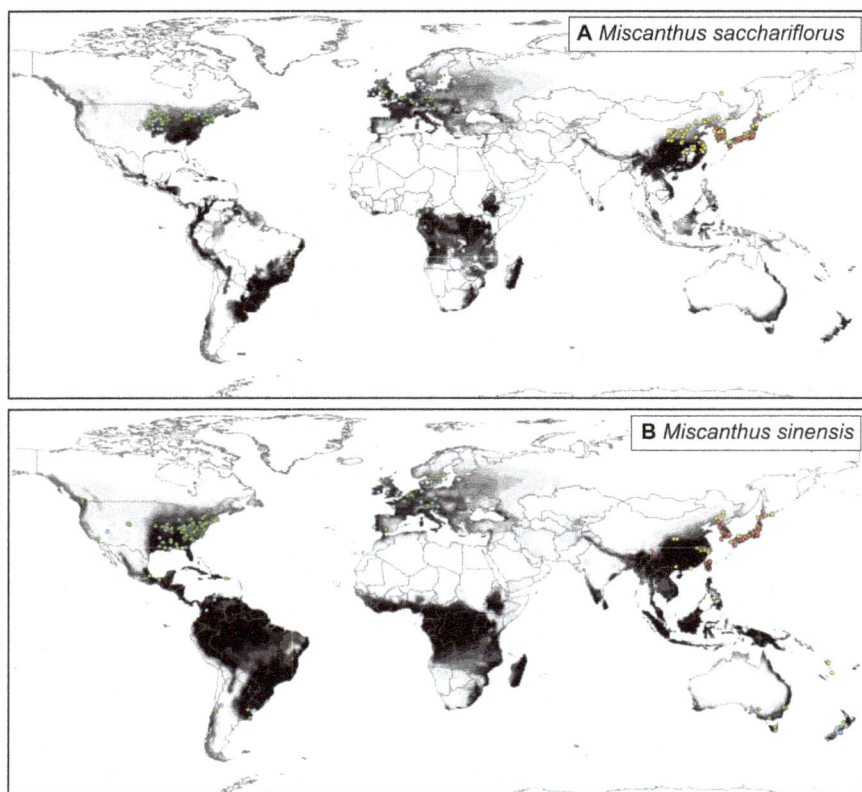

Figure 1. Observed (circles) and modelled (shaded) native and introduced plant distributions. Circles indicate native (red), native inferred (yellow), introduced (green), and introduced inferred (blue) geolocations. The density of shading represents the bioclimatic suitability of a region for plant population persistence, with white areas indicating no suitability, lighter grey areas indicating low suitability, and darker grey areas indicating high suitability. (a) *Miscanthus sacchariflorus*. Geolocations: native (n = 94), native inferred (n = 171), introduced (n = 119), and introduced inferred (n = 43). (b) *Miscanthus sinensis*. Geolocations: native (n = 335), native inferred (n = 52), introduced (n = 297), and introduced inferred (n = 81).

upper temperature parameters tended to decrease EI towards less favourable values, whereas an increase tended to increase EI towards more favourable values for both species. The heat stress temperature threshold followed a similar pattern. The *M. sacchariflorus* model was more sensitive than the *M. sinensis* model to changes in these parameters. For *M. sacchariflorus*, a decrease in the hot-wet temperature threshold also decreased EI towards less favourable values, whereas an increase had only a small effect.

Sensitive soil moisture parameters all related to upper soil moistures and the wet stress threshold. Changes in upper soil moisture parameters had weaker effects than for temperature. A decrease in the upper soil moisture parameters tended to decrease the most favourable EI values but had little effect on the unsuitable climate area. An increase in these parameters had little effect, except for slight increases in EI towards more favourable values for *M. sacchariflorus*. For *M. sacchariflorus*, a decrease in the wet stress moisture threshold tended to decrease the suitable climate area.

The only rate parameter that exhibited sensitivity was the dry stress rate for the *M. sacchariflorus* model. An increase in this parameter increased EI values from unsuitable towards more favourable values, with the largest change in marginal climate area (Table 2).

Range Shifts Under Climate Change Projections

Projected changes in the potential area occupied by *M. sacchariflorus* and *M. sinensis* are generally large, ranging from global decreases in potential area by 2080 of 4 to 6%, depending

on the species, model, and scenario chosen (Table 3, World; Fig. 2). In all cases, the area of climatically suitable locations (EI >10) continues to decrease over time.

Limiting this analysis to North America reveals some important regional differences (Table 3, North America). Projected percent changes in climatically suitable area are smaller for North America than for the world and are often different in sign. The direction of change also differs between the two models: For both species by 2080, regardless of scenario, the BCM projects reductions in the total suitable area and the CGCM projects increases in suitable area. There are also large changes in the relative proportions of habitat categories. In all projections, including those in which the area of climatically suitable habitat increases, highly favourable area (EI >30) decreases and favourable area (EI 20–30) increases.

Comparing the future potential distributions of *Miscanthus* species to their baseline potential distributions provides an indication of how rapidly shifts in suitable range might occur. This analysis indicates that projected suitability shifts are moderately large for both species and under both emissions scenarios (Table 4). The smallest projected shifts in suitable range occur under the B1 emissions scenario, with *M. sacchariflorus* projected to have larger shifts than *M. sinensis*. The mean overlap between the baseline and 2080s potentially suitable areas is only 61% for *M. sacchariflorus* and 78% for *M. sinensis*. Very little range contraction is projected to occur in the native range, except for some projections for *M. sacchariflorus*. In contrast, range expansion is projected in northern parts of the native range for both species.

Table 2. Sensitivity analysis results for *Miscanthus sacchariflorus* and *M. sinensis* indicating the proportional change in area from that of the baseline model global distribution for each category of ecoclimatic index (EI) modelled using CLIMEX software.

| | *Miscanthus sacchariflorus* | | | | | | *Miscanthus sinensis* | | | | | |
| | | Proportional change in area from baseline (%) | | | | | | Proportional change in area from baseline (%) | | | | |
Parameter	Value[a]	EI = 0	EI = 1–10	EI = 10–20	EI = 20–30	EI = 30–100	Value[a]	EI = 0	EI = 1–10	EI = 10–20	EI = 20–30	EI = 30–100
DV0	4 (20)	−0.61	0.98	1.17	−1.08	2.50	4 (20)	−0.71	0.36	1.81	−0.76	1.09
	6 (20)	0.69	−1.39	−1.28	2.13	−3.01	6 (20)	0.85	−1.16	−1.75	1.88	−1.30
DV1	14 (6.7)	−0.14	−2.13	0.19	−1.66	3.68	14 (6.7)	−0.19	−2.45	0.79	−0.88	1.55
	16 (6.7)	0.12	2.15	0.11	2.67	−4.32	16 (6.7)	0.19	2.12	−0.79	2.61	−1.83
DV2	27 (3.6)	0.06	**5.31**	**9.94**	−0.27	**−11.68**	29 (3.3)	0.11	1.77	**8.97**	**16.08**	**−7.08**
	29 (3.6)	−0.07	**−5.62**	**−7.63**	−2.98	**12.25**	31 (3.3)	−0.13	−1.01	**−3.92**	**−13.90**	**5.10**
DV3	31 (3.1)	0.54	**10.64**	−0.84	**−7.20**	**−9.72**	34 (2.9)	0.26	0.46	**5.41**	**5.22**	**−3.28**
	33 (3.1)	−0.15	**−5.80**	**−4.26**	3.08	**7.80**	36 (2.9)	−0.12	−0.46	−1.62	**−5.13**	2.11
SM0	0.225	−0.77	1.62	1.62	−0.10	2.03	0.225	−1.46	2.97	3.59	−1.03	1.11
	0.275	0.71	−1.43	−1.67	0.57	−2.10	0.275	1.41	−2.80	−3.76	1.52	−1.17
SM1	0.72	−0.28	−3.06	−0.42	−2.27	6.18	0.72	−0.54	−3.30	1.42	−4.28	3.36
	0.88	0.24	4.14	−1.59	2.13	−5.68	0.88	0.38	4.48	−3.69	5.13	−3.26
SM2	1.08	0.03	4.32	7.60	7.37	**−12.94**	0.9	0.00	1.13	2.74	**10.08**	**−3.65**
	1.32	−0.01	−3.89	−6.71	−4.50	**10.39**	1.1	−0.01	−0.91	−3.23	−8.07	3.18
SM3	1.62	0.17	7.86	7.60	1.22	**−14.21**	2.25	0.00	1.01	4.39	**13.26**	**−4.76**
	1.98	−0.05	−5.12	−5.40	1.93	7.81	2.75	−0.01	−0.50	−2.77	−7.37	2.71
TTCS	−6 (20)	0.76	−1.86	−2.26	−0.91	−0.83	−4 (20)	−1.10	2.60	2.67	0.88	0.26
	−4 (20)	−0.78	2.32	2.42	0.37	0.67	−6 (20)	1.00	−1.66	−4.25	−0.36	−0.25
THCS	−0.00022	0.66	−2.56	−1.45	−0.44	−0.31	−0.00022	0.81	−1.96	−2.80	−0.24	−0.08
	−0.00018	−0.83	3.34	2.03	0.44	0.19	−0.00018	−1.07	3.18	2.61	0.24	0.11
DTCS	11 (8.3)	−0.36	0.95	1.20	0.10	0.40	13 (7.1)	−0.42	0.61	1.48	−0.21	0.31
	13 (8.3)	0.38	−1.15	−0.53	−0.41	−0.57	15 (7.1)	0.44	−0.63	−1.78	0.03	−0.23
DHCS	−0.000055	0.26	−0.69	−0.39	−0.17	−0.55	−0.000055	0.33	−0.23	−1.62	−0.12	−0.17
	−0.000045	−0.27	0.71	0.95	0.07	0.31	−0.000045	−0.42	0.68	1.32	−0.12	0.28
TTHS	31 (3.1)	**8.79**	**−13.96**	**−22.28**	**−21.78**	**−13.97**	35 (2.8)	2.29	**−3.12**	−2.01	**−7.37**	−1.03
	33 (3.1)	**−8.15**	**26.66**	**16.04**	**8.12**	**8.21**	37 (2.8)	−1.82	**3.30**	**3.99**	**5.01**	0.11
THHS	0.054	−0.21	0.46	0.72	0.30	0.21	0.045	−0.05	0.10	0.16	0.09	0.00
	0.066	0.18	−0.46	−0.47	−0.41	−0.14	0.055	0.08	−0.10	−0.43	−0.06	0.00
SMDS	0.09	−0.12	−0.15	−1.64	0.98	1.39	0.09	−0.42	0.28	−2.84	0.55	1.25
	0.11	−0.36	−1.91	−6.35	−0.91	8.48	0.11	0.43	−0.02	1.39	0.21	−1.25
HDS	−0.022	−0.36	−1.91	−6.38	−0.88	8.48	−0.022	0.15	0.45	0.53	0.09	−0.66
	−0.018	**−17.00**	**88.48**	6.10	7.51	5.97	−0.018	−0.25	0.05	−1.65	0.76	0.68

Table 2. Cont.

Parameter	Miscanthus sacchariflorus						Miscanthus sinensis					
	Value[a]	Proportional change in area from baseline (%)					Value[a]	Proportional change in area from baseline (%)				
		EI=0	EI=1-10	EI=10-20	EI=20-30	EI=30-100		EI=0	EI=1-10	EI=10-20	EI=20-30	EI=30-100
SMWS	1.62	1.03	-2.96	**-10.05**	-2.71	4.37	2.7	0.01	0.23	-0.20	0.00	-0.08
	1.98	-0.63	-2.77	-5.35	0.57	9.55	3.3	-0.01	0.02	-0.03	0.03	0.02
HWS	0.018	-0.39	-2.08	-6.04	-0.78	8.57	0.045	0.00	0.00	-0.03	0.00	0.01
	0.022	-0.32	-1.94	-6.52	-1.05	8.45	0.055	0.00	0.00	0.00	0.00	0.00
TTHW	30 (3.2)	**3.67**	**-5.43**	**-6.91**	**-9.77**	**-7.37**						
	32 (3.2)	-1.15	2.87	**3.93**	0.95	1.12						
MTHW	0.9	0.14	-0.05	-0.58	-0.07	-0.40						
	1.1	-0.19	0.52	0.11	0.51	0.28						
PHW	0.009	-0.09	0.21	0.28	0.10	0.14						
	0.011	0.05	-0.02	-0.22	0.00	-0.15						
PDD	540	-0.60	-0.53	-6.35	-0.95	8.52	540	-0.27	1.21	0.03	0.00	0.00
	660	-0.13	-3.11	-6.74	-0.95	8.52	660	0.27	-1.14	-0.16	0.00	0.00

[a]Parameter value used in the sensitivity analysis. The percent change from the original model parameter value was 10%, except for temperature variables, which were changed by 1 °C and for which percent change is indicated in parentheses. Changing temperature variables by 10% gave qualitatively similar results (data not shown).
Parameter values were each decreased and increased by 10% from the baseline value (Table 1) unless indicated otherwise[a]. Changes in area that exceed the percent change in parameter value are shown in boldface font; positive and negative proportions respectively indicate increases and decreases in area.

Table 3. Percent change in climatically suitable area for the World and for North America projected using the BCM and CGCM models and the A2 and B1 emissions scenarios relative to the baseline projections.

Model	Year, scenario	World						North America					
		EI 0-1	EI >1-10	EI >10-20	EI >20-30	EI >30	EI >10	EI 0-1	EI >1-10	EI >10-20	EI >20-30	EI >30	EI >10
Miscanthus sacchariflorus													
BCM	2050 (A2)	+4.1	+6.9	+15.3	−1.7	−20.6	**−10.4**	−1.6	+1.5	+15.4	+74.5	−32.4	**+3.9**
	2080 (A2)	+12.4	+14.7	+8.1	−31.8	−38.6	**−28.9**	−4.4	+18.5	−2.7	+16.7	−18.8	**−6.7**
	2050 (B1)	+3.2	+2.1	+12.6	−0.2	−14.6	**−6.9**	−1.1	−0.3	+6.6	+58.6	−18.4	**+4.4**
	2080 (B1)	+6.5	+5.4	+9.3	−4.6	−24.3	**−14.3**	−1.1	+4.3	+4.8	+49.8	−26.2	**−1.4**
CGCM	2050 (A2)	+8.7	+7.4	−6.3	−12.0	−58.2	**−31.1**	−7.2	+32.8	+27.0	+5.0	−51.5	**+0.6**
	2080 (A2)	+13.0	+10.0	−11.8	−41.7	−71.9	**−46.1**	−13.9	+55.9	+47.7	+21.0	−70.1	**+9.9**
	2050 (B1)	+7.0	+6.4	−9.7	−9.9	−44.3	**−25.4**	−6.5	+26.0	+23.0	+18.2	−42.3	**+5.0**
	2080 (B1)	+8.6	+6.5	−3.8	−13.9	−57.5	**−30.5**	−6.7	+27.2	+27.6	+23.8	−59.9	**+4.3**
Miscanthus sinensis													
BCM	2050 (A2)	+4.0	+8.3	−7.4	−14.4	−5.0	**−6.9**	+1.9	−2.8	−6.5	+23.1	−7.1	**−1.9**
	2080 (A2)	+9.7	+19.1	−8.5	−32.4	−14.9	**−16.5**	−5.4	+28.9	−22.0	+10.3	−9.3	**−11.4**
	2050 (B1)	+1.8	+4.9	+0.7	−11.4	−2.7	**−3.5**	+1.6	+1.8	−11.4	+15.7	−6.8	**−5.0**
	2080 (B1)	+5.1	+12.8	−12.9	−16.1	−6.9	**−9.4**	+0.0	+17.9	−34.6	+17.4	−6.3	**−14.3**
CGCM	2050 (A2)	+8.9	−5.6	+8.3	−0.4	−24.5	**−13.4**	−8.2	+24.3	+13.0	+61.5	−42.1	**+5.3**
	2080 (A2)	+10.9	+6.9	+15.6	+5.9	−42.2	**−21.6**	−17.2	+44.2	+44.7	+101.6	−71.5	**+18.2**
	2050 (B1)	+7.1	−4.1	+6.5	−3.3	−19.0	**−10.9**	−6.7	+14.5	+5.1	+68.0	−22.3	**+9.9**
	2080 (B1)	+8.0	−5.2	+10.3	−1.0	−22.7	**−12.0**	−9.4	+22.0	+17.6	+78.0	−38.6	**+12.2**

The column EI>10 (bold) indicates a composite analysis combining all area classified as suitable, favourable, and highly favourable for each species.

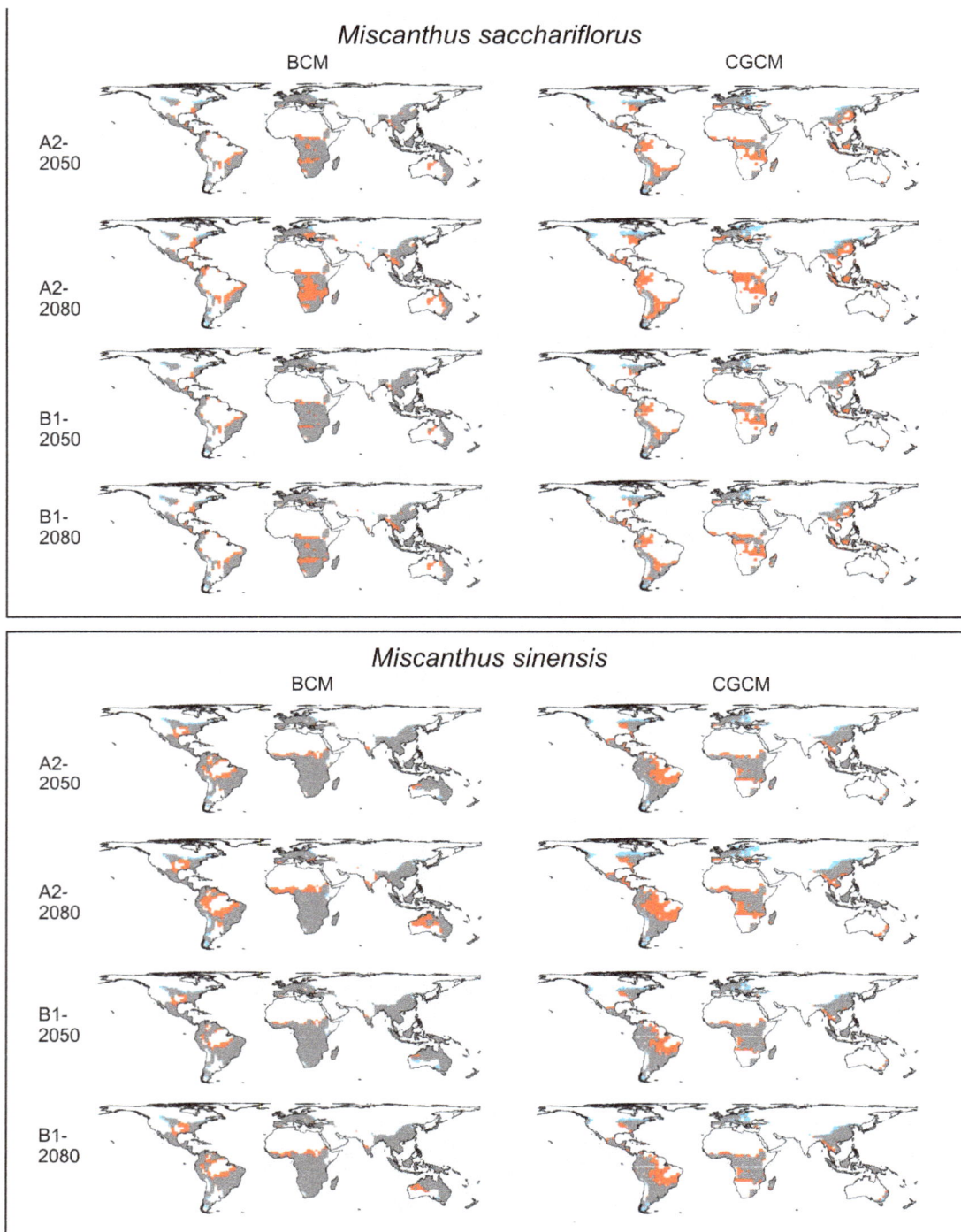

Figure 2. The potential future distribution of *Miscanthus sacchariflorus* **and** *M. sinensis* **(for all area classified as suitable, favourable, and highly favourable, or EI >10) as modeled by CLIMEX under the BCM and CGCM models run using the A2 and B1 emissions scenarios.** All potential ranges are shown relative to their baseline potential distributions. Areas shown in red indicate area where the range is projected to contract; grey indicates the species are potentially present under both baseline and future conditions; and blue indicates areas of range expansion.

In the non-native range, contraction is projected mainly in areas where the species have not yet been introduced such as South America, Africa, and parts of Australia for both species, as well as islands of Southeast Asia for *M. sacchariflorus*. Contraction of the potential range is also projected to occur in southern parts of the United States, but more so for *M. sinensis* than for *M. sacchariflorus*. The majority of range expansion is projected to occur in northern areas of North America, eastern Europe, and Scandinavia.

Table 4. Projected changes in the global climatically suitable area for *Miscanthus sacchariflorus* and *M. sinensis* under the BCM and CGCM models run using the A2 and B1 scenarios.

Model	Year, scenario	Expansion[a] (km×10⁶)	Contraction (km×10⁶)	No change (km×10⁶)	Overlap[b] (%)
		Miscanthus sacchariflorus			
BCM	2050 (A2)	1.54	5.74	34.81	86%
	2080 (A2)	2.28	13.99	26.56	65%
	2050 (B1)	1.10	3.88	36.67	90%
	2080 (B1)	1.35	7.16	33.38	82%
CGCM	2050 (A2)	3.08	11.94	16.51	58%
	2080 (A2)	5.08	18.20	10.25	36%
	2050 (B1)	2.33	9.56	18.89	66%
	2080 (B1)	2.73	11.41	17.04	60%
		Miscanthus sinensis			
BCM	2050 (A2)	2.08	5.85	48.85	89%
	2080 (A2)	2.78	11.79	42.91	78%
	2050 (B1)	1.95	3.85	50.84	93%
	2080 (B1)	1.60	6.72	47.98	88%
CGCM	2050 (A2)	3.68	9.36	33.10	78%
	2080 (A2)	6.01	15.17	27.29	64%
	2050 (B1)	2.84	7.47	35.00	82%
	2080 (B1)	3.51	8.60	33.86	80%

[a]Potential future suitable area is defined as EI>10.
[b]Overlap is calculated as the portion of the baseline range that overlaps with the projected range.

Sensitivity of Results to Choice of Climate Model and Scenario

By overlaying the results for the different models and scenarios, we can identify patterns of agreement and disagreement between the various projections (Fig. 3). This analysis helps to identify changes in the suitable range that are highly likely (e.g., robust to the selection of model or scenario), versus those that are more speculative (e.g., those that differ depending on the model or scenario chosen). This analysis also allows us to identify whether the projections differ because of differences between the climate models or between the scenarios. These results indicate that the choice of climate model accounts for more difference in the results than the choice of scenario and that the two *Miscanthus* species differ in their sensitivity to the selection of model and scenario (Table 5). For the model sensitivity analysis, agreement values are all either <1.0 (greater area of disagreement than agreement between models; five out of eight comparisons), or very slightly > 1.0 (maximum value 1.22). For the scenario sensitivity analysis, all agreement values are substantially >1.0. Different scenarios are more similar within models than are the same scenarios between models for these species.

Discussion

Potential Global Distribution

Niche and bioclimatic envelope models in general provide a coarse-scale indication of areas where the climate might be suitable for the species of interest to establish. The limitations of such models are well discussed elsewhere [34], [54], [55], [56]. Here, we note two associated factors that require consideration in interpreting our models: the species' ecology and their likely non-equilibrium distribution in the introduced ranges.

Little is known of the species' comparative ecology in the native and introduced ranges because most research to date has focused on determining optimal conditions for agricultural production (but see [57], [42], [50]). Parameterizing a model based solely on the native range distribution assumes similar biotic interactions in the introduced and native ranges. However, potential release from suppressive interactions such as competition, herbivory, parasitism, and disease (e.g., [58], [59], [60]), or a lack of mutualist organisms (e.g., [61]) could result in different realized distributions or niche shifts in the introduced compared to the native range (e.g., [62], [63], but see [64], [65]). We parameterized our model using both the native range distribution and some physiological data, which could improve model estimations compared to strictly correlative methods [66], [67]. Model validation using occurrences in the introduced range indicates that the models fit the current introduced distributions well. Determining whether an introduced species will establish beyond the modelled range requires further fine-scale assessment and/or field experiments [54], [68].

It is highly likely that *M. sinensis* and *M. sacchariflorus* are still spreading in the introduced ranges. Indeed, the species only became naturalized in North America in the mid-1900s [12], [11], and time since introduction is a well-known correlate of plant escape and abundance in the introduced range (e.g., [69], [70]). The bioclimatic models suggest that there are additional moderate to large amounts of climatically suitable area in Central America, South America, and Africa, as well as small parts of Australia and New Zealand, where these species have not yet established. The likelihood of spread beyond the modelled potential distributions is unknown, but given the increase in cultivation of these species as

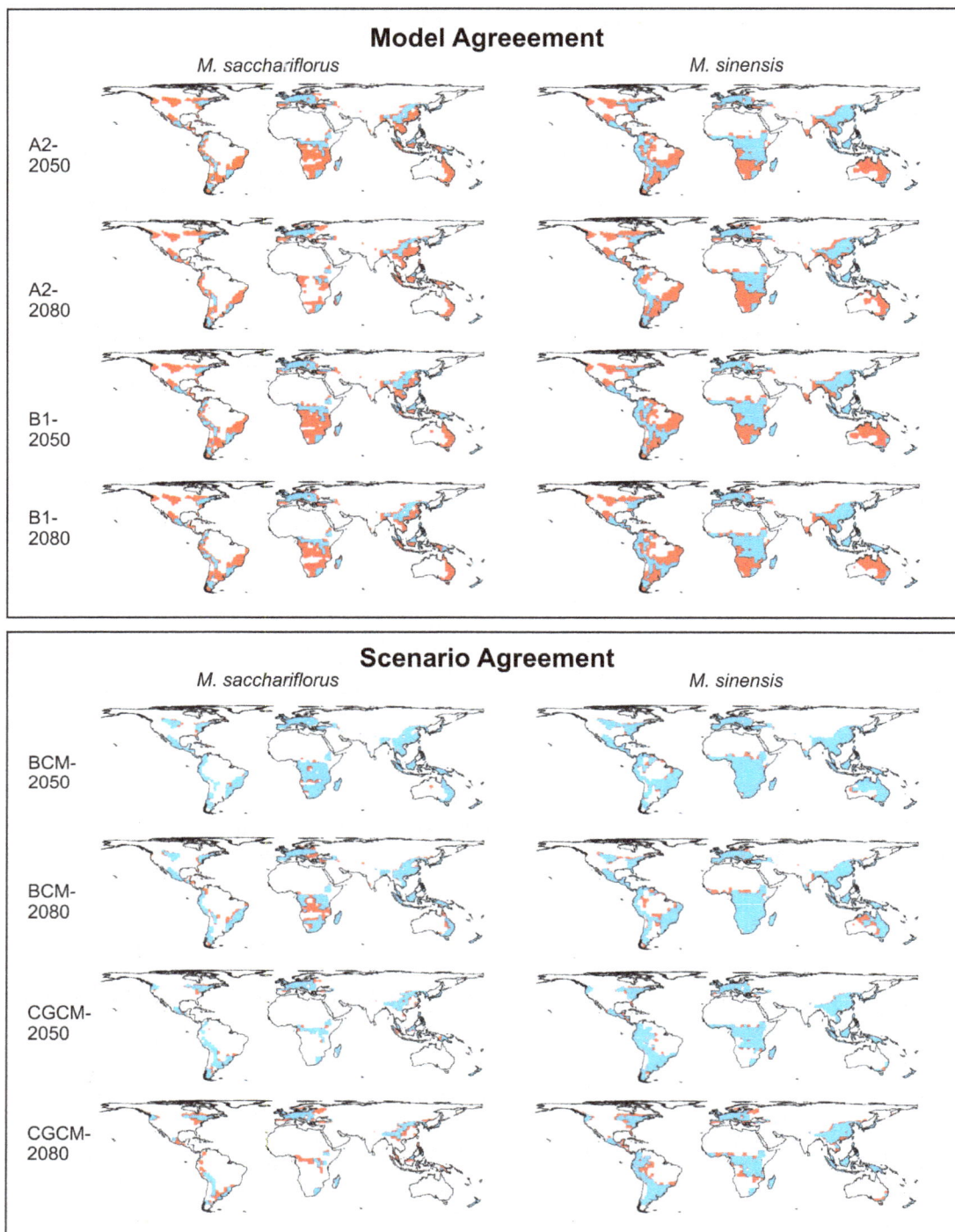

Figure 3. Areas of agreement and disagreement in the projected potential distribution of *Miscanthus sacchariflorus* **and** *M. sinensis* **(all area classified as suitable, favourable, and highly favourable, or EI >10) as modeled by CLIMEX.** The upper model shows areas of agreement between the BCM and CGCM models for a given scenario and year. The lower panel shows areas of agreement between the A2 and B1 scenarios for a given model and year. Areas shown in blue indicate suitable climate under both projections; areas shown in red indicate that only one of the two projections predicts suitable climate.

both horticultural and agricultural materials, which reduces dispersal limitations, and the potential for plant breeding programs to introduce new genetic material with a wider range of trait variation, proliferation within and beyond the introduced area should be monitored closely.

According to the native distributions and our models, *M. sinensis* has a wider range and greater climatically suitable area than *M. sacchariflorus* [37] (Fig. 1). Although these characteristics could make *M. sinensis* attractive for cultivation across a wide area, species that have wider native distributions and occur in more

Table 5. The index of agreement[a] between different models run under the same scenario and between different scenarios run within the same model, for projected distributions of *M. sacchariflorus* and *M. sinensis* in 2050 and 2080.

Model agreement:	2050	2080
M. sacchariflorus–A2	0.58	0.44
M. sacchariflorus–B1	0.64	0.63
M. sinensis–A2	1.17	0.96
M. sinensis–B1	1.21	1.22
Scenario agreement:		
M. sacchariflorus–BCM	13.68	3.23
M. sacchariflorus–CGCM	5.28	1.33
M. sinensis–BCM	16.35	6.25
M. sinensis–CGCM	11.74	3.25
Model average:	0.90	0.81
Scenario average:	11.76	3.51

[a]The index is calculated as the ratio of the area of overlap between the two projections to the area of non-overlap.

habitats and climate zones are likely to be more successful as invaders, regardless of other biological traits [71]. Thus, *M. sinensis* might also have greater potential to become a weedy or invasive plant than does *M. sacchariflorus*. Rather than developing one or two cultivars suitable for widespread production, a best management practice might be to develop regionally restricted cultivars to minimize widespread escape and invasion of novel species from the agriculture and horticulture trades.

Consequences of Parameter Sensitivity

The smaller native distribution and therefore narrower environmental tolerances of *M. sacchariflorus* likely contribute to its greater sensitivity to changes in model parameters than for *M. sinensis*. If the extent of the native distribution is correlated with model sensitivity across a range of species, this would have implications for modelling and interpreting models for both invasive species and rare species of conservation concern. For potential invaders, this might mean that some priority is given to species with wide distributions. For rare species, i.e., those with small native distributions that are not due to anthropogenically caused local extinctions, obtaining accurate model results could require greater accuracy in parameter estimation.

The two *Miscanthus* species models showed sensitivity to similar parameters, which might not be surprising, given that their distributions overlap in temperate areas. However, the main sets of parameters exhibiting sensitivity were those for which there are the least data. The most sensitive parameters were related to upper temperatures and heat tolerance, but most studies of temperature-related growth for these species have examined cold tolerance because of interest in their cultivation at northern latitudes (e.g., [39], [38]). Physiological heat thresholds remain to be explored for these species to improve the confidence of lower-latitude thresholds for growth in the northern hemisphere, where they have been introduced, as well as potential range contractions at lower latitudes under climate change.

Similarly, although weakly sensitive and thus potentially of lesser importance the upper temperature parameters, upper soil moistures and moisture tolerance have rarely been examined. Most studies of soil moisture effects for these species examine drought, rather than saturation (e.g., [45], but see [50]). The

accuracy of these parameter estimates could be important in predicting potential invasion of these species into drainage ditches, riparian areas, and wetlands.

Most stress rate parameters were relatively insensitive to changes in value. CLIMEX determines stress as the annual exponential accumulation of weekly population reduction when a stress threshold is exceeded [34]. Stress accumulation rate is difficult to estimate empirically without extensive field or laboratory trials under various stress thresholds, and the magnitude of the accumulation rate could depend on the threshold value chosen; for many species, few data of this type likely exist [72]. However, the minimal sensitivity of the stress rate parameters implies that their accuracy is less influential than that of other, more easily estimated parameters.

Two previous tests of sensitivity using CLIMEX have some similarities. A study of the invasive tropical/subtropical shrub *Lantana camara* identified model sensitivity to limiting low and high temperatures and limiting low soil moisture [27]. A study of the invading pathogen *Phytophthora ramorum* identified model sensitivity to optimal high temperature and limiting and optimal low soil moisture [26]. However, neither study tested model sensitivity to the stress rate parameters. Nevertheless, both our and their models show high sensitivity to some of the limiting upper or lower temperature and moisture parameters. Sensitivity analyses should be performed for additional species to determine whether some parameters are consistently more sensitive than others. If sensitivity to specific parameters is consistent within biomes and species types (forb, shrub, etc.), researchers could focus their efforts on measuring those specific environmental tolerances to maximize model estimation accuracy.

Future Climate Projections

Although the climatically suitable area for the two *Miscanthus* species is projected to decrease globally with climate change, areas of North America, eastern Europe, and Scandinavia are projected to experience some future increase in suitable climate. This could be beneficial for cultivating these species as bioenergy crops in these regions if suitable habitat is available. However, it could also place these regions at greater risk of invasion through increases in the area of suitable climate outside of cultivation. These regions, in particular, are projected to be future hotspots of invasion for 99 of the worst invaders globally [73]. Areas of range contraction for the two *Miscanthus* species also coincide with areas where future invasion is projected to decrease [73].

Additionally, climate niche projections do not account for the potential that rapid evolution in introduced species and their recipient communities could allow species to become invasive beyond their current tolerances (e.g., [74], [75], [76], [77]). In these *Miscanthus* species, rapid evolution could be aided by the introduction of new horticultural genotypes from widely separated populations in the native range [78]. These species are obligate out-crossers [79], so isolated populations composed of a single clone do not produce seed. Introduction of different genotypes could increase the probability of sexual reproduction and long-distance spread via the wind-dispersed seed [80]. Similarly, if these plants are developed for biomass production, intensive plant breeding programs will aim to improve their performance under a variety of conditions, including resistance to pests and disease, drought and heat tolerance, cold tolerance, and possibly salinity tolerance [81]. These efforts will potentially expand the plants' realized distributions as well as their invasive potential.

Variation in the area of future projected climate was greater among climate models than among emissions scenarios. This has also been found previously, when quantified, for both plant (e.g., [82], [83], [84]) and insect (e.g., [31], [32]) species under a number of different model-scenario combinations. Coupled with the observation that some currently observed climate changes might be greater than those predicted by even the highest emissions scenarios [85], this result suggests that future bioclimatic envelope model projections should focus efforts on increasing the number of models compared using one high-emissions scenario to develop composite projections of future suitable climate areas.

Acknowledgments

We thank Emily Robinson-Berzitis and Filip Cybula for formulating the climate model data. We also thank two referees for helpful comments on an earlier version of the manuscript.

Author Contributions

Conceived and designed the experiments: HAH JAN. Performed the experiments: HAH SES. Analyzed the data: HAH SES ZG. Wrote the paper: HAH SES.

References

1. Mack R, Erneberg M (2002) The United States naturalized flora: largely the product of deliberate introductions. Ann Mo Bot Gard 89: 176–189.
2. Williamson M, Fitter A (1996) The varying success of invaders. Ecology 77: 1661–1666.
3. Pheloung PC, Williams PA, Halloy SR (1999) A weed risk assessment model for using as a biosecurity tool evaluating plant introductions. J Environ Manage 57: 239–251.
4. IPPC (2007) International Standards for Phytosanitary Measures: Framework for Pest Risk Analysis. ISPM No 2. Rome: Food and Agriculture Organization.
5. Koop A, Fowler L, Newton L, Caton B (2012) Development and validation of a weed screening tool for the United States. Biol Invasions 14: 273–294.
6. Baker RHA, Benninga J, Bremmer J, Brunel S, Dupin M, et al. (2012) A decision-support scheme for mapping endangered areas in pest risk analysis. Bulletin OEPP/EPPO Bulletin 42: 65–73.
7. Ewel JJ, O'Dowd DJ, Bergelson J, Daehler CC, D'Antonio CM, et al. (1999) Deliberate introductions of species: research needs - benefits can be reaped, but risks are high. Bioscience 49: 619–630.
8. Raghu S, Anderson RC, Daehler CC, Davis AS, Wiedenmann RN, et al. (2006) Adding biofuels to the invasive species fire? Science 313: 1742–1742.
9. Venette RC, Kriticos DJ, Magarey RD, Koch FH, Baker RHA, et al. (2010) Pest risk maps for invasive alien species: a roadmap for improvement. Bioscience 60: 349–362.
10. Atkinson CJ (2009) Establishing perennial grass energy crops in the UK: a review of current propagation options for *Miscanthus*. Biomass Bioenergy 33: 752–759.
11. Quinn LD, Allen DJ, Stewart JR (2010) Invasiveness potential of *Miscanthus sinensis*: implications for bioenergy production in the United States. Glob Change Biol Bioenergy 2: 310–320.
12. Pohl RW (1963) Phytogeographic notes on *Rottboellia*, *Paspalum*, and *Miscanthus* (Gramineae). Rhodora 65: 146–147.
13. Dore WG, McNeill J (1980) Grasses of Ontario. Agriculture Canada Monograph 26. Hull, Quebec, Canada: Canadian Government Publishing Centre.
14. Nishiwaki A, Mizuguti A, Kuwabara S, Toma Y, Ishigaki G, et al. (2011) Discovery of natural *Miscanthus* (Poaceae) triploid plants in sympatric populations of *Miscanthus sacchariflorus* and *Miscanthus sinensis* in southern Japan. Am J Bot 98: 154–159.
15. Dougherty RF, Quinn LD, Endres AB, Voigt TB, Barney JN (2014) Natural history survey of the ornamental grass *Miscanthus sinensis* in the introduced range. Invasive Plant Sci Manag 7: in press.
16. Clifton-Brown JC, Lewandowski I, Andersson B, Basch G, Christian DG, et al. (2001) Performance of 15 *Miscanthus* genotypes at five sites in Europe. Agron J 93: 1013–1019.
17. Jorgensen U, Mortensen J, Kjeldsen JB, Schwarz KU (2003) Establishment, development and yield quality of fifteen *Miscanthus* genotypes over three years in Denmark. Acta Agric Scand B Soil Plant Sci 53: 190–199.
18. Pyter R, Heaton E, Dohleman F, Voigt T, Long S (2009) Agronomic experiences with *Miscanthus x giganteus* in Illinois, USA. In Mielenz J, editor. Biofuels: methods and protocols. New York: Springer. 41–52.
19. Barney JN, DiTomaso JM (2008) Nonnative species and bioenergy: are we cultivating the next invader? Bioscience 58: 64–70.
20. Holdridge LR (1947) Determination of world plant formations from simple climatic data. Science 105: 367–368.
21. Woodward FI (1987) Climate and plant distribution. Cambridge: Cambridge University Press.
22. Gordon DR, Gantz CA (2008) Screening new plant introductions for potential invasiveness: a test of impacts for the United States. Conservation Letters 1: 227–235.
23. Scott JK, Panetta FD (1993) Predicting the Australian weed status of southern African plants. J Biogeog 20: 87–93.
24. Sutherst RW, Maywald GF (1985) A computerised system for matching climates in ecology. Agric Ecosyst Environ 13: 281–299.
25. Beerling DJ, Huntley B, Bailey JP (1995) Climate and the distribution of *Fallopia japonica*: use of an introduced species to test the predictive capacity of response surfaces. J Veg Sci 6: 269–282.
26. Venette RC, Cohen SD (2006) Potential climatic suitability for establishment of *Phytophthora ramorum* within the contiguous United States. For Ecol Manage 231: 18–26.
27. Taylor S, Kumar L (2012) Sensitivity analysis of CLIMEX parameters in modelling the potential distribution of *Lantana camara* L. Plos One 7(7): 1–16.
28. Sutherst RW, Baker RHA, Coakley SM, Harrington R, Kriticos DJ, et al. (2007) Pests under global change - meeting your future landlords? In: Canadell JG, Pataki DE, Pitelka LF, editors. Terrestrial ecosystems in a changing world. Berlin: Springer. 211–223.
29. Walther G, Roques A, Hulme PE, Sykes MT, Pysek P, et al. (2009) Alien species in a warmer world: risks and opportunities. Trends Ecol Evol 24: 686–693.
30. Dukes JS, Mooney HA (1999) Does global change increase the success of biological invaders? Trends Ecol Evol 14: 135–139.
31. Mika AM, Weiss RM, Olfert O, Hallett RH, Newman JA (2008) Will climate change be beneficial or detrimental to the invasive swede midge in North America? Contrasting predictions using climate projections from different general circulation models. Glob Change Biol 14: 1721–1733.
32. Mika AM, Newman JA (2010) Climate change scenarios and models yield conflicting predictions about the future risk of invasive species in North America. Agric For Entomol 12: 213–221.
33. Commonwealth Scientific and Industrial Research Organisation (2004) Dymex simulator application 2.0. South Yarra, Australia: Hearn Scientific Software.
34. Sutherst RW, Maywald GF, Kriticos DJ (2006) CLIMEX Version 3 User's Guide. Melbourne: CSIRO.
35. New M, Hulme M, Jones P (1999) Representing twentieth-century space-time climate variability. Part I: development of a 1961–90 mean monthly terrestrial climatology. J Clim 12: 829–856.
36. Sutherst RW (2003) Prediction of species geographical ranges. J Biogeog 30: 805–816.
37. Hodkinson TR, Chase MW, Lledo MD, Salamin N, Renvoize SA (2002) Phylogenetics of *Miscanthus*, *Saccharum* and related genera (Saccharinae, Andropogoneae, Poaceae) based on DNA sequences from ITS nuclear ribosomal DNA and plastic trnL intron and trnL-F intergenic spacers. J Plant Res 115: 381–392.
38. Yan J, Chen W, Lou F, Ma H, Meng A, et al. (2011) Variability and adaptability of *Miscanthus* species evaluated for energy crop domestication. Glob Change Biol Bioenergy 4: 49–60.

39. Farrell AD, Clifton-Brown JC, Lewandowski I, Jones MB (2006) Genotypic variation in cold tolerance influences the yield of *Miscanthus*. Ann Appl Biol 149: 337–345.

40. Naidu SL, Long SP (2004) Potential mechanisms of low-temperature tolerance of C4 photosynthesis in *Miscanthus* x *giganteus*: an in vivo analysis. Planta 220: 145–155.

41. Farage PK, Blowers D, Long SP, Baker NR (2006) Low growth temperatures modify the efficiency of light use by photosystem II for CO2 assimilation in leaves of two chilling-tolerant C4 species, *Cyperus longus* L. and *Miscanthus* x *giganteus*. Plant Cell Environ 29: 720–728.

42. Quinn LD, Stewart JR, Yamada T, Toma Y, Saito M, et al. (2012) Environmental tolerances of *Miscanthus sinensis* in invasive and native populations. Bioenerg Res 5(1): 139–148.

43. Naidu SL, Moose SP, Al-Shoaibi AK, Raines CA, Long SP (2003) Cold tolerance of C$_4$ photosynthesis in *Miscanthus* x *giganteus*: adaptation in amounts and sequence of C$_4$ photosynthetic enzymes. Plant Physiol 132: 1688–1697.

44. Clifton-Brown J, Lewandowski I (2000) Overwintering problems of newly established *Miscanthus* plantations can be overcome by identifying genotypes with improved rhizome cold tolerance. New Phytol 148: 287–294.

45. Clifton-Brown JC, Lewandowski I, Bangerth F, Jones MB (2002) Comparative responses to water stress in stay-green, rapid- and slow senescing genotypes of the biomass crop, *Miscanthus*. New Phytol 154: 335–345.

46. Daubenmire RF (1974) Plants and environment: a textbook of plant autecology. Sydney, Australia: Wiley.

47. Kriticos DJ, Sutherst RW, Brown JR, Adkins SW, Maywald GF (2003) Climate change and biotic invasions: a case history of a tropical woody vine. Biol Invasions 5: 145–165.

48. Hastings A, Clifton-Brown J, Wattenbach M, Mitchell CP, Smith P (2009) The development of MISCANFOR, a new *Miscanthus* crop growth model: towards more robust yield predictions under different climatic and soil conditions. Glob Change Biol Bioenergy 1: 154–170.

49. Yamasaki S (1990) Potential dynamics in overlapping zones of *Phragmites australis* and *Miscanthus sacchariflorus*. Aquat Bot 36: 367–377.

50. Li F, Qin X, Xie Y, Chen X, Hu J, et al. (2013) Physiological mechanisms for plant distribution pattern: responses to flooding and drought in three wetland plants from Dongting Lake, China. Limnology 14: 71–76.

51. Sutherst RW, Maywald GF (2005) A climate model of the red imported fire ant, *Solenopsis invicta* Buren (Hymenoptera: Formicidae): implications for invasion of new regions, particularly Oceania. Environ Entomol 34(2): 317–335.

52. Randall DA, Wood RA, Bony S, Colman R, Fichefet T, et al. (2007) Climate models and their evaluation. In Solomon S, Qin D, Manning M, Chen Z, Marquis M, Averyt KB, Tignor M, Miller HL, editors. Climate Change 2007: The Physical Science Basis. Contribution of Working Group I to the Fourth Assessment Report of the Intergovernmental Panel on Climate Change. Cambridge: Cambridge University Press. 589–662.

53. IPCC (2000) IPCC Special Report: emissions scenarios. Summary for policy makers. Geneva: Intergovernmental Panel on Climate Change.

54. Guisan A, Thuiller W (2005) Predicting species distribution: offering more than simple habitat models. Ecol Lett 8: 993–1009.

55. Heikkinen RK, Luoto M, Araujo MB, Virkkala R, Thuiller W, et al. (2006) Methods and uncertainties in bioclimatic envelope modelling under climate change. Prog Phys Geogr 30: 751–777.

56. Araújo MB, Peterson AT (2012) Uses and misuses of bioclimatic envelope modeling. Ecology 93: 1527–1539.

57. Horton JL, Fortner R, Goklany M (2010) Photosynthetic characteristics of the C4 invasive exotic grass *Miscanthus sinensis* Andersson growing along gradients of light intensity in the southeastern United States. Castanea 75: 52–66.

58. Keane R, Crawley M (2002) Exotic plant invasions and the enemy release hypothesis. Trends Ecol Evol 17: 164–170.

59. Ross JL, Ivanova ES, Severns PM, Wilson MJ (2010) The role of parasite release in invasion of the USA by European slugs. Biol Invasions 12: 603–610.

60. Schaffner U, Ridenour WM, Wolf VC, Bassett T, Mueller C, et al. (2011) Plant invasions, generalist herbivores, and novel defense weapons. Ecology 92: 829–835.

61. Parker MA (2001) Mutualism as a constraint on invasion success for legumes and rhizobia. Divers Distrib 7: 125–136.

62. Broennimann O, Treier UA, Mueller-Schaerer H, Thuiller W, Peterson AT, et al. (2007) Evidence of climatic niche shift during biological invasion. Ecol Lett 10: 701–709.

63. Gallagher RV, Beaumont LJ, Hughes L, Leishman MR (2010) Evidence for climatic niche and biome shifts between native and novel ranges in plant species introduced to Australia. J Ecol 98: 790–799.

64. Guo Q, Sax DF, Qian H, Early R (2012) Latitudinal shifts of introduced species: possible causes and implications. Biol Invasions 14: 547–556.

65. Petitpierre B, Kueffer C, Broennimann O, Randin C, Daehler C, et al. (2012) Climatic niche shifts are rare among terrestrial plant invaders. Science 335: 1344–1348.

66. Kearney M, Porter W (2009) Mechanistic niche modelling: combining physiological and spatial data to predict species' ranges. Ecol Lett 12: 334–350.

67. Kearney MR, Wintle BA, Porter WP (2010) Correlative and mechanistic models of species distribution provide congruent forecasts under climate change. Conserv Lett 3: 203–213.

68. Flory SL, Lorentz KA, Gordon DR, Sollenberger LE (2012) Experimental approaches for evaluating the invasion risk of biofuel crops. Environ Res Lett 7: 045904.

69. Pysek P, Sadlo J, Mandak B, Jarosik V (2003) Czech alien flora and the historical pattern of its formation: what came first to Central Europe? Oecologia 135: 122–130.

70. Huang QQ, Qian C, Wang Y, Jia X, Dai XF, et al. (2010) Determinants of the geographical extent of invasive plants in China: effects of biogeographical origin, life cycle and time since introduction. Biodivers Conserv 19: 1251–1259.

71. Pysek P, Jarosik V, Pergl J, Randall R, Chytry M, et al. (2009) The global invasion success of Central European plants is related to distribution characteristics in their native range and species traits. Divers Distrib 15: 891–903.

72. Sutherst RW, Maywald GF, Bourne AS (2007) Including species interactions in risk assessments for global change. Global Change Biol 13: 1843–1859.

73. Bellard C, Thuiller W, Leroy B, Genovesi P, Bakkenes M, et al. (2013) Will climate change promote future invasions? Global Change Biol 19: 3740–3748.

74. Broennimann O, Treier UA, Mueller-Schaerer H, Thuiller W, Peterson AT, et al. (2007) Evidence of climatic niche shift during biological invasion. Ecol Lett 10: 701–709.

75. Whitney KD, Gabler CA (2008) Rapid evolution in introduced species, 'invasive traits' and recipient communities: challenges for predicting invasive potential. Divers Distrib 14: 569–580.

76. Xu C, Julien MH, Fatemi M, Girod C, Van Klinken RD, et al. (2010) Phenotypic divergence during the invasion of *Phyla canescens* in Australia and France: evidence for selection-driven evolution. Ecol Lett 13: 32–44.

77. Sultan SE, Horgan-Kobelski T, Nichols LM, Riggs CE, Waples RK (2013) A resurrection study reveals rapid adaptive evolution within populations of an invasive plant. Evol Appl 6: 266–278.

78. Lavergne S, Molofsky J (2007) Increased genetic variation and evolutionary potential drive the success of an invasive grass. Proc Natl Acad Sci U S A 104: 3883–3888.

79. Heaton EA, Dohleman FG, Miguez AF, Juvik JA, Lozovaya V, et al. (2010) *Miscanthus*: a promising biomass crop. In: Kader JC, Delseny M, editors. Advances in botanical research. Volume 56. London: Academic Press. 76–138.

80. Quinn LD, Matlaga DP, Stewart JR, Davis AS (2011) Empirical evidence of long-distance dispersal in *Miscanthus sinensis* and *Miscanthus* x *giganteus*. Invasive Plant Sci Manag 4: 142–150.

81. Chapman SC, Chakraborty S, Dreccer MF, Howden SM (2012) Plant adaptation to climate change-opportunities and priorities in breeding. Crop Pasture Sci 63: 251–268.

82. Barney JN, DiTomaso JM (2010) Bioclimatic predictions of habitat suitability for the biofuel switchgrass in North America under current and future climate change scenarios. Biomass Bioenergy 34: 124–133.

83. Kriticos DJ, Watt MS, Potter KJB, Manning LK, Alexander NS, et al. (2011) Managing invasive weeds under climate change: considering the current and potential future distribution of *Buddleja davidii*. Weed Res 51: 85–96.

84. Bourdôt G, Lamoureaux S, Watt M, Manning L, Kriticos D (2012) The potential global distribution of the invasive weed *Nassella neesiana* under current and future climates. Biol Invasions 14: 1545–1556.

85. Rahmstorf S, Cazenave A, Church JA, Hansen JE, Keeling RF, et al. (2007) Recent climate observations compared to projections. Science 316: 709–709.

7

Avoided Heat-Related Mortality through Climate Adaptation Strategies in Three US Cities

Brian Stone Jr[1]*, Jason Vargo[2], Peng Liu[3], Dana Habeeb[1], Anthony DeLucia[4], Marcus Trail[3], Yongtao Hu[3], Armistead Russell[3]

1 School of City and Regional Planning, Georgia Institute of Technology, Atlanta, Georgia, United States of America, **2** Center for Sustainability and the Global Environment, University of Wisconsin-Madison, Madison, Wisconsin, United States of America, **3** School of Civil and Environmental Engineering, Georgia Institute of Technology, Atlanta, Georgia, United States of America, **4** Quillen College of Medicine, East Tennessee State University, Johnson City, Tennessee, United States of America

Abstract

Heat-related mortality in US cities is expected to more than double by the mid-to-late 21st century. Rising heat exposure in cities is projected to result from: 1) climate forcings from changing global atmospheric composition; and 2) local land surface characteristics responsible for the urban heat island effect. The extent to which heat management strategies designed to lessen the urban heat island effect could offset future heat-related mortality remains unexplored in the literature. Using coupled global and regional climate models with a human health effects model, we estimate changes in the number of heat-related deaths in 2050 resulting from modifications to vegetative cover and surface albedo across three climatically and demographically diverse US metropolitan areas: Atlanta, Georgia, Philadelphia, Pennsylvania, and Phoenix, Arizona. Employing separate health impact functions for average warm season and heat wave conditions in 2050, we find combinations of vegetation and albedo enhancement to offset projected increases in heat-related mortality by 40 to 99% across the three metropolitan regions. These results demonstrate the potential for extensive land surface changes in cities to provide adaptive benefits to urban populations at risk for rising heat exposure with climate change.

Editor: Igor Linkov, US Army Engineer Research and Development Center, United States of America

Funding: This research was made possible by a grant from the US Centers for Disease Control and Prevention (CDC) through project #5U01EH000432-02. Although the research described in the article has been funded wholly or in part by the CDC, it has not been subjected to any CDC review and therefore does not necessarily reflect the views of the Center, and no official endorsement should be inferred. The funders had no role in study design, data collection and analysis, decision to publish, or preparation of the manuscript.

Competing Interests: The authors have declared that no competing interests exist.

* Email: stone@gatech.edu

Introduction

Human health effects associated with rising temperatures are expected to increase significantly by mid-to-late century. A large body of work now estimates an increase in mean global temperature from pre-industrial averages of more than 2°C by late century under mid-range emissions scenarios [1]. A smaller but growing body of work has sought to estimate the effects of projected warming on heat-related mortality. Employing health impact functions derived from epidemiological studies of historical warm season mortality rates, recent work projects an increase in annual heat-related mortality of between 3,500 and 27,000 deaths in the United States by mid-century [2]. Studies focused on individual cities estimate an increase in annual heat-related mortality by a factor of 2 to 7 by the mid-to-late 21st century [3–5].

The urban heat island effect compounds the potential effects of global scale climate change on heat-related mortality among urban populations. Time series analyses of climatic trends in cities find large urbanized regions to be warming at a higher rate than proximate rural areas, with many cities warming at more than twice the mean global rate [6,7]. The combined effects of urban heat island formation and the global greenhouse effect are projected to significantly increase the number of extreme heat events in urbanized regions [8]. At present, the extent to which the urban heat island effect may further increase heat-related mortality is not well established.

Here we examine the potential for urban heat island mitigation as a climate adaptation strategy to reduce projected heat-related mortality in three large US cities by mid-century. Future year climate and seasonal mortality are modeled across the metropolitan statistical areas (MSAs) of Atlanta, Georgia, Philadelphia, Pennsylvania, and Phoenix, Arizona to capture a wide continuum of climatic, geographic, and demographic characteristics known to underlie population vulnerability to extreme heat. Using coupled global and regional scale climate models together with an environmental health effects model, we project the number of heat-related deaths expected for these regions in 2050 in response to a "business as usual" (BAU) and an array of urban heat management scenarios characterized by variable land cover modifications. Employing separate health impact functions responsive to temperature change and derived from prior epidemiological studies, referred to herein as "heat response functions" (HRFs), we find different combinations of heat management strategies to offset projected increases in heat-related mortality across the three MSAs by a range of 40 to 99%.

Our work builds on previous studies of climate change and heat-related mortality in three respects. First, we develop a set of climate projections responsive not only to future changes in atmospheric composition but to changes in land cover characteristics as well to capture the influence of heat island formation on heat-related health outcomes. Second, in addition to estimating changes in heat-related mortality resulting from future year climatic conditions, we further model the influence of alternative heat management strategies on health outcomes. Third, we introduce a modeling approach that enables health outcomes resulting from mean warm season temperatures and shorter-term heat wave events to be estimated by employing multiple HRFs.

Methods

Land cover modeling

The influence of local climate modification on heat-related mortality was estimated through the integration of separate land cover, climate, and human health effects models. To account for separate global and regional climate forcings on future climate, our approach made use of a land cover modeling routine responsive to historical rates of land cover change. As presented in an earlier paper [9], historical land cover change rates by urbanization class were developed for each metropolitan region from the National Land Cover Database [10] and projected forward based on population projections for each decade from 2010 to 2050.

Employing this approach, the type and area of vegetative (tree canopy, grass, shrubland, agriculture) and impervious (building roofing, street paving, other surface paving) land cover were estimated across the three metropolitan regions for a 2050 BAU scenario linking land cover change to population growth at the census tract level. Under the BAU scenario, historical rates of land cover change per unit of population growth, such as the area of forestland lost with each new 10,000 residents added to a census tract, are expected to continue into the future. Based on projected population growth per census tract between 2010 and 2050, we estimated how seven classes of land cover, including forest, grass, barren land, water, wetlands, agriculture, and impervious surfaces, are expected to change by 2050. These tract-level land cover values were then aggregated to a 4 km model grid resolution and used as land surface inputs to a mesoscale meteorological model.

County level population data were obtained for all counties in the Atlanta, Philadelphia, and Phoenix MSAs from the economic forecasting firm *Woods & Poole* for the years 2010 to 2040 and then extrapolated to 2050 with a statistical routine employing ordinary least squares. County level population estimates stratified by age cohorts and race/ethnicity were then disaggregated to the census tract level with a spatial weighting algorithm. Figure 1 (top panel) illustrates the percent change in population between 2010 and 2050 in each MSA. These maps project Atlanta and Phoenix to grow more rapidly than Philadelphia, and for expected growth to continue to largely occur in suburban zones. The population projections used in this study are publicly available through the following URL: http://doi.org/10.5061/dryad.14g40.

Global and regional climate modeling

Future year climatic conditions were simulated though the coupling of the Weather Research Forecasting (WRF) mesoscale meteorological model to the Goddard Institute for Space Studies (GISS) Global Atmosphere-Ocean ModelE (the GISS-WRF model system) [11–13]. The coarse resolution ($2° \times 2.5°$) meteorological fields of the GISS ModelE global circulation model (GCM) were dynamically downscaled to 36 km, after which each metropolitan study region was nested and downscaled to 12 and 4 km to capture fine scale land use impacts and temperature

variations. Energetic parameters in WRF were set to correspond with either base year (2010) or projected 2050 land cover attributes in each region, aggregated to the 4 km grid resolution. Simulated meteorological fields of interest included hourly ambient temperature (2 m), surface skin temperature, ambient humidity (2 m), surface heat fluxes (latent, sensible, and radiation), precipitation, wind velocities, and turbulence.

The GCM projections used for this study were obtained from a previous study by Trail and others [14]. Through this previous study, the performance of the GISS model in reproducing observed warm season temperatures across the continental US over the period of 2006 to 2010 was assessed and found to have a moderate cool bias of $-1.9°C$, with regional mean biases of $-0.2°C$, $-2.0°C$, and $-4.2°C$, in the south, northeast, and west, respectively. These results compare favorably with other studies making use of the GISS model [15,16].

Meteorological fields corresponding to one emissions and eight land cover scenarios were simulated through WRF for the full year of 2050. In order to assess the extent to which modeled temperature changes fall beyond the range of natural variability captured in the WRF model, we generated a five-member ensemble of the BAU scenario by modifying the initial conditions in each simulation. The results of this ensemble modeling yielded standard errors averaging $0.112°C$ across the three temperature metrics in Atlanta, $0.018°C$ in Philadelphia, and $0.004°C$ in Phoenix. Supplemental Figure S1 and Table S1 summarize in more detail the results of these ensemble simulations.

While it is a conventional practice to average future year climate estimates over successive years, we made use of a single year of simulated meteorology in this study to capture heat wave episodes that may have been moderated or lost from a statistical smoothing of temperature projections across multiple years. A comparison of GISS model results across the five years of 2048 through 2052 did not find 2050 to be a statistically anomalous year for any of the meteorological fields of interest [14]. In each case, the Intergovernmental Panel on Climate Change (IPCC) RCP4.5 emissions scenario was used for the GISS model runs, corresponding to "middle of the road" emissions assumptions. As our interest in this study is on the influence of alternative land cover change scenarios on future regional climates, rather than on the influence of alternative global emissions scenarios, we make use of a single emissions forecast and then assess how temperature and heat-related mortality vary in response to changing land cover conditions at the metropolitan scale. The RCP4.5 emissions scenario was selected for the GCM projection as it represents middle range assumptions pertaining to emissions growth and radiative forcing.

The influence of alternative heat management scenarios was assessed through separate WRF runs parameterized around variable land cover assumptions. These WRF runs included two scenarios focused on vegetation enhancement differentiated by property type (private vs. public land parcels), two scenarios focused on albedo enhancement differentiated by material type (roofing vs. surface paving), and three scenarios employing varying combinations of these heat management strategies. Table 1 presents a detailed overview of each heat management scenario included in the study. All WRF model output is publicly available through the following URL: http://doi.org/10.5061/dryad.14g40.

To date, numerous observational and modeling studies have found vegetative cover and high albedo roofing and paving materials – referred to generally as "cool materials" – to be associated with lower surface and near-surface air temperatures than sparsely vegetated areas with low albedo impervious

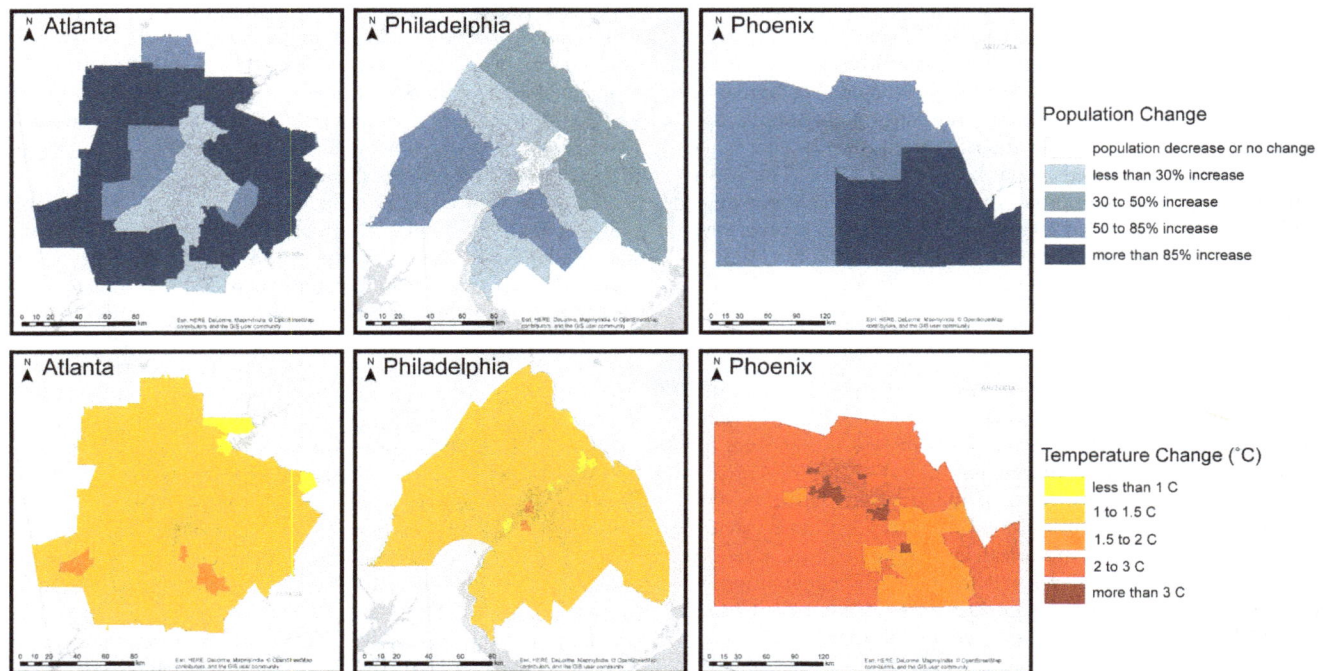

Figure 1. Change in population and average warm season (May – Sept) temperature under the BAU scenario between 2010 and 2050.

materials [17–19]. In some studies the cooling benefits of vegetative and high albedo materials are shown to extend beyond the zones in which these materials are found [20]. The climatic benefits of these land cover types result from either an increase in local rates of evapotranspiration, which offsets sensible heating at the surface, or through an increase in shortwave reflection, reducing the absorption of solar energy [21]. The heat management strategies outlined in Table 1 are designed to increase

Table 1. Description of 2050 heat management WRF simulations.

Scenario definition	Type of land cover modified	Type of modification
Private Greening (PRG)		
trees, grass, and/or shrubs added to private property to achieve minimum 80% green cover for residential parcels and minimum 50% green cover for non-residential parcels	building roof and surface paving (e.g., driveways and sidewalks) on private parcels	1. All commercial roofs converted to grass. 2. Surface paving and building roofs overlaid with tree canopy or converted to grass/shrubs to achieve green area minimum by land use class
Public Greening (PUG)		
trees, grass, and/or shrubs added to public property to achieve or approach green area minimum of 80% for publicly owned land	street surfaces, parkland, and other publicly owned parcels	1. 50% of all roadway surfaces overlaid by tree canopy. 2. Grass or barren land in public parcels converted to tree canopy (ATL and PHL). 3. Barren or agricultural land in public parcels converted to a grass/shrub mix (PHX).
Building Albedo Enhancement (BAE)		
increase the albedo of building roof surfaces	all building roofs	converted to high albedo (0.9) impervious surfaces
Road Albedo Enhancement (RAE)		
increase the albedo of paved surfaces	all roads, parking lots, and other surface paving	converted to moderate albedo (0.45) impervious surfaces
Combined Green Strategies (GREEN)		
combination of PRG and PUG scenarios	see PRG and PUG scenarios	see PRG and PUG scenarios
Combined Albedo Strategies (ALBEDO)		
combination of BAE and RAE scenarios	see BAE and RAE scenarios	see BAE and RAE scenarios
All Strategies Combined (ALL)		
combination of PRG, PUG, BAE, RAE scenarios	see PRG, PUG, BAE, and RAE scenarios	for areas subject to either vegetation or albedo enhancement, vegetation enhancement is prioritized

vegetative and high albedo materials separately and in combination to assess the relative impacts of each type of intervention, as well as to establish feasible land cover change goals achievable through municipal land use policies. Consistent with environmental zoning policies recently adopted in Seattle, WA and Washington, DC [22], several of our scenarios set minimum green area targets per parcel to bring about land cover changes associated with urban heat island mitigation.

To assess the potential for municipal governments to reduce urban temperatures through strategies focused on publicly managed land alone, we further stratify our vegetation and albedo enhancement scenarios by land ownership class. As such, the Public Greening (PUG) and Road Albedo Enhancement (RAE) scenarios modify the land cover characteristics of publicly owned street surfaces, parks, and other publicly owned parcels, while the Private Greening (PRG) and Building Albedo Enhancement (BAE) scenarios modify the land cover characteristics of privately owned parcels across each region. The combined scenarios of GREEN, ALBEDO, and ALL enable the assessment of vegetation and albedo enhancement alone or in combination across all land ownership types. Figure 2 illustrates the percent area of each census tract modified through the heat management strategies included in the ALL scenario, showing a concentration of these strategies in the highest density tracts where impervious cover is greatest.

Health effects modeling

The implications of each land cover change scenario for heat-related mortality were assessed with an environmental health effects model developed by the US Environmental Protection Agency, the *Environmental Benefits Mapping and Analysis Program* (BenMAP). Constructed as a damage function estimation tool, the BenMAP model estimates excess mortality resulting from heat exposure through the following function:

$$\Delta y = y_o \cdot (e^{\beta \cdot \Delta x} - 1) \cdot Pop$$

where Δy is the estimated change in mortality, y_o is the baseline incidence rate for a specified cause of mortality, β is a measure of relative risk obtained from surveillance studies of heat mortality, Δx is the estimated change in heat exposure, and Pop is the size of the exposed population.

Initially developed for the purpose of estimating health benefits associated with air pollution control strategies, the BenMAP model

has recently been adapted to treat heat exposure as a pollutant through the integration of published impact functions for temperature [2]. In this study, we employ an approach similar to that developed by Voorhees and others [2], which estimated national excess heat-related mortality associated with projected global temperature change between the present period and 2050. The BenMAP model offers two important benefits to our study. First, its use for the estimation of climate change-related health effects has been validated through an initial "proof of concept" paper focused on the same period of projection. Second, the BenMAP model provides a standardized set of present-day and future year estimates of baseline incidence (y_o – the number of deaths expected to occur independent of changes in temperature) for all causes of mortality at the county level across the US. As such, use of the BenMAP model for climate and health research enhances the comparability of different studies focused on the health risks of temperature change over time.

We selected from the epidemiology literature three published measures of relative risk (β) quantifying excess mortality above baseline incidence rates per unit change in temperature or per heat wave day. Separate HRFs for average warm season mortality were modeled to account for differing temperature-mortality associations established in response to alternative temperature metrics. As presented in Table 2, HRFs developed by Medina-Ramon and Schwartz [23] and Zanobetti and Schwartz [24] are responsive to differences between scenarios in either average daily minimum temperatures ("Medina-Ramon HRF") or average daily mean apparent temperatures ("Zanobetti HRF") over a May through September warm season.

Prior work has found HRFs constructed from seasonal or annual health surveillance data to underestimate heat-related mortality during episodic heat wave conditions [25,26]. To capture the additive effect of heat wave conditions on mortality rates, we make use of a secondary HRF developed by Anderson and Bell [26], which measures the additional risk of heat exposure during consecutive days of extreme temperatures. The "Anderson HRF" measures the increase in mortality per heat wave day, defined as one of two or more consecutive days in which the mean temperature exceeds the 95th percentile of the long-term (1987–2005) daily mean temperature for each MSA. Mean temperature thresholds of 28.7°C, 28.8°C, and 37.3°C were used to identify heat wave periods in Atlanta, Philadelphia, and Phoenix, respectively (temperature thresholds provided by Dr. Brooke Anderson, December 2013).

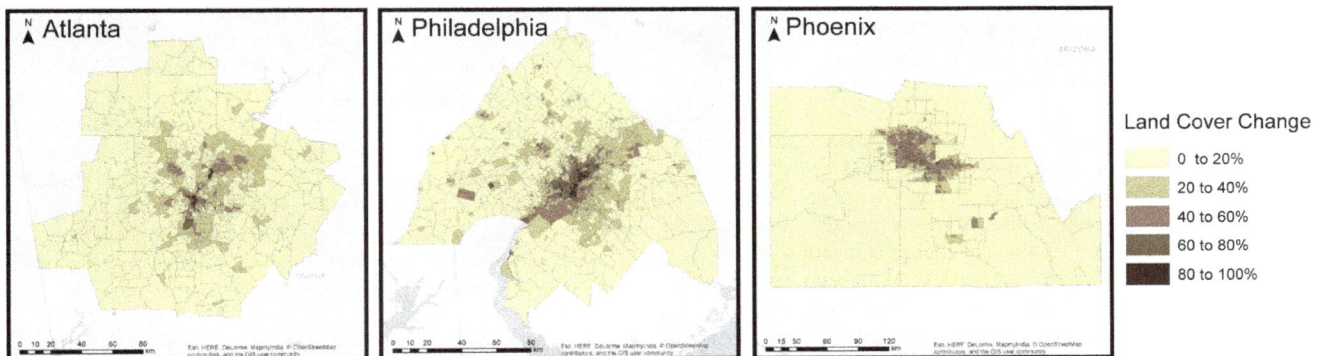

Figure 2. Percent of census tract area modified through the ALL heat management scenario. Land cover changes include the addition of new tree canopy, grass, or shrubs; conversion of roofing areas to greenroofs or high albedo materials; and the conversion of streets and other surface paving to moderate albedo materials (see Table 1).

Table 2. Description of heat response functions.

Study	Temperature metric	Relative risk	Mortality type	Study population
Anderson and Bell, 2011	heat wave periods classified as 2 or more days with mean daily T above 95th percentile of 1987–2005 average for May-Sept	1.0367 (1.0295, 1.0439) per heat wave day	non-accidental	all ages in 43 US cities (1987–2005)
Medina-Ramon and Schwartz, 2007	minimum daily T (May-Sept) above 17°C; measured as 2-day cumulative T	1.0043 (1.0024, 1.0061) per 1°C (O_3 adjusted)	all cause	all ages in 42 US cities (1989–2000)
Zanobetti and Schwartz, 2008	mean daily apparent temperature (May-Sept)	1.018 (1.0109, 1.025) per 5.55°C (O_3 and $PM_{2.5}$ adjusted)	non-accidental	all ages in 9 US cities (1999–2002)

Confidence intervals (95%) for estimates of relative risk in parentheses.

Results

Temperature change was first modeled for the 2050 BAU scenario in reference to base year (2010) conditions in each metropolitan region. Three measures of temperature change were derived for each MSA to correspond with the HRFs used to assess health outcomes, including average temperature (AvgT), average apparent temperature (AvgapT), and minimum temperature (MinT). Under the BAU scenario, warm season temperatures increase from base year temperatures by an average of 1.2°C in the Atlanta and Philadelphia MSAs, and by an average 2.2°C in the Phoenix MSA (see Figure 1, bottom panel).

Figure 3 presents the results of the heat management scenario modeling relative to BAU conditions. Across the MSA-temperature metric combinations, the influence of variable heat management strategies on BAU temperatures was found to range from an increase in mean warm season apparent temperature of 0.06°C to a reduction in minimum temperature of 0.57°C. In Atlanta and Philadelphia, maximum temperature reductions were achieved through either vegetation enhancement or a combination of vegetation and albedo enhancement. By contrast, albedo enhancement alone was associated with the greatest temperature reductions in Phoenix. Fifty-eight of the 63 scenarios presented in Figure 3 were associated with reductions in average MSA temperatures relative to BAU during the warm season. In addition, the multi-strategy ALL, ALBEDO, and GREEN scenarios were found to produce a cooling effect beyond the uncertainty ranges estimated through the ensemble modeling discussed above (Figure S1 and Table S1).

The variable effectiveness of the heat management strategies in lowering metro-wide, average warm season temperatures is driven in large part by differences in the area of land conversions and rates of soil moisture availability. In the eastern US, where annual precipitation rates are high, both the spatial extent and species mix of vegetation are conducive to higher rates of evapotranspiration than found in the arid climate of Phoenix, promoting in these areas more regional cooling through the latent heat flux. In Atlanta and Phoenix, albedo enhancement, on average, is found to be more effective than the combined green strategies due to the fact that more land area is available for modification. In these cities, the total area of land converted to high albedo materials is about one-third greater than that total land area subject to vegetation enhancement, yielding a greater cooling effect in most scenario/temperature metric combinations.

Consistent with reported temperature change, trends in heat-related mortality are reported in Figure 4 as differences in the number of deaths relative to the BAU scenario (deaths under BAU scenario minus deaths under heat management scenario). The total number of avoided (or increased) deaths reported for each scenario accounts for non-heat wave and episodic heat wave periods during the 2050 warm season in each region. The "BASE" scenario reports the projected number of deaths resulting from changes in climate between the base year of 2010 and the 2050 BAU scenario, holding regional population characteristics constant at 2050 levels. The projected increase in annual heat-related mortality in response to warming over this 40-year period ranges from a low of 53 additional deaths in Philadelphia to a high of 132 additional deaths in Phoenix. Relative to base year levels, these 2050 projections represent an increase in heat-related mortality of 55% in Phoenix, 77% in Atlanta, and 319% in Philadelphia.

Heat management strategies offset projected increases in heat-related mortality within a 95% confidence level in 37 of 42 scenario runs. Mirroring the temperature change results, vegetation enhancement or a combination of vegetation and albedo enhancement resulted in the greatest reductions in BAU mortality in Atlanta and Philadelphia, while albedo enhancement in Phoenix was found to have the most significant effect on heat-related mortality in response to the Zanobetti/Anderson HRFs. The most effective heat management strategies in each region were found to offset projected increases in heat-related mortality by between a low of 40% in Atlanta and Philadelphia (Zanobetti/Anderson HRFs) to a high of 99% in Atlanta (Medina-Ramon/Anderson HRFs), with an average reduction across all MSA and HRF combinations of 57%. We find vegetation and albedo enhancement in the Atlanta region to almost fully offset the projected increase in heat-related mortality associated with changes in minimum temperature between 2010 and 2050.

Figure 5 illustrates the change in heat-related mortality by census tract associated with the Medina-Ramon/Anderson HRFs, chosen due to evidence finding minimum temperatures to be the strongest predictor of heat-related health outcomes [27,28]. The greatest concentration of avoided mortality is seen in the urban core of each metropolitan region, where population densities are high and the proportion of the land surface impacted by either albedo or vegetation enhancement is greatest. Heat-related mortality is shown to marginally increase in a small number of zones in the central districts of Philadelphia and Phoenix, an outcome attributed to the mixed thermal effects of tree planting along roadways and in public parks, as discussed below.

Discussion

The results of our study support climate adaptation strategies designed to lessen the risk of heat exposure through mitigation of the urban heat island effect. Heat management strategies were found to be effective in offsetting mortality during both heat wave

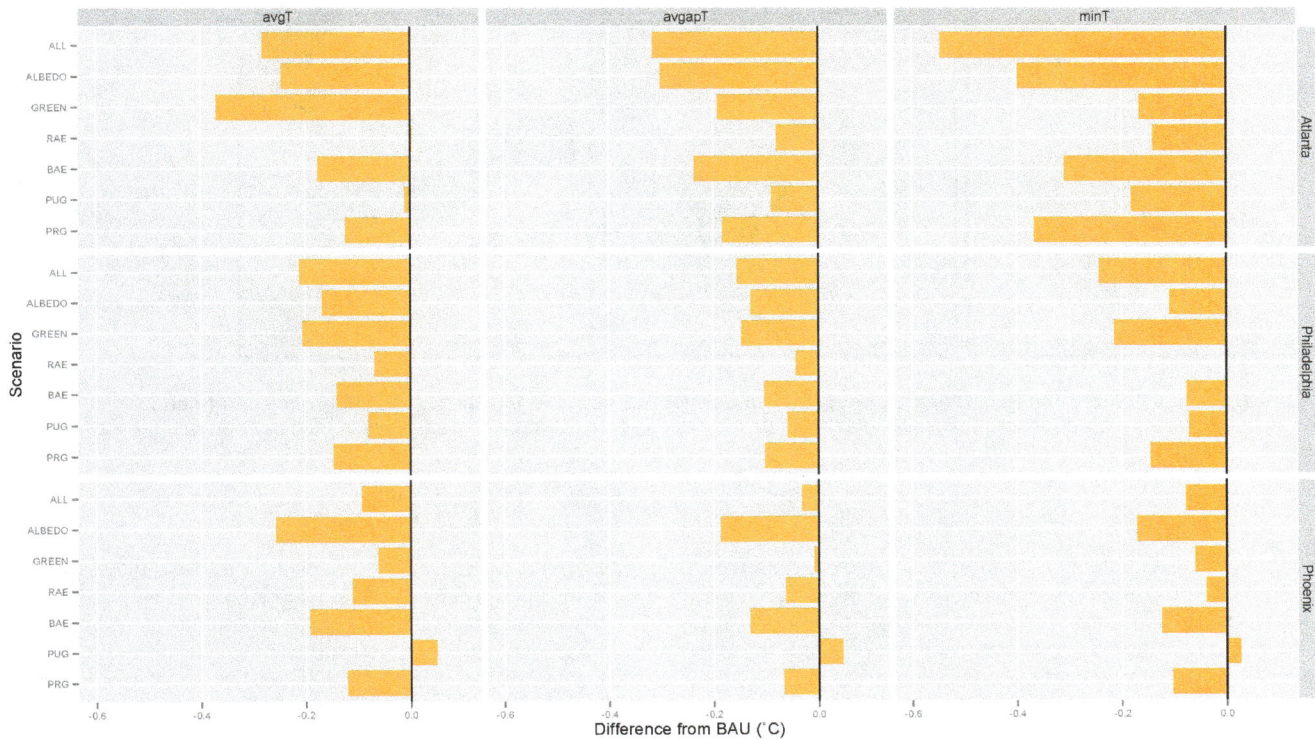

Figure 3. Differences in warm season temperature from BAU by heat management scenario, temperature metric, and MSA.

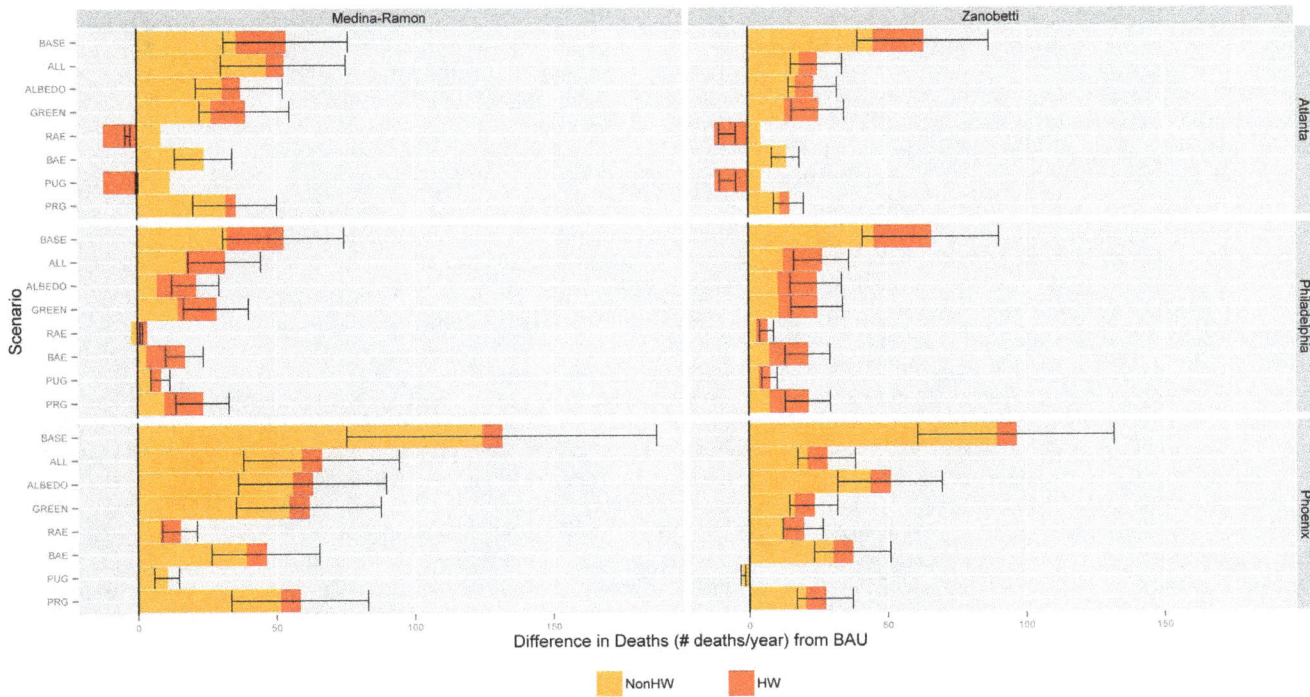

Figure 4. Difference in mortality relative to BAU by heat management scenario, HRF, and MSA. Bars report estimated difference in mortality relative to BAU in response to either the Medina-Ramon minimum temperature (minT) or Zanobetti average apparent temperature (avgapT) HRFs (orange shading) combined with the Anderson average temperature (avgT) HRF for heat wave conditions (red shading). Positive results denote a reduction in mortality relative to the BAU scenario; negative results denote an increase in mortality relative to the BAU scenario. Error bars report 95% confidence intervals.

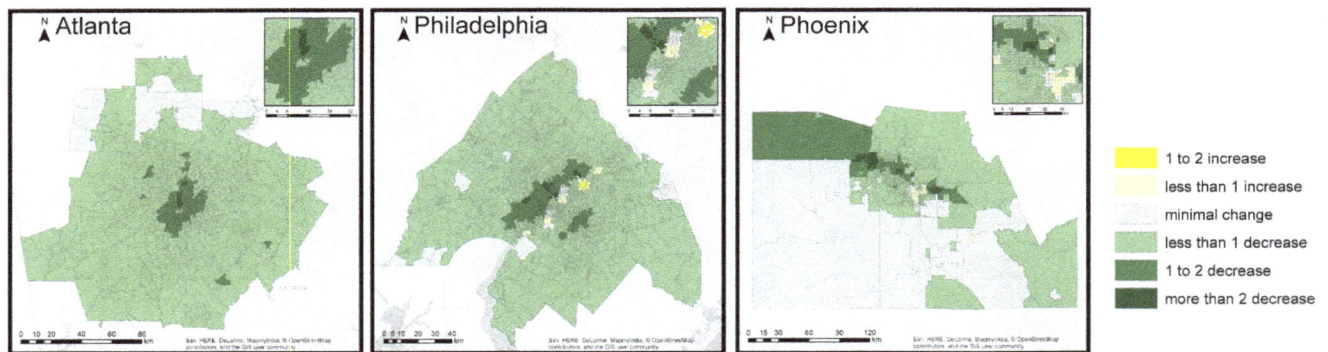

Figure 5. Change in mortality (per 100,000 population) under the ALL scenario in Atlanta, Philadelphia, and Phoenix. 2050 mortality changes are estimated in response to the Medina-Ramon/Anderson HRFs and are based on the difference in mortality between the BAU and ALL scenarios.

and non-heat wave conditions. Our results suggest that measures of relative risk for heat-related mortality based on average warm season temperatures only may significantly underestimate the potential for heat deaths during extreme heat events spanning two or more days. When accounting for both average warm season and heat wave conditions, the estimated number of avoided deaths due to the various heat management strategies was found to be 33% higher, on average, than model runs responsive to average warm season temperatures only.

The additional health benefits found to result from heat island mitigation during periods of extreme temperatures highlights the potential for established climate modeling protocols to systematically underestimate health risks associated with climate change. While a multi-year smoothing of climate projections may reduce the uncertainty of a median-year temperature estimate, it further carries the potential to obscure climate-related impacts associated with enhanced temperature variability. For this reason, standard climate modeling protocols that serve to statistically smooth annual and seasonal temperature variability may require modification when employed in health impact studies.

The heat management strategies most effective in offsetting mortality vary by region. Accounting for both warm season and heat wave deaths, vegetative strategies were found to have protective benefits greater than or comparable to albedo enhancement in Atlanta and Philadelphia, while albedo enhancement was found to be more protective in Phoenix. While the combined vegetation and albedo enhancement scenario (ALL) was generally found to be more effective in offsetting heat-related mortality, albedo strategies alone were found to be most protective of health in Phoenix in response to the apparent temperature response function (Zanobetti/Anderson HRF). The greater effectiveness of albedo strategies in arid climates reflects the limitations of vegetation enhancement in regions characterized by low soil moisture availability. The variable effect of heat management strategies by region demonstrates the need for heat abatement approaches to be tailored to the unique climatic conditions of different urban environments. Our findings further demonstrate the need to associate heat management strategies with a health endpoint directly, as some strategies found to be highly effective in reducing temperatures were less effective in offsetting heat-related mortality.

We find limited evidence that an increase in tree canopy on public property (PUG) in Atlanta and Phoenix may serve to marginally increase mortality – an effect that is not found when public tree planting is combined with other vegetation enhancement strategies. An examination of WRF model output for these scenario runs suggests the displacement of grass through tree planting may serve to elevate sensible heating through a dampening of the latent heat flux relative to BAU conditions. We conclude this outcome to result from an increase in leaf stomatal resistance with a shift from grass to tree canopy through this scenario, a physiological change that can decrease transpiration rates. Additionally, the PUG and RAE scenarios in Atlanta produce a larger number of heat wave days while simultaneously reducing average temperatures during heat wave episodes relative to BAU. As the heat wave response function is sensitive to changes in the number of heat wave days but not to changes in temperatures between scenarios, the modeled increase in heat wave mortality is in part an artifact of the heat wave response function employed in this study.

An additional limitation to our approach entails the use of historical temperature-mortality HRFs for the estimation of future year heat-related mortality [29]. If urban populations successfully exhibit physiological and/or behavioral (e.g., greater prevalence of air conditioning) adaptations to rising temperatures over time, the rate of mortality per degree rise in temperature or per additional heat wave day may be lower than reflected through the HRFs employed in this study.

Conclusions

We examined the potential for urban heat management strategies to offset projected increases in heat-related mortality in three large US metropolitan regions by mid-century using a set of global/regional climate and human health effects models. Variable combinations of heat management strategies involving vegetation and albedo enhancement were estimated to offset projected heat-related mortality by a range of 40 to 99%, depending on the metropolitan region and health impact function applied. These results highlight the potential for extensive land surface changes in cities to provide adaptive benefits to urban populations at risk for rising heat exposure with climate change.

We believe the study findings can inform the development of urban heat adaptation plans through which municipal governments can moderate the extremity of ambient temperatures during heat wave events, in concert with the implementation of emergency operations plans designed to protect public health once such events are underway. In selecting among alternative heat management strategies, urban planners and public health officials will want to consider the extent to which vegetation and albedo enhancement are consistent with a range of other climate adaptation objectives – including stormwater and air quality management — as well as other stakeholder preferences [30]. Future work will estimate the air quality implications of the modeled land cover change in Atlanta, Philadelphia, and Phoenix, as well as the economic costs and benefits of the various heat management strategies evaluated herein.

Supporting Information

Figure S1   Uncertainty in WRF model initializations by temperature metric and MSA. Error bars display confidence intervals around mean warm season temperatures (AvgT, AvgapT, and MinT) for five-member ensemble runs in which initial conditions were modified for the BAU scenario.

Table S1   Mean warm season (May-Sept) standard errors (°C) from ensemble simulations by temperature metric and MSA.

Author Contributions

Conceived and designed the experiments: BS JV PL DH AD MT YH AR. Performed the experiments: BS JV PL DH AD MT YH AR. Analyzed the data: BS JV PL DH AD MT YH AR. Contributed reagents/materials/analysis tools: BS JV PL DH AD MT YH AR. Wrote the paper: BS JV PL DH AD MT YH AR.

References

1. IPCC (2013) Climate change 2013: The physical science basis. New York: Cambridge University Press, New York.
2. Voorhees A, Fann N, Fulcher C, Dolwick P, Hubbell B, et al. (2011) Climate change-related temperature impacts on warm season heat mortality: A proof-of-concept methodology using BenMAP. Environ Sci Technol 45: 1450–1457.
3. Hayhoe K, Cayan D, Field C, Frumhoff P, Maurer E, et al. (2004) Emissions pathways, climate change, and impacts on California. Proc Natl Acad Sci USA 101: 12422–12427.
4. Peng R, Bobb J, Tebaldi C, McDaniel L, Bell M, et al. (2011) Toward a quantitative estimate of future heat wave wave mortality under global climate change. Environ Health Perspect 119: 701–706.
5. Sheridan S, Allen M, Lee C, Kalkstein L (2012) Future heat vulnerability in California, Part I: Projecting future weather types and heat events. Clim Chang 115: 291–309.
6. Zhou L, Dickinson R, Tian Y, Fang J, Li Q, et al. (2004) Evidence for a significant urbanization effect on climate in China. Proc Natl Acad Sci USA 101: 9540–9544.
7. Stone B Jr (2007) Urban and rural temperature trends in proximity to large US cities: 1951–2000. Int J Clim 27: 1801–1807.
8. McCarthy M, Best M, Betts R (2010) Climate change in cities due to global warming and urban effects. Geophys Res Lett 37: L09705.
9. Vargo J, Habeeb D, Stone B Jr (2013) The importance of land cover change across urban-rural typologies for climate modeling. J Environ Manage 114: 243–252.
10. Fry J, Coan M, Homer C, Meyer D, Wickman J (2009) Completion of the national land cover database (NLCD) 1992/2001 land cover change retrofit product. US Geological Survey open-file report 1379. 18p.
11. Schmidt G, Ruedy R, Hansen J, Aleinov I, Bell N, et al. (2006) Present-day atmospheric simulations using GISS ModelE: Comparison to in situ, satellite, and reanalysis data. J Clim 19: 153–192.
12. Skamarock W, Klemp J (2008) A time-split nonhydrostatic atmospheric model for weather research and forecasting applications. J Comput Phys 227: 3465–3485.
13. Liu P, Tsimpidi A, Hu Y, Stone B, Russell A, et al. (2012) Differences between downscaling with spectral and grid nudging using WRF. Atmos Chem Phys 12: 3601–3610.
14. Trail M, Tsimpidi A, Liu P, Tsigardis K, Hu Y, et al. (2013) Downscaling a global climate model to simulate climate change impacts on US regional and urban air quality. Geosci Mod Dev 6: 1429–1445.
15. Racherla P, Shindell D, Faluvegi G (2012) The added value to global model projections of climate change by dynamical downscaling: A case study over the continental U.S. using the GISS-ModelE2 and WRF models. J Geophys Res 117: D20118, doi:10.1029/2012JD018091.
16. Lynn B, Rosenzweig C, Goldberg R, Rind D, Hogrefe C, et al. (2010) Testing GISS-MM5 physics configurations for use in regional impacts studies. Climatic Change 99: 567–587.
17. Spronken-Smith R, Oke T (1998) The thermal regime of urban parks in two cities with different summer climates. Int J Remote Sens 19: 2085–2104.
18. Hart M, Sailor D (2009) Quantifying the influence of land-use and surface characteristics on spatial variability in the urban heat island. Theor Appl Climatol 95: 397–406.
19. Zhou Y, Shepherd M (2010) Atlanta's urban heat island under extreme heat conditions and potential mitigation strategies. Natural Hazards 52: 639–668.
20. Taha H, Akbari H, Rosenfeld A (1991) Heat island and oasis effects of vegetative canopies: Micro-meteorological field-measurements. Th App Clim 44: 123–138.
21. Oke T (1987) Boundary layer climates. Routledge.
22. Alpert D (2012) DC looks to the past to fix its zoning code. The Atlantic Cities, May 5.
23. Medina-Ramon M, Schwartz J (2007) Temperature, temperature extremes, and mortality: A study of acclimatisation and effect modification in 50 US cities. Occup Environ Med 64: 827–833.
24. Zanobetti A, Schwartz J (2008) Temperature and mortality in nine US cities. Epidemiol 19: 563–570.
25. Hajat S, Armstrong B, Baccini M, Biggeri A, Bisanti L, et al. (2006) Impact of high temperatures on mortality: Is there an added heat wave effect? Epidemiol. 17: 632–638.
26. Anderson G, Bell M (2011) Heat waves in the United States: Mortality risk during heat waves and effect modification by heat wave characteristics in 43 US communities. Environ Health Perspect 119: 210–218.
27. Kalkstein L, Davis R (1989) Weather and human mortality: An evaluation of demographic and interregional responses in the United States. Ann Assoc Am Geogr 79: 44–64.
28. Hajat S, Kovats R, Atkinson R, Haines A (2002) Impact of hot temperatures on death in London: A time series approach. J Epi Com Health 56: 367–372.
29. Knowlton K, Lynn B, Goldberg R, Rosenzweig C, Hogrefe C, et al. (2007) Projecting heat-related mortality impacts under a changing climate in the New York City region. Am J Public Health 97: 2028–2034.
30. Convertino M, Foran C, Keisler J, Scarlett L, LoSchiavo A, et al. (2013) Enhanced adaptive management for Everglades in response to climate change. Nature Scientific Reports 3: 2922.

8

Combining Inferential and Deductive Approaches to Estimate the Potential Geographical Range of the Invasive Plant Pathogen, *Phytophthora ramorum*

Kylie B. Ireland[1,2], **Giles E. St. J. Hardy**[1,2], **Darren J. Kriticos**[1,3]*

1 Cooperative Research Centre for National Plant Biosecurity, Canberra, Australian Capital Territory, Australia, **2** Centre for Phytophthora Science and Management, School of Veterinary and Life Sciences, Murdoch University, Perth, Western Australia, Australia, **3** Commonwealth Scientific and Industrial Research Organisation (CSIRO) Ecosystem Sciences, Canberra, Australian Capital Territory, Australia

Abstract

Phytophthora ramorum, an invasive plant pathogen of unknown origin, causes considerable and widespread damage in plant industries and natural ecosystems of the USA and Europe. Estimating the potential geographical range of *P. ramorum* has been complicated by a lack of biological and geographical data with which to calibrate climatic models. Previous attempts to do so, using either invaded range data or surrogate species approaches, have delivered varying results. A simulation model was developed using CLIMEX to estimate the global climate suitability patterns for establishment of *P. ramorum*. Growth requirements and stress response parameters were derived from ecophysiological laboratory observations and site-level transmission and disease factors related to climate data in the field. Geographical distribution data from the USA (California and Oregon) and Norway were reserved from model-fitting and used to validate the models. The model suggests that the invasion of *P. ramorum* in both North America and Europe is still in its infancy and that it is presently occupying a small fraction of its potential range. *Phytophthora ramorum* appears to be climatically suited to large areas of Africa, Australasia and South America, where it could cause biodiversity and economic losses in plant industries and natural ecosystems with susceptible hosts if introduced.

Editor: Zhengguang Zhang, Nanjing Agricultural University, China

Funding: The authors would like to acknowledge the support of the Australian Government's Cooperative Research Centres Program and the Cooperative Research Centre for National Plant Biosecurity in particular. We also acknowledge the financial support of the Australian Government Department of Sustainability, Environment, Water, Population and Communities and the School of Biological Sciences at Murdoch University. The funders had no role in study design, data collection and analysis, decision to publish, or preparation of the manuscript.

Competing Interests: The authors have declared that no competing interests exist.

* E-mail: Darren.Kriticos@csiro.au

Introduction

Phytophthora ramorum is an invasive plant pathogen causing considerable and widespread damage in nurseries, gardens, natural woodland and plantation forest ecosystems of the USA and Europe [1,2]. It is internationally recognized as a plant biosecurity threat in many regions. Australia, Canada, the Czech Republic, Mexico, New Zealand, South Korea and Taiwan have established specific quarantine policies and protocols to prevent the spread of contaminated plant materials from areas known to have the disease [3]. Outbreaks of the pathogen into areas of the west coast of the USA containing remnant native vegetation, particularly in central coastal California, have decimated populations of the keystone species tanoak (*Notholithocarpus densiflorus*) and coast live oak (*Quercus agrifolia*) [4,5]. Similarly in England, *P. ramorum* has caused mortality of important ecological and commercial species such as *Vaccinium myrtillus* in heathlands [6] and Japanese larch (*Larix kaempferi*) in plantations [2]. While eradication efforts have been undertaken in natural areas of Oregon, USA [7] and parts of Europe [8], *P. ramorum* continues to invade new forest sites in these regions and in coastal California where eradication efforts have not been attempted.

While the geographical centre of origin for *P. ramorum* remains unknown, both molecular and biological evidence suggest that it is exotic to both North America and Europe. *Phytophthora ramorum* populations in Europe and North America are dominated by different mating types [9] and significant genotypic and phenotypic differences exist between these populations [10,11,12,13]. At present, four distinct clonal lineages (NA1, NA2, EU1 and EU2) have been identified from North America and Europe [11,14]. NA1, NA2 and EU1 are all found in North American nurseries, while NA1 is the only genotype present in Californian and Oregon forests [11]. In Europe, EU1 is the dominant genotype in nurseries and forests, while EU2 is a newly discovered genotype only known at present from infected *L. kaempferi* in Northern Ireland and Southern Scotland [14]. Although the disease emerged around the same time in nurseries in Europe and woodlands in California (the early to mid-1990s), molecular evidence suggests these lineages diverged at least 150 000 years ago and are most probably independent introductions from its native range [15]. The current geographical range of the pathogen in North American native vegetation ecosystems extends over 850 km from south of Big Sur in California to Curry County in Southwest Oregon [16]. The pathogen has also been recorded in streams as far north as King County in Washington state in the Pacific Northwest of the USA and associated with streams with inlet water from nurseries in Alabama, Georgia, Florida, Mississippi and North Carolina [17]. In Europe, natural outbreaks have largely been limited to southern England and south Wales in the UK [2,18], with smaller outbreaks in public greens and woodlands of Belgium, Denmark, France, Germany, Ireland, Luxembourg, the Netherlands, Norway, Slovenia, Spain and Switzerland [8]. Infected nursery stock has been detected in 22 European countries, Canada and in numerous states in the USA where it predominates in the west coast states of Washington, Oregon and California [8,19].

Movement of nursery stock has been highlighted as the primary factor spreading the pathogen both within the USA and globally [20]. This is of particular concern given that the geographical centre of origin for *P. ramorum* remains unknown, the trade in mature plants for horticulture and landscaping continues to grow [21] and biosecurity agencies rely on knowledge of the distribution of unwanted invasive alien organisms to ensure the application of effective quarantine regulations.

In order to justify both phytosanitary measures associated with international trade and domestic biosecurity actions aimed at slowing or preventing the spread of an established pest, under the International Standards for Phytosanitary Measures (ISPM), it is necessary to estimate the *endangered area* and the value of the impacts that might arise if the pest were to spread throughout that range [22]. Previously, a variety of models have been used to answer questions related to the probability and consequences of an invasion of *P. ramorum*. Techniques applied include niche modelling using programmes such as CLIMEX [23] and Genetic Algorithm for Rule-set Production (GARP) [24], as well as fine-scale tactical models and maps to address immediate management issues using support vector machines [25], rule-based expert elicitation approaches incorporating both climatic and host data [26,27] and most recently, spatio-temporal, stochastic epidemiological modelling in combination with realistic geographical modelling to estimate spread [28]. Despite the importance of understanding the potential geographical range of *P. ramorum*, most studies have modelled its potential effects and spread where it already exists in the United States and Europe [8,23,25,29,30,31,32,33]. Only two models have explored the potential for disease establishment and outbreaks globally [24,27], and neither of these approaches addressed the question of the endangered area in terms of the ISPM framework.

While the various geographical models of *P. ramorum* were developed with the best available information at the time, many have failed to produce robust pest risk maps due to either a lack of suitable information to parameterise the model or the incorrect application or interpretation of a particular modelling method for simulating an emerging infectious disease. Distribution data used to inform models such as Guo et al. [25] and Kluza et al. [24] remain incomplete, as it is clear that the pathogen has not filled its ecological niche in its invaded range. In basing their projections on the known Californian distribution data at the time, these are best described as tactical models, estimating the invaded environment at that point in time, rather than the potential distribution of the pathogen. As a result, both of these models face challenges to the extrapolation of their outputs beyond that point in time and beyond the locations they are based upon.

CLIMEX [34,35] and NAPPFAST [27] are software packages designed to deal with the piecemeal nature of information regarding invasive species. Both are capable of incorporating knowledge of an organism's response to environmental variables, particularly climatic data, gleaned from direct observations or inferred from phenological observations. Both models can use information regarding the species' potential to grow as a function

of climatic factors to estimate the potential for the species to grow at any site for which suitable climatic data are available. Where NAPPFAST has a more detailed treatment of the potential for infection, CLIMEX is also able to use distribution data or experimental data to assess the potential for the species to survive inclement seasonal conditions. CLIMEX has been used previously to model potential distribution and relative disease risk of important plant pathogens on both continental [23,36,37] and global scales [38,39].

Both the Venette and Cohen [23] CLIMEX and Magarey et al. [27] NAPPFAST models were based on the best ecophysiological data available at the time to define their parameters. The CLIMEX model [23] was built using some parameter values for *P. cinnamomi*, a congeneric species which is understood to be a species adapted to warmer climates [40], to define soil moisture and stress parameters for *P. ramorum*. Whilst informative at that time, laboratory and field-based studies have since revealed more information about the ecological and climatic factors necessary for the infection, transmission and persistence of *P. ramorum* [41,42,43,44,45,46]. This new knowledge can now be used to inform an improved niche model specific to *P. ramorum*, and hence improve the understanding of the geographical risks posed by this organism.

The objective of this study was to estimate the global climate suitability patterns for the establishment of *P. ramorum* by developing a climatic niche model using the Compare Locations model in CLIMEX [34,35]. Parameters were defined using revised growth and stress parameters for *P. ramorum*, based on the best available experimentally-derived ecophysiological responses and site-level phenological factors of transmission and disease persistence associated with climate data from the field. Independent distribution data from the USA (California and Oregon) and Norway were used to validate the model. The results of the model are discussed and related to quarantine and management implications for international plant biosecurity.

Materials and Methods

The Compare Locations function in CLIMEX 3.0 [34] was used to develop a simulation model to estimate the climate suitability for the establishment of *P. ramorum* populations. The CliMond CM10_1975H_V1 interpolated climate surface [47] was used for all modelling. It is a fine scale (10 arc minute) dataset of long-term monthly climate means centred on 1975 for precipitation, maximum temperature, minimum temperature and relative humidity at 9 am and 3 pm. CLIMEX interpolates the monthly means to weekly values prior to calculating growth and stress indices.

The Compare Locations function in CLIMEX calculates an annual index of climatic suitability, the Ecoclimatic Index (EI), which reflects the combined potential for population growth during favourable periods and survival during stressful periods (Equation 1). The annual growth index (GI_A) describes the potential for growth of the host and pathogen as a function of average weekly soil moisture (Moisture Index; MI) and temperature (Temperature Index; TI) during favourable conditions (Equation 2; Table 1; Weekly Thermo-hydrological Growth Index, $TGI_W = TI \times MI$). Stress indices describing cold stress (CS), wet stress (WS), heat stress (HS), and dry stress (DS) (Table 1) and their interactions with one another can be used to describe the species response to climatically unfavourable conditions. The individual components of stress are combined into a stress index (SI) and a stress interaction index (SX) (Equations 3 and 4; CDX = Cold-Dry Stress, CWX = Cold-Wet Stress, HDX = Hot-

Dry Stress and HWX = Hot-Wet Stress) [34].

$$EI = GI_A \times SI \times SX \qquad (1)$$

$$GI_A = 100 \sum_{i=1}^{52} TGIW/52 \qquad (2)$$

$$SI = (1 - CS/100)(1 - DS/100)(1 - HS/100)(1 - WS/100) \qquad (3)$$

$$SX = (1 - CDX/100)(1 - CWX/100)(1 - HDX/100)(1 - HWX/100) \qquad (4)$$

The EI ranges from 0 for locations at which the species is not able to persist to 100 at locations that are optimal for the species year round. To downplay the implied modelling precision of the percentile values of EI provided by CLIMEX, the EI was classified into four arbitrary classes: unsuitable (EI = 0), marginal (EI = 1–5), moderately favourable (EI = 6–25) and highly favourable (EI>25), as described by Kriticos et al. [48].

Phenological observations, and relevant laboratory and field-based biological information were used to inform the selection of

Table 1. CLIMEX parameter values used to model eco-climatic suitability of Phytophthora ramorum.

Parameter	Description	Value
Temperature Index (TI)		
DV0	Lower temperature threshold for growth	0°C
DV1	Lower optimum for growth	18°C
DV2	Upper optimum for growth	22°C
DV3	Upper temperature threshold for growth	30°C
Moisture Index (MI)		
SM0	Lower soil moisture threshold for growth	0.2[a]
SM1	Lower optimum for growth	0.7[a]
SM2	Upper optimum for growth	1.3[a]
SM3	Upper soil moisture threshold for growth	2.0[a]
Cold stress (CS)		
TTCS	Temperature threshold for cold stress	−8°C
THCS	Cold stress accumulation rate	−0.02 week[−1]
Heat Stress (HS)		
TTHS	Temperature threshold for heat stress	31°C
THHS	Heat stress accumulation rate	0.03 week[−1]
Dry Stress (DS)		
SMDS	Soil moisture threshold for dry stress	0.2[a]
HDS	Dry stress accumulation rate	−0.005 week[−1]
Wet Stress (WS)		
SMWS	Soil moisture threshold for wet stress	2.0[a]
HWS	Wet stress accumulation rate	0.002 week[−1]

[a]Expressed as a proportion of soil moisture holding capacity, where 0 = oven dry and 1 = field capacity (saturation).

relevant parameters for growth and stress of *P. ramorum* (Table 1). Data were combined for the three main *P. ramorum* genotypes, NA1, NA2 and EU1, to produce a composite model of the potential geographic range of the pathogen as insufficient data exists at this point in time to treat each lineage separately. The stress indices were fitted in such a way as to conform to the guidance of Kriticos et al. [49], so that the stresses and growth should not occur at the same time and hence the thresholds for stresses occur outside the limits for growth.

Temperature Index

Temperature parameters were estimated from laboratory studies of the pathogen *in vitro* [41,42,45,50], *in vivo* [51,52] and from field-based studies of natural forest infections [53]. The lower temperature threshold for growth (DV0) was set to 0°C as infection and lesion growth can barely occur at this temperature [52], and below 0°C chlamydospore production and germination is impaired [45]. The lower and upper optimum temperatures for growth (DV1 and DV2) were set at 18 and 22°C, respectively, based on *in vitro* laboratory studies by Werres et al. [41], Tooley et al. [51] and Englander et al. [42] and optimum conditions for sporangia production and transmission under natural conditions [53]. The upper temperature threshold for growth (DV3) was set at 30°C based on the work of Werres et al. [41] and Brasier et al. [50]. Isolate growth and disease progression have not been observed above 30°C [50,51,53].

Moisture Index

The soil moisture index in CLIMEX is used as a proxy for moisture availability, relying on observed correlations between soil moisture and relative humidity and/or rainfall [34,38]. Moisture parameters for *P. ramorum*, not based on soil moisture, such as extended periods of leaf wetness (24 to 48 hours), high levels of rainfall and extended rainfall, have been associated with increased disease in the laboratory [51] and under natural field conditions [53]. These studies and those by Fichtner et al. [43] on the effect of soil drying on pathogen recovery, were used to infer and refine soil moisture parameter estimates. Extra drying of the substrate for the lower limit for growth (SM0), which was set to 0.2, was allowed for (expressed as a proportion of soil moisture holding capacity, where 1 = saturation and >1 indicates excess moisture, i.e. run-off). The lower soil moisture value for optimal population growth (SM1) was taken from Brasier and Scott [54], as modified by Sutherst et al. [55] and subsequently applied by Venette and Cohen [23] to *P. ramorum*. Following iterative sensitivity analyses, the upper soil moisture value for optimal population growth (SM3) from these models was reduced from 3 to a more realistic value of 2, in order to indicate that wet conditions favour *P. ramorum* transmission and infection, but recognising that it was possible to have excessive amounts of rainfall and soil moisture.

Cold Stress

Cold stress parameters were based on the results of recovery of chlamydospores (an asexual survival structure) at extreme cold temperatures [45,52]. Tooley et al. [45] observed reduced recovery of free chlamydospores after exposure to temperatures of 0°C for 24 hours and little or no recovery at −10°C and −20°C, while Turner et al. [52] found they were able to survive at −2°C for up to four hours in the laboratory. Hyphal colonies on the other hand have been shown to have no reduction in recovery after exposure to −5°C for 24 hours, but reductions at or below −10°C [56]. Despite the direct effects of cold on the pathogen, chlamydospores were able to survive at least one week in infected leaf tissue after exposure to a continuous −10°C, indicating that survival in infected plant tissue provides a more robust method of pathogen survival during cold periods [45]. Turner et al. [52] also found that chlamydospores inside infected leaf tissue were able to survive mild winters with minimum temperatures of $-9°C$ over 16 weeks under field conditions in the UK. At least 50% of chlamydospores survived for 16 weeks in the study by Turner et al. [52], with 80% chlamydospore survival in leaves buried 5 cm beneath the soil surface. In CLIMEX, the threshold cold stress due to damaging cold temperatures (TTCS) was set to begin accumulating at $-8°C$, at a slow enough rate (THCS $= .-0.02$ week^{-1}) to allow for survival for at least two months at $-10°C$. This accumulation rate was chosen to incorporate survival in the coldest location where the pathogen is known to survive winter, in naturally infected nursery plants kept outside in the south of Finland (A. Rytkŏnen, Finnish Forest Research Institute, pers. comm.)

Heat Stress

Heat stress indices were fitted based on experiments that indicated that isolate growth and disease on plant material was impaired at temperatures above $30°C$ [41,45,50,51,53]. Brasier et al. [50] found that 37% of EU1 isolates and 80% of NA1 isolates grew at $30°C$, but that neither lineage of isolates grew at $31°C$. Similarly, Tooley et al. [45] found that chlamydospores of *P. ramorum* held in moist sand showed a high rate of recovery at $30°C$, but at $35°C$ recovery declined steadily with time, and over a seven-day period there was no recovery of the pathogen at $40°C$. Likewise, when Tooley et al. [45] tested recovery of *P. ramorum* from infected *Rhododendron* tissue they found high recoveries at 20 and $30°C$ after seven days, but at a constant $35°C$ recoveries declined within two days, and there was no recovery by four days. However, Tooley et al. [45] also found that *P. ramorum* was able to survive several weeks of maximum temperatures of $31°C$ and $32°C$ in a variable temperature growth chamber experiment, designed to represent temperature minima and maxima of an average summer in Lewisburg, Tennessee in the USA. Collectively, the results of these experiments indicate that heat stress (TTHS) for *P. ramorum* is likely to begin accumulating at or above approximately $31°C$. The heat stress accumulation rate (THHS) was therefore set to accumulate at a rate of 0.03 week^{-1}, to allow for a survival rate of at least 50% of *P. ramorum* under the conditions described by Tooley et al. [45], as those representative of a summer in Lewisburg, TN. A survival rate at or near 100%, as reported by Tooley et al. [45], was not calculated for as the methods used to break potential dormancy of *P. ramorum* in the Tooley et al. [45] study (where dry leaf disks were rehydrated for one hour prior to plating) are not considered likely to occur systematically and regularly under natural field conditions.

Dry Stress

Phytophthora ramorum has been shown to be sensitive to drought, as highlighted by studies showing that free sporangia and chlamydospores were killed by drying at 30% relative humidity at room temperature for 30 minutes [57] and relative humidity below 50% significantly affected growth and germ tube elongation of zoospores [58]. However, sporangia have been recorded to survive up to six hours in moisture free conditions [52] and studies indicate that the pathogen can survive temperature and moisture stresses much more effectively within infected plant tissue [46]. Dry stress parameters were altered to begin accumulating below the lower soil moisture threshold for growth (SM0 $= 0.2$). The dry stress accumulation

rate (HDS $= -0.005$ week^{-1}) was selected to reflect the pathogen's apparent sensitivity to drought.

Wet Stress

Wet stress (SMWS) was set to occur when the soil moisture exceeds SM3 (2), with a relatively low stress accumulation rate (0.002 week^{-1}).

Model Run and Validation

The model was run on a world-wide scale using the CliMond historical climate data [47]. Geographical distribution data (both stream bait and infected tree confirmed positives) from California and Oregon [16] and Norway [59] were used to validate the model. The locations of ten positive waterways detected in the 2010 National *P. ramorum* Early Detection Survey of Forests [60] in Alabama, Georgia, Florida, Mississippi, North Carolina and Washington were also assessed in relation to the risk areas identified by the model.

Results

Model Fit and Projections

The majority of Mediterranean and maritime temperate climates were projected to be favourable for *P. ramorum* and the model indicates that the pathogen could extend into some continental climates with warm or cool summers (e.g. in the western USA and south-eastern Canada) as well as some sub-tropical climates, such as Virginia and North Carolina in the USA and coastal northern New South Wales and southeast Queensland in Australia (Fig. 1). The modelled climate suitability fits the known occurrences within California and Norway and, as might be expected for a new invader, extends significantly beyond the known current distribution (Fig. 2).

North America

The model suggests that the invasion of *P. ramorum* in North America is still in its infancy and it is presently occupying a small fraction of its available range (Figs. 2a and 3). All but one of the known positive detections (2696 total) of the pathogen in California and Oregon from the SODMAP data set [16] fell into the moderate to highly favourable risk areas (Fig. 2a). This one detection, NA-1, associated with infected *Umbellularia californica* trees, was located in Lake county, within 10 km of areas defined as being of moderate risk (Fig. 2a) and was a confirmed positive detection by PCR in 2006 [16].

Coastal areas with a Mediterranean climate along the west coast of the USA, extending into coastal areas of British Columbia (Canada), and continental areas of the East Coast, encompassing most of the Appalachians and extending into the Great Lakes region, appear suitable. Stream-associated nursery finds in Georgia and North Carolina fell into areas modelled as having highly favourable climates for pathogen establishment and survival, while those in Alabama, Florida and Mississippi fell into areas modelled as climatically unsuitable for persistence due to excessive heat stress (Fig. 3a; Supplementary Fig. S1). This does not preclude the species from epidemics in these areas, but indicates that populations are unlikely to be able to over-summer there. This is apparent in the high GI values throughout the south-eastern USA (Supplementary Fig. S2).

Within Central America and Mexico, areas classified as favourable were predominantly located at high elevations along the central mountain ranges. Large areas were modelled as climatically unsuitable for the establishment and persistence of *P. ramorum*, primarily due to a lack of adequate moisture for growth

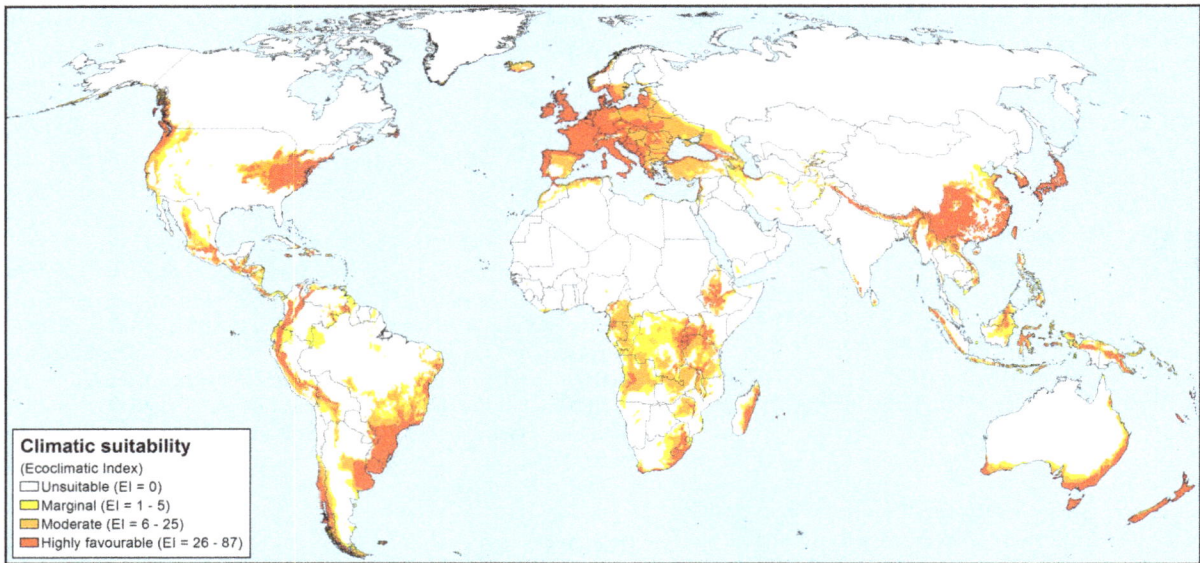

Figure 1. Global eco-climatic suitability for *Phytophthora ramorum* **under the 1961–1990 climate normals, as modelled using CLIMEX.**

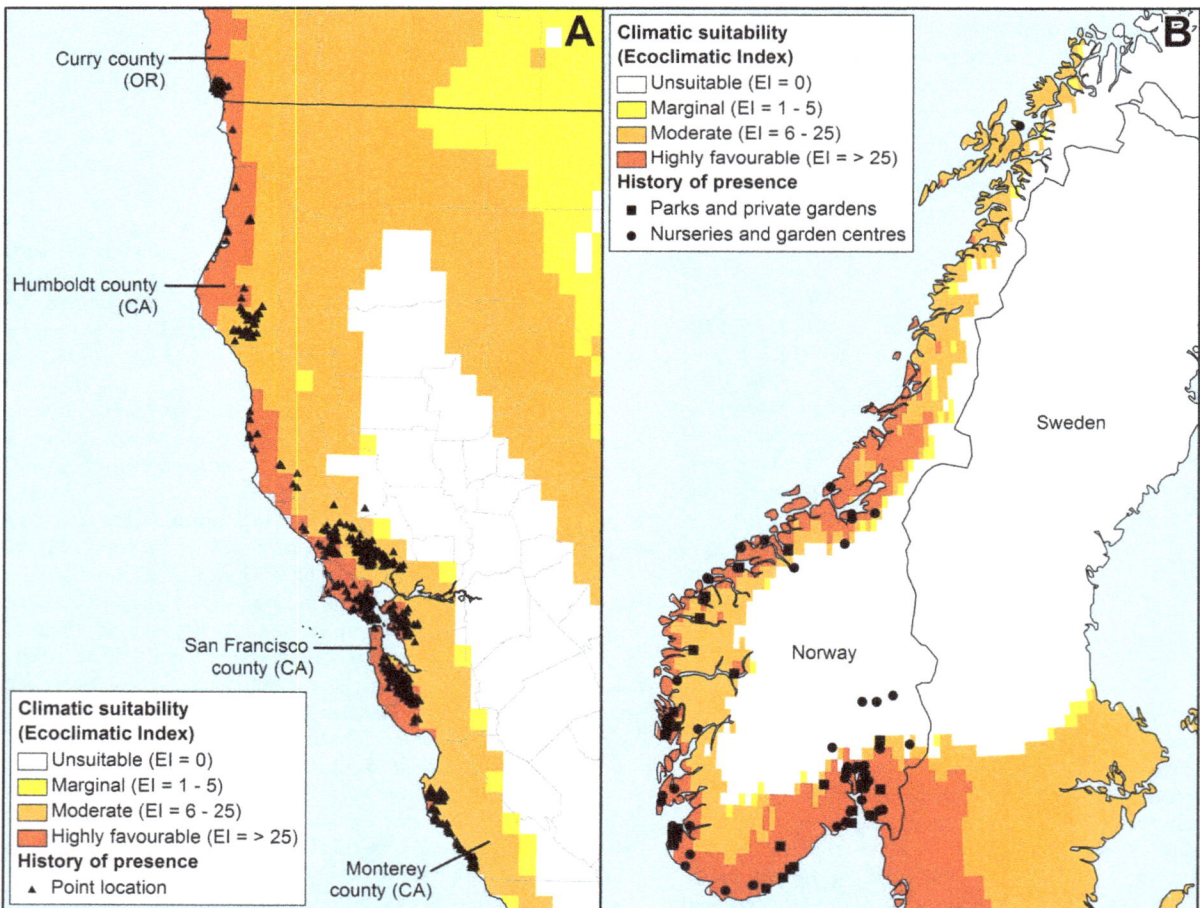

Figure 2. Known distribution of *Phytophthora ramorum* **in northern California and Oregon, USA (a) and Norway (b) and eco-climatic suitability.** Projected using the 1961–1990 climate normals, as modelled using CLIMEX.

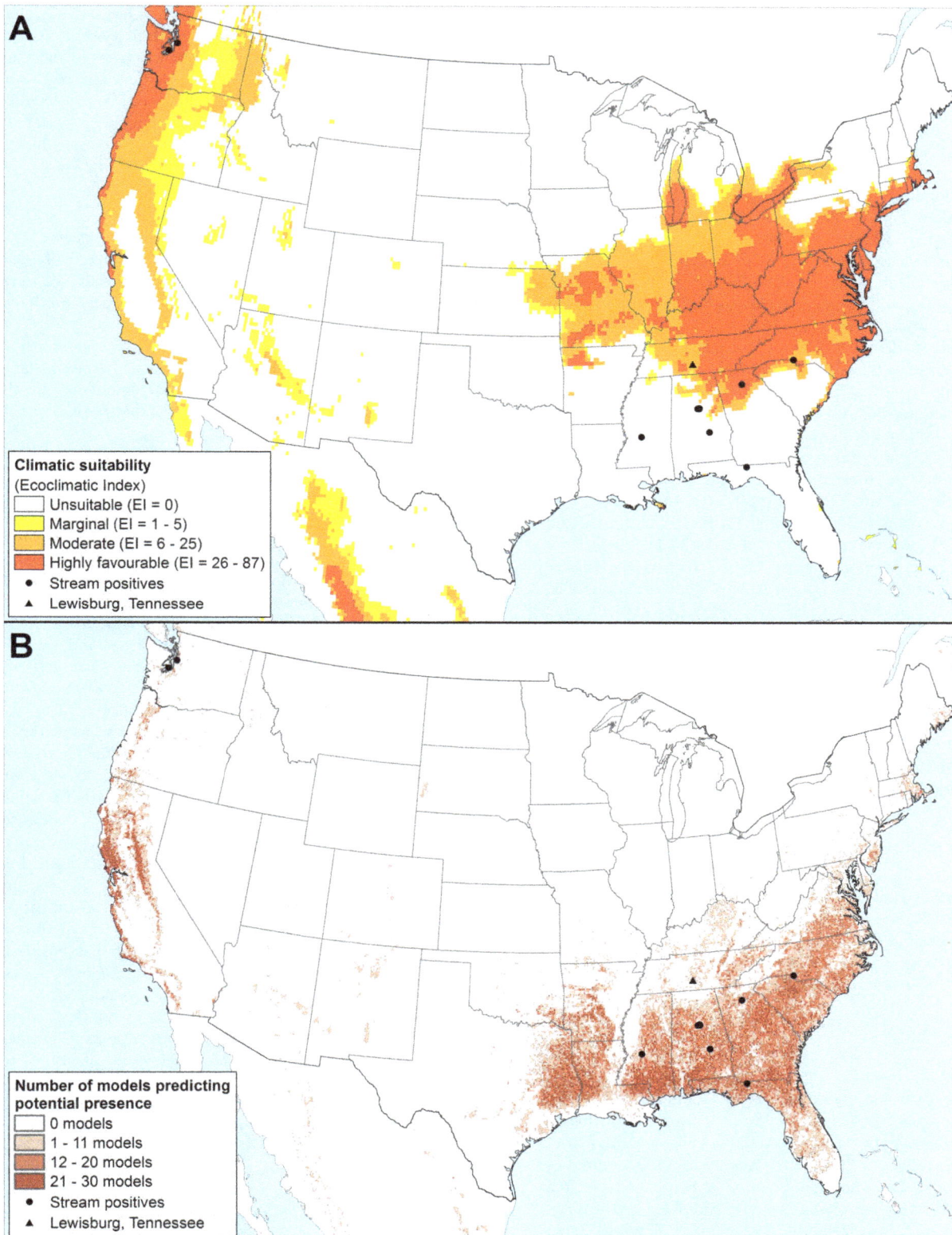

Figure 3. *Phytophthora ramorum* **positive streams and models of potential range of the pathogen in the USA.** Stream positive detections in the eastern USA and Washington state are those known at July 2010 [60]. As modelled using CLIMEX and the composite model parameters of our model under 1961–1990 climate normals (a); and as generated using a Genetic Algorithm for Rule-set Production (GARP) based on known Californian occurrences and three ecological datasets by Kluza et al. [24] (b).

throughout the continental USA as well as being limited by temperature extremes in the far north and in the south toward the equator (Supplementary Figs. S1 and S3–S5).

Europe

In Europe, it would appear that approximately half of the climatically favourable countries have already been invaded (22/45), predominantly within nurseries, though there is still considerable scope for further international spread and within country range expansion (Fig. 4). According to our model, all known countries in Europe where the pathogen has been detected have regions of moderate to highly favourable climate suitability for *P. ramorum* establishment. In Norway, all known places where the pathogen exists in parks and private gardens (59) and the majority (122/137) of locations of infected nursery and garden centres fell into areas with moderate to highly favourable climate (Fig. 2b). Almost all of Western Europe, apart from some high altitude areas in the Swiss Alps and the Carpathian Mountains in Romania (too cold; Supplementary Fig. S5) and southern regions of the Iberian Peninsula (too hot; Supplementary Fig. S1) were projected to have moderate to highly favourable climates for *P. ramorum* (Fig. 4). In Scandinavia, only the most southern coastal regions of Norway and Sweden and a small coastal port area of Finland were considered to be climatically favourable for the establishment and persistence of *P. ramorum* under historical climates. The majority of the Russian Federation and north-eastern regions of Belarus, Estonia, Latvia and Ukraine appear to be climatically unsuitable, due primarily to extreme cold weather conditions and inadequate soil moisture (Supplementary Figs. S4 and S5).

Asia

Throughout Asia *P. ramorum* is projected to be primarily restricted to the humid sub-tropics (Figs. 1 and 5a), with some areas classed as having a continental climate with warm summers projected as favourable in China i.e., the central south eastern provinces of Yunnan and Sichuan, extending west across the Himalayas. Much of northern Asia, including Mongolia, western China and the Russian Federation appear too cold for *P. ramorum* (Supplementary Fig. S5). The Middle East is too hot and dry (Supplementary Figs. S1 and S4), and low-lying areas within tropical Asia are projected to be unsuitable for *P. ramorum* due to heat stress (Supplementary Fig. S1). In contrast, the majority of southern Japan and all of Taiwan were projected to be climatically moderate to highly favourable for *P. ramorum* (Figs. 1 and 5a).

South America

The majority of South America was projected to be unsuitable for *P. ramorum* (Fig 1), largely due to conditions not being conducive to pathogen growth (Supplementary Fig. S2) and heat stress (Supplementary Fig. S1). Moderate to highly favourable conditions occur at high altitudes in the Andes from northern Colombia to northern Argentina, and south of Concepción in Chile, where conditions are cooler. East of the Andes, the only region projected to be favourable was within a band on the East Coast, from Rio de Janeiro in Brazil to Bahia Blanca in Argentina.

Africa

Moderate to highly favourable conditions for *P. ramorum* in Africa were largely restricted to elevated sites within Ethiopia, Kenya and along the borders of Uganda, Rwanda, Burundi and the Democratic Republic of Congo (Fig. 1). The majority of the continent was projected to be unsuitable for *P. ramorum* due to heat or dry stress (Supplementary Figs. S1 and S4). Central coastal subtropical areas of Angola, coastal subtropical areas of the East Coast and a small portion of the region of the Cape region of South Africa with a Mediterranean climate were also projected to have marginal to highly favourable climatic conditions for *P. ramorum* (Fig. 1). The model also projected moderate to highly favourable climates for the pathogen in eastern Madagascar.

Australasia

In Australia, climatically favourable areas for *P. ramorum* were confined to the temperate moist periphery, predominantly in New South Wales, Victoria, Tasmania and south-west Western Australia, with coastal areas of the south east of Queensland and South Australia also projected as being favourable (Fig. 1). The pathogen was primarily restricted by hot, arid conditions in Australia (Supplementary Figs. S1 and S4). All of New Zealand was projected to have climates either moderate or highly favourable for *P. ramorum*, including the southern Alps, which extend above the tree-line (Fig. 1).

Discussion

The CLIMEX model developed in this paper suggests that *P. ramorum* is presently occupying a small fraction of its available range in both North America and Europe. With continued global spread *P. ramorum* could potentially invade and establish in climatically favourable areas of Asia, Australasia, Africa and South America, where it could cause detrimental biodiversity loss and severe economic losses in plant industries and natural ecosystems with suitable hosts. It has been suggested by pathologists and modellers that the absence of *P. ramorum* from climatically favourable areas in North America is likely due to lack of historical opportunity, rather than intrinsic factors preventing its establishment [24,25,30,33,61,62]. This hypothesis is supported by the model presented here, fuelling the debate as to why the pathogen has not established in the eastern USA. Early detection of *P. ramorum* in horticultural shipments to eastern USA, combined with effective quarantine measures on materials in infested areas and increased awareness of the disease [3], may have resulted in lack of establishment of the pathogen in these areas. Conversely, the pathogen may be in a lag phase of invasion in these areas or may have entered a state of equilibrium with the ecosystem and may not cause disease of epiphytotic proportions [63]. It has been suggested that *P. ramorum* may have been present in California for some years before the disease that it caused was noticed [1], so vigilance should be maintained when surveying in the eastern USA and maintaining and updating quarantine policies. Further research into this lag phase of *P. ramorum* invasion is crucial to understanding the aetiology of the pathogen's disease cycle and could then be used to inform and formulate better risk models.

Given that all but one of the confirmed positive locations in California used to validate the model fell into areas that were moderate to highly favourable according to the model presented in this paper, as well as all known locations in Curry County, Oregon, where attempted eradication of the pathogen has been taking place [64]; our model suggests that surrounding areas identified as being climatically favourable areas are at high risk. Similarly, our model suggests that native vegetation communities in the Appalachian Mountains, for which potentially susceptible species have been identified in laboratory studies [65], are also at risk. Promisingly, the majority of stream associated nursery finds in the south-eastern USA fell into areas that experience prolonged periods of temperatures that are apparently too hot for *P. ramorum*

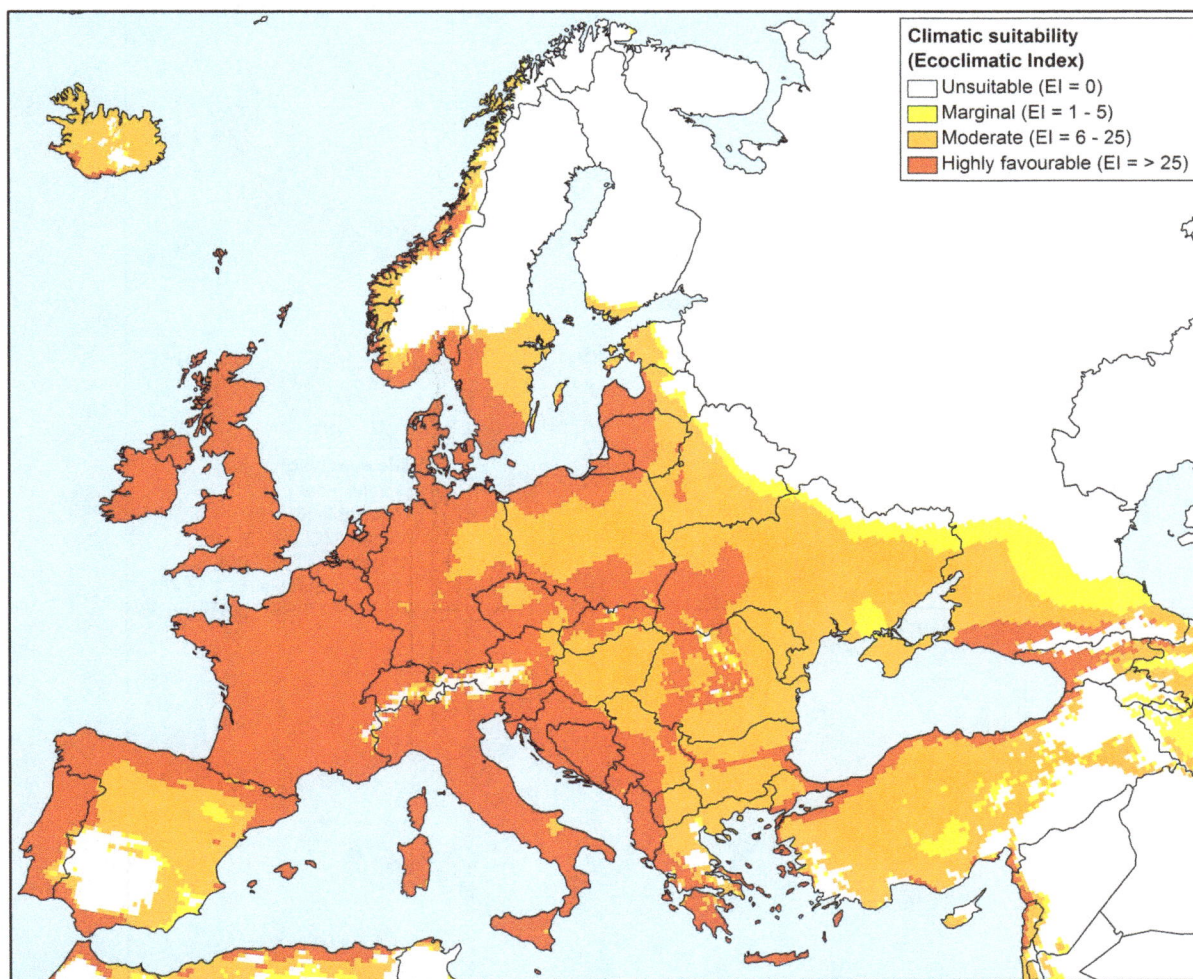

Figure 4. Known European distribution and eco-climatic suitability for *Phytophthora ramorum*. Projected using the 1961–1990 climate normals, as modelled using CLIMEX.

(Supplementary Fig. S1) and may therefore present a reduced risk to native ecosystems; unless infected nursery stock growing in a protected environment provides a recurrent source of inoculum during periods favourable for infection and population growth of *P. ramorum*.

While the stream positive detection originating from a nursery in Mississippi was associated with infected streamside vegetation in 2010 [60], it is unclear whether this association represents a consistent microclimate that allows summer temperatures to be moderated sufficiently for persistence, or whether it is simply an ephemeral population that established between Autumn and Spring of that year, when both the CLIMEX (Supplementary Fig. S2) and NAPPFAST [27] models indicate that conditions are likely to be favourable for population growth. Subsequent surveys in this area over the past three years have failed to detect infected streamside vegetation again, despite continued stream positive detections (Steven Oak, USDA Forest Service, personal communication). While this may indicate an ephemeral streamside establishment of the pathogen associated with a suitable microclimate for that particular point in time, it may also simply be a reflection of sampling effort, chance infection and/or detection. This uncertainty illustrates one of the challenges of relying upon single field detections or laboratory observations recorded under

constant or simplified diurnal temperature regimes to define stress parameters. Stress parameters are generally best-defined by fitting them to geographical distribution data based on consistent and/or well-designed sampling methodology, using mechanisms that are informed by laboratory experiments or theoretical expectations [66]. Irrespective of the explanation for this apparent discrepancy, one would expect that *P. ramorum* would be unlikely to establish a persistent population away from the streamside vegetation in this unsuitably hot region unless a favourable microclimate existed.

The Kluza et al. [24] model differed significantly from both the CLIMEX and NAPPFAST [27] model projections for the USA (Fig. 3; data not shown for NAPPFAST model). It is likely that the model of Kluza et al. [24] suffers from lack of sufficient input data, particularly given that it is unlikely that *P. ramorum* has reached its full expansion in the region upon which the model was based, in northern California. Using the Kluza et al. [24] model, vegetation surrounding the stream associated nursery finds in the southeastern USA would be considered at a much greater risk of *P. ramorum* invasion (Fig. 3).

In Europe it would appear that approximately half of the climatically favourable countries (22/45) have recorded positive nursery finds, but few natural ecosystems, parks or private gardens have been invaded, leaving considerable scope for further within-

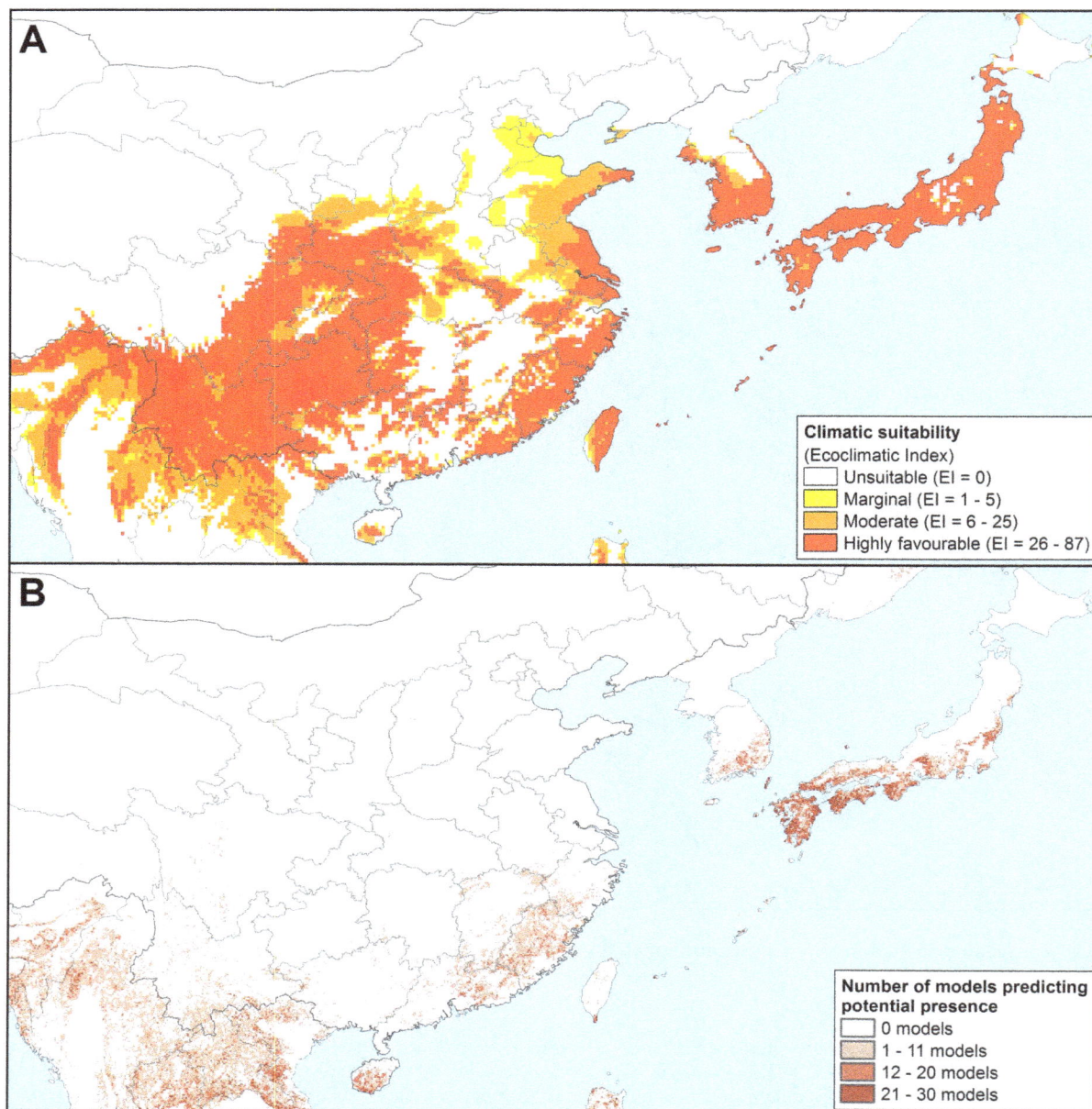

Figure 5. Projected potential distribution of *Phytophthora ramorum* **in Eastern Asia using two different niche-models.** As modelled using CLIMEX and the composite model parameters of our model under 1961–1990 climate normals (a); and as generated using a Genetic Algorithm for Rule-set Production (GARP) based on known Californian occurrences and three ecological datasets by Kluza et al. [24] (b).

country invasion. In Norway, which contains some of the most frigid conditions in which the pathogen may be able to survive, all natural infections in parks and private gardens and 89% of nurseries and garden centres which have been found to have positively infected stock [59] fell into areas modelled as being climatically suitable. Within-country invasion of areas outside of nurseries has been limited to only 11 European countries [8]. This may be a function of the pathogen not being spread to susceptible hosts in natural environments, lack of detection as a function of lack of surveillance, or may reflect the favourability of the artificial nursery environment, where temperatures may be regulated and plants are regularly watered, providing ideal conditions for the transmission of *Phytophthora* species. This lack of infestation in

natural areas or gardens of climatically favourable countries may also be due to the lag phase of the pathogen invasion [63], as it was only discovered in the mid-1990s, and is likely to be at the beginning of its invasive spread. Therefore, continued and regular surveys for the pathogen outside the nursery environment may be warranted in these countries. The recent rapid spread of *P. ramorum* on *L. kaempferi* [18] in the UK highlights the potential for rapid spread elsewhere in climatically favourable regions.

Examination of the stress indices used in the models indicated that the distribution of the pathogen is largely restricted by extreme hot and dry conditions in the mid-latitudes, and cold stress in the high northern latitudes (Supplementary Figs. S1, S4 and S5). Nowhere in the southern hemisphere (excluding

Antarctica) appears to be too cold for *P. ramorum*. Our models suggest that warm summer conditions may not be severe enough to reduce the risk of the pathogen establishing and spreading throughout most of the Mediterranean. It should be noted that the experimental data which underpin the selection of the heat stress parameters for *P. ramorum* in our model [45,50,51] are far from conclusive at this point in time, rarely exploring the conditions to break any potential dormancy of the pathogen. Our methodology and results reflect the best reasonable inference which could be derived at this point in time. Further experiments exploring the survival of *P. ramorum* propagules within smaller temperature increments (such as one degree increments between 30°C and 40°C, coupled with a range of dormancy breaking methods such as post-hydration of leaf discs or propagules [43]) would be invaluable in determining heat and heat/wet interaction stress parameters for the pathogen, and ultimately for informing the potential area of climatic suitability for *P. ramorum* in high risk areas such as the south-eastern USA.

Moisture, rainfall and days of consecutive rainfall are considered to play a crucial role in the development and spread of disease caused by *P. ramorum*, especially in years of high tree mortality, which historically follow very wet springs [53,67]. This hypothesis has been supported by an experimental irrigation nursery study, which showed that the highest concentrations of infective propagules occurred when stream sampling was preceded by about two months with low minimum daily temperatures and by four days of high rainfall, indicating that cold wet conditions are highly conducive to disease development [44]. However, persistence and production of chlamydospores in forest and nursery soils and leaf litter over the summer contribute to the disease cycle by providing an inoculum reserve at the onset of the Autumn disease cycle [43,46,68], allowing the pathogen to survive otherwise non-conducive conditions for growth and survival. In the case of Greece, our model suggests that the pathogen would be able to survive the hot summer in this manner, contributing to the disease cycle at the onset of the rains in the winter, when conditions are ideal for *P. ramorum* transmission. However, should these moist conditions not occur following these initial dry conditions where the pathogen can survive, the risk of disease outbreaks and pathogen establishment may be negligible.

While the effects of weather cannot be explored in the CLIMEX Compare Locations function, further work utilising weather data from specific locations within CLIMEX and its sister program DYMEX (which models population dynamics) may be able to shed some light on the nature and influence of extreme weather events on *P. ramorum* disease incidence and severity. In this case, further questions as to the particular stages of the pathogen's life cycle and their relationships to climatic variables and behaviour within populations would need to be addressed in much further detail to build such a model. At present, spread and potential establishment modelling utilising host data has been developed for California by Meentemeyer et al. [28,30] to address some of the concerns associated with indicating where epidemics are most likely to occur in the near future. Models of this nature are valuable in a different way to our global establishment model, as in these regions the estimation of epidemic outbreaks, rather than simply the potential presence of the pathogen, is more important when making considered management decisions with finite financial resources.

Presently, CLIMEX is limited in not being able to use moisture parameters based upon relative humidity and leaf wetness, which are likely to be more influential to disease development than soil moisture alone, for aerial plant pathogens such as *P. ramorum*. Nevertheless, soil moisture is a relatively stable index that is very useful in defining the condition of the host plant and is influential in disease development, especially as it relates to dry stress and the permanent wilting point of the host plant [37,38]. It is also regarded that soil moisture is a valid proxy for moisture availability when developing CLIMEX models for aerial pathogens [38]. Correlations between soil moisture and relative humidity and/or rainfall inform the selection of moisture related parameters when using the Compare Locations function in CLIMEX [34]. Detail on the interpretation of the relationship between rainfall and evaporation on the one hand and subsequent soil moisture parameter estimation on the other is outlined by Sutherst et al. [34], and it is recommended that new CLIMEX users make themselves familiar with the aforementioned relationships when estimating moisture based parameters.

Many opportunities exist in the future to further understand the underlying biology of *P. ramorum* and how this biology is associated with climatic and weather events. Study into changes of the potential range of *P. ramorum* under the effects of climate change would be valuable, especially as to how frequency of severe weather events such as particularly wet spring rains will influence disease spread, incidence and severity in areas where the pathogen is well established, such as California. As the pathogen becomes more widely established it may also be possible to apply methods of attaching severity ratings to the Ecoclimatic Index, as has been demonstrated as an effective modelling tool to plan future management in forestry by Pinkard et al. [37] for Tasmanian *Eucalyptus globulus* forests under the threat of Mycosphaerella leaf disease. This method also has applicability for understanding underlying mechanisms of pathogen behaviour and infective potential seasonally and across years [37].

This study adds to the body of work modelling the potential global distribution of *P. ramorum* [24,69]. In particular, the model presented here builds upon the previous model developed by Venette and Cohen [23], who used CLIMEX to assess the potential climatic suitability for establishment of *P. ramorum* within the contiguous United States. Stress and soil moisture parameters of the surrogate *Phytophthora* species, *P. cinnamomi*, were replaced with those derived from the published literature on *P. ramorum*, much of which became available after the Venette and Cohen [23] paper was written. The modelled growth indices of our model and that of Venette and Cohen [23] are remarkably similar (Supplementary Fig. S6c and S6d), differing mostly in the ability to grow under dry conditions (data not shown). There were, however, significant differences in the potential range as indicated by the Ecoclimatic Indices (growth indices with stress indices applied; Supplementary Fig. S6a and S6b), with a significantly increased area at risk under our model in the west coast states and also the north eastern states, and a projected decreased risk in the southern (stream positive) states of Mississippi, Alabama and Georgia. This comparison highlights the importance of using and updating parameter sets providing the basis for models. Pest Risk Modelling has been described as the 'art of the possible' [70]. Risk models should be based on the best knowledge of an organism's distribution and ecology at the time the model is developed, with scope to update and improve the model as new evidence becomes available.

Although our model clearly highlights areas at risk of invasion by *P. ramorum*, it should be noted that optimal climatic conditions for establishment do not necessarily translate into frequent or severe outbreaks. Not only must the pathogen be transported to suitable locations, and an infectious (sporulating) host be present in sufficient numbers to support infection in the long term at that location, but it must also be under the influence of local scale factors that may support severe outbreaks of the disease.

Microclimatic factors play a big role in the presence of the disease in Northern California [30], and this may explain why one of the known presences used to validate the model in Northern California fell into an unsuitable area. Favourable microclimates for pathogen establishment and survival are unlikely to be apparent on the CliMond 10′ grid [66]. Notably, this positive detection of *P. ramorum* located in an area projected to be unsuitable by the model is located within 10 km of suitable climatic envelopes and associated with the highly favourable host, *U. californica*, lending weight to the idea that a suitable microclimate exists in this area. Nevertheless, climatic suitability is a necessary condition for any outbreak of the disease and the model projections provide a useful indication of potential outbreak severity. Given this, the risk maps provided in this paper could be used to provide guidance on areas to target for the early detection and monitoring of the pathogen globally.

Notwithstanding uncertainties about host range, no host implies no problem [71]. Furthermore, in the case of *P. ramorum*, no infectious host implies less of a problem, as rapid expansion of the pathogen's range has been linked to "super-sporulating" hosts such as *U. californica* (California bay laurel) in California [72] and *L. kaempferi* (Japanese larch) in the United Kingdom [2,73]. Potential host range studies of species from areas considered to be at risk of *P. ramorum* invasion have indicated that susceptible and infectious hosts exist on the east coast of the USA, Asia and Australasia [2,74,75,76,77,78]. Knowledge of these hosts combined with knowledge of the potential geographical distribution is important when concentrating efforts for early detection of the pathogen, or when attempting to uncover the origins of invasive species such as *P. ramorum*. The origin of *P. ramorum* has been hypothesised by many to be in eastern Asia, based upon climatic variables [24] and origin of hosts such as *Rhododendron* [74]. Expeditions to regions of Yunnan Province in China, the centre of diversity for *Rhododendron* species, and Taiwan have not yet recovered *P. ramorum* [79,80,81]. Our results contrast very strongly with that of Kluza et al. [24] with regards to eastern Asia (Fig. 5). While both models agree that the Yunnan province in South-west China is climatically favourable, elsewhere the models disagree strongly. Where the model of Kluza et al. [24] indicates that the warmer humid tropical and sub-tropical climates are highly favourable for *P. ramorum* (e.g., Guangdong, Fujian, Zhejiang and Jiangxi provinces in China), our model indicates that it has a more temperate climatic preference (e.g. Guizhou, Hubei and Hunnan provinces in China). Perhaps targeting stream surveys for *P. ramorum* in the cooler regions of China, Taiwan (where the closely related *P.lateralis* has been discovered [81]) or Japan may be more fruitful.

Additional regions highlighted by the models, including temperate regions of China, elevated Andean locations in central South America and central Chile, the highly biodiverse Mediterranean region of the cape of South Africa, coastal Australia and the entirety of New Zealand, should be considered at risk of invasion by *P. ramorum*, or as potential origins of the pathogen, particularly where susceptible hosts naturally occur and/or are planted and traded. It is hoped the projections from this model will provide useful guidance on areas to target for early detection and monitoring of the pathogen, particularly in novel environments.

Supporting Information

Figure S1   Heat Stress (HS) for *Phytophthora ramorum* as modelled using CLIMEX with the CliMond dataset of historical climate normals centred on 1975. Where HS = 0, heat does not limit the distribution of *P. ramorum* and where HS >0 heat stress is represented by a factor of 1000, with increasing limitation as HS increases.

Figure S2   Annual Growth Index (GI; climatic suitability without stress) for *Phytophthora ramorum* as modelled using CLIMEX with the CliMond dataset of historical climate normals centred on 1975. Climatic conditions are classified as being unfavourable for growth when GI = 0, marginally favourable when GI = 1–5, moderately favourable when GI = 6–25 and highly favourable when GI >25. The GI does not factor in climatic stress and therefore does not represent the potential distribution of *P. ramorum*, only growth during non-stressful periods of the year.

Figure S3   Wet Stress (WS) for *Phytophthora ramorum* as modelled using CLIMEX with the CliMond dataset of historical climate normals centred on 1975. Where WS = 0, soil moisture does not limit the distribution of *P. ramorum* and where WS >0 wet stress is represented by a factor of 1000, with increasing limitation as WS increases.

Figure S4   Dry Stress (DS) for *Phytophthora ramorum* as modelled using CLIMEX with the CliMond dataset of historical climate normals centred on 1975. Where DS = 0, soil dryness does not limit the distribution of *P. ramorum* and where DS >0 dry stress is represented by a factor of 1000, with increasing limitation as DS increases.

Figure S5   Cold Stress (CS) for *Phytophthora ramorum* as modelled using CLIMEX with the CliMond dataset of historical climate normals centred on 1975. Where CS = 0, cold does not limit the distribution of *P. ramorum* and where CS >0 cold stress is represented by a factor of 1000, with increasing limitation as CS increases.

Figure S6   Ecoclimatic Index and suitability and Annual Growth Index for *Phytophthora ramorum* in the USA. As modelled using the CLIMEX parameters of Venette et al. [23] (a and c) and our model (b and d), with the CliMond dataset of historical climate normals centred on 1975.

Acknowledgments

The authors would like to thank A. Kanaskie, R. Meentemeyer, T. Rafoss, A. Rytkönen and T. Václavík for assistance with obtaining point location data for *P. ramorum*. Thanks also to N. Brouwers for assistance with ArcGIS and my PhD supervisors Daniel Húberli, Bernard Dell and Ian Smith for valuable editorial support. The authors would like to acknowledge the support of the Cooperative Research Centre for National Plant Biosecurity, especially for supporting key collaborative opportunities. We would also like to acknowledge the efforts of three anonymous reviewers and thank them for their valuable feedback.

Author Contributions

Conceived and designed the experiments: KBI GESH DJK. Performed the experiments: KBI. Analyzed the data: KBI DJK. Contributed reagents/materials/analysis tools: GESH DJK. Wrote the paper: KBI GESH DJK.

References

1. Rizzo DM, Garbelotto M, Davidson JM, Slaughter GW, Koike ST (2002) *Phytophthora ramorum* as the cause of extensive mortality of *Quercus* spp. and *Lithocarpus densiflorus* in California. Plant Disease 86: 205–214.

2. Brasier C, Webber J (2010) Plant Pathology: Sudden larch death. Nature 466: 824–825.

3. Kliejunas JT (2010) Sudden oak death and *Phytophthora ramorum*: a summary of the literature. Gen. Tech. Rep. PSW-GTR-234 Albany, CA: U.S. Department of Agriculture, Forest Service, Pacific Southwest Research Station.

4. Rizzo DM, Garbelotto M (2003) Sudden oak death: endangering California and Oregon forest ecosystems. Frontiers in Ecology and the Environment 1: 197–204.

5. Meentemeyer RK, Rank NE, Shoemaker DA, Oneal CB, Wickland AC, et al. (2008) Impact of sudden oak death on tree mortality in the Big Sur ecoregion of California. Biological Invasions 10: 1243–1255.

6. Webber J, Denman S. *Phytophthora ramorum* and *P. kernoviae*: An update on distribution, policy and management in Europe; 2009 15–18 June 2009; Santa Cruz, CA. Available: http://www.cnr.berkeley.edu/comtf/sodsymposium4/pdf/presentations/1.sodiv_webber%20final.pdf. Accessed 2011 April 1.

7. Hansen EM, Kanaskie A, Prospero S, McWilliams M, Goheen EM, et al. (2008) Epidemiology of *Phytophthora ramorum* in Oregon tanoak forests. Canadian Journal of Forest Research 38: 1133–1143.

8. Sansford CE, Inman AJ, Baker R, Brasier C, Frankel S, et al. (2009) Report on the risk of entry, establishment, spread and socio-economic loss and environmental impact and the appropriate level of management for *Phytophthora ramorum* for the EU. Deliverable Report 28. EU Sixth Framework Project RAPRA: Risk Analysis for *Phyophthora ramorum*. Available at: http://rapra.csl.gov.uk/pra/index.cfm. Accessed 18 March 2011.

9. Werres S, Kaminski K (2005) Characterisation of European and North American *Phytophthora ramorum* isolates due to their morphology and mating behaviour in vitro with heterothallic *Phytophthora* species. Mycological Research 109: 860–871.

10. Brasier C (2003) Sudden oak death: *Phytophthora ramorum* exhibits transatlantic differences. Mycological Research 107: 258–259.

11. Ivors K, Garbelotto M, Vries IDE, Ruyter-Spira C, Hekkert BT, et al. (2006) Microsatellite markers identify three lineages of *Phytophthora ramorum* in US nurseries, yet single lineages in US forest and European nursery populations. Molecular Ecology 15: 1493–1505.

12. Grünwald NJ, Goss EM, Ivors K, Garbelotto M, Martin FN, et al. (2009) Standardizing the nomenclature for clonal lineages of the Sudden Oak Death pathogen, *Phytophthora ramorum*. Phytopathology 99: 792–795.

13. Elliott M, Sumampong G, Varga A, Shamoun SF, James D, et al. (2011) Phenotypic differences among three clonal lineages of *Phytophthora ramorum*. Forest Pathology 41: 7–14.

14. Brasier C (2012) EU2, a fourth evolutionary lineage of *Phytophthora ramorum*. Sudden Oak Death 5th Science Symposium. Petaluma, California, USA. 19–22 June 2012. Available at: http://ucanr.org/sites/sod5/files/147414.pdf. Accessed 2013 Feb 16.

15. Goss EM, Carbone I, Grünwald NJ (2009) Ancient isolation and independent evolution of the three clonal lineages of the exotic sudden oak death pathogen *Phytophthora ramorum*. Molecular Ecology 18: 1161–1174.

16. University of California Berkeley Forest Pathology and Mycology Laboratory (2012) SODMAP. Available at: http://nature.berkeley.edu/garbelotto/english/sodmap.php. Accessed 23 September 2012.

17. COMTF (2011) California oak mortality task force newsletter. April 2011 ed: Available: http://www.suddenoakdeath.org/wp-content/uploads/2010/03/COMTF_Report_April_20111.pdf. Californian Oak Mortality Task Force. Accessed 2011 April 9.

18. Forestry Commission (2011) *Phytophthora ramorum* in larch trees - Update. Forestry Commission Great Britain. Available: http://www.forestry.gov.uk/forestry/INFD-8EJKP4. Accessed 2011 May 6.

19. Tsopelas P, Paplomatas E, Tjamos S, Soulioti N, Elena K (2011) First report of *Phytophthora ramorum* on *Rhododendron* in Greece. Plant Disease 95: 223.

20. Brasier C (2008) The biosecurity threat to the UK and global environment from international trade in plants. Plant Pathology 57: 792–808.

21. Dehnen-Schmutz K, Holdenrieder O, Jeger MJ, Pautasso M (2010) Structural change in the international horticultural industry: Some implications for plant health. Scientia Horticulturae 125: 1–15.

22. FAO (2006) International standards for phytosanitary measures. 1–24. Rome: Food and Agriculture Organisation of the United Nations. Available: http://www.fao.org/docrep/009/a0450e/a0450e00.htm. Accessed 2011 July 8.

23. Venette RC, Cohen SD (2006) Potential climatic suitability for establishment of *Phytophthora ramorum* within the contiguous United States. Forest Ecology and Management 231: 18–26.

24. Kluza DA, Vieglais DA, Andreasen JK, Peterson AT (2007) Sudden oak death: geographic risk estimates and predictions of origins. Plant Pathology 56: 580–587.

25. Guo QH, Kelly M, Graham CH (2005) Support vector machines for predicting distribution of sudden oak death in California. Ecological Modelling 182: 75–90.

26. USDA (2004) Sudden oak death: protecting America's woodlands from *Phytophthora ramorum*: State and Private Forestry, USDA Forest Service. Publication no. FS-794.

27. Magarey RD, Fowler GA, Borchert DM, Sutton TB, Colunga-Garcia M (2007) NAPPFAST: An internet system for the weather-based mapping of plant pathogens. Plant Disease 91: 336–345.

28. Meentemeyer RK, Cunniffe NJ, Cook AR, Filipe JAN, Hunter RD, et al. (2011) Epidemiological modeling of invasion in heterogeneous landscapes: spread of sudden oak death in California (1990–2030). Ecosphere 2: art17.

29. Kelly M, Meentemeyer RK (2002) Landscape dynamics of the spread of sudden oak death. Photogrammetric Engineering and Remote Sensing 68: 1001–1009.

30. Meentemeyer R, Rizzo D, Mark W, Lotz E (2004) Mapping the risk of establishment and spread of sudden oak death in California. Forest Ecology and Management 200: 195–214.

31. Guo QH, Kelly M, Gong P, Liu DS (2007) An object-based classification approach in mapping tree mortality using high spatial resolution imagery. Geoscience & Remote Sensing 44: 24–47.

32. Václavík T, Kanaskie A, Hansen EM, Ohmann JL, Meentemeyer RK (2010) Predicting potential and actual distribution of sudden oak death in Oregon: Prioritizing landscape contexts for early detection and eradication of disease outbreaks. Forest Ecology and Management 260: 1026–1035.

33. Václavík T, Meentemeyer RK (2009) Invasive species distribution modeling (iSDM): Are absence data and dispersal constraints needed to predict actual distributions? Ecological Modelling 220: 3248–3258.

34. Sutherst RW, Maywald GF, Kriticos DJ (2007) CLIMEX Version 3: User's Guide: Hearne Scientific Software Pty Ltd. Available: http://www.hearne.com.au/attachments/ClimexUserGuide3.pdf. Accessed 2011 April 3.

35. Sutherst RW, Maywald GF (1985) A computerised system for matching climates in ecology. Agriculture, Ecosystems & Environment 13: 281–299.

36. Scherm H, Yang XB (1999) Risk assessment for sudden death syndrome of soybean in the north-central United States. Agricultural Systems 59: 301–310.

37. Pinkard EA, Kriticos DJ, Wardlaw TJ, Carnegie AJ, Leriche A (2010) Estimating the spatio-temporal risk of disease epidemics using a bioclimatic niche model. Ecological Modelling 221: 2828–2838.

38. Watt MS, Kriticos DJ, Alcaraz S, Brown AV, Leriche A (2009) The hosts and potential geographic range of Dothistroma needle blight. Forest Ecology and Management 257: 1505–1519.

39. Yonow T, Kriticos DJ, Medd RW (2004) The potential geographic range of *Pyrenophora semeniperda*. Phytopathology 94: 805–812.

40. Hardham AR (2005) *Phytophthora cinnamomi*. Molecular Plant Pathology 6: 589–604.

41. Werres S, Marwitz R, Man in't Veld WA, De Cock AWAM, Bonants PJM, et al. (2001) *Phytophthora ramorum* sp. nov., a new pathogen on *Rhododendron* and *Viburnum*. Mycological Research 105: 1155–1165.

42. Englander L, Browning M, Tooley PW (2006) Growth and sporulation of *Phytophthora ramorum in vitro* in response to temperature and light. Mycologia 98: 365–373.

43. Fichtner EJ, Lynch SC, Rizzo DM (2007) Detection, distribution, sporulation, and survival of *Phytophthora ramorum* in a California redwood-tanoak forest soil. Phytopathology 97: 1366–1375.

44. Tjosvold SA, Chambers DL, Koike ST, Mori SR (2008) Disease on nursery stock as affected by environmental factors and seasonal inoculum levels of *Phytophthora ramorum* in stream water used for irrigation. Plant Disease 92: 1566–1573.

45. Tooley PW, Browning M, Berner D (2008) Recovery of *Phytophthora ramorum* following exposure to temperature extremes. Plant Disease 92: 431–437.

46. Fichtner EJ, Lynch SC, Rizzo DM (2009) Survival, dispersal, and potential soil-mediated suppression of *Phytophthora ramorum* in a California redwood-tanoak forest. Phytopathology 99: 608–619.

47. Kriticos DJ, Webber BL, Leriche A, Ota N, Macadam I, et al. (2011) CliMond: global high-resolution historical and future scenario climate surfaces for bioclimatic modelling. Methods in Ecology and Evolution 3: 53–64.

48. Kriticos DJ, Sutherst RW, Brown JR, Adkins SW, Maywald GF (2003) Climate change and biotic invasions: A case history of a tropical woody vine. Biological Invasions 5: 147–165.

49. Kriticos DJ, Yonow T, McFadyen RE (2005) The potential distribution of *Chromolaena odorata* (Siam weed) in relation to climate. Weed Research 45: 246–254.

50. Brasier C, Kirk S, Rose J (2006) Differences in phenotypic stability and adaptive variation between the main European and American lineages of *Phytophthora ramorum*. In: Brasier C, Jung T, Oβswald W, editors. Progress in Research on *Phytophthora* Diseases of Forest Trees. Farnham: Forest Research. 166–173.

51. Tooley PW, Browning M, Kyde KL, Berner D (2009) Effect of temperature and moisture period on infection of *Rhododendron* 'Cunningham's White' by *Phytophthora ramorum*. Phytopathology 99: 1045–1052.

52. Turner J, Jennings P, Humphries G (2005) *Phytophthora ramorum* epidemiology: sporulation potential, dispersal, infection, latency and survival. Defra Project Report PH0194. Department for Environment, Food and Rural Affairs. Available at: http://randd.defra.gov.uk/Document.aspx?Document=PH0194_2004_FRP.pdf. Accessed 8 July 2011.

53. Davidson JM, Wickland AC, Patterson HA, Falk KR, Rizzo DM (2005) Transmission of *Phytophthora ramorum* in mixed-evergreen forest in California. Phytopathology 95: 587–596.

54. Brasier CM, Scott JK (1994) European oak declines and global warming: a theoretical assessment with special reference to the activity of *Phytopthora cinnamomi*. EPPO Bulletin 24: 221–232.

55. Sutherst RW, Maywald GF, Bottomley W, Bourne A (2004) CLIMEX v2 User's Guide. Melbourne: Hearne Scientific Software.

56. Browning M, Englander L, Tooley PW, Berner D (2008) Survival of *Phytophthora ramorum* hyphae after exposure to temperature extremes and various humidities. Mycologia 100: 236–245.

57. Davidson JM, Rizzo DM, Garbelotto M, Tjosvold S, Slaughter GW (2002) *Phytophthora ramorum* and sudden oak death in California: II. Transmission and survival. In: Standiford RB, McCreary D, Purcell KL (tech. coords). Proceedings of the fifth symposium on oak woodlands: oaks in California's changing landscape 2001 October 22–25; San Diego, CA General Technical Report - PSW-GTR-184. Albany, CA: U.S. Department of Agriculture, Forest Service, Pacific Southwest Research Station. 741–749.

58. Turner J, Jennings P, Thorp G, MacDonough S, Beales P, et al. (2008) Management and containment of *Phytophthora ramorum* infections in the UK. Defra Project Report PH0308. http://randd.defra.gov.uk/Document. aspx?Document = ph0308_7379_FRP.pdf: Department for Environment, Food and Rural Affairs. Available at: http://randd.defra.gov.uk/Document. aspx?Document = ph0308_7379_FRP.pdf. Accessed 8 July 2011.

59. Sundheim L, Herrero ML, Rafoss T, Toppe B (2009) Pest risk assessment of *Phytophthora ramorum* in Norway. Opion of the Panel on Plant Health of the Norwegian Scientific Committee for Food Safety, 08/907–3 final. Oslo, Norway: Norwegian Science Committee (VKM).

60. COMTF (2010) California oak mortality task force newsletter. July 2010 ed: Available: http://www.suddenoakdeath.org/wp-content/uploads/2010/03/ COMTF_Report_July_2010.pdf. Californian Oak Mortality Task Force. Accessed 2011 April 9.

61. Maloney PE, Lynch SC, Kane SF, Jensen CE, Rizzo DM (2005) Establishment of an emerging generalist pathogen in redwood forest communities. Journal of Ecology 93: 899–905.

62. Rizzo DM, Garbelotto M, Hansen EA (2005) *Phytophthora ramorum*: Integrative research and management of an emerging pathogen in California and Oregon forests. Annual Review of Phytopathology 43: 309–335.

63. Sakai AK, Allendorf FW, Holt JS, Lodge DM, Molofsky J, et al. (2001) The population biology of invasive species. Annual Review of Ecology and Systematics 32: 305–332.

64. Kanaskie A, Goheen E, Osterbauer N, McWilliams M, Hansen E, et al. (2008) Eradication of *Phytophthora ramorum* from Oregon forests: status after 6 years. In: Frankel SJ, Shea PJ, Haverty MI (tech. coords). Proceedings of the sudden oak death third science symposium General Technical Report PSW-GTR-214. Albany, CA: U.S. Department of Agriculture, Forest Service, Pacific Southwest Research Station. 15–17.

65. Tooley PW, Kyde KL (2007) Susceptibility of some Eastern forest species to *Phytophthora ramorum*. Plant Disease 91: 435–438.

66. Kriticos DJ, Leriche A (2010) The effects of climate data precision on fitting and projecting species niche models. Ecography 33: 115–127.

67. Davidson JM, Patterson HA, Wickland AC, Fichtner EJ, Rizzo DM (2011) Forest type influences transmission of *Phytophthora ramorum* in California oak woodlands. Phytopathology 101: 492–501.

68. Tjosvold SA, Chambers DL, Fichtner EJ, Koike ST, Mori SR (2009) Disease risk of potting media infested with *Phytophthora ramorum* under nursery conditions. Plant Disease 93: 371–376.

69. Fowler G, Magarey R, Colunga M (2006) Climate-host mapping of *Phytophthora ramorum*, causal agent of sudden oak death. In: Frankel SJ, Shea PJ, Haverty MI (tech. coords). Proceedings of the sudden oak death second science symposium: the state of our knowledge General Technical Report PSW-GTR-196. Albany, CA: U.S. Department of Agriculture, Forest Service, Pacific Southwest Research Station. 329–332.

70. Sutherst RW (2003) Prediction of species geographical ranges. Journal of Biogeography 30: 805–816.

71. McKenney D, Hopkin AA, Campbell KL, Mackey BG, Foottit R (2003) Opportunities for improved risk assessments of exotic species in Canada using bioclimatic modeling. Environmental Monitoring and Assessment 88: 445–461.

72. Swiecki TJ, Bernhardt E (2007) Influence of local California Bay distribution on the risk of *Phytophthora ramorum* canker (Sudden Oak Death) in Coast Live Oak. Vacaville: Phytosphere Research. Available at: http://phytosphere.com/ publications/influence_bay_dist_SOD.htm. Accessed 20 May 2011.

73. Webber JF, Mullet M, Brasier CM (2010) Dieback and mortality of plantation Japanese larch (*Larix kaempferi*) associated with infection by *Phytophthora ramorum*. New Disease Reports 22: 19.

74. Brasier C, Denman S, Brown A, Webber J (2004) Sudden Oak Death (*Phytophthora ramorum*) discovered on trees in Europe. Mycological Research 108: 1108–1110.

75. Ireland KB, Hüberli D, Dell B, Smith IW, Rizzo DM, et al. (2011) Potential susceptibility of Australian flora to a NA2 isolate of *Phytophthora ramorum* and pathogen sporulation potential. Forest Pathology 42: 305–320.

76. Ireland KB, Hüberli D, Dell B, Smith IW, Rizzo DM, et al. (2012) Potential susceptibility of Australian native plant species to branch dieback and bole canker diseases caused by *Phytophthora ramorum*. Plant Pathology 61: 234–246.

77. Hüberli D, Lutzy B, Voss B, Calver M, Ormsby M, et al. (2008) Susceptibility of New Zealand flora to *Phytophthora ramorum* and pathogen sporulation potential: an approach based on the precautionary principle. Australasian Plant Pathology 37: 615–625.

78. Tooley PW, Browning M (2009) Susceptibility to *Phytophthora ramorum* and inoculum production potential of some common eastern forest understory plant species. Plant Disease 93: 249–256.

79. Goheen EM, Kubisiak TL, Zhao W (2006) The search for the origin of *Phytophthora ramorum*: A first look in Yunnan Province, People's Republic of China. In: Frankel SJS, Patrick J; Haverty MI (tech. coords). Proceedings of the sudden oak death second science symposium: the state of our knowledge 2005 January 18–21; Monterey, CA Gen Tech Rep PSW-GTR-196. Albany, CA, U.S.A.: Pacific Southwest Research Station, Forest Service, U.S. Department of Agriculture. 113–115.

80. Vannini A, Brown A, Brasier C, Vettraino A (2009) The search for *Phytophthora* centres of origin: *Phytophthora* species in mountain ecosystems in Nepal. In: Goheen EM, Frankel SJ (tech. coords). Proceedings of the Fourth Meeting of the International Union of Forest Research Organisations (IUFRO) Working Party S070209: Phytophthoras in Forest and Natural Ecosystems Gen Tech Rep PSW GTR-221. Albany, California, U.S.A.: U.S. Department of Agriculture, Forest Service, Pacific Southwest Research Station. 54–55.

81. Brasier CM, Vettraino AM, Chang TT, Vannini A (2010) Phytophthora lateralis discovered in an old growth Chamaecyparis forest in Taiwan. Plant Pathology 59: 595–603.

9

Direct Fitness Correlates and Thermal Consequences of Facultative Aggregation in a Desert Lizard

Alison R. Davis Rabosky[1,2,3]*, Ammon Corl[1,4], Heather E. M. Liwanag[1,5], Yann Surget-Groba[1,6], Barry Sinervo[1]

1 Department of Ecology and Evolutionary Biology, University of California Santa Cruz, Santa Cruz, California, United States of America, 2 Department of Integrative Biology and Museum of Vertebrate Zoology, University of California, Berkeley, California, United States of America, 3 Department of Ecology and Evolutionary Biology, University of Michigan, Ann Arbor, Michigan, United States of America, 4 Department of Evolutionary Biology, Evolutionary Biology Centre, Uppsala University, Uppsala, Sweden, 5 Department of Biology, Adelphi University, Garden City, New York, United States of America, 6 Ecological Evolution Group, Xishuangbanna Tropical Botanical Garden, Menglun, Mengla, Yunnan, P. R. China

Abstract

Social aggregation is a common behavioral phenomenon thought to evolve through adaptive benefits to group living. Comparing fitness differences between aggregated and solitary individuals in nature – necessary to infer an evolutionary benefit to living in groups – has proven difficult because communally-living species tend to be obligately social and behaviorally complex. However, these differences and the mechanisms driving them are critical to understanding how solitary individuals transition to group living, as well as how and why nascent social systems change over time. Here we demonstrate that facultative aggregation in a reptile (the Desert Night Lizard, *Xantusia vigilis*) confers direct reproductive success and survival advantages and that thermal benefits of winter huddling disproportionately benefit small juveniles, which can favor delayed dispersal of offspring and the formation of kin groups. Using climate projection models, however, we estimate that future aggregation in night lizards could decline more than 50% due to warmer temperatures. Our results support the theory that transitions to group living arise from direct benefits to social individuals and offer a clear mechanism for the origin of kin groups through juvenile philopatry. The temperature dependence of aggregation in this and other taxa suggests that environmental variation may be a powerful but underappreciated force in the rapid transition between social and solitary behavior.

Editor: Jane M. Waterman, University of Manitoba, Canada

Funding: This research was funded by grants from the American Museum of Natural History, the American Society of Ichthyologists and Herpetologists, the United States Department of Education, and NSF (DBI-09060346) to ARDR. The funders had no role in study design, data collection and analysis, decision to publish, or preparation of the manuscript.

Competing Interests: The authors have declared that no competing interests exist.

* E-mail: ardr@umich.edu

Introduction

Few topics within evolutionary biology have inspired more excitement and controversy than the origin and evolution of social behavior [1,2,3,4]. The tendency for individuals to form conspicuous groups is surprisingly widespread and easy to observe across both vertebrates and invertebrates, with repeated independent origins throughout many taxa [5,6,7,8,9,10]. Sociality can take many forms, from the simple clustering of individuals in space to complex forms of cooperative breeding and reciprocal altruism [11]. However, one of the most poorly understood aspects of the evolution of any social system is the origin of group interaction in a population of solitary individuals [5]. Regardless of the eventual complexity of a social system, the evolutionary transition from solitary living to any form of sociality must begin with the simple act of initiating and maintaining contact with conspecifics, which is best described as aggregation.

Social aggregation is thought to evolve though adaptive benefits to group living, but hypothesized benefits of communal behavior can be difficult to quantify, compare, and interpret in natural systems. Most empirical research on social evolution has focused on taxa with obligate sociality (eusocial insects, most birds and mammals) and highly complex social interactions, precluding comparison to solitary individuals and limiting insights to the maintenance of sociality rather than its origins [12]. Thus, the mechanisms and directionality of fitness advantages to group living remain contentious but critical to understanding why solitary individuals transition to communal living and how group formation can change over time [13,14,15,16].

Comparative studies across social insects [17], spiders [18], birds [7], mammals [6], fish [19], and lizards [8,9] have revealed that many groups form specifically through delayed dispersal of offspring, creating highly related kin groups [20]. The broad taxonomic distribution of this pattern suggests that extending the interaction of parents and offspring is a simple, common process by which solitary animals become social [8,20]. However, the ecological mechanisms promoting long-term juvenile philopatry are rarely tested empirically (but see evidence in cooperatively-breeding birds [21] and fish [22]). Hypothesized mechanisms tend to be species-specific and invoke complicated group cooperation scenarios, but several suggest dependence on resource availability or environmental conditions (biotic or abiotic) that are inherently unstable [20,21,23]. Directly testing these mechanisms offers a powerful way to understand how kin groups form by explaining why juveniles may drive the transition from solitary to group living and by predicting when and how this behavior may change as environmental conditions vary. This integration of ultimate fitness consequences of social behavior with the proximate mechanisms promoting aggregation among kin [13] is key to understanding how and why social systems may form and change over time.

Here we show direct empirical evidence that social group participation in a reptile (the Desert Night Lizard, *Xantusia vigilis*) provides both survival and reproductive success advantages, test a physiological mechanism driving these fitness benefits, and predict changes to social behavior with changing environmental conditions. *Xantusia vigilis* is a very small (1.5 g), long-lived (at least 8–10 years), viviparous lizard that lives at high densities under fallen logs (*Yucca sp.*) in the Mojave Desert of California, USA [24,25]. Night lizards are highly secretive and rarely seen away from cover objects. Despite their common name, night lizards are diurnally active, especially in the winter [26].

Every winter between November and February, night lizards aggregate into groups of 2–20 individuals underneath fallen logs (Figure S1). Molecular analyses and field studies have previously shown that these aggregations are often highly related family groups produced through delayed juvenile dispersal, although groups without juveniles can contain kin or unrelated individuals [8]. Cross-fostering manipulations show that juveniles remain philopatric and aggregate specifically when with kin as opposed to unrelated individuals, so these family groups do not form simply as a by-product of globally infrequent dispersal [27]. Furthermore, groups of related individuals are stable across years despite aggregation being a winter-restricted phenomenon, as individuals will re-select former aggregation partners from a high density pool after eight months of solitary living during the spring, summer, and early fall [8]. These groups form outside of the mating (May–June) or birthing (August–September) seasons and independently of resource distribution and habitat quality, as there are no environmental differences between sites containing aggregations as opposed to solitary lizards (Figure S2–S3; Text S1; [8]). However, only about 2/3 of the population participates in aggregations each winter [8], and this facultative behavior allows the rare comparison between aggregated and solitary individuals and a test of mechanisms promoting social group formation.

We used this natural variation in social behavior for three specific tests. First, we compared fitness measurements between naturally aggregated and solitary individuals in the field to test for adaptive benefits encouraging the formation of social groups. Second, we examined the temperature-dependence of aggregation behavior and compared rates of heat loss between solitary lizards and aggregations to test a thermal mechanism driving juveniles to join groups. Third, we used climate models to predict how this temperature-dependent aggregation will respond to changing environmental conditions, informing how social behavior can change over time.

Results

To test for reproductive success and survival consequences of aggregation, we combined a mark-recapture field study of 2,332 lizards in 441 social groups with a DNA microsatellite analysis of parentage. For both males and females, we found that lizards caught in winter aggregations had higher reproductive success in the following summer than solitary individuals (females: $t = 2.70$, df = 32, $P = 0.011$; males: x = 24, n = 24, $P < 0.001$; Figure 1A). In males, this reproductive skew was extreme, and no winter-solitary male was ever found to sire offspring in a consecutive summer,

while several winter-aggregated males sired offspring by multiple females within a single reproductive season. We also found higher survival of aggregated individuals in adult females ($\chi^2 = 4.31$, df = 1, $P = 0.038$) and a trend in the same direction for juveniles ($\chi^2 = 2.82$, df = 1, $P = 0.093$) and adult males ($\chi^2 = 2.20$, df = 1, $P = 0.138$; Figure 1B), which are significant when combined with a weighted Z-test ([28]; $P = 0.0116$; Table S1). Additionally, we found that aggregated lizards had better body condition (plumper per unit body length) than solitary lizards, although this effect was not significant in adult males (females: $F_{1,638} = 4.07$, $P = 0.04$; males: $F_{1,387} = 0.398$, $P = 0.53$; juveniles: $F_{1,505} = 7.15$, $P = 0.008$; Figure 1C).

As aggregation was only observed during winter, we next examined the role of environmental temperature in the prevalence of this behavior (see Text S1 for tests and discussion of other non-significant environmental variables). We found that group formation was strongly temperature-dependent, with high levels of aggregation occurring only on cold winter days (Figure 2 inset, with winter only observations; $\rho = -0.63$, N = 17, $P = 0.007$; sigmoidal function with both winter and summer points was only used to generate predictions of aggregation levels at unobserved temperatures, see Methods). Daily observed aggregation levels did not depend on the total number of lizards caught ($P = 0.21$) and were not an artifact of small-scale fluctuations in daytime temperature (Figure S4).

Due to the basic principles of convective heat transfer and the fact that huddling physically alters surface area to volume ratios, there are unavoidable thermal consequences to winter aggregation in night lizards. We quantified these consequences by comparing heat flux in solitary individuals, natural aggregations, and experimentally isolated lizards. We found that, as expected, aggregations cool more slowly than solitary lizards (Figure 3A; decay time for solitary lizards: $F_{1,51} = 5.07$, $P = 0.029$, Figure 3B; decay time for aggregations: $F_{1,53} = 12.16$, $P < 0.001$, Figure 3C; decay rate for aggregations: $F_{1,52} = 14.08$, $P < 0.001$, Figure 3D). This relationship scales positively with mass, such that small juveniles track environmental temperature very closely and derive the greatest increase in thermal stability from social behavior (Figure 3A; note comparison to blank [empty arena] showing environmental temperature). Small-massed juveniles joining aggregations take 30–60% longer to reach equilibrium with environmental temperature than solitary juveniles (Figure 3A–C). When comparing solitary neonates to aggregations under natural field conditions, this heat loss difference translates into an average time lag to temperature equilibrium of about 6.5 hours during a typical night of cooling, and aggregations only experience the coldest temperatures for a limited amount of time (Figure 3E; solitary mean rate = -0.484, aggregation mean rate = -0.329). As winter temperatures in this habitat regularly drop below freezing (Table S2), even underneath fallen logs (Figure S3), the thermal buffer of aggregation is critical to avoiding mortality from freezing and the metabolically expensive tissue repair associated with even mild freeze events (Supporting Information; [29]).

Given the strong temperature-dependence of aggregation, we were able to use four climate projection models to generate predictions about how social behavior will respond to long-term changes in average temperature. By modeling monthly aggregation levels from 1950–2099, we found that the predicted proportion of the population aggregating each winter decreased over time both in peak magnitude and in the number of months per year with high levels of aggregation (Figure 4A). By 2099, annual aggregation is predicted to decline by a minimum of 20% to more than 50% of 1950 levels (four model mean = 33%), depending on the severity of the climate model (Figure 4B).

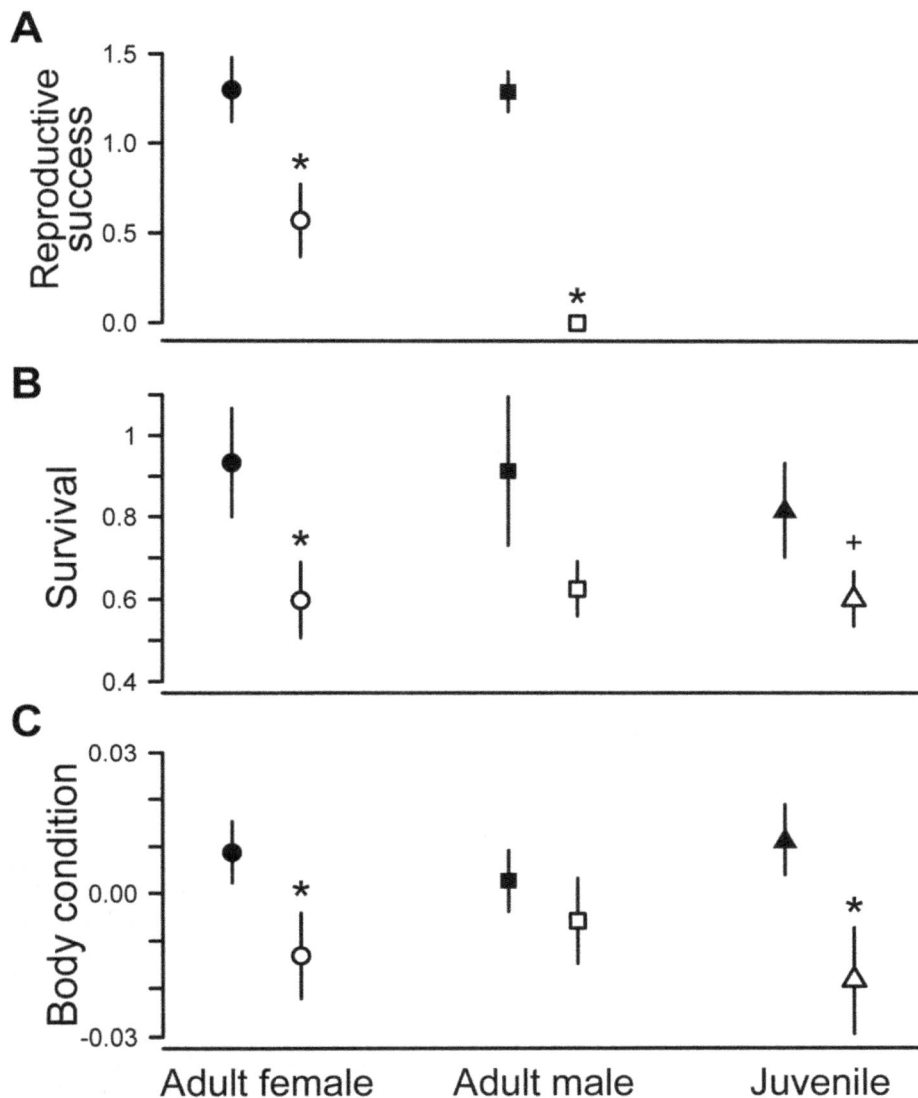

Figure 1. Fitness comparisons of aggregated and solitary night lizards. Higher fitness of aggregated (filled symbols) versus solitary (open symbols) lizards shows clear adaptive benefits to winter aggregation (circles = females, squares = males, triangles = juveniles). (**A**) Reproductive success (# offspring) of lizards that were aggregated or solitary in the winter directly preceding summer reproduction ($N_F = 34$, $N_M = 24$; see text). (**B**) Multi-state model estimates of survival associated with aggregated and solitary social states ($N_F = 723$, $N_M = 443$, $N_J = 1166$). (**C**) Body condition (residual mass on body size) of all winter-collected lizards by aggregation state ($N_F = 640$, $N_M = 389$, $N_J = 507$). Asterisks and cross denote $P < 0.05$ and $P < 0.1$ significance levels, respectively, and error bars are ± 1SEM.

Discussion

Our study provides strong empirical evidence that direct fitness benefits favor social aggregation and offers a clear mechanistic explanation for the formation of kin groups through juvenile philopatry [20]. Juvenile lizards are disproportionately favored to join winter groups because of their small size, but the fitness and thermal benefits of aggregation are evident in all age and sex classes. However, our results also suggest that aggregated males, females, and juveniles may have multiple ways of benefiting in this system. Our finding that aggregated lizards have higher fitness could be explained either by social interaction driving fitness benefits or by the preferential aggregation of otherwise high fitness individuals. The former interpretation better fits the observed data because the ability of aggregations to increase thermal buffering

and reduce metabolic costs uniquely accounts for both the temperature dependence of group living and the absence of body condition differences in males. Both juveniles and females have special body condition needs not experienced by adult males; juveniles are challenged by their tiny size (average mass = 0.25 g) and high ratio of surface area to volume, and viviparous females by their high reproductive investment in litters up to half their body mass [26]. However, males may also use aggregations as a form of mate-guarding and be better able to retain female aggregation mates, accounting for the extreme reproductive skew towards social males. These two processes are not necessarily mutually exclusive, and they may well form a positive feedback loop promoting the maintenance of aggregative behavior even after the initial benefit to philopatric juveniles.

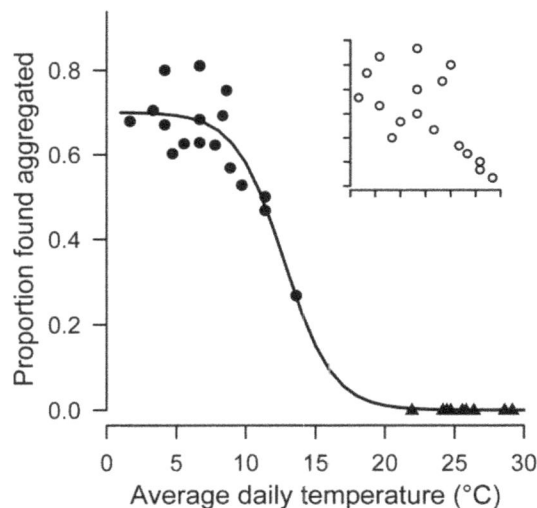

Figure 2. Social aggregation as a function of environmental temperature. The proportion of total lizards found aggregated each field collection day shows strong temperature-dependence ($N = 17$ winter [circles] and $N = 9$ summer [triangles] days; sigmoidal function with both winter and summer points was only used to generate predictions of aggregation levels at unobserved temperatures, see Methods and Figure 4). Inset shows Spearman correlation among ranked winter days only ($\rho = -0.63$, $P = 0.007$).

This research has implications for appreciating the potentially large ecological consequences of environmental variation that can occur when animal behavior is mediated by temperature. Potential and realized effects of climate change on species diversification [30], extinction [31], geographic range or niche shifts [32], and reproductive phenology [33] have been thoroughly discussed, but purely behavioral responses to climate change have received less attention (except for migration [34]). However, temperature-mediated behavior is present in many animal systems, of which huddling at cold temperatures is especially common [35,36,37], suggesting that our results are part of a more general phenomenon. In line with what we describe here in *Xantusia*, aggregation is considered beneficial in other winter huddling species because of the thermal advantages of buffering against extreme temperatures, although exact mechanisms vary slightly due to the differences in metabolic profiles between endotherms and ectotherms. Indeed, winter aggregation in some lizard species is also suggested to be mediated by thermal buffering [38,39], although it should be noted that other species seem to have quite extensive social interactions without significant thermal pressures (especially in the Australian skink genus *Egernia*, reviewed in [9]).

The major question for species with temperature-mediated social aggregation is the ecological and behavioral response to broad scale changes in climate. Although our data suggest that there may be additional benefits to aggregation other than simple thermal buffering at cold temperatures (*e.g.*, male reproductive success), the strong temperature dependence of social behavior (Figure 2) means that a mechanism is already in place that yields less aggregation at warmer temperatures. For annual aggregation to be maintained at current levels under future climate scenarios, this established relationship between temperature and group formation would have to break down. Moreover, several recent studies have implicated a strong environmental effect on macroevolutionary patterns of sociality and facultative social behavior [23,40,41], and changes in either behavior or ecology

should be expected when the benefits of sociality are dependent on particular environmental conditions. Recognizing that simple changes in temperature may have profound and cascading effects on higher-order processes like social evolution is critical to appreciating the ramifications of future environmental change.

From a macroevolutionary perspective, investigating the transition between solitary and group living may inform the ramifications of extended conspecific interactions in the broader context of the evolution of sociality. The popular perception of social evolution is one of a unidirectional "evolutionary trajectory," a progressive process in which species get locked into obligate sociality. Although this model is likely true for some systems (particularly eusocial insects), our data support the idea that social group formation may be a very dynamic process in species that maintain reproductive independence of individuals [16], including repeated bidirectional transitions between social and solitary states [42]. Especially in the early stages of sociality, transitions between social and solitary behavior may occur easily and rapidly by tracking changes in an ecological or environmental variable like temperature. Behavioral modifications are likely to be the first changes seen when environmental conditions vary through time, and this study adds to a growing body of work highlighting the importance of social plasticity and the role of environment in social evolution. Broadening such perspectives can guide both theoretical and empirical research by generating testable predictions about how, why, when, and in which species social behavior may arise and change over time and space.

Methods

Ethics Statement

All methods were approved by the Chancellor's Animal Research Committee at the University of California, Santa Cruz (Sine00.02-1) and California Department of Fish and Game (SC-05560 to ARDR).

Field Collection

We conducted a mark-recapture survey of 2,332 lizards from 441 social groups every summer and winter from August 2003 - January 2008 on a 36 hectare plot in the western Mojave Desert, approximately 16 km from Pearblossom, CA, USA (UTM coordinates easting 434561, northing 3816412, zone 11N). We hand-captured lizards by turning every fallen Joshua tree (*Yucca brevifolia*) log in this plot once per season.

Each winter, we classified any lizards found within a radius of approximately 30 cm of each other as aggregated. Although the vast majority of aggregated lizards were found in direct physical contact and unambiguously intertwined with the other group members, this guideline was occasionally necessary to account for the fleeing of lizards due to the unavoidable effects of sampling disturbance incurred by the rolling of logs (see Figure S1; [8]). For reference, this distance corresponds to approximately 3.5 times the total length of an adult *X. vigilis* and to the distance an adult can sprint in 0.5–1.5 seconds, even at cold body temperatures ([43]; Supporting Information). Although we think it is important to report that this guideline was used, we rarely needed to employ it and do not believe that it had a significant effect on our results or interpretation.

At each capture or birth, we measured the mass, snout-vent length (SVL), and tail condition (broken, regrown, intact) of each lizard. We sexed each lizard by shining a light through the base of the tail to visualize hemipenes in males [44]. We then toe-clipped each new individual for future identification and took a small piece of tail tissue (stored in 95% ethanol) for genetic analysis of paternity.

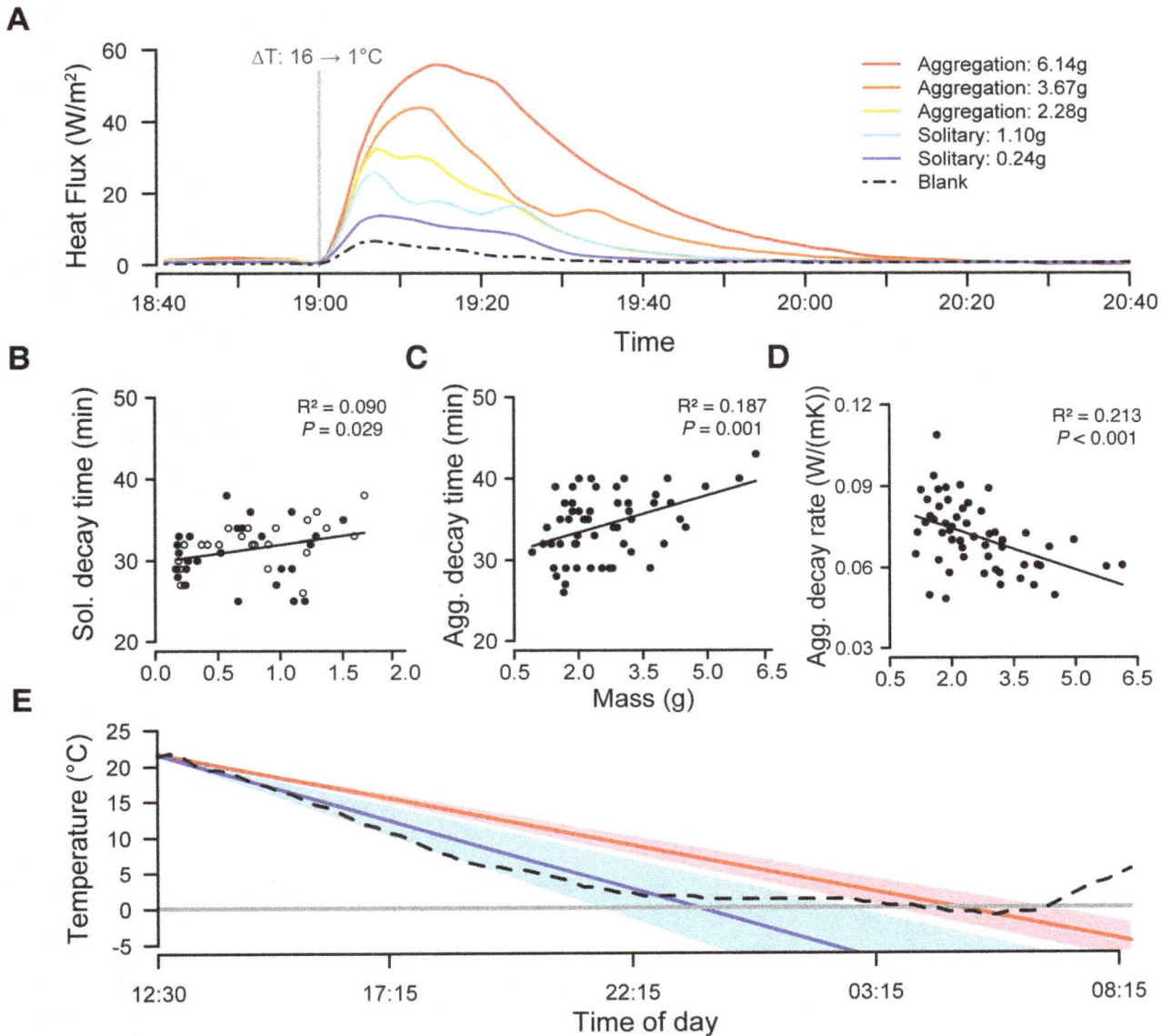

Figure 3. Rates of heat loss in solitary and aggregated night lizards. (**A**) Representative heat loss curves measured for two solitary lizards and three aggregations of different mass show that thermal stability increases with mass because lizards in large aggregations take longer than solitary lizards to return to equilibrium with environmental temperature. Properties of each curve are represented by single points in panels B–D. (**B**) Thermal decay time (two half-lives) for solitary lizards increases with mass ($F_{1,51} = 5.07$, $P = 0.029$). There was no difference between naturally solitary (closed circles) and experimentally isolated (open circles) lizards ($F_{1,50} = 0.59$, $P = 0.45$). (**C**) Thermal decay time (two half-lives) for lizard aggregations increases with mass ($F_{1,53} = 12.16$, $P < 0.001$). (**D**) Rates of thermal decay (heat loss) decrease as aggregation mass increases ($F_{1,52} = 14.08$, $P < 0.001$). (**E**) Rates of heat loss under natural conditions, as predicted from measured laboratory rates in (D), show that solitary neonates (blue) reach equilibrium with microclimate temperature (dashed black line; intersection) much earlier than aggregations greater than 4g (red) and spend more time at cold temperatures. Shading denotes ±1 SD, and gray line at 0°C shows environmental freezing point.

Reproductive Success

To assess the reproductive success of social and solitary lizards, we analyzed only adults for which we knew aggregation status in the winter directly preceding summer reproduction ($N = 34$ females, 24 males, all from separate aggregations). Summer-recaptured females were kept in the laboratory until parturition to determine litter size and then returned with their offspring to their exact log of capture. We used general linear models (GLMs) to test for the potential confounding effect of body size (SVL) on litter size ($P = 0.56$) and to assess the effect of winter sociality on litter size.

To assign paternity, we genotyped 369 females, 249 males, and 624 juveniles at seven unlinked, highly polymorphic microsatellite loci [8]. We assigned paternity to candidate sires with pair or trio critical LOD scores above the 80% confidence level in CERVUS v3.0.3 and included genotypes of known mothers when available. Although we were able to confidently assign paternity of 230 juveniles to 123 sires, only 24 of these males were also captured in the winter directly preceding offspring birth. Because all 24 of these sires were aggregated males, we used a binomial test with probabilities equal to the observed frequencies of the two aggregation states (69% of winter-captured males were aggregated)

A

B

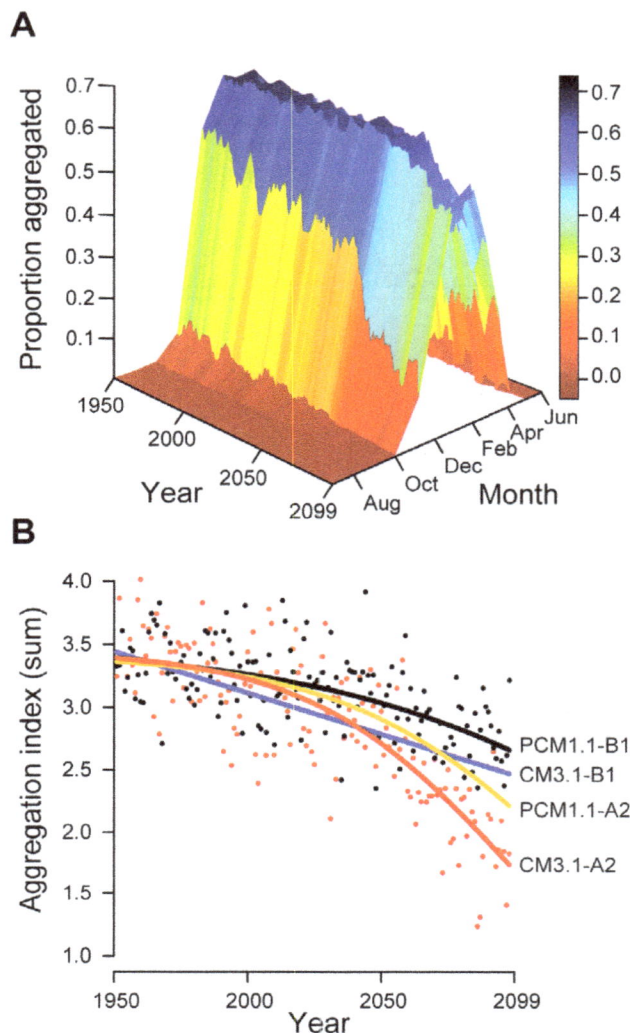

Figure 4. Predicted changes in night lizard aggregation under projected climate scenarios. (**A**) Proportion of the population predicted to aggregate under a climate projection model (CNRM CM3.1-A2). Over time, this proportion decreases both in peak magnitude and in number of months per year with high levels of aggregation. (**B**) Sum of monthly aggregation proportions by year (aggregation index) under two climate models (CNRM CM3.1 and NCAR PCM1.1) and two emissions projections (A2 and B1). Predicted aggregation declines by 20% under mild climate projections and more than 50% under more severe projections (four model mean = 33%). Points generating the curves are shown only for the two most disparate models.

to assess the effect of winter sociality on male reproductive success. There was no significant effect of male body size (SVL) on reproductive success ($P = 0.36$).

Survival

To estimate survival of social and solitary lizards, we used multistate models in MARK v6.0 by coding encounter histories as A (solitary state), B (aggregated state), or 0 (not captured) over the 10 sampling occasions ($N = 723$ females, 443 males, 1166 juveniles). We constructed four main models with constant but state-specific survival (Φ_A, Φ_B), time-dependent state transition probabilities (Ψ_A, Ψ_B), and all four combinations of constant and time-

dependent capture probabilities by aggregation state (pA, pB). To minimize over-parameterizing models, we independently analyzed adult males, adult females, and juveniles. Our sampling occasions were not equally spaced across the year, as winter observations were taken in December/January and summer observations in August/September during late stages of female pregnancy (see Methods for Reproductive Success). We accounted for these unequal time intervals by calculating the fraction of one year that had passed between the last day of the earlier sampling occasion and the first day of the later sampling occasions, and then scaling them by dividing each by the longest duration between sampling occasions. To account for aggregation (State B) only being present in the winter, we fixed eight of the possible 18 state transition probabilities (Ψ): $\Psi_{A \to B}$ was fixed to 0 and $\Psi_{B \to A}$ was fixed to 1 for all four winter to summer transitions.

We ran all models with the logit link function, 2ndPart variation estimation, and identity design matrix options. We then weight-averaged the real survival parameter estimates from our four models to obtain robust estimates of survival (Table S1) and compared solitary and aggregated survival estimates and error using CONTRAST [45]. We also performed likelihood ratio tests comparing the model with the lowest AIC score to the most reduced model (all parameters constant), and in all cases, the best fit model was significantly better than the reduced model (females: $\chi^2 = 113.3$, df = 11, $P < 0.0001$; males: $\chi^2 = 42.6$, df = 3, $P < 0.0001$; juveniles: $\chi^2 = 101.5$, df = 9, $P < 0.0001$).

Body Condition

To compare body condition of aggregated and solitary lizards, we log transformed (for linearity) mass and SVL of all winter-caught lizards at time of capture and regressed log mass against log SVL by age/sex class ($N = 640$ females, 389 males, 507 juveniles). We then used residuals from class-specific regressions as an index of body condition and used linear models to compare aggregated and solitary lizards within each class. All individuals with incomplete tails were excluded from analysis, and pregnancy does not occur in the winter.

Environmental Temperature

We obtained historical temperature data for Pearblossom, CA, USA by downloading daily and monthly temperatures for the Pearblossom weather station (PWS; #046773) from the NOAA National Climatic Data Center (NCDC) database for all available years (1986–2009). We compiled summary statistics for annual number of nights below 0°C and annual extreme T_{min} for both the full 24 year period for which data were available and the 2003–2008 period during which we conducted our study (Table S2).

To assess the effect of field temperature on aggregation, we averaged maximum and minimum daily temperature data for each collection day from the PWS. We analyzed only days for which we caught more than 30 lizards to avoid small sample biases ($N = 17$ winter and 9 summer days). We then fit a self-starting three parameter logistic model to generate predicted aggregation values by temperature and used a Spearman rank test on the 17 winter collection days to assess the effect of temperature on winter aggregation.

Laboratory Measurements of Heat Flux

To quantify the thermal stability of social groups, we measured heat flux of 26 naturally solitary lizards, 55 natural aggregations, and 27 experimentally isolated lizards originally found in aggregations. We placed lizards inside 59 ml cylindrical airtight plastic arenas filled across the bottom by a 2 cm diameter heat flux disk (Thermonetics, Inc.; Figure S5). We then placed these arenas inside an environmental chamber for at least two hours at 16°C before the temperature dropped to 1°C for 12 hours to simulate natural daytime and nighttime winter temperatures. Each heat flux disk was connected to an external data logger (Hydra®) that scanned all arenas once every six seconds and recorded heat loss curves for each arena each night. We also weighed lizards before and after each run to measure evaporative water loss (Text S1; Fig S3B). To compare cooling rates of aggregated and solitary lizards, we averaged measurements from each arena over one minute intervals, calculated the half-lives and rate of decay (by linearizing) for each of these classic first-order decay curves (Figure S5), and used GLMs to assess the effect of mass and aggregation. Although the cooling rate of this experiment is much faster than the real rate of environmental cooling, we could combine these lab-measured heat loss rates (lizard and environmental) with field-measured cooling rates (environmental; Text S1) to extrapolate natural rates of lizard heat loss and calculate the magnitude (mean and s.d.) of thermal buffering derived by juveniles (<0.3 g) joining aggregations (>4 g) in nature.

Climate Models

We used LLNL-Reclamation-SCU downscaled climate projections data derived from the World Climate Research Programme's (WCRP's) Coupled Model Intercomparison Project phase 3 (CMIP3) multimodel dataset, stored and served at the LLNL Green Data Oasis. As per the setup of this online database's access capabilities, we queried the closest geographic approximation of our field site, which was a 156 km^2 rectangle encompassing our site in the middle of the rectangle's eastern side. This downscaling method predicts average monthly temperatures for each of the four vertices of this rectangle, and we then averaged the values from the two northernmost vertices that were unambiguously in the desert instead of the foothills of the nearby San Gabriel Mountains (see Figure S1B) so as not to bias our temperature models towards elevations higher than our field site. We obtained projected mean monthly temperatures for two climate models (Centre National de Recherches Météorologiques (CNRM) CM3.1 and National Center for Atmospheric Research (NCAR) PCM1.1) under two different emissions scenarios (A2 and B1), yielding a total of four models.

Modeling Social Behavior on Climate Projections

After acquiring average monthly temperature projections for the four climate models, we used the three parameter logistic model generated from observed aggregation by mean temperature (Figure 2) to predict monthly aggregation proportions for each model from 1950–2099. We then summed monthly aggregation proportions by year (area under yearly curves, Figure 4A) to create an Aggregation Index and fit logistic and linear models as appropriate to assess change in aggregation over time for each climate model (Figure 4B). Climate model CM3.1-B1 produced a linear relationship instead of a logistic curve. All four of these statistical models showed a significant negative relationship between cumulative annual aggregation and time as compared to model specific zero-slope lines (PCM1.1-B1: $\Delta AIC = 62.78$; CM3.1-B1: $\Delta AIC = 83.71$; PCM1.1-A2: $\Delta AIC = 115.15$; CM3.1-A2: $\Delta AIC = 170.86$).

Statistical Analysis

Unless otherwise stated, we performed all statistical tests in R v2.13.1 and assessed significance at $P<0.05$. For parametric analyses, normality of residuals was assessed using Shapiro-Wilk tests, linearity by visual assessment of residual by predicted plots, autocorrelation with Durbin-Watson tests, and homogeneity of variance with Levene's test for all relevant analyses. Nonparametric tests were chosen where described for data that violated assumptions of parametric analyses. Unless otherwise noted, only significant effects are reported.

Supporting Information

Figure S1 Night lizard aggregation and Joshua tree habitat. (A) *In situ* night lizard (*Xantusia vigilis*) aggregation of three adults and two juveniles demonstrates winter huddling behavior. The lizard above is walking away from the aggregation after being disturbed by the rolling of the cover log. **(B)** Joshua tree (*Yucca brevifolia*) habitat at the field site shows both living trees and fallen logs, the site of winter aggregation.

Figure S2 Fallen log microhabitat characteristics do not predict aggregation. (A) None of the following habitat variables vary between sites with aggregations versus solitary lizards (left to right): Under log temperature (residual from regression with air temperature), buffer from air temperature, log decomposition (*p*-value from Chi square test of count data, excluding logs with no lizards), log size, nearest fallen log, or nearest living tree. **(B)** Evaporative water loss (EWL) rates are the same in aggregated and solitary lizards, and EWL rates do not scale with aggregation mass as expected if water loss needs were driving social behavior. Comparison graphs show means ±1 s.d.

Figure S3 Daily temperature profiles underneath fallen logs. Microhabitat temperatures measured by HOBO data loggers underneath five fallen logs are remarkably similar to environmental temperatures at the Pearblossom weather station, especially underneath preferred logs of medium decomposition (top row). Sheltering logs can weakly buffer lizards against extreme temperatures, but the maximum effect of this buffer is no more than a few degrees C and subzero temperatures are encountered as late as March.

Figure S4 Lack of aggregation capture bias during daily temperature fluctuation. (A) Typical temperature profile from sunrise to sunset (3 March 2002) from a HOBO data logger underneath fallen log of medium decomposition. (B) Group size by daily capture order (sunrise to sunset) show that lizards caught at the beginning of the day (the coldest temperatures) were not more likely to be aggregated. Aggregations were found throughout each collection day. Collection dates and average daily temperatures are at the top of each graph.

Figure S5 Arenas for laboratory measurements of heat flux. We measured heat flux in 26 naturally solitary lizards, 55 natural aggregations, and 27 experimentally isolated lizards originally found in aggregations by placing them inside plastic arenas filled across the bottom by a heat flux disk. We then placed these arenas inside an environmental chamber at 16°C before the temperature dropped to 1°C for 12 hours to simulate natural daytime and nighttime winter temperatures. We recorded changes in heat flux over time to generate heat loss curves for which we then calculated decay rate and half-life to compare the thermal stability of solitary and aggregated lizards.

Table S1 Multi-strata survival model output from MARK. All four main models for each class (female, male, juvenile), with constant survival (Φ) for each state (A = solitary, B =

aggregated), time-dependent transition probabilities between states ($\Psi_{A \to B}$, $\Psi_{B \to A}$), but variable capture probability (p) parameters (t = time-dependent, . = constant). The weighted model averages are bolded, and the fully time-constant and time-dependent models are italicized for comparison. Estimates of survival are remarkably robust to changes in model structure; in all cases, the aggregated state is associated with higher survival than the solitary state.

Table S2 Historical weather data for Pearblossom, CA, USA. Annual average and extreme minimum temperatures and number of nights below freezing from 1986–2009 (subset 2003–2008 is the duration of this study). Source is NOAA National Climatic Data Center (NCDC) for the Pearblossom weather station (#046773).

Acknowledgments

We thank S. Adolph, J. Lighton, B. Lyon, D. Rabosky, P. Raimondi, M. Taborsky, T. Williams, D. Wake, the Sinervo Lab, the McGuire Lab, and anonymous reviewers for assistance and comments on the manuscript.

Author Contributions

Conceived and designed the experiments: ARDR AC HEML YSG BS. Performed the experiments: ARDR AC HEML YSG. Analyzed the data: ARDR. Contributed reagents/materials/analysis tools: ARDR YSG. Wrote the paper: ARDR.

References

1. Hamilton WD (1964) Genetical evolution of social behaviour I & II. Journal of Theoretical Biology 7: 1–52.
2. Wilson EO (1975) Sociobiology: The New Synthesis. Cambridge, MA: Harvard University Press.
3. Nowak MA, Tarnita CE, Wilson EO (2010) The evolution of eusociality. Nature 466: 1057–1062.
4. Abbot P, Abe J, Alcock J, Alizon S, Alpedrinha JAC, et al. (2011) Inclusive fitness theory and eusociality. Nature 471: E1–E4.
5. Agnarsson I, Aviles L, Coddington JA, Maddison WP (2006) Sociality in Theridiid spiders: Repeated origins of an evolutionary dead end. Evolution 60: 2342–2351.
6. Clutton-Brock T (2009) Structure and function in mammalian societies. Philosophical Transactions of the Royal Society B-Biological Sciences 364: 3229–3242.
7. Hatchwell BJ (2009) The evolution of cooperative breeding in birds: kinship, dispersal and life history. Philosophical Transactions of the Royal Society B-Biological Sciences 364: 3217–3227.
8. Davis AR, Corl A, Surget-Groba Y, Sinervo B (2011) Convergent evolution of kin-based sociality in a lizard. Proceedings of the Royal Society B-Biological Sciences 278: 1507–1514.
9. Chapple DG (2003) Ecology, life-history, and behavior in the Australian Scincid genus *Egernia*, with comments on the evolution of complex sociality in lizards. Herpetological Monographs 17: 145–180.
10. Hughes WOH, Oldroyd BP, Beekman M, Ratnieks FLW (2008) Ancestral monogamy shows kin selection is key to the evolution of eusociality. Science 320: 1213–1216.
11. Krause J, Ruxton GD (2002) Living in Groups. Oxford, UK: Oxford University Press. 224 p.
12. Silk JB (2007) The adaptive value of sociality in mammalian groups. Philosophical Transactions of the Royal Society B-Biological Sciences 362: 539–559.
13. West SA, Griffin AS, Gardner A (2007) Social semantics: altruism, cooperation, mutualism, strong reciprocity and group selection. Journal of Evolutionary Biology 20: 415–432.
14. Wilson DS, Wilson EO (2007) Rethinking the theoretical foundation of sociobiology. Quarterly Review of Biology 82: 327–348.
15. Okasha S (2010) Altruism researchers must cooperate. Nature 467: 653–655.
16. Helms Cahan S, Blumstein DT, Sundstrom L, Liebig J, Griffin A (2002) Social trajectories and the evolution of social behavior. Oikos 96: 206–216.
17. Hunt JH (1999) Trait mapping and salience in the evolution of eusocial vespid wasps. Evolution 53: 225–237.
18. Jones TC, Parker PG (2002) Delayed juvenile dispersal benefits both mother and offspring in the cooperative spider *Anelosimus studiosus* (Araneae : Theridiidae). Behavioral Ecology 13: 142–148.
19. Dierkes P, Heg D, Taborsky M, Skubic E, Achmann R (2005) Genetic relatedness in groups is sex-specific and declines with age of helpers in a cooperatively breeding cichlid. Ecology Letters 8: 968–975.
20. Emlen ST (1995) An evolutionary theory of the family. Proceedings of the National Academy of Sciences of the United States of America 92: 8092–8099.
21. Eikenaar C, Richardson DS, Brouwer L, Komdeur J (2007) Parent presence, delayed dispersal, and territory acquisition in the Seychelles warbler. Behavioral Ecology 18: 874–879.
22. Heg D, Rothenberger S, Schürch R (2011) Habitat saturation, benefits of philopatry, relatedness, and the extent of co-operative breeding in a cichlid. Behavioral Ecology 22: 82–92.
23. Jetz W, Rubenstein DR (2011) Environmental uncertainty and the global biogeography of cooperative breeding in birds. Current Biology 21: 72–78.
24. Stebbins RC (2003) A Field Guide to Western Reptiles and Amphibians. New York: Houghton Mifflin Publishing Company. 560 p.
25. Zweifel RG, Lowe CH (1966) The ecology of a population of *Xantusia vigilis*, the Desert Night Lizard. American Museum Novitates 2247: 1–57.
26. Miller MR (1951) Some aspects of the life history of the Yucca Night Lizard, *Xantusia vigilis*. Copeia: 114–120.
27. Davis AR (2012) Kin presence drives philopatry and social aggregation in juvenile Desert Night Lizards (*Xantusia vigilis*). Behavioral Ecology 23: 18–24.
28. Whitlock MC (2005) Combining probability from independent tests: the weighted Z-method is superior to Fisher's approach. Journal of Evolutionary Biology 18: 1368–1373.
29. Voituron Y, Storey JM, Grenot C, Storey KB (2002) Freezing survival, body ice content and blood composition of the freeze-tolerant European common lizard, *Lacerta vivipara*. Journal of Comparative Physiology B-Biochemical Systemic and Environmental Physiology 172: 71–76.
30. Rabosky DL, Sorhannus U (2009) Diversity dynamics of marine planktonic diatoms across the Cenozoic. Nature 457: 183–186.
31. Sinervo B, Mendez-de-la-Cruz F, Miles DB, Heulin B, Bastiaans E, et al. (2010) Erosion of lizard diversity by climate change and altered thermal niches. Science 328: 894–899.
32. Tingley MW, Monahan WB, Beissinger SR, Moritz C (2009) Birds track their Grinnellian niche through a century of climate change. Proceedings of the National Academy of Sciences USA: 19637–19643.
33. Fitter AH, Fitter RSR (2002) Rapid changes in flowering time in British plants. Science 296: 1689–1691.
34. Cotton PA (2003) Avian migration phenology and global climate change. Proceedings of the National Academy of Sciences of the United States of America 100: 12219–12222.
35. Ancel A, Visser H, Handrich Y, Masman D, LeMaho Y (1997) Energy saving in huddling penguins. Nature 385: 304–305.
36. Berteaux D, Bergeron JM, Thomas DW, Lapierre H (1996) Solitude versus gregariousness: Do physical benefits drive the choice in overwintering meadow voles? Oikos 76: 330–336.
37. Arnold W (1988) Social thermoregulation during hibernation in alpine marmots (*Marmota marmota*). Journal of Comparative Physiology B: Biochemical, Systemic, and Environmental Physiology 158: 151–156.
38. Elfstrom BEO, Zucker N (1999) Winter aggregation and its relationship to social status in the tree lizard, *Urosaurus ornatus*. Journal of Herpetology 33: 240–248.
39. Shah B, Shine R, Hudson S, Kearney M (2003) Sociality in lizards: Why do thick-tailed geckos (*Nephrurus milii*) aggregate? Behaviour 140: 1039–1052.
40. Avilés L, Agnarsson I, Salazar P, Purcell J, Iturralde G, et al. (2007) Altitudinal patterns of spider sociality and the biology of a new midelevation social *Anelosimus* species in Ecuador. The American Naturalist 170: 783–792.
41. Field J, Paxton RJ, Soro A, Bridge C (2010) Cryptic plasticity underlies a major evolutionary transition. Current Biology 20: 2028–2031.
42. Wcislo WT, Danforth BN (1997) Secondarily solitary: the evolutionary loss of social behavior. Trends in Ecology & Evolution 12: 468–474.
43. Kaufmann JS, Bennett AF (1989) The effect of temperature and thermal acclimation on locomotor performance in *Xantusia vigilis*, the Desert Night Lizard. Physiological Zoology 62: 1047–1058.
44. Davis AR, Leavitt DH (2007) Candlelight *vigilis*: A noninvasive method for sexing small, sexually monomorphic lizards. Herpetological Review 38: 402–404.
45. Hines JE, Sauer JR (1989) Program CONTRAST– a general program for the analysis of several survival or recovery rate estimates. US Fish and Wildlife Technical Report 24: 1–7.

10

Time Series Analysis of Hand-Foot-Mouth Disease Hospitalization in Zhengzhou: Establishment of Forecasting Models Using Climate Variables as Predictors

Huifen Feng[1,2], Guangcai Duan[1]*, Rongguang Zhang[1], Weidong Zhang[1]

1 Department of Epidemiology, College of Public Health, Zhengzhou University, Zhengzhou, Henan, China, 2 Department of Infectious Diseases, the Fifth Affiliated Hospital of Zhengzhou University, Zhengzhou, Henan, China

Abstract

Background: Large-scale outbreaks of hand-foot-mouth disease (HFMD) have occurred frequently and caused neurological sequelae in mainland China since 2008. Prediction of the activity of HFMD epidemics a few weeks ahead is useful in taking preventive measures for efficient HFMD control.

Methods: Samples obtained from children hospitalized with HFMD in Zhengzhou, Henan, China, were examined for the existence of pathogens with reverse-transcriptase polymerase chain reaction (RT-PCR) from 2008 to 2012. Seasonal Autoregressive Integrated Moving Average (SARIMA) models for the weekly number of HFMD, Human enterovirs 71(HEV71) and CoxsackievirusA16 (CoxA16) associated HFMD were developed and validated. Cross correlation between the number of HFMD hospitalizations and climatic variables was computed to identify significant variables to be included as external factors. Time series modeling was carried out using multivariate SARIMA models when there was significant predictor meteorological variable.

Results: 2932 samples from the patients hospitalized with HFMD, 748 were detected with HEV71, 527 with CoxA16 and 787 with other enterovirus (other EV) from January 2008 to June 2012. Average atmospheric temperature ($T\{avg\}$) lagged at 2 or 3 weeks were identified as significant predictors for the number of HFMD and the pathogens. $SARIMA(0,1,0)(1,0,0)_{52}$ associated with $T\{avg\}$ at lag 2 ($T\{avg\}$-Lag 2) weeks, $SARIMA(0,1,2)(1,0,0)_{52}$ with $T\{avg\}$-Lag 2 weeks and $SARIMA(0,1,1)(1,1,0)_{52}$ with $T\{avg\}$-Lag 3 weeks were developed and validated for description and predication the weekly number of HFMD, HEV71-associated HFMD, and Cox A16-associated HFMD hospitalizations.

Conclusion: Seasonal pattern of certain HFMD pathogens can be associated by meteorological factors. The SARIMA model including climatic variables could be used as an early and reliable monitoring system to predict annual HFMD epidemics.

Editor: Benjamin J. Cowling, University of Hong Kong, Hong Kong

Funding: Grant numbers 81172740, http://isisn.nsfc.gov.cn. The funders had no role in study design, data collection and analysis, decision to publish, or preparation of the manuscript.

Competing Interests: The authors have declared that no competing interests exist.

* E-mail: gcduan@yeah.net

Introduction

Hand-foot-mouth disease (HFMD) is a common infectious illness in young children, particularly those less than 5 years old. Numerous large outbreaks of HFMD have occurred in Eastern and Southeastern Asian countries, including Singapore, Malaysia, Japan, and China since 1997 [1,2,3,4,5], which have caused death and neurological sequelae, and have become a growing public health threat. HFMD is an acute enterovirus infection. Human enterovirus 71 (HEV71) and Coxsackievirus A16 (CoxA16) are the major causative agents of this disease [6,7]. HFMD usually resolves spontaneously, but severe complications can arise, particularly when HEV71 is the causative agent [8,9]. Currently, neither vaccine nor effective drug against HEV71 is available for human use [10,11]. Thus, epidemiological surveillance of HFMD and its pathogens is important to take the proper and timely public health interventions to prevent its outbreaks. Early warning of

HFMD outbreaks could improve the efficiency of control campaigns and help to take prevention actions to delay the epidemic, thus reducing its impact on health system. HFMD morbidity and mortality would be minimized through earlier and efficient public health response.

Many mathematical models have been developed to predict the occurrence of outbreaks using a combined environmental approach. Seasonal Autoregressive Integrated Moving Average (SARIMA) models allow the integration of external factors, such as climatic variables, to increase their predictive power. This approach has been successfully used to predict the evolution of infectious diseases, such as dengue, vibrio cholera, malaria, and deaths due to influenza [12,13,14,15]. Many studies have indicated that meteorological conditions are the most important factors of HFMD outbreaks [16,17,18]. A study in China showed that the weekly number of HFMD cases in the 0–14 years age group increase by 1.86% for every 1°C increase in temperature and by 1.42% for every 1% increase in relative humidity [19].

In this study, we proposed to develop SARIMA models using time series analysis of the number of laboratory-confirmed HFMD hospitalizations. The goals of this study were to characterize whether climatic factors are associated with HFMD epidemics among children and whether inclusion of such factors is useful to predict epidemics with higher precision. This predicable model would be used to facilitate efficient HFMD control.

Materials and Methods

Ethics Statement

The study was approved by the Life Sciences Institutional Review Board of Zhengzhou University. It was also approved by the Ethics Committee of each participating hospital: the Ethics Committee of the Fifth Affiliated Hospital of Zhengzhou University, the Ethics Committee of the Third Affiliated Hospital of Zhengzhou University, the Ethics Committee of the Sixth People's Hospital of Zhengzhou and the Ethics Committee of the Zhengzhou Children's Hospital. Written informed consent was obtained from the parent of every child participant enrolled in this study.

Study Area

Zhengzhou, the capital of Henan Province, locates in the central of China, is situated at 34°16′–34°58′ north latitude and 112°42′–114°13′ east longitude, and the total area is 7446.2 square kilometers. The total population of the city amounted to 9.1 million by the end of 2012 (data from the Henan Bureau of Statistics). The population density is the second in China. China lies mainly in the north-temperate zone, characterized by a warm climate and distinctive seasons. In terms of temperature, the nation can be sectored from south to north into equatorial, tropical, subtropical, warm-temperate, temperate, and cold-temperate zones. The average temperature and the average annual precipitation vary greatly from place to place. Zhengzhou lies in the north warm-temperate zone, characterized by a warm climate and distinctive seasons, with a draught spring (March-May) and a hot and rainy summer (June-September). The annual average temperature is 14~14.3°C, the average annual precipitation is 640.9 mm, and the total sunshine is 2400 hours. The number of clinically diagnosed HFMD reported to the Centers Disease Control (CDC, Zhengzhou, China) was highest in 2009(17,792 cases), and lowest in 2008(1778 cases). Most of the cases occurred from March to June. The average annual incidence was from 26.21/100,000 to 260.56/100,000.

Meteorological Data

Average atmospheric temperature (T{avg}), maximum atmospheric temperature (T{max}), minimum atmospheric temperature (T{min}), relative humidity (RH), duration of sunshine (SS) and vapor pressure (VP) were routinely measured at the Zhengzhou Meteorological Administration. Daily diurnal variation in temperature was calculated by subtracting the maximum and minimum temperature. These data were available for the period from January 2008 to June 2012 without any missing values, and aggregated on a weekly basis which comprised a total of 234 weeks period.

Hospitalizations Information of Children with HFMD

The patients were identified according to the diagnostic criteria defined by Ministry of Health. Clinical diagnosis HFMD is characterized by oral vesicular exanthema/ulcers plus vesicular lesions on the hands, and/or feet, and/or buttocks. Laboratory-confirmed cases were clinically diagnosed HFMD with enterovirus-positive, and/or HEV71-positive, and/or Cox A16-positive.

The criteria for HFMD hospitalization included one of the following conditions: total duration of fever≥3 days, peak temperature≥38.5°C on examination, toxic and ill in appearance, recurrent vomiting (at least twice), tachycardia (heart rate≥150/min), breathlessness, poor perfusion (cold clammy skin), reduced consciousness (lethargy, drowsiness, coma), limb weakness, meningitis (neck stiffness or positive Kernig's sign), seizures.

All HFMD cases are the sentinel hospital-based clinical and laboratory HFMD surveillance scheme in China. All children with HFMD in this region hospitalized at the Fifth Affiliated Hospital of Zhengzhou University, the Third Affiliated Hospital of Zhengzhou University, the Sixth People's Hospital of Zhengzhou and the Zhengzhou Children's Hospital were eligible for participation in this study. These hospitals host the large-scale dedicated HFMD clinics and wards in Zhengzhou. Almost all of hospitalized HFMD cases were treated in these hospitals. Stool specimens were collected from each child hospitalized with clinical diagnosis HFMD. Participation in the study was voluntary and was proposed to all eligible patients until the target sample was reached. Samples not taken or refusal of participation rate was approximately 12%. Pan-enterovirus, HEV71, and Cox A16 were routinely detected in the clinical laboratory of each sentinel hospital. Moreover, the samples from severe cases were sent to the Centers for Disease Control (CDC, Zhengzhou, China) or the Molecular Laboratory of Zhengzhou University for virus identification. The data collection mechanism has been stable over time, and this routinely collected data can be used for analyzing factors affecting the occurrence of HFMD. Weekly and monthly cases of HFMD were obtained from sentinel hospitals for early detection and measurement of magnitude of epidemics, weekly confirmed cases of HFMD from all laboratories were verified for circulating virus. Data from 2932 samples tested with RT-PCR were subjected to statistical analysis from January 2008 to June 2012.

Laboratory Analysis

Stool specimens were collected from each child enrolled in this study. These samples were transported immediately at 4°C to the clinical laboratory of each sentinel hospital, the Centers for Disease Control (CDC, Zhengzhou, China) or the Molecular Laboratory of Zhengzhou University and then kept at −70°C until for the detection of HEV71, CoxA16 and universal enterovirus (EV) using the QIGEN Viral RNA kit (QIAGEN, Germany) according to the manufacturer's instructions. Briefly, The OneStep RT-PCR Kit (QIAGEN, Germany) was used for RT-PCR with a 50 μl reaction mixture containing 3 μl of RNA sample, 5 μl 10× buffer, 2.0 μl dNTP mix (25 mM), 1.0 μl enzyme mix, 0.5 μl RNase inhibitor (40 U/μl), 1.0 μl forward primer, and 1.0 μl reverse primer. The reactions were carried on 7500 fast PCR instrument (Applied Biosystems), with an initial reverse transcription step at 50°C for 45 min, followed by PCR activation at 95°C for 3 min and 35 cycles of amplification (95°C for 30 s, 50°C for 30 s, 65°C for 60 s). A final extension at 65°C for 10 min was performed. PCR products were observed in 2% agarose electrophoresis. **Table 1** shows the nucleotide sequences of the specific primers and Taq Man probes used in this study.

Statistical Analysis

Number of hospitalized children with HFMD and the mean values of the meteorological parameters were calculated for intervals of 7 consecutive days, which are maximal coverage of current weather forecast. Description was performed by time series

Table 1. Nucleotide sequences of the specific primers and Taq Man probes.

	Primer (5'-3')	Probe (5'-3')
Pan-EV	Forward: GCAAGTCTGTGGCGGAACC	(FAM)-AATAACAGGAAACACGGACACCCAAAGTA(TAMRA)
	Reverse: TGTCACCATAAGCAGCCATGATA	
HEV71	Forward: GTTCACCTACATGCGCTTTGA	(VIC)-TCTTGCGTGCACACCCACCG(TAMRA)
	Reverse: TGGAGCAATTGTGGGACAAC	
CoxA16	Forward: CCTAAAGACTAATGAGACCACCC	(TEXASRED)-CTTGTGCTTTCCAGTGTCGGTGCA(TAMRA)
	Reverse: CTAAAGGCAGCACACAATTCG	

diagrams. Inferential statistics included Spearman rank correlations, partial and cross correlations, univariate and multiple time series analysis. The number of children hospitalized with HFMD, HEV71-associated HFMD and CoxA16-associated HFMD were considered as the dependent variables. The meteorological data and a seasonal component were considered as the independent variables.

With the goal of predicting the number of HFMD hospitalizations and the major enterovirus infection, SARIMA models were developed. SARIMA models (Box and Jenkins models) have the flexibility to control the autocorrelation of time series data. Four steps were undertaken in the modeling of the number of HFMD and the climate variables. First, using the mean range plot to determine whether the time series of the children hospitalized with HFMD and the climate variables is in a stationary or non-stationary condition. If non-stationary, it has to be transformed into a stationary time series by applying an appropriate transformation (logarithmic, square root, inverse transformation or differencing). Since both HFMD and the climate variables exhibited strong seasonal variation and fluctuations in their yearly means, we adjusted for seasonality by first seasonally differencing the series in the analysis. Second, the temporal structure of seasonal and non-seasonal autoregressive parameters $(AR)(P,p)$, moving average parameters $(MV)(Q,q)$ of the series were determined by assessing the analysis of autocorrelation function (ACF) and partial autocorrelation function (PACF). Once the model was specified, parameters of the model were estimated by using the maximum likelihood method. Third, the goodness-of-fit of the models were determined for appropriate modeling, using the Ljung-Box test measures both ACF and PACF of the residuals, and checking the normality of the residuals. The significance of the parameters should be statistically different from zero. The normalized Bayesian Information Criteria (BIC) and stationary R square (R^2) were also conducted to compare the goodness-of-fit among SARIMA models. The lowest BIC and the highest stationary R^2 values was considered good model. Finally, the models developed were verified by dividing the data file into two data sets: the data from the 1st calendar week of 2008 to the 52nd calendar week of 2011 (estimation period) were used to construct a SARIMA model and those between the 1st calendar week to the 26th calendar week of 2012 (evaluation period) were used to validate the model.

We further evaluated whether alternative SARIMA models incorporating climate variables as external regressors have greater predictive power. To facilitate selection of climate variables to be used as external regressors, we computed the association between the number of HFMD hospitalizations and meteorological parameters using spearman rank correlations. Pearson's or Spearman's rank correlation was used to further test any correlation among the meteorological parameters. Cross-correla-

tion analysis was used to assess associations between HFMD cases and covariates over a range of time lags (a time lag was defined as the time span between climatic observation and the incidence of HFMD). The time lags chosen for the final model were outcomes of the cross-correlation analysis. To overcome the autocorrelation within each individual series, the correlation coefficients between the number HFMD and climate variables were computed after pre-whitening. Pre-whitening was performed by modeling each time series individually using the SARIMA model. Climatic variables significantly associated to the number of HFMD cases were tested as predictors in multivariate SARIMA model. Similar to the univariate SARIMA model, we estimate the coefficients of multivariate SARIMA associated with the lagged climate variable. The comparison of the SARIMA with and without climatic variables was conducted. The predictive validity of the models was evaluated by calculating the root mean square error (RMSE), which measures the amount by which the fitted values differ from the observed values. The smaller the RMSE, the better the model is for forecasting.

All statistical tests were 2-tailed, and P value <0.05 were considered to be statistically significant in terms of an explorative data analysis. For statistical analysis we used SPSS software, version 19 (SPSS).

Results

Classification of Pathogens in the Patients with HFMD

Of the 3380 subjects admitted to the isolation wards for treatment between January 2008 and June 2012, 48 were excluded from the protocol analysis for failing to meet inclusion criteria with respect definition of HFMD. 3332 hospitalized with HFMD cases, 2932 children provided stool samples for testing, 201 were severe and 5 died of HFMD. 93.5% patients were under 5 years old, the youngest was 5 months old and the oldest was 12.5 years old. In 2062(69.18%) of the 2932 stool samples tested for HFMD from January 2008 to June 2012, at least one kind of HFMD pathogen was detected. HEV71 (748[36.28%], CoxA16 (527[25.56%]) and other EV (787[38.17%]), were the most common pathogens detected in these samples.

The number of clinical diagnosis HFMD cases (**Figures 1A**) and the classification of the pathogens (**Figures 1B**) were shown in Figure 1. HEV71, CoxA16, and other EV were detected all year round, whereas HEV71, CoxA16 and other EV, showed distinctive spring and early summer peaks (**Figures 1**).

Bivariate Analysis

T{avg}, T{max}, T{min}, RH, SS and VP were significantly correlated with the overall number of HFMD hospitalizations. HEV71 was most strongly correlated with T{avg}, then the CoxA16. We found statistically significant but weaker correlations

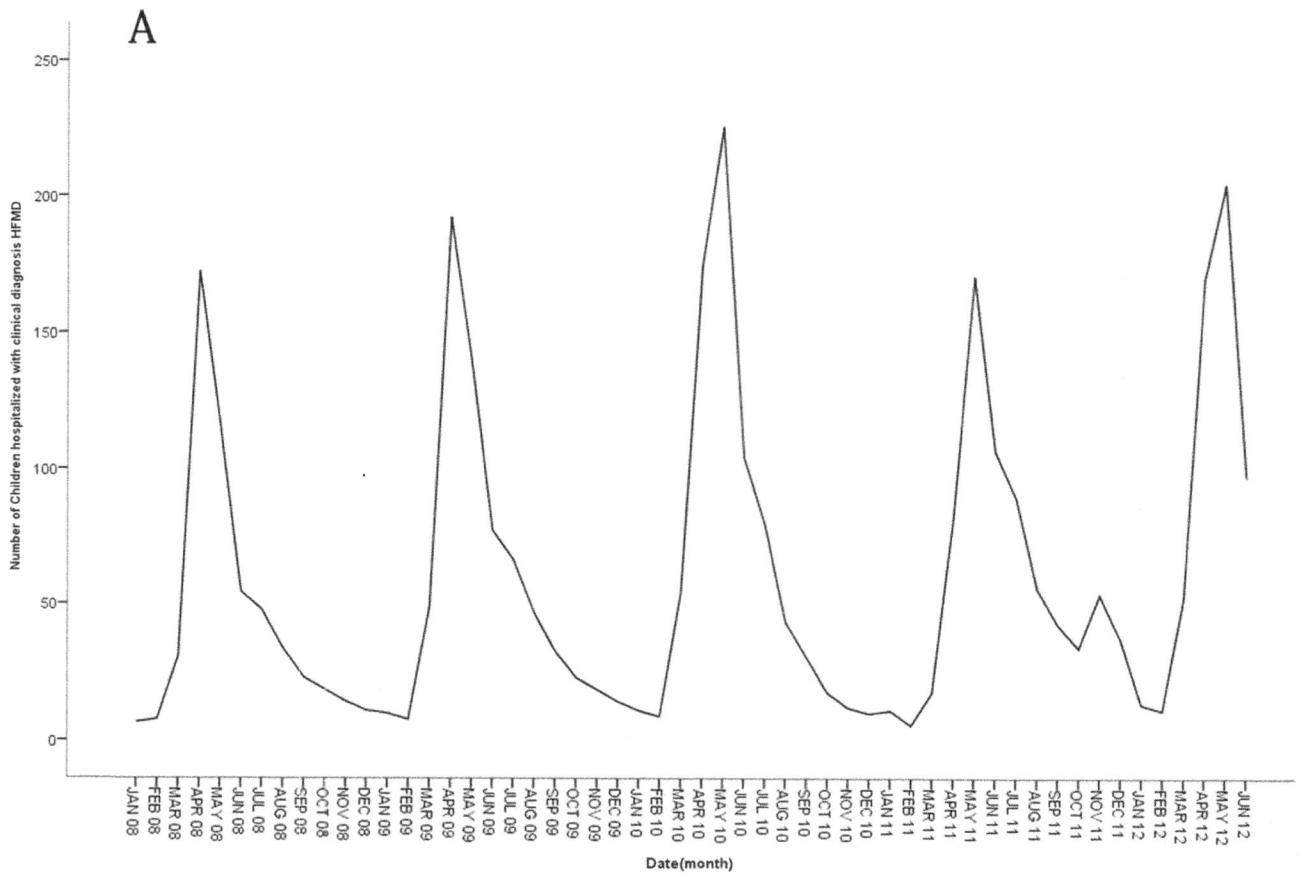

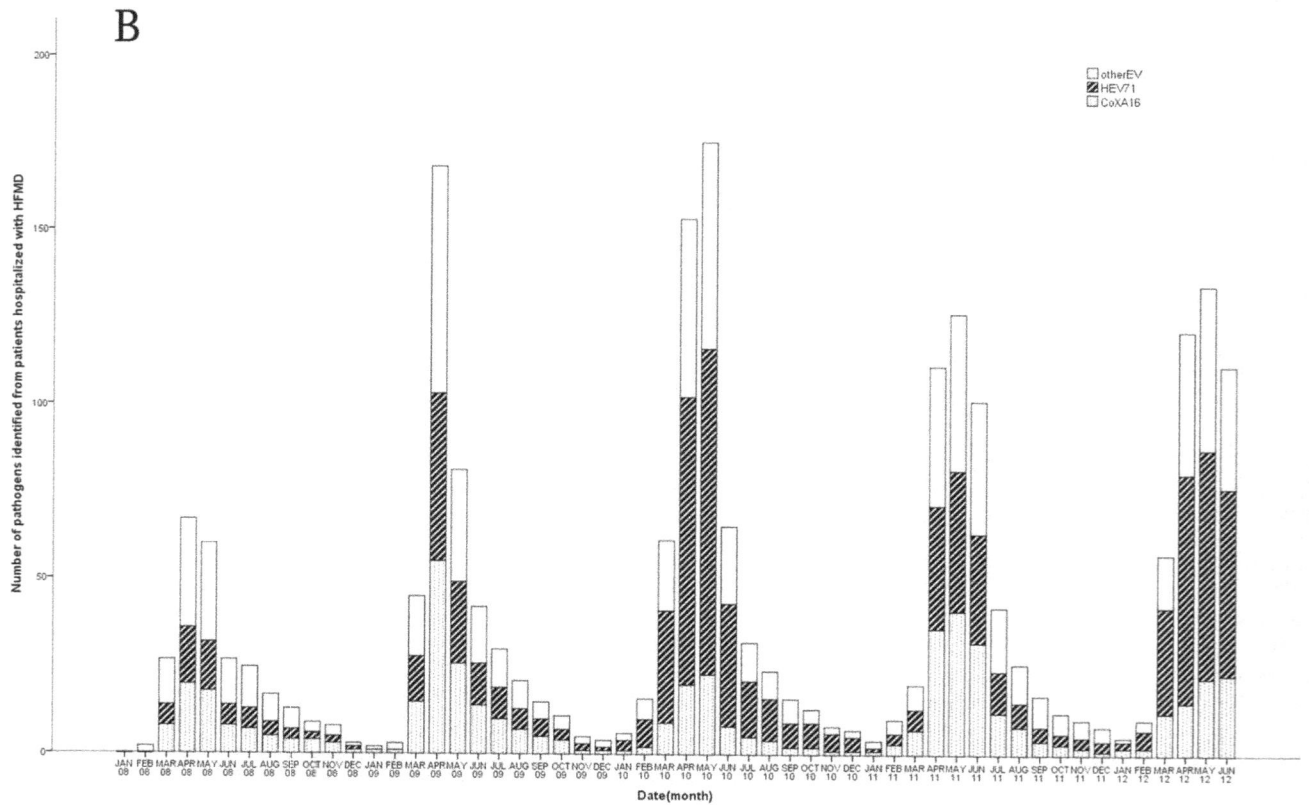

Figure 1. The number of clinical diagnosis cases and the pathogens hospitalized with hand-foot-mouth disease (HFMD) in Zhengzhou, China from 2008 to 2012. The 3 most frequent pathogens leading to hospitalized children with HFMD in Zhengzhou from 2008 to 2012 were, in order, other enterovirus (other EV), Human enterovirus 71 (HEV71) and CoxsackievirusA16 (CoxA16).

for the association between RH, SS and these 2 pathogens (**Table 2**).

Because different meteorological parameters may also be correlated with each other, we analyzed the relationship among these parameters. In fact, average atmospheric temperature was inversely correlated with vapor pressure ($r_s = -0.930$; $P < 0.001$), but correlated with duration of sunshine ($r_s = 0.178$; $P = 0.006$), relative humidity ($r_s = 0.259$; $P < 0.001$).

Accounting for these intercorrelations, associations between meteorological factors and the number of HFMD hospitalization were then analyzed using partial correlations: detection of any of the pathogens was associated with average atmospheric temperatures (**Table 3**). The figures also demonstrated temperature and hospitalization caused by the most common pathogens detected over time, showing association of increased activity of HFMD with atmospheric temperatures (**Figure 2**).

Multiple Analysis

In the first step of the HFMD time series analysis, a square root transformation was performed to stabilize the variance of the series. Then we calculated one time regular differencing for the variable to ensure the time series stationary. The plots of auto correlation function (ACF) and partial auto correlation function (PACF) (**Figures 3A and 3B**) showed the temporal dependence of the number of cases hospitalized with HFMD and confirmed the need to use a SARIMA model with seasonal (P, D, Q) and non-seasonal (p, d, q) parameters. Upon checking ACF and PACF, after differencing, a significant cut offs at one week lag and another at lag 52 weeks were observed on the plot ACF (**Figure 3C**). These two cut offs were less marked on the plot PACF (**Figure 3D**) and evolve more gradually over the time, compared to the plot ACF. The analysis from the correlograms of the series suggests that p value should be equal to 1 or 2 and q value equal to 0 or 1 of moving average parameters. We fitted the data with several univariate SARIMA (p,d,q)(P,D,Q)s with different orders and excluded the models in which the residual is not likely to be white noise. Among these models, the univariate SARIMA $(1,1,1)(1,0,0)_{52}$ model had both lowest BIC and highest R^2 values and appeared the best to fit the cases hospitalized with HFMD (**Table 4**). The analyses of residuals on ACF and PACF plots assessed the absence of persistent temporal correlation (**Figure 4**). The Ljung-Box test confirmed that the residuals of time series

were statistically not dependent (**Table 4**). The selected SARIMA model fitted observed data from 2008 to 2011. Furthermore, the model was used to forecast the number of HFMD hospitalizations between January and June 2012, and was then validated by the actual observations. The validation analyses indicate that the model had reasonable accuracy over the predictive period (RMSE = 0.377). **Table 4** also described the characteristics of the SARIMA models for the number of HEV71-associated HFMD and Cox A16-associated HFMD hospitalizations.

To include climatic variables (time series) as external variables in the univariate model, a SARIMA model was applied to the time series. We first removed trend and seasonal components of each time series through SARIMA modeling. A regular differencing and a seasonal differencing were applied to the atmospheric temperature. Climatic variables identified as the most interconnected to the number of HFMD were accounted one by one, due to their strong interconnection. The results of the cross-correlations show that the number of children hospitalized with HFMD were significantly positively associated with T{avg} (coefficients = 0.296, $P < 0.05$), T{max}(coefficients = 0.207, $P < 0.05$) at lag 2 weeks. HEV71-associated HFMD positively associated with T{avg}at lag 2 weeks (coefficients = 0.182, $P < 0.05$) and T{max}at lag1 week (coefficients = 0.211, $P < 0.05$), while CoxA16-associated HFMD only positively associated with T{avg}at lag 3 weeks (coefficients = 0.190, $P < 0.05$) and T{max} (coefficients = 0.183, $P < 0.05$) at lag 3 weeks.

The identification of climate variables that significantly correlated with HFMD hospitalizations were tested with univariate SARIMA models, which were carried out by including external independent variables. Average atmospheric temperature at lag 2 weeks (T{avg}-Lag 2) was the only independent covariate that significantly associated with the number of HFMD hospitalizations in the multiple time series analysis. Overall models the SARIMA $(0,1,0)(1,0,0)_{52}$ associated with T{avg} -Lag 2 weeks is the most appropriate, which has the best fit and highest R^2. The model estimated with the T{avg} -Lag 2 weeks was a better fit than the model without the variable (Stationary R-squared (Stationary R^2) increased, while the BIC decreased). (**Table 4, Table 5**). The model was used to predict the number of HFMD hospitalizations from January to June 2012, and validated by the actual observations (**Figure 5**). The validation analyses indicate that the model increased with the inclusion of T{avg}-Lag 2 weeks

Table 2. Spearman rank correlation coefficients for the associations between meteorological parameters and hospitalizations of children with HFMD.

Parameters	HFMD		other EV		HEV71		CoxA16	
	r_s	P	r_s	P	r_s	P	r_s	P
VP	−0.654	0.000	−0.619	0.000	−0.553	0.000	−0.561	0.000
T{avg}	0.647	0.000	0.611	0.000	0.533	0.000	0.531	0.000
T{max}	0.627	0.000	0.595	0.000	0.517	0.000	0.523	0.000
T{min}	0.622	0.000	0.579	0.000	0.510	0.000	0.496	0.000
RH	−0.137	0.025	0.125	0.033	−0.172	0.022	−0.151	0.0392
SS	0.235	0.002	0.229	0.002	0.177	0.007	0.272	0.001

Table 3. Partial correlations between meteorological parameters and hospitalizations of children with HFMD for the adjustment of correlations among other meteorological parameters.

Parameters	HFMD		other EV		HEV71		CoxA16	
	r_s	P	r_s	P	r_s	P	r_s	P
VP	−0.048	0.468	−0.052	0.431	−0.011	0.873	−0.091	0.172
T{avg}	0.442	0.001	0.418	0.000	0.349	0.000	0.374	0.000
RH	−0.115	0.085	−0.096	0.153	−0.107	0.109	−0.075	0.259
SS	0.015	0.827	0.017	0.795	−0.032	0.634	−0.075	0.258

(RMSE = 0.352) compared with the model without this variable (RMSE = 0.377).

Multiple time series analysis was also performed for the climate variables on the number of hospitalizations due to HEV71 and Cox A 16 infections. T{avg}-Lag 2 weeks and T{avg} at lag 3 (T{avg}-Lag 3) weeks were the independent covariate that significantly associated with the number of HEV71-associated HFMD and the Cox A 16-associated HFMD hospitalizations in

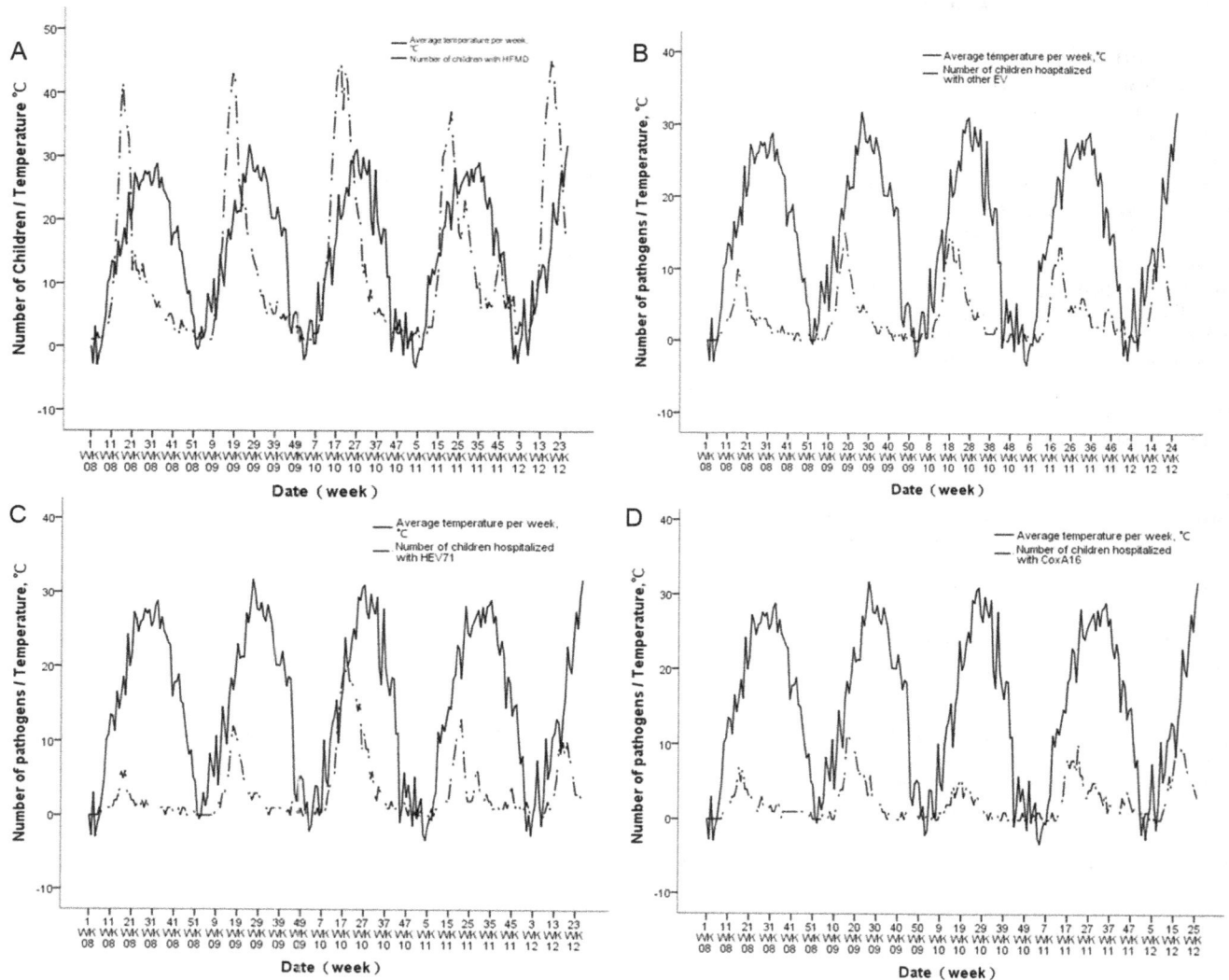

Figure 2. Weekly numbers of hospitalized children with HFMD in Zhengzhou, China from January 2008 to June 2012 compared to crude meteorological variables for the same period. An alternate course is seen between temperature and the pathogens. HFMD (A), other EV(B), HEV71(C) and CoxA16(D).

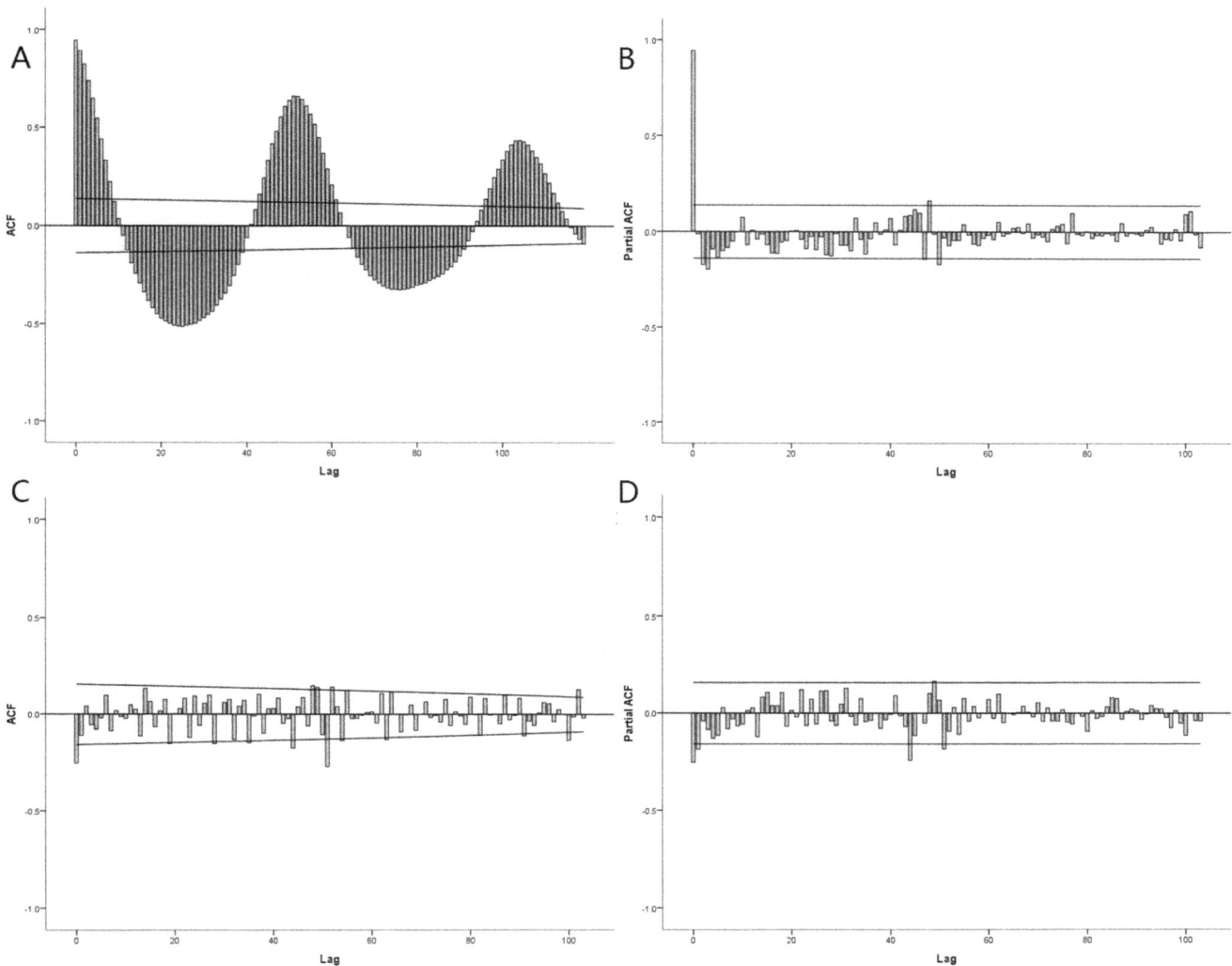

Figure 3. Autocorrelation function (ACF) and Partial ACF (PACF) plot of original and integrated the number of HFMD hospitalizations. A and B) shows ACF and PACF plot of original HFMD hospitalizations. C and D) ACF and PACF plot of integrated HFMD hospitalizations.

the multiple time series analysis, respectively. Models of SARIMA $(0,1,2)(1,0,0)_{52}$, SARIMA $(0,1,1)(1,1,0)_{52}$ shows the fitted models of HEV71-associated HFMD with T{avg}-Lag 2 weeks and Cox A16-associated HFMD with T{avg}-Lag 3 weeks. HEV71-associated HFMD model with T{avg} -Lag 2 weeks was better fit and validity than the univariate model, while the Cox A16-associated HFMD model with T{avg}-Lag 3 weeks didn't show difference (**Table 5, Figure 5**).

Discussion

It was observed from this study that HFMD was prevalent year round in this region and peaked between April and July during spring and early summertime. In August, the activity of HFMD fell sharply. However, in 2011 the peak season was in May, one month later than that seen in previous years, followed by a second smaller and unusual epidemic wave of HFMD was observed in middle autumn and winter. Moreover, we also found that the pathogens of HFMD, such as HEV71 and CoxA16, presented a specific annual or biannual specific pattern (**Figure 1**). Our

findings are in agreement with the incidence of HFMD that has been reported to exhibit seasonal variation in a number of different areas [4,5]. Epidemiologists have been perplexed by the causes and consequences of seasonal infectious disease for long time, and there is no theory that can alone explain this phenomenon [20,21]. Environment changes, particularly changes in weather, have been mostly implicated. Annual variation in climate has been proposed to result in annual or more complex peaks in disease incidence, depending on the influence of climatic variables [22,23]. Many studies suggested that HFMD consultation rates were positively associated with temperature and humidity [16,17,18,19]. Herein, we report that HFMD and the pathogens are significantly associated with meteorological parameters (**Figure 2**). This study provided confirmatory evidence for the notion that mean temperature, among various climate variables is the key contributor to HFMD outbreak, which is consistent with results from other studies.

Temperatures and other climatic factors may influence the survival and spread of infectious pathogen in the environment, exposure probability, and the host susceptibility [24,25,26]. On

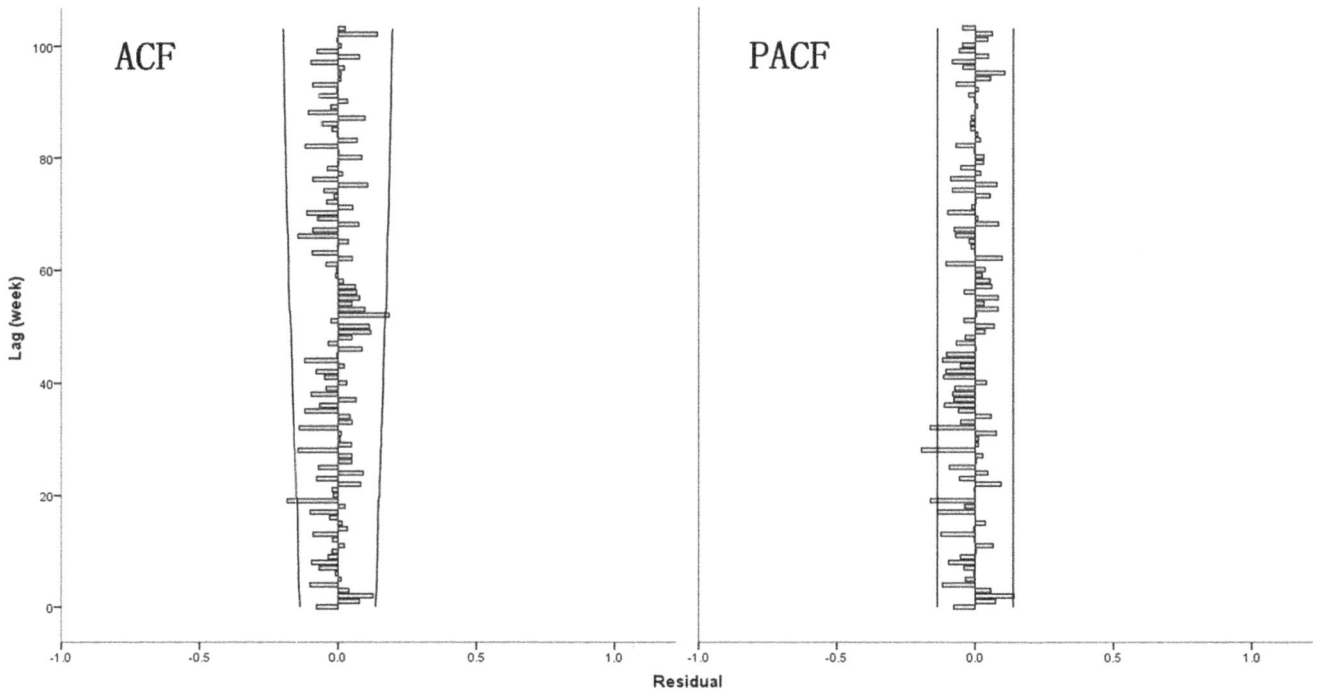

Figure 4. Autocorrelation function (ACF) and Partial ACF (PACF) plot of residuals after applying a SARIMA (1, 1, 1) (1, 0, 0)$_{52}$ model. The x-axis gives the number of lags in weeks and, the y-axis, the value of the correlation coefficient comprised between −1 and 1. Dotted lines indicate 95% confidence interval.

the one hand, virus survival under certain climatic conditions could play a role. The survival of the pathogenic organisms outside a host depends on the characteristics of the environment, particularly temperature, humidity, exposure to sunlight, pH and salinity [27,28].Experimental studies have shown the stability of enteric viruses is influenced by environmental factors such as temperature and relative humidity, which could survive for at least 45 days on nonporous fomites [29]. These findings are supported by epidemiological studies. For example, in the tropics, seasonal peaks in the incidence of enteric viruses have been found to correlate with temperature and relative humidity [16]. This is present study also showed that the activity of HFMD and the pathogens pattern are associated with average atmospheric temperature and the maximum temperature. However, a complicated relationship exists between the micro-environment and enteric viruses, which depends on temperature, salinity and overall levels of water in the environment [28]. It is difficult to predict the incidence of HFMD only on climate since it may peak once or twice a year due to local environment alterations. On the other

hand, the probability of transmission of HFMD pathogens might be changed due to host behavior in different seasons. Children are more likely to go outside for playing or swimming during summer than in winter. A lot of previous studies have shown that the summer peaks of polio and other enteric viruses were associated to swimming [30,31,32]. Additionally open and weeping skin vesicles, direct contact of contaminated toys and environmental non-hygienic surfaces are other approaches for the spread of enteric viruses infection with the fecal-oral route. In winter time children stay indoor longer, resulting in more contact opportunity and higher transmission among household members. This in turn facilitates transmission of enteric viruses through respiratory droplets. Enterovirus transmitted mainly via faecal-oral, in temperate climates, enteroviral infection occurs primarily in the summer. Therefore, the changes of host behavior, particular patterns of movement and contact, have a potent impact on the seasonality of HFMD.

The time series analysis used in this study produced similar results to previous studies, which made it possible to develop a

Table 4. Characteristics of SARIMA models for the number of cases hospitalized with HFMD, HEV71-associated HFMD, Cox A16-associated HFMD.

Variables	SARIMA model	AR	MA1	MA2	SAR1	R^2	BIC	P	RMSE
HFMD	(1,1,1)(1,0,0)$_{52}$	0.754	0.623	–	0.375	0.198	2.905	0.339	0.377
HEV71	(0,1,2)(1,0,0)$_{52}$	–	–	−0.234	0.291	0.162	0.529	0.177	1.269
Cox A 16	(0,1,1)(1,1,0)$_{52}$	–	0.563	–	−0.551	0.417	0.488	0.329	1.236

SARIMA: Seasonal Autoregressive Integrated Moving Average model, AR: autoregressive, MA: moving average, SAR: seasonal autoregressive, R^2: Stationary R-squared, BIC: Bayesian information criteria, P: Ljung-Box test, RMSE: Root Mean Square Error.

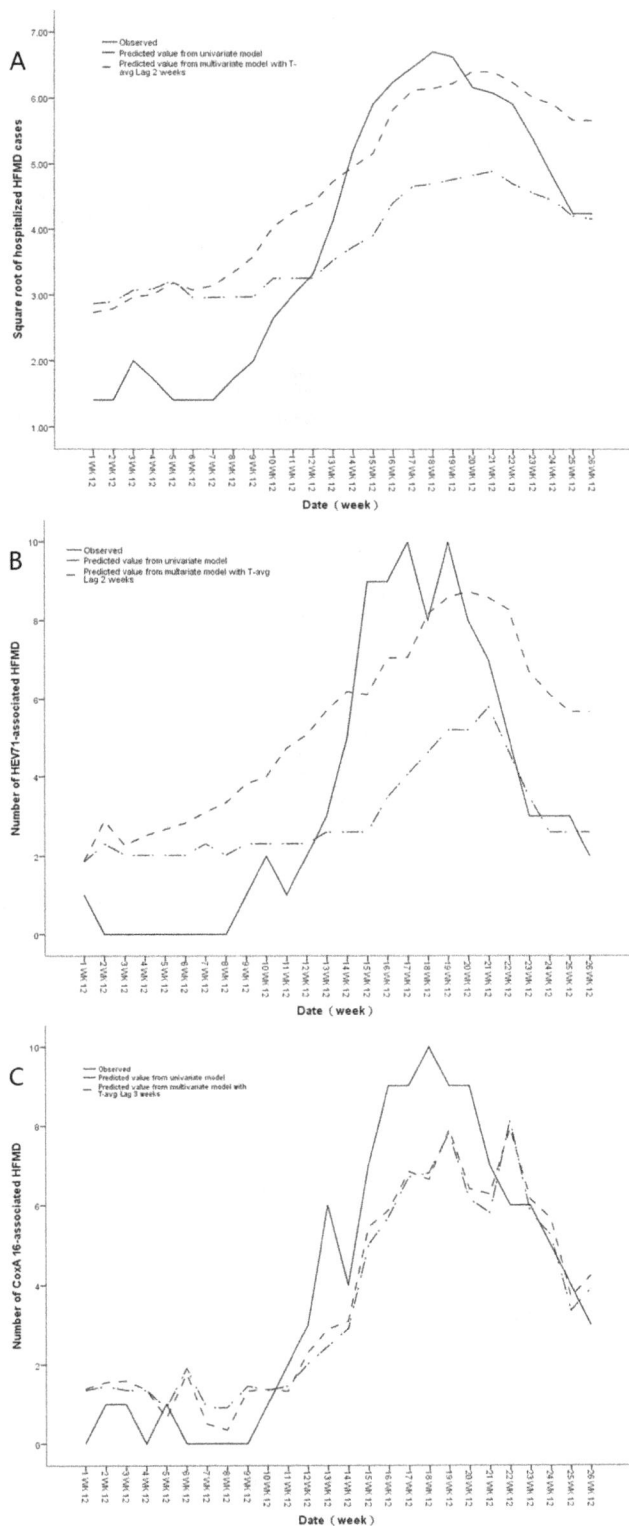

of HFMD hospitalizations, B: the number of HEV71-associated HFMD hospitalizations, C: the number of CoxA16-associated HFMD hospitalizations.

temporal structure model, especially for seasonal infections. The SARIMA modeling is a useful tool for interpreting and applying surveillance data in disease control and prevention. The model allows the integration of external factors, such as climatic variables, therefore increasing its predictive power [13,33]. In Japan, HFMD prevalence was positively correlated with the temperature and humidity at lag 0–3 weeks [16]. In Hong Kong, relative humidity, mean temperature, difference in diurnal temperature at 2 weeks' lag time was positively associated with HFMD consultation rates [17]. And in the city of Guangzhou in China, temperature and relative humidity were significantly associated with HFMD infection with one week lag [19]. We have shown that the increase in average atmospheric temperature was a determining factor in predicting changes of the HFMD incidence. On the contrary, the relative humidity did not appear to play a significant role in this aspect. This study developed a climate-based forecasting model using HFMD hospitalization data collected from 2008 to 2011 in this region, to predict the onset of HFMD of 2012 (**Figure 5**). Average atmospheric temperature was identified as a significant predictor for the occurrence of HFMD and the pathogens. After the introduction of the average atmospheric temperature at lag 2 weeks increased the SARIMA models of HFMD and HEV71's predictive power, which might be implemented in routine surveillance of HFMD and useful for the evaluation of new intervention strategies introduced into this region. However, including weather parameters the prediction model of Cox A 16 could not accurately predict the actual diseases occurrence. Nevertheless, producing accurate predictions using climate data remains a challenge. This study first analyzes the relationship between the most common known HFMD pathogens in children and different meteorological parameters for 5 years, and develops a model for prediction of the number of HFMD hospitalizations on the basis of weather variables in an SARIMA model. The majority of HFMD cases were clinically diagnosed but

Figure 5. Prediction of square root transformation of the number of HFMD hospitalizations, the number of HEV71-associated and CoxA16-associated HFMD hospitalizations on the basis of a seasonal autoregressive integrated moving average model (SARIMA) model with average atmospheric temperature as the covariate for 2012. Solid line: observed values during the period, dashed line: predicted values for 2012 with and without climatic variables. A: Square root transformation of the number

Table 5. Characteristics of multivariate SARIMA models using climate variables for the number of cases hospitalized with HFMD, HEV71-associated HFMD, Cox A16-associated HFMD.

Parameters	HFMD	HEV71	CoxA16
SARIMA model	$(0,1,0)(1,0,0)_{52}$	$(0,1,2)(1,0,0)_{52}$	$(0,1,1)(1,1,0)_{52}$
MA1	–	–	0.529 ± 0.074
MA2	–	-0.227 ± 0.071	–
SAR1	0.369 ± 0.079	-0.251 ± 0.088	-0.490 ± 0.099
T{avg}-Lag2 weeks	0.019 ± 0.005	0.079 ± 0.026	–
T{avg}-Lag3 weeks	–	–	0.091 ± 0.037
R^2	0.229	0.232	0.402
BIC	1.871	0.543	0.627
P	0.356	0.585	0.664
RMSE	0.352	1.230	1.297

SARIMA: Seasonal Autoregressive Integrated Moving Average model, AR: autoregressive, MA: moving average, SAR: seasonal autoregressive. β: Coefficient, SE: Standard Error, R^2: Stationary R-squared, BIC: Bayesian information criteria, P*: Ljung-Box test, RMSE: Root Mean Square Error, T{avg}-Lag2 weeks: average atmospheric temperature at lag 2 weeks, T{avg}-Lag3 weeks: average atmopheric temperature at lag 3 weeks.

only a small proportion were laboratory-confirmed in the earlier studies. An early warning of HFMD outbreaks could improve the efficiency of control campaigns and help to take preventive measures. In addition, it provides insight into the local etiology of HFMD, and is helpful in designing preventive strategies. Such early interventions could delay the epidemic, thus reducing its impact on health. Health facilities could adjust their response in terms of availability of beds and mobilization of human and material resources. HFMD morbidity and mortality would be minimized through earlier and proper public health response.

The limitation of this study is that we failed to detect other serotypes of enterovirus except HEV71 and CoxA16 and monitor pathogens in outpatients with HFMD. There is potential to develop an early warning system for HFMD in this region using a predictive model, which would give public health authorities sufficient time to prepare medical equipments and staff in the event of an outbreak. Prediction of outbreaks is imperative in order to develop efficient and cost-effective prevention strategies for HFMD control. However, more work is needed to refine such a model before it is ready for routine use.

Acknowledgments

We are indebted to Sanjing Li of the Sixth People's Hospital of Zhengzhou (China), Xinhong Wang of the Zhengzhou Children's Hospital for collecting clinical samples. We are also grateful to Shuling Li of Zhengzhou Meteorological Administration and Hongjun Li of Zhengzhou Center of Diseases Control (China) for giving the meteorological and the hospitalizations of children with HFMD data.

Author Contributions

Conceived and designed the experiments: HF GD RZ WZ. Performed the experiments: HF GD RZ. Analyzed the data: HF. Contributed reagents/materials/analysis tools: HF WZ. Wrote the paper: HF.

References

1. Ang LW, Koh BK, Chan KP, Chua LT, James L, et al. (2009) Epidemiology and control of hand, foot and mouth disease in Singapore, 2001–2007. Ann Acad Med Singapore 38: 106–112.
2. Chan LG, Parashar UD, Lye MS, Ong FG, Zaki SR, et al. (2000) Deaths of children during an outbreak of hand, foot, and mouth disease in sarawak, malaysia: clinical and pathological characteristics of the disease. For the Outbreak Study Group. Clinical Infectious Diseases 31: 678–683.
3. Fujimoto T, Chikahira M, Yoshida S, Ebira H, Hasegawa A, et al. (2002) Outbreak of central nervous system disease associated with hand, foot, and mouth disease in Japan during the summer of 2000: detection and molecular epidemiology of enterovirus71. Microbiol Immunol 46: 621–627.
4. Chen KT, Chang HL, Wang ST, Cheng YT, Yang JY (2007) Epidemiologic features of hand-foot-mouth disease and herpangina caused by enterovirus71 in Taiwan, 1998–2005. Pediatrics 120 e244–252.
5. Yang F, Ren L, Xiong Z, Li J, Xiao Y, et al. (2009) Enterovirus 71 outbreak in the People's Republic of China in 2008. Journal of Clinical Microbiology 47: 2351–2352.
6. Schmidt NJ, Lennette EH, Ho HH (1974) An apparently new enterovirus isolated from patients with disease of the central nervous system. Journal of Infectious Diseases 129: 304–309.
7. Robinson CR, Frances WD, Rhodes AJ (1958) Report of outbreak of febrile illness with pharyngeal lesions and exanthema: Toronto summer 1957 isolation of group A Coxsackie virus. Canadian Medical Association Journal 79: 615–621.
8. Chang LY, Huang LM, Gau SS, Wu YY, Hsia SH, et al. (2007) Neurodevelopment and cognition in children after enterovirus71 infection. Engl Med 356: 1226–1234.
9. Chen SC, Chang HL, Yan TR, Cheng YT, Chen KT (2007) An eight-year study of epidemiologic features of enterovirus 71 infection in Taiwan. Am J Trop Med Hyg 77: 188–191.
10. Lee BY, Wateska AR, Bailey RR, Tai JH, Bacon KM (2010) Forecasting the economic value of an enterovirus 71(EV71) vaccine. Vaccine 28: 7731–7736.
11. Wu KX, Ng MM, Chu JJ (2010) Developments towards antiviral therapies against enterovirus 71. Drug Discov Today 15: 1041–1051.
12. Rajendran K, Sumi A, Bhattacharjya MK, Manna B, Sur D, et al. (2011) Influence of relative humidity in Vibrio cholerae infection: a time series model. Indian J Med Res 133: 138–145.
13. Luz PM, Mendes BV, Codeco CT, Struchiner CJ, Galvani AP (2008) Time series analysis of dengue incidence in Rio de Janeiro, Brazil. Am J Trop Med Hyg 79: 933–939.
14. Zhang Y, Bi P, Hiller JE (2010) Meteorological variables and malaria in a Chinese temperate city: A twenty-year time-series data analysis. Environ Int 36: 439–445.
15. Quénel P, Dab W (1999) Influenza A and B epidemic criteria based on time-series analysis of surveillance data. European Journal of Epidemiology 14: 275–285.
16. Onozuka D, Hashizume M (2011) The influence of temperature and humidity on the incidence of hand, foot, and mouth disease in Japan. Science of the Total Enviroment 410–411: 119–125.
17. Ma E, Lam T, Wong C, Chuang SK (2010) Is hand, foot and mouth disease associated with meteorological parameters? Epidemiol Infect 138: 1779–1788.
18. Wang Y, Feng Z, Yang Y, Self S, Gao Y, et al. (2011) Hand, Foot and Mouth Disease in China: Patterns of Spread and Transmissibility during 2008–2009. Epidemiology 22: 781–792.
19. Huang Y, Deng T, Yu S, Gu J, Huang G, et al. (2013) Effect of meteorological variables on the incidence of hand, foot, and mouth disease in children: a time-series analysis in Guangzhou, China. BMC Infect Dis 13: 134.
20. Fares A (2013) Factors Influencing the Seasonal Patterns of Infectious Diseases. Int J Prev Med 4: 128–132.
21. Nicholas CG, Christophe F (2006) Seasonal infectious disease epidemiology. Proc R Soc B 273: 2541–2550.
22. Tini G, Neil MF, Azra CG (2013) Estimating Air Temperature and Its Influence on Malaria Transmission across Africa. Plos One 8: e56487.
23. Jean-Baptist du P, Wolfram P, Britta G, Markus K, Josef AI, et al. (2009) Are Meteorological Parameters Associated with Acute Respiratory Tract Infections? Clinical Infectious Diseases 49: 861–868.
24. Pirtle EC, Beran GW (1991) Virus survival in the environment. Rev Sci Tech 10: 733–748.
25. Si Y, Wang T, Skidmore AK (2010) Environmental factors influencing the spread of the highly pathogenic avian influenza H5N1 virus in wild birds in Europe. Ecol Soc 15: 26.
26. Dowell SF (2001) Seasonal variation in host susceptibility and cycles of certain infectious diseases. Emerg Infect Dis 7: 369–374.
27. Brown JD, Goekjian G, Poulson R, Valeika S, Stalkncht DE (2009) Avian influenza virus in water: infectivity is dependent on pH, salinity and temperature. Vet Microb 136: 20–26.
28. Rzezutka A, Cook N (2004) Survival of human enteric viruses in the environment and food. FEMS Microbiol Rev 28: 441–453.
29. Fischer TK, Steinsland H, Valentiner-Branth P (2002) Rotavirus particles can survive storage in ambient tropical temperatures for more than 2 months. J Clin Microbiol 40: 4763–4764.
30. Bashiardes S, Koptides D, Pavlidou S (2011) Analysis of enterovirus and adenovirus presence in swimming pools in Cyprus from 2007–2008. Water Sci Technel 63: 2674–2684.
31. D'Alessio D, Minor T, Allen C, Tsiatis A, Nelson D (1981) A study of the proportions of swimmers among well controls and children with enterovirus-like illness shedding or not shedding an enterovirus. Am J Epidemiol 113: 533–541.
32. Hawley H, Morin D, Geraghty M, Tomkow J, Phillips C (1973) Coxsackievirus B epidemic at a boys' summer camp: isolation of virus from swimming water. JAMA 226: 33–36.
33. Nobre FF, Monteiro AB, Telles PR, Williamson GD (2001) Dynamic linear model and SARIMA: a comparison of their forecasting performance in epidemiology. Stat Med 20: 3051–3069.

11

Validating Predictions from Climate Envelope Models

James I. Watling[1]*, David N. Bucklin[1], Carolina Speroterra[1], Laura A. Brandt[2], Frank J. Mazzotti[1], Stephanie S. Romañach[3]

1 Ft Lauderdale Research and Education Center, University of Florida, Ft Lauderdale, Florida, United States of America, 2 U.S. Fish and Wildlife Service, Ft Lauderdale, Florida, United States of America, 3 Southeast Ecological Science Center, U.S. Geological Survey, Ft Lauderdale, Florida, United States of America

Abstract

Climate envelope models are a potentially important conservation tool, but their ability to accurately forecast species' distributional shifts using independent survey data has not been fully evaluated. We created climate envelope models for 12 species of North American breeding birds previously shown to have experienced poleward range shifts. For each species, we evaluated three different approaches to climate envelope modeling that differed in the way they treated climate-induced range expansion and contraction, using random forests and maximum entropy modeling algorithms. All models were calibrated using occurrence data from 1967–1971 (t_1) and evaluated using occurrence data from 1998–2002 (t_2). Model sensitivity (the ability to correctly classify species presences) was greater using the maximum entropy algorithm than the random forest algorithm. Although sensitivity did not differ significantly among approaches, for many species, sensitivity was maximized using a hybrid approach that assumed range expansion, but not contraction, in t_2. Species for which the hybrid approach resulted in the greatest improvement in sensitivity have been reported from more land cover types than species for which there was little difference in sensitivity between hybrid and dynamic approaches, suggesting that habitat generalists may be buffered somewhat against climate-induced range contractions. Specificity (the ability to correctly classify species absences) was maximized using the random forest algorithm and was lowest using the hybrid approach. Overall, our results suggest cautious optimism for the use of climate envelope models to forecast range shifts, but also underscore the importance of considering non-climate drivers of species range limits. The use of alternative climate envelope models that make different assumptions about range expansion and contraction is a new and potentially useful way to help inform our understanding of climate change effects on species.

Editor: Raphaël Arlettaz, University of Bern, Switzerland

Funding: Funding for this work was provided by the U.S. Fish and Wildlife Service (http://www.fws.gov/), Everglades and Dry Tortugas National Park through the South Florida and Caribbean Cooperative Ecosystem Studies Unit (http://www.nps.gov/ever/index.htm), and USGS Greater Everglades Priority Ecosystem Science (http://access.usgs.gov/). The views in this paper do not necessarily represent the views of the U.S. Fish and Wildlife Service. Use of trade, product, or firm names does not imply endorsement by the U.S. Government. The funders had no role in study design, data collection and analysis, decision to publish, or preparation of the manuscript.

Competing Interests: The authors have declared that no competing interests exist.

* E-mail: watlingj@ufl.edu

Introduction

Climate change is one of the major conservation issues of the twenty-first century. Because the effects of increasing greenhouse gas are expected to exacerbate climate change over the course of the twenty-first century and beyond [1], models are an important tool for anticipating potential future effects of climate change and identifying proactive mitigation and adaptation strategies. Climate envelope models (CEMs) establish species-climate relationships that can be extrapolated in space and time [2]. Because climate is one of the major filters determining broad patterns of species distribution [3], [4] and because models can be constructed using relatively simple statistical models and data inputs [2], CEMs have become a widely-used tool for forecasting climate change effects on species distributions [5], [6]. However, CEMs have been criticized as lacking a sound theoretical foundation, making unrealistic assumptions about species-climate relationships (e.g., assuming niche conservatism [7]), and too-easily leading to unjustified conclusions [3], [4], [7]. In some cases, empirical data refute the importance of climate change in underlying contemporary range shifts, even for species presumed to be vulnerable to climate change [8]. Here, we use independent data on changes in the distribution of selected breeding birds in North America from 1967–71 to 1998–2002 to evaluate the ability of three alternative models, including one with no climate change, to correctly classify species and absence.

If CEMs are to be used as a robust natural resource management tool, their ability to accurately forecast species' distributional shifts [9] or population trends [10] needs to be evaluated with field data. Relatively few studies that have evaluated CEMs by calibrating models with historical data (i.e., an initial time period, t_1) and evaluating them with data from a future time period for which there are empirical data on climate and species occurrence (t_2). Those studies that have been conducted have differed in their assessments of CEM performance, with one study suggesting that CEMs are capable of making predictions that are of fair to good performance [9], one indicating relatively poor predictive performance [11], and another showing mixed results [8]. To some degree, the determination of a model's ability to accurately forecast a species' future distribution depends on the metric used to evaluate model performance [12], which depends in part on the relative importance of omission and commission errors [13]. Although some have suggested that when projecting future climate change effects, omission errors (i.e., failing to predict a known occurrence) are more serious than commission errors (predicting species presence in areas where it is not known to occur; [13], [14]), the decision of how to balance omission versus commission errors is highly case-specific.

To determine the ability of CEMs to forecast geographic range shifts presumed to have occurred in response to recent climate change, we evaluated performance of CEMs using metrics describing both omission and commission error. We compared three alternative approaches to model construction and evaluation (Figure 1) that differed in the way they described areas of expansion and contraction of the climate envelope. The first approach incorporated climate change between t_1 and t_2 by calibrating a model with the t_1 occurrences and t_1 climate data, extrapolating the model into t_2 climate conditions, and evaluating model classification with the t_2 occurrence data. We refer to this as the 'dynamic' approach to climate envelope modeling. Under a dynamic model, the climate envelope was allowed to both contract and expand in response to changing climate. The second approach calibrated a model with the t_1 occurrence and t_1 climate data and evaluated the ability of that model to correctly classify t_2 occurrences. In other words, the second 'static' approach tested the ability of a model that described no change in climate suitability (e.g., neither expansion nor contraction of the climate envelope) to classify the t_2 occurrences. A third 'hybrid' approach calibrated a model with the t_1 occurrences and t_1 climate data and projected the model into t_2 climate conditions. We then identified those portions of the map that changed from being outside of the climate envelope in t_1 to within the climate envelope in t_2 (i.e., the areas in which the climate envelope expanded between the two time steps, Figure 2), appended those areas of expansion to the t_1 climate envelope, and evaluated the ability of the model to correctly classify the t_2 occurrences. This approach explicitly assumes that areas of climate suitability at t_1 will remain suitable at t_2, while also considering newly suitable areas when classifying t_2 occurrences. We did not eliminate areas where the climate envelope contracted between 1967–71 and 1998–2002 because recent work suggests the potential for long-term persistence of sink populations experiencing negative growth rates [15].

Materials and Methods

Species occurrences for model calibration and evaluation were drawn from the Breeding Bird Survey (BBS) dataset [16]. Breeding Bird Survey data are collected annually by thousands of volunteers who record species observations along fixed survey routes, and are a key source of long-term population data for North American breeding birds. To define the pool of species for which models would be constructed, we searched the primary literature for studies of latitudinal range shifts in birds; three studies presented data for multiple species and were used to create our species pool ([17], [18], [19]). We created models for species known on the basis of previously published data to have experienced a poleward distributional shift (either north or south). Hitch & Leberg [17] tested for significant distributional shifts among species included in their study so we included species for which their tests were significant at $\alpha \leq 0.05$. In the remaining two studies, the significance of range shifts was not tested for individual species, and different metrics were used to describe range shifts. In lieu of a significance test, we developed operational criteria for including species in our study. For species reported on in La Sorte & Thompson [18] we included species for which the slope of the

relationship describing movement of the northern range boundary through time was $5 < $ slope < -5 (e.g., the species that had experienced the most dramatic range shifts in their study). For species reported on in Zuckerberg et al. [19], we included species for which the northern or southern range boundary shifted >50 km. In all cases, migratory species were excluded from consideration because fine temporal resolution climate data for evaluating models are not available for much of the Neotropics. Modeled species therefore had to be resident in and restricted to the contiguous United States (two species, the Wild turkey *Meleagris gallopavo* and Gambel's quail, *Callipepla gambelii* also occur in parts of Mexico, but because most of their range is within the contiguous United States, they were included in our analysis) and have experienced a significant poleward distributional shift according to criteria described above. Across the three studies, our selection criteria identified 12 species for modeling (Table 1).

Following Hitch & Leberg [17], we used two five-year time periods for model development, the first for calibration and the second for evaluation. Models were calibrated using t_1 observations made on BBS routes (e.g., between the years 1967–1971). Repeat observations of a species from a survey route were removed such that if a species was observed at any point during the five year t_1 period, it was counted as a single presence. Latitude and longitude coordinate data for routes were obtained from the BBS database ([16]; coordinates are expressed as a single latitude and longitude observation for each 24.5 mile route). Using the coordinate data for each survey route, we extracted the values of seven climate variables at routes known to be occupied by each species. We also selected 1000 random routes from which focal species were unobserved and assumed to be absent, and extracted the values of the same seven variables. Climate variables included were: annual precipitation, precipitation of the driest month, precipitation of the wettest month, mean annual temperature, temperature annual range, maximum mean monthly temperature and minimum mean monthly temperature. Models were evaluated with occurrence data from t_2 (1998–2002), which were compiled from BBS survey routes as described for the t_1 data. Climate data were obtained from the PRISM dataset (PRISM Climate Group, Oregon State University, August 2011, and average values for all variables were calculated for each of the two five year periods used in our study. Because the PRISM data were downloaded at a resolution of 4×4 km, we resampled the climate grids to 40 km $\times$ 40 km to approximate the resolution of the BBS survey routes (24.5 miles = ~ 39.4 km).

Models were constructed using two different algorithms: the random forest algorithm, which classifies observations (e.g., species presence/absence) based on an iterative, recursive partitioning of observations into the most homogeneous subsets possible [20], and maximum entropy, which calculates a species' probability of occurrence based on knowledge of environmental conditions at sites known to be occupied by the species and background environmental conditions [21], [22]. Here, we used the 1000 random absences for each species to calculate the environmental background for maximum entropy modeling. Random forest models and all other statistical analyses were conducted in R [23] and maximum entropy modeling was done using the MaxEnt software package [21], [22] using default settings.

We evaluated performance of all CEMs by constructing models with calibration data (t_1) and testing them with t_2 occurrences. Four criteria were used to assess model performance: the area under the receiver-operator curve (AUC), the true skill statistic, sensitivity and specificity [12]. The AUC metric ranges from 0–1 and measures the tendency for a random presence point to have a higher predicted probability of climate suitability than a random

'Dynamic' approach

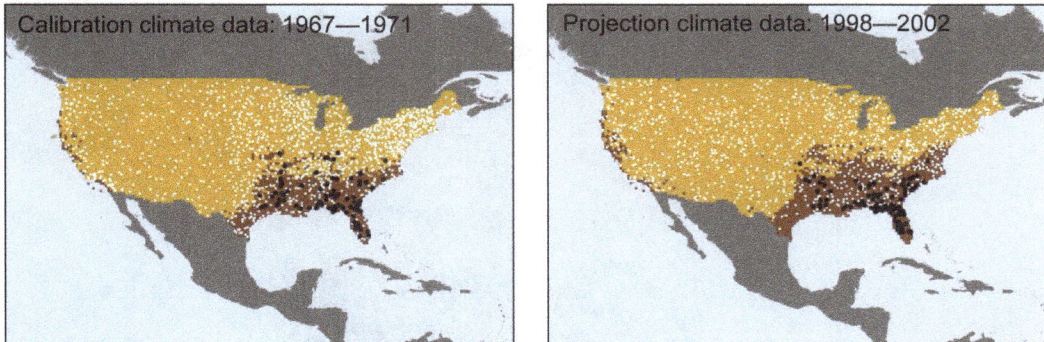

'Static' approach

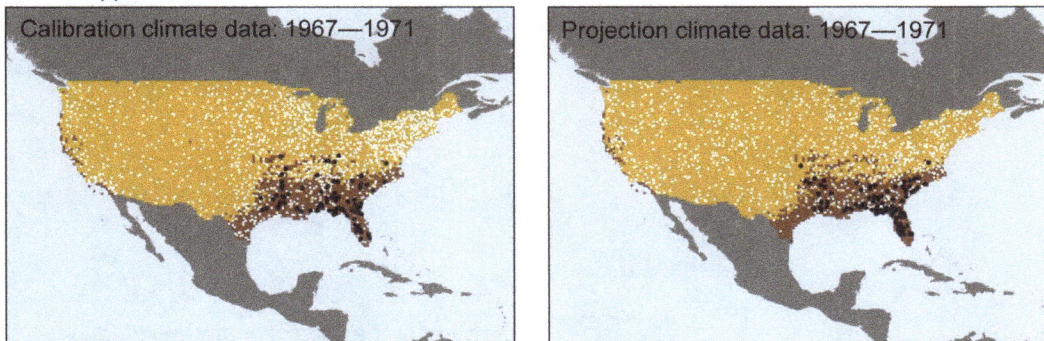

'Hybrid' approach

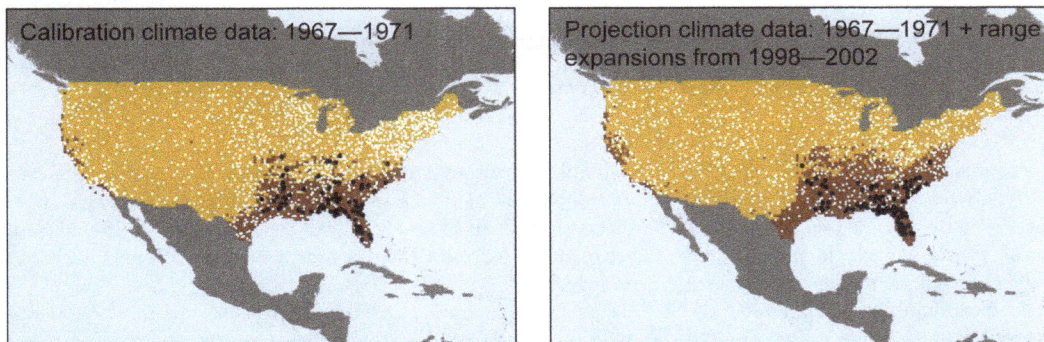

Figure 1. Examples of the three approaches to construction and validation of climate envelope models. All models are calibrated with occurrence data and climate conditions from 1967–1971 (left-hand panels), and validated with occurrence data from 1998–2002 (right hand panels). Black circles indicate species presence and white dots indicate species absence. The prediction map against which occurrences are validated using the dynamic approach represent the climate envelope under 1998–2002 conditions and for the static approach the prediction map represents the 1967–1971 climate envelope. Under the hybrid approach, areas of range expansion between 1967–1971 and 1998–2002 are merged with the 1967–1971 climate envelope to create a third prediction map.

background point. In addition to using AUC to evaluate the extrapolated model (based on t_2 climate and occurrences), we also calculated AUC using a static model in which the t_2 occurrences were evaluated against t_1 climate conditions. We expected that species whose ranges were shifting in response to climate change would have greater AUC values using the t_2 climate data compared with the static AUC calculation. Like AUC, the true skill statistic also ranges from 0–1, but is independent of species prevalence [24]. Sensitivity measures the proportion of correctly classified presences in the test dataset, whereas specificity measures the proportion of correctly classified absences; both metrics range from 0–1. Sensitivity is a measure of omission error (high

sensitivity = low omission), and specificity is a measure of commission error (high specificity = low commission). A number of authors have suggested that the 'best' models should achieve low rates of omission (i.e., they should accurately classify presences) even if commission error is relatively high, because at least some commission error is not truly error but rather reflects our incomplete knowledge of species distributions or the identification of environmentally suitable area that is inaccessible to species because of dispersal barriers, species interactions or other factors [13], [14]).

Because two of our performance metrics describe the ability of a model to correctly classify presences and absences, they require the

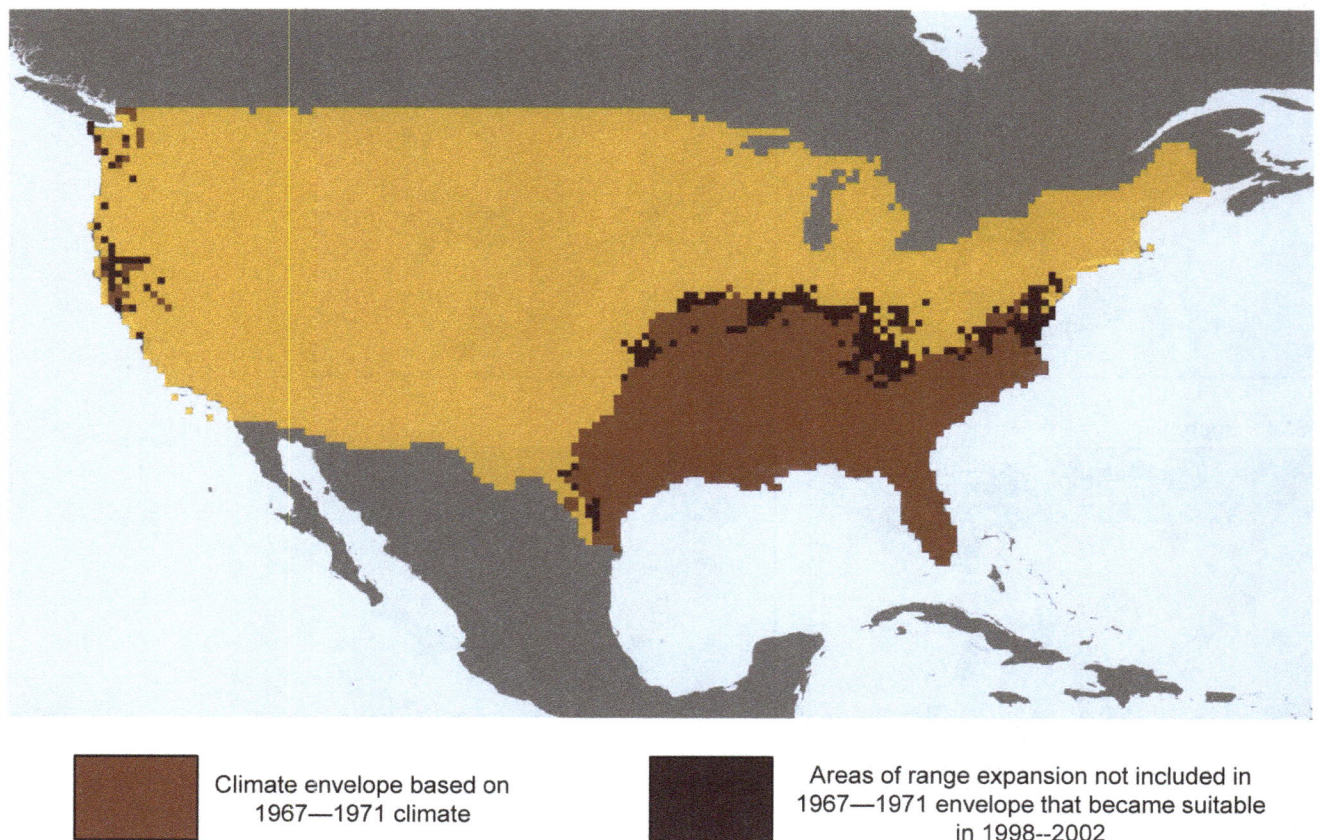

Figure 2. Example map illustrating a 'hybrid' approach to climate envelope model construction. Areas indicated in black are included in the 1998–2002 projection, but were not part of the initial climate envelope. These areas range expansion between 1967–1971 and 1998–2002 are merged with the initial 1967–1971 climate envelope to create a hybrid prediction map.

user define the threshold probabilities at which presence is differentiated from absence. We used two alternative criteria to determine that threshold. One criterion converted continuous probabilities into a categorical prediction by identifying the threshold that maximized Cohen's kappa, a model performance metric that measures overall classification ability [25]. To identify this threshold, we ran five replicate model runs using random subsets of the species occurrence data in the calibration dataset (1967–1971) for each 0.01 unit change in threshold between .01 and 0.99 and calculated kappa for each randomization (using a 75–25% training-testing partition of the occurrence data). We calculated the average kappa for each incremental change in the threshold to identify the threshold at which kappa was maximized. A second criterion used a prevalence-based approach to defining the threshold used to calculate kappa [25]. We calculated the prevalence of each species as number of occurrences/(number of occurrences +1000) because all CEMs used 1000 absence points, and used the estimate of prevalence as the threshold for converting probability into categorical predictions. We calculated each threshold (maximum kappa and prevalence) once for each species, and report the threshold that resulted in the greatest model sensitivity (e.g., best classified species presences at t_2). We calculated all performance metrics using a 75–25% training-testing split on 100 random partitions of the occurrence data, and tested for significant effects of algorithm and approach on AUC, the true skill statistic, sensitivity and specificity using generalized linear mixed-effects models [26] with a binomial distribution and a

logit link. Algorithm and approach were tested as fixed effects, and species were treated as a random effect. The significance of fixed effects and their interaction was tested as the likelihood ratio between the full model and a model with the effect being tested removed.

The dynamic, static and hybrid approaches to model evaluation differ in the extent to which they treat range expansion and contraction (see above). To understand how model performance varied as a function of classification specifically in areas of range change, we calculated the proportion of t_2 presences and absences of each species that occurred in areas of range expansion or contraction between 1967–71 and 1998–2002 using the dynamic approach to model evaluation. We expected that the hybrid approach would result in increased sensitivity compared with the dynamic approach because the hybrid approach assumes no range contraction and therefore maximizes the area of predicted suitability. We further expected that the hybrid approach would improve sensitivity the most for those species for which the dynamic approach resulted in the greatest number of misclassified presences. In other words, models assuming no range contraction should yield the biggest gains in sensitivity for species that continue to persist in areas where range contraction is predicted under the dynamic approach. Therefore, we used linear regression to determine whether species-by-species differences in sensitivity between the dynamic and hybrid approaches were associated with proportions of misclassified presences (i.e., those occurring in areas of range contraction) using the dynamic approach. We also

Table 1. Summary statistics for climate envelope models for twelve species of resident North American breeding birds.

Species	Common Name	No. presences 1967-1971/ 1998-2002	Range shift	Threshold criterion	AUC (static)	AUC (dynamic)	True skill statistic	Sensitivity (dynamic approach)	Specificity (dynamic approach)	Sensitivity (static approach)	Specificity (static approach)	Sensitivity (hybrid approach)	Specificity (hybrid approach)
Phasianidae													
Meleagris gallopavo	Wild turkey	73/967	North	Prevalence									
RF					0.577±0.030	0.613±0.032	0.159±0.051	0.425±0.043	0.734±0.036	0.295±0.028	0.844±0.033	0.591±0.041	0.610±0.044
Maxent					0.650±0.029	0.652±0.032	0.091±0.047	0.957±0.022	0.134±0.056	0.904±0.034	0.204±0.051	0.995±0.007	0.031±0.020
Centrocercus urophasianus	Greater sage grouse	20/60	South	Prevalence									
RF					0.943±0.020	0.927±0.021	0.721±0.060	0.951±0.055	0.768±0.033	0.955±0.054	0.794±0.033	0.997±0.016	0.591±0.041
Maxent					0.951±0.016	0.944±0.015	0.652±0.047	0.999±0.007	0.653±0.048	0.993±0.020	0.610±0.058	0.999±0.001	0.398±0.067
Tympanuchus cupido	Greater prairie chicken	19/35	North	Prevalence									
RF					0.827±0.077	0.838±0.067	0.479±0.158	0.749±0.153	0.745±0.038	0.751±0.140	0.779±0.030	0.934±0.093	0.577±0.042
Maxent					0.875±0.041	0.910±0.032	0.095±0.091	0.999±0.001	0.095±0.091	0.999±0.001	0.082±0.071	0.999±0.001	0.013±0.021
Callipepla gambelii	Gambel's quail	27/78	North	Prevalence									
RF					0.943±0.021	0.938±0.021	0.723±0.058	0.953±0.051	0.768±0.033	0.919±0.059	0.836±0.026	0.996±0.013	0.625±0.043
Maxent					0.955±0.015	0.948±0.013	0.669±0.034	0.999±0.001	0.670±0.034	0.997±0.018	0.710±0.038	0.999±0.001	0.463±0.039
Picidae													
Melanerpes erythrocephalus	Red-headed woodpecker	624/852	South	Maximum kappa									
RF					0.834±0.015	0.814±0.018	0.402±0.041	0.929±0.017	0.472±0.038	0.910±0.016	0.581±0.030	0.994±0.005	0.215±0.029
Maxent					0.877±0.016	0.870±0.016	0.613±0.035	0.905±0.019	0.709±0.032	0.877±0.021	0.743±0.029	0.987±0.007	0.420±0.031
Melanerpes carolinus	Red-bellied woodpecker	599/830	North	Maximum kappa									
RF					0.776±0.020	0.766±0.019	0.311±0.037	0.912±0.017	0.398±0.033	0.754±0.028	0.579±0.031	0.977±0.009	0.201±0.026
Maxent					0.852±0.017	0.806±0.019	0.466±0.036	0.852±0.022	0.614±0.029	0.748±0.030	0.768±0.026	0.962±0.012	0.397±0.031
Corvidae													
Corvus ossifragus	Fish crow	153/404	North	Prevalence									
RF					0.933±0.014	0.912±0.015	0.605±0.051	0.755±0.050	0.851±0.022	0.736±0.044	0.931±0.016	0.938±0.025	0.741±0.026
Maxent					0.938±0.012	0.921±0.013	0.702±0.034	0.919±0.027	0.783±0.026	0.879±0.031	0.849±0.021	0.989±0.011	0.605±0.033
Paridae													
Poecile carolinensis	Carolina chickadee	468/791	North	Maximum kappa									
RF					0.911±0.013	0.914±0.011	0.686±0.030	0.956±0.016	0.730±0.024	0.474±0.034	0.974±0.009	0.980±0.011	0.662±0.029
Maxent					0.935±0.011	0.877±0.015	0.674±0.029	0.957±0.015	0.716±0.027	0.499±0.043	0.966±0.011	0.978±0.011	0.653±0.030
Troglodytidae													
Thryothorus ludovicianus	Carolina wren	535/689	North	Maximum kappa									

Table 1. Cont.

Species	Common Name	No. presences 1967-1971/ 1998-2002	Range shift	Threshold criterion	AUC (static)	AUC (dynamic)	True skill statistic	Sensitivity (dynamic approach)	Specificity (dynamic approach)	Sensitivity (static approach)	Specificity (static approach)	Sensitivity (hybrid approach)	Specificity (hybrid approach)
RF					0.815±0.018	0.868±0.014	0.451±0.030	0.991±0.007	0.459±0.030	0.828±0.023	0.602±0.025	0.999±0.002	0.248±0.022
Maxent					0.861±0.017	0.875±0.018	0.662±0.032	0.960±0.017	0.702±0.028	0.847±0.023	0.766±0.027	0.994±0.006	0.454±0.032
Thryomanes bewickii	Bewick's wren	246/440	South	Prevalence									
RF					0.780±0.024	0.725±0.023	0.332±0.041	0.824±0.030	0.507±0.031	0.783±0.034	0.658±0.032	0.961±0.020	0.307±0.032
Maxent					0.774±0.024	0.717±0.024	0.335±0.044	0.881±0.028	0.454±0.041	0.829±0.037	0.611±0.033	0.982±0.012	0.249±0.033
Emberizidae													
Aimophila aestivalis	Bachman's sparrow	81/95	South	Prevalence									
RF					0.962±0.012	0.954±0.015	0.762±0.041	0.976±0.035	0.785±0.026	0.955±0.042	0.859±0.021	0.999±0.004	0.657±0.031
Maxent					0.965±0.010	0.957±0.013	0.716±0.031	0.999±0.004	0.716±0.031	0.986±0.021	0.772±0.028	0.999±0.001	0.536±0.037
Icteridae													
Agelaius tricolor	Tricolored blackbird	15/29	North	Prevalence									
RF					0.942±0.044	0.901±0.054	0.654±0.127	0.866±0.113	0.786±0.043	0.919±0.098	0.833±0.035	0.984±0.049	0.659±0.052
Maxent					0.954±0.019	0.940±0.030	0.670±0.065	0.976±0.079	0.694±0.069	0.983±0.065	0.727±0.064	0.997±0.020	0.515±0.121
Average ±1 SD	RF												
	Maxent												

Three approaches to climate envelope model evaluation were compared, a dynamic approach in which a model calibrated on conditions at t_1 (here 1967–1971) was projected to predict occurrences at t_2 (here 1998–2002), a static approach in which t_2 occurrences are predicted using the t_1 climate data, and a hybrid approach in which t_2 occurrences are predicted using a model that allows for range expansion, but not range contraction at t_2.

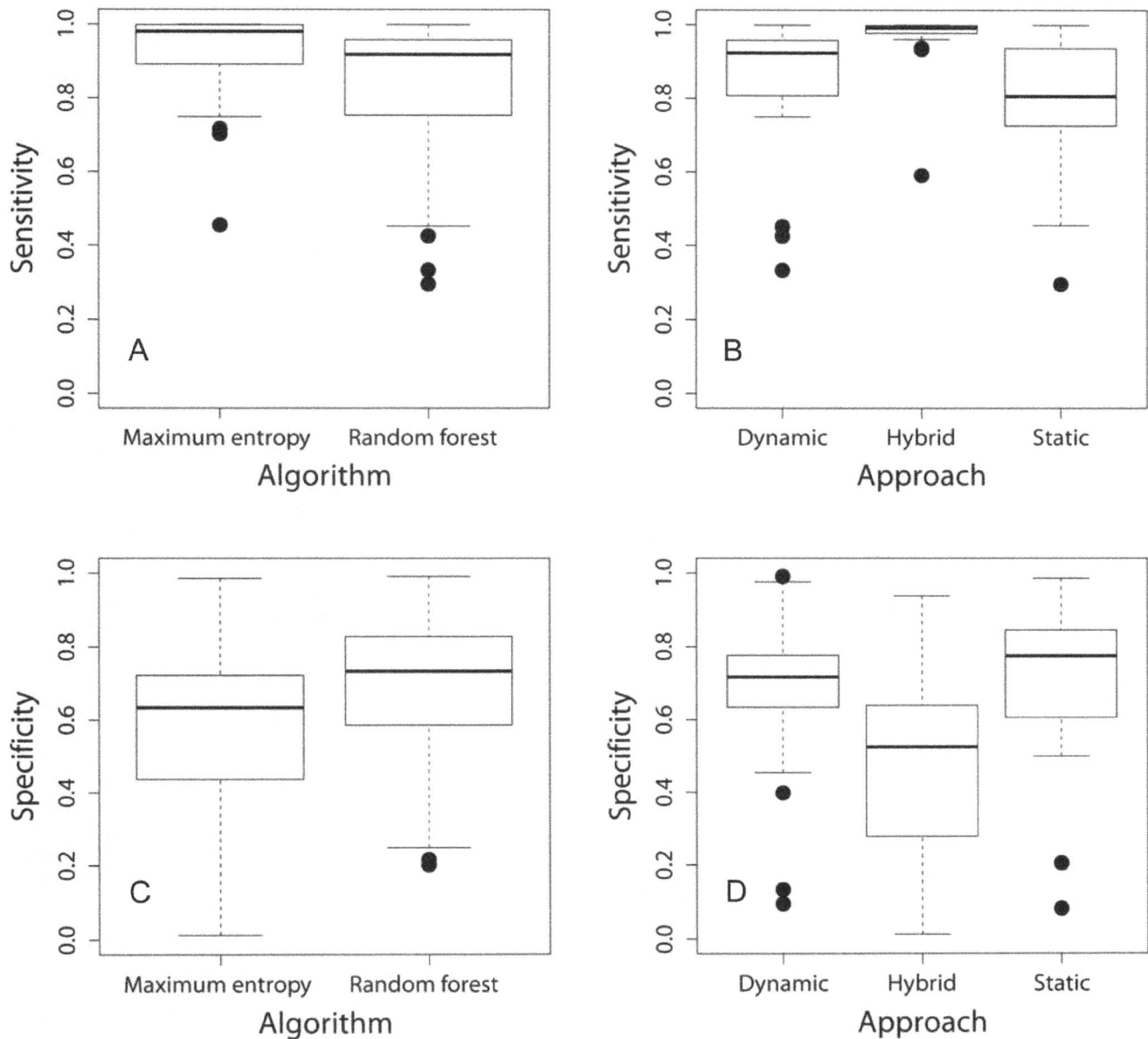

Figure 3. Box plots illustrating differences in climate envelope model sensitivity between models constructed with the maximum entropy and random forest algorithms (A), differences in sensitivity between models using three approaches that differ in the way they treat range expansion and contraction (B), differences in specificity between algorithms (C) and differences in specificity among approaches (D).

wanted to determine whether species-specific gains in model sensitivity under a hybrid approach were related to species traits. We reasoned that if the gain in sensitivity achieved using the hybrid approach is indeed greatest for species that maintain populations in areas deemed unsuitable by a dynamic CEM, there may be a positive effect of niche breadth on such resistance, much as has been described for the relatively generalist species that persist in fragmented landscapes [27]. We counted the number of habitat categories to which species were assigned in the Zip Code Zoo database (www.zipcodezoo.com) as an index of habitat niche breadth. To test whether habitat generalists showed the greatest improvement in sensitivity using the hybrid approach, we used linear regression to determine whether differences in sensitivity

between the dynamic and hybrid approaches were positively associated with habitat niche breadth. In contrast, we expected the static approach to result in increased specificity relative to the dynamic approach (because the range expansion predicted using the dynamic approach may increase the number of misclassified absences). Therefore, we used linear regression to determine whether differences in specificity between the dynamic and static approaches were associated with proportions of absences in areas of range expansion. We reasoned that species for which the dynamic approach most overestimated range expansion may be dispersal limited, and unable to track changing climate [28]. Although we searched for information on known dispersal distances for species using online databases and literature searches,

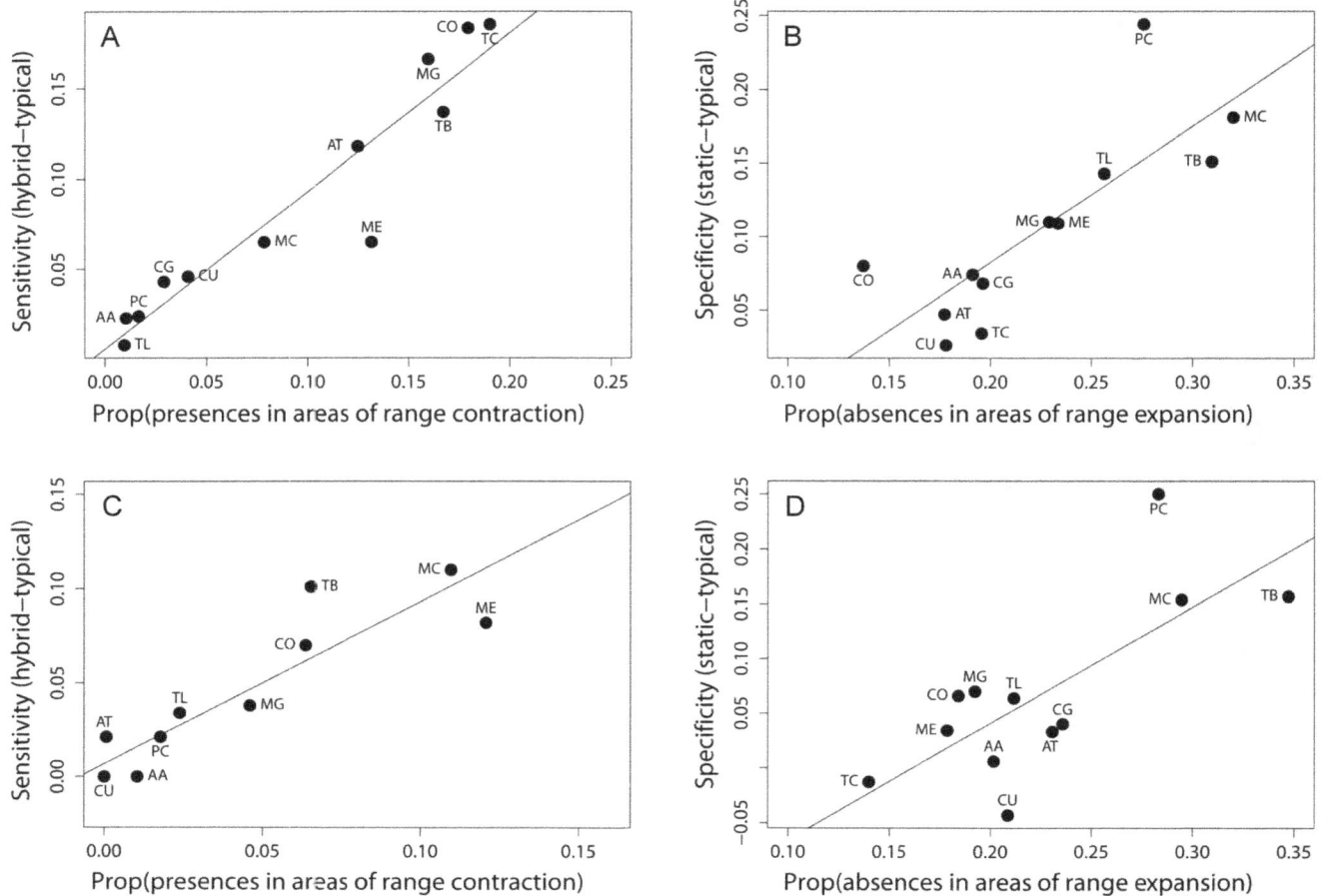

Figure 4. Relationships between differences in model sensitivity as a function of the proportion of presences in 1998–2002 that occurred in areas of range expansion (A and C for random forest and maximum entropy models, respectively) and differences in model specificity as a function of the proportion of absences in 1998–2002 that occurred in areas of range expansion (B and D for random forest and maximum entropy models, respectively).

we were only able to obtain dispersal data for seven of our twelve study species. Because body size is positively correlated with dispersal distance for active dispersers [29], we used body size as a proxy for dispersal ability. We obtained data on maximum body mass from online databases (Animal Diversity Web, and Zip Code Zoo), which we log-transformed prior to analysis. We used linear regression to determine whether differences in specificity between the static and dynamic approaches were greatest for the species with the smallest body mass (i.e., the species expected to be most dispersal limited).

Results

The PRISM data describe a warmer and slightly wetter climate across the contiguous United States in 1998–2002 compared with the 1967–1971 period. Annual precipitation in 1998–2002 averaged 768.5 mm compared with 763.3 mm in 1967–1971. Precipitation of the driest month was slightly greater in 1998–2002 (25.9 mm) than in 1967–1971 (25.4 mm), although precipitation of the wettest month was slightly lower (118.7 mm in 1998–2002 compared with 125.2 mm in 1967–1971). Temperature annual mean, maximum mean monthly temperature and minimum mean monthly temperature were all warmer in 1998–2002 compared with 1967–1971 (11.6°C vs 10.7°C, 30.8°C vs 30.3°C, −5.7°C vs

−7.8°C, respectively), and the temperature annual range was lower in 1998–2002 (23.4°C) than in 1967–1971 (24.6°C).

Of the 12 species included in the study, previously published data suggest that eight experienced a northward range shift and four experienced a southward range shift (Table 1). In general, model sensitivity was greatest when presences were differentiated from absences using a prevalence criterion for the rarest species in the analysis (those represented by 262 or fewer occurrences in the calibration dataset, Table 1), whereas for more common species, model sensitivity was greatest when presence and absence was differentiated using the threshold that maximized kappa in the calibration dataset (Table 1). Although all 12 species have been suggested to have shifted their range in response to changing climate, static AUC values were higher than projected AUC values for at least one algorithm in nine out of 12 species (Table 1), suggesting that not all range shifts are consistent with a climate change model.

Average (dynamic) AUC values for the 12 random forest CEMs were 0.848 ± 0.103 and for the maximum entropy CEMs average AUC was 0.868 ± 0.097 (Table 1). The difference in AUC between algorithms was significant ($\chi^2 = 5.021$, df = 1, P = 0.025). Values of the true skill statistic averaged 0.524 ± 0.196 for random forest CEMs and 0.551 ± 0.258 for maximum entropy CEMs, but this difference was not statistically significant ($\chi^2 = 0.324$, df = 1,

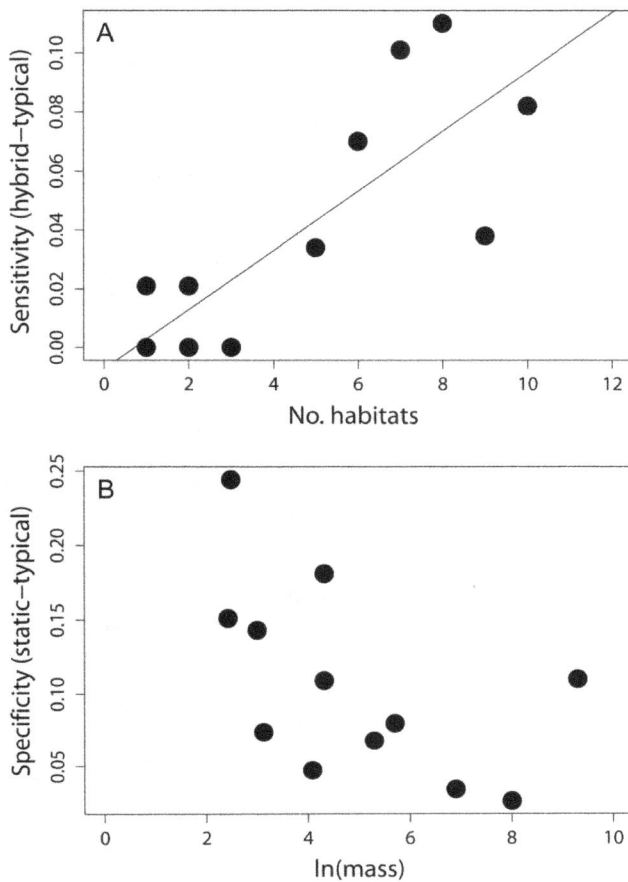

Figure 5. The improvement in sensitivity using a hybrid approach to climate envelope model evaluation that assumed no range contraction relative to a dynamic model in which ranges expanded and contracted in response to climate change was greatest for species that have been reported from relatively more habitat types (A), whereas differences in specificity between the dynamic approach and a static approach assuming no climate change were not significantly associated with body size (B).

Table 2. Body size and habitat niche breadth for 12 species of resident North American breeding birds.

Common Name	Number of land cover types used	ln (mass)
Wild turkey	9	9.31
Greater sage grouse	1	8.01
Greater prairie chicken	2	6.91
Gambel's quail	3	5.30
Red-headed woodpecker	10	4.32
Red-bellied woodpecker	8	4.32
Fish crow	6	5.71
Carolina chickadee	1	2.49
Carolina wren	5	2.99
Bewick's wren	7	2.43
Bachman's sparrow	2	3.12
Tricolored blackbird	2	4.09

Trait data were obtained from online natural history databases.

varied in specificity ($\chi^2 = 21.059$, df = 2, P<0.001), with the hybrid approach having lower specificity (0.47 ± 0.228) than either dynamic (0.67 ± 0.219) or static approaches (0.71 ± 0.218; Table 1, Figure 3D). Prediction maps for all species using the two algorithms and three approaches to model construction are included as supplementary figures (Figures S1–S6).

Both random forest and maximum entropy models indicated that difference in sensitivity between dynamic and hybrid approaches increased with the proportion of presences occurring in areas of range contraction between t_1 and t_2 ($F_{1,10} = 89.80$, P<0.001 and $F_{1,10} = 49.36$, P<0.001; Figure 4A and 4C for random forest and maximum entropy models, respectively). The difference in specificity between dynamic and static approaches increased with the proportion of absences in areas of range expansion for both random forest and maximum entropy models ($F_{1,10} = 18.24$, P = 0.002 and $F_{1,10} = 12.41$, P = 0.006 for random forest and maximum entropy models, respectively; Figure 4B & 4D). For tests investigating the effect of niche breadth on changes in sensitivity between hybrid and dynamic approaches, we focused on results from maximum entropy models because sensitivity was greater, on average, than for random forest models (Table 1, Table 2). As hypothesized, species for which the hybrid approach yielded the greatest increase in sensitivity have been reported from more habitat types than species for which the hybrid approach had little effect on sensitivity ($F_{1,10} = 17.98$, P = 0.002; Figure 5A). We investigated whether small body size was associated with changes in specificity between static and dynamic approaches for random forest models, because specificity was greater than for maximum entropy models (Table 1, Table 2). There was suggestive, but not significant relationship indicating that differences in specificity between static and dynamic approaches were greatest for the smallest-bodied species ($F_{1,10} = 4.30$, P = 0.065; Figure 5B).

Discussion

The choice of modeling algorithm, rather than the approach to model evaluation, had the greatest overall effect on sensitivity, whereas specificity was affected by both algorithm and approach. The hybrid approach to CEM evaluation did not result in a significant overall increase in model sensitivity, but did result in

P = 0.569). Generalized linear mixed effects models describing effects of algorithm and approach on CEM sensitivity did not differ with or without interaction terms ($\chi^2 = 0.865$, df = 2, P = 0.649), so the significance of fixed effects was tested against the full model without interaction terms. Although maximum entropy models had greater sensitivity (0.92 ± 0.122) than random forest models (0.82 ± 0.202; Table 1, Figure 3A), the effect of algorithm on sensitivity was not significant ($\chi^2 = 1.624$, df = 1, P = 0.203). Mean sensitivity of the hybrid approach (0.97 ± 0.082) was greater than either the dynamic (0.85 ± 0.189) or static approach (0.80 ± 0.183), but this difference was not statistically significant ($\chi^2 = 3.902$, df = 2, P = 0.142; Figure 3B).

Like the test for sensitivity, tests of all fixed effects on CEM specificity did not differ with or without interaction terms ($\chi^2 = 1.132$, df = 2, P = 0.568), so the significance of fixed effects was again tested against the full model without interaction terms. The effect of algorithm on specificity was significant ($\chi^2 = 7.806$, df = 1, P = 0.005), with random forest models having greater specificity (0.68 ± 0.207) than maximum entropy models (0.56 ± 0.263; Table 1, Figure 3C). The different approaches also decreased specificity relative to dynamic and static approaches (Figure 3B & 3D). These general trends, however, obscure substantial species-specific responses that illustrate how users can create models that vary in their assumptions about climate change effects on range expansion and contraction depending on characteristics of the species in question and the relative importance of omission and commission error. For example, we were able to correctly almost 30% more presences for some species using the hybrid approach compared with the dynamic approach (Table 1), although at a cost of reduced specificity. Improved sensitivity was the result of an increase in presences that were correctly classified using the hybrid approach, but misclassified using the dynamic approach because they occurred in areas of range contraction between t_1 and t_2 (Figure 4A & 4C). Species that experienced the greatest improvement in sensitivity using the hybrid approach were reported from more land cover types than species for which dynamic and hybrid approaches differed little in sensitivity. Overall, specificity was similar using both the dynamic and static approaches, although specificity tended to be greatest using the static approach for species experiencing the greatest number of absences in areas of range expansion. However, for three species evaluated at least one algorithm indicated that sensitivity of the static approach was at least as good as with a dynamic approach, and for nine species the static AUC was greater than the dynamic AUC, suggesting that climate change may not necessarily always underlie purported climate-induced range shifts.

Although differences in model sensitivity between approaches were not statistically significant overall, our use of alternative approaches to CEM evaluation provides a useful framework for evaluating alternative explanations about climate change effects on species. Comparing CEMs that make competing predictions about climate-induced range expansion and contraction may have important implications for natural resource management decisions made on the basis of CEMs. If managing for an endangered or invasive species, for example, for which the priority is to identify all possible suitable areas (e.g., a prioritization of reduced omission error at the expense of increased commission error), a hybrid model that includes the possibility of range expansion but not contraction may be preferred. On the other hand, increased specificity (which may be desirable, for example, when identifying areas unlikely to be suitable for a problematic invasive species) may be achieved using the random forest algorithm (an observation consistent with the random forest's tendency to reduce overprediction, [20]).

The dynamic approach to climate envelope modeling tended to overestimate the extent of range contraction that species experienced (Figure 3A & 3C). We found that species for which the dynamic CEM approach most overestimated range contractions have been reported from many different land cover types. It has been suggested that habitat generalists are buffered from the negative effects of habitat fragmentation [27], and our work suggests that generalists may also be buffered to some degree from climate-induced range contractions. Alternatively, it has been suggested that generalist species may be better able to track changing climate than habitat specialists [30]. Given the potential for complex interactions between species traits and changing climate, we suggest that more work relating species traits to the extinction debt [31] accumulated as a result of climate change is needed.

Although specificity showed little overall difference between dynamic and static approaches, there were differences between the two approaches for individual species, and a tendency for the static approach to perform best when the dynamic approach overesti-

mated the number of absences in areas of range expansion. We found a marginally non-significant relationship suggesting that smaller-bodied species experienced that greatest improvement in specificity using the static CEM approach. Although smaller body size generally equates to dispersal limitation [29], there is substantial error in this generalization. Unfortunately, dispersal distances are unreported for many of the species reported here so we are unable to test for direct effects of dispersal limitation on differences in specificity. Although transplant experiments in which species successfully establish themselves in areas of climate suitability attest to the potential importance of dispersal limitation as a factor that prevents species from tracking changing climate [30], little work is available to suggest that the most dispersal limited species are least able to track climate change. Interpreting model specificity is also complicated by the uncertain nature of absences [32], because species are likely present but unobserved at some locations categorized as absences. However, the suggestive relationship between body size and improved specificity using a static approach may indicate that small bodied species are not able to track changing climate as efficiently as larger-bodied species with greater dispersal abilities.

Although climate may be an important determinant of species distributions at broad spatial scales [4], it is not necessarily the most important factor in circumscribing species' geographic ranges. For many species, habitat loss [33] or other factors (e.g., competition and dispersal, [34]) may be as or more important than climate in determining current or future geographic distributions. Some of the species included in our analysis have experienced range expansion at least partly because of factors other than climate change (e.g., the Wild turkey has been the target of reintroduction efforts in parts of its range, [35]). Furthermore, for three species in our analysis, at least one algorithm showed that a static approach assuming no climate change classified at least as many t_2 presences as the dynamic approach assuming climate change effects, and for one species (the Tricolored blackbird), the static approach had greater sensitivity using both algorithms (Table 1). Indeed, the relative improvement in sensitivity between dynamic and static approaches may be a useful metric of the magnitude of climate change effects on species. It may also be expected that performance of some models would be improved by adding data describing non-climate environmental conditions (e.g., land cover) in addition to climate data. For example, the Fish crow, *Corvus ossifragus*, is associated with wetland habitats in the eastern United States, suggesting that models that do not include the distribution of wetland habitat may underperform relative to models that include habitat. Consistent with this expectation, experimental niche models for the fish crow that included land cover data had greater sensitivity than the models we report on here (J. I. Watling, unpublished). We also acknowledge that non-climate driven spatial variation in population dynamics (i.e., metapopulation structure) may play a role in driving the range shifts we describe here.

Our results suggest cautious optimism when using predictions from CEMs to infer climate change effects on species, and we demonstrate how model construction may be manipulated to best suit alternative model needs. We suggest that the use of alternative CEMs that make different assumptions about range expansion and contraction can help inform an understanding of climate change effects on species. However, our results also suggest that CEMs do not unambiguously implicate climate change as a driver of observed species range shifts in many cases, underscoring the importance of considering additional factors when considering species range shifts through time.

Supporting Information

Figure S1   Figure panels with binary prediction maps indicating areas of suitable (brick red) and unsuitable (dark yellow) climate for 12 species of resident North American breeding birds. Models were calibrated on climate conditions for the 1967–1971 period and projected using climate conditions for 1998–2002. Presences (dark circles) and absences (white circles) from 1998–2002 surveys are indicated. Illustrated are predictions from a random forest model using the dynamic approach described in the text.

Figure S2   Figure panels with binary prediction maps indicating areas of suitable (brick red) and unsuitable (dark yellow) climate for 12 species of resident North American breeding birds. Models were calibrated on climate conditions for the 1967–1971 period and projected using climate conditions for 1998–2002. Presences (dark circles) and absences (white circles) from 1998–2002 surveys are indicated. Illustrated are predictions from a maximum entropy model using the dynamic approach described in the text.

Figure S3   Figure panels with binary prediction maps indicating areas of suitable (brick red) and unsuitable (dark yellow) climate for 12 species of resident North American breeding birds. Models were calibrated on climate conditions for the 1967–1971 period and projected using climate conditions for 1998–2002. Presences (dark circles) and absences (white circles) from 1998–2002 surveys are indicated. Illustrated are predictions from a random forest model using the static approach described in the text.

Figure S4   Figure panels with binary prediction maps indicating areas of suitable (brick red) and unsuitable (dark yellow) climate for 12 species of resident North American breeding birds. Models were calibrated on climate conditions for the 1967–1971 period and projected using climate

conditions for 1998–2002. Presences (dark circles) and absences (white circles) from 1998–2002 surveys are indicated. Illustrated are predictions from a maximum entropy model using the static approach described in the text.

Figure S5   Figure panels with binary prediction maps indicating areas of suitable (brick red) and unsuitable (dark yellow) climate for 12 species of resident North American breeding birds. Models were calibrated on climate conditions for the 1967–1971 period and projected using climate conditions for 1998–2002. Presences (dark circles) and absences (white circles) from 1998–2002 surveys are indicated. Illustrated are predictions from a random forest model using the hybrid approach described in the text.

Figure S6   Figure panels with binary prediction maps indicating areas of suitable (brick red) and unsuitable (dark yellow) climate for 12 species of resident North American breeding birds. Models were calibrated on climate conditions for the 1967–1971 period and projected using climate conditions for 1998–2002. Presences (dark circles) and absences (white circles) from 1998–2002 surveys are indicated. Illustrated are predictions from a maximum entropy model using the hybrid approach described in the text.

Acknowledgments

The views in this paper do not necessarily represent the views of the U.S. Fish and Wildlife Service. Use of trade, product, or firm names does not imply endorsement by the US Government.

Author Contributions

Conceived and designed the experiments: JIW DNB CS LAB SSR FJM. Performed the experiments: JIW DNB CS. Analyzed the data: JIW DNB CS. Contributed reagents/materials/analysis tools: JIW DNB CS. Wrote the paper: JIW DNB CS LAB SSR FJM.

References

1. Solomon S, Qin D, Manning M, Chen Z, Marquis M, et al. (2007) Climate Change 2007: The physical Science Basis. Contribution of Working Group I to the Fourth Assessment Report of the Intergovernmental Panel on Climate Change. Cambridge, UK and New York, USA : Cambridge University Press. 996 p.
2. Franklin J (2009) Mapping species distributions: spatial inference and prediction. New York: Cambridge University Press. 320 p.
3. Toledo M, Peña-Claros M, Bongers F, Alarcón A, et al. (2012) Distribution patterns of tropical woody species in response to climatic and edaphic gradients. J Ecol 100: 253–263.
4. Pearson RG, Dawson TP (2003) Predicting the impacts of climate change on the distribution of species: are bioclimate envelope models useful? Glob Ecol Biogeogr 12: 361–371.
5. Thomas CD, Cameron A, Green RE, Bakkenes M, Beaumont LJ, et al. (2004) Extinction risk from climate change. Nature 427: 145–148.
6. Lawler JJ, Shafer SL, White D, Kareiva P, Maurer EP, et al. (2009) Projected climate-induced faunal change in the Western Hemisphere. Ecology 90: 588–597.
7. Wiens JA, Stralberg D, Jongsomjit D, Howell CA, Snyder MA (2009) Niches, models, and climate change: assessing the assumptions and uncertainties. Proc Nat Acad Sci USA 106: 19729–19736.
8. Rubidge EM, Monahan WB, Parra JL, Cameron SE, Brashares JS (2011) The role of climate, habitat, and species co-occurrence as drivers of change in small mammal distributions over the past century. Glob Change Biol 17: 696–708.
9. Araújo MB, Pearson RG, Thuiller W, Erhard M (2005) Validation of species-climate impact models under climate change. Glob Change Biol 11: 1504–1513.
10. Green RE, Collingham YC, Willis SG, Gregory RD, Smith KW, et al. (2008) Performance of climate envelope models in retrodicting recent changes in bird population size from observed climate change. Biol Lett 4: 599–602.

11. Mitikka V, Heikkinen RK, Luoto M, Araújo MB, Saarinen K, et al. (2008) Predicting range expansion of the map butterfly in Northern Europe using bioclimate models. Biodiver Conserv 17: 623–641.
12. Fielding AH, Bell JF (1997) A review of methods for the assessment of prediction errors in conservation presence/absence models. Environ Conserv 24: 38–49.
13. Anderson RP, Lew D, Townsend Peterson A (2003) Evaluating predictive models of species' distributions: criteria for selecting optimal models. Ecol Model 162: 211–232.
14. Peterson AT, Papeş M, Soberón J (2008) Rethinking receiver operator characteristic analysis applications in ecological niche modeling. 2008. Ecol Model 213: 63–72.
15. Matthews DP, Gonzalez A (2007) The inflationary effects of environmental fluctuations ensure the persistence of sink metapopulations. Ecology 88: 2848–2856.
16. Sauer JR, Hines JE, Fallon JE, Pardieck KL, Ziolkowski D Jr, et al. (2011) The North American Breeding Bird Survey, Results and Analysis 1966–2009. Version 3.23.2011. USGS Patuxent Wildlife Research Center, Laurel, MD.
17. Hitch AT, Leberg PL (2007) Breeding distributions of North American bird species moving north as a result of climate change. Conserv Biol 21: 534–539.
18. La Sorte FE, Thompson III FR (2007) Poleward shifts in winter ranges of North American birds. Ecology 88: 1803–1812.
19. Zuckerberg B, Woods AM, Porter WF (2009) Poleward shifts in breeding bird distributions in New York state. Glob Change Biol 35: 1866–1883.
20. Cutler DR, Edwards Jr, TC, Beard KH, Cutler A, Hess KT, et al. (2007) Random forests for classification in ecology. Ecology 88: 2783–2792.
21. Phillips SJ, Anderson RP, Schapire RE (2006) Maximum entropy modeling of species geographic distributions. Ecol Model 190: 231–259.
22. Elith J, Phillips SJ, Hastie T, Dudík M, Chee YE, et al. (2011) A statistical explanation of MaxEnt for ecologists. Div Distrib 17: 43–57.
23. R project website. Available: www.R-project.org. Accessed 2013 1 April.

24. Allouche O, Tsoar A, Kadmon R (2006) Assessing the accuracy of species distribution models: prevalence, kappa and the true skill statistic (TSS). J Appl Ecol 43: 1223–1232.

25. Freeman EA, Moisen GG (2008) A comparison of the performance of threshold criteria for binary classification in terms of predicted prevalence and kappa. Ecol Model 217: 48–58.

26. Bolker BM, Brook ME, Clark CJ, Geange SW, Poulson JR, et al. (2008) Generalized linear mixed models: a practical guide for ecology and evolution. Trends Ecol Evol 24: 127–135.

27. Swihart RK, Gehring TM, Kolozsvary MB, Nupp TE (2003) Responses of 'resistant' vertebrates to habitat loss and fragmentation: the importance of niche breadth and range boundaries. Div Distrib 9: 1–18.

28. Schloss CA, Nuñez TA, Lawler JJ (2012) Dispersal will limit ability of mammals to track climate change in the Western Hemisphere. Proc Nat Acad Sci USA 109: 8606–8611.

29. Jenkins DG, Brescarin CR, Duxbury CV, Elliott JA, Evans JA et al. (2007) Does size matter for dispersal distance? Glob Ecol Biogeogr 16: 415–425.

30. Menéndez R, González Megías A, Hill JK, Braschler B, Willis SC, et al. (2006) Species richness changes lag behind climate change. Proc Royal Soc B: Biol. Sci. 273: 1465–1470.

31. Tilman D, May RM, Lehman CL, Nowak MA (1994) Habitat destruction and the extinction debt. Nature 371: 65–66.

32. Lobo JM, Jiménez-Valverde A, Hortal J (2010) The uncertain nature of absences and their importance in species distribution modeling. Ecography 33: 103–114.

33. Fahrig L (2003) Effects of habitat fragmentation on biodiversity. Ann Rev Ecol Evol Syst 34: 487–515.

34. Urban MC, Tewksbury JJ, Sheldon KS (2012) On a collision course: competition and dispersal differences create no-analogue communities and cause extinctions during climate change. Proc Royal Soc B Biol Sci: 1–9.

35. Mitchell MD, Kimmel RO, Snyders J (2011) Reintroduction and range expansion of eastern wild turkeys in Minnesota. Geogr Rev 101: 269–284.

12

Severe Loss of Suitable Climatic Conditions for Marsupial Species in Brazil: Challenges and Opportunities for Conservation

Rafael D. Loyola[1]*, **Priscila Lemes**[2], **Frederico V. Faleiro**[2], **Joaquim Trindade-Filho**[2], **Ricardo B. Machado**[3]

1 Department of Ecology, Universidade Federal de Goiás, Goiânia, Goiás, Brazil, **2** Graduate Program in Ecology and Evolution, Universidade Federal de Goiás, Goiânia, Goiás, Brazil, **3** Departament of Zoology, Universidade de Brasília, Brasília, Distrito Federal, Brazil

Abstract

A wide range of evidences indicate climate change as one the greatest threats to biodiversity in the 21st century. The impacts of these changes, which may have already resulted in several recent species extinction, are species-specific and produce shifts in species phenology, ecological interactions, and geographical distributions. Here we used cutting-edge methods of species distribution models combining thousands of model projections to generate a complete and comprehensive ensemble of forecasts that shows the likely impacts of climate change in the distribution of all 55 marsupial species that occur in Brazil. Consensus projections forecasted range shifts that culminate with high species richness in the southeast of Brazil, both for the current time and for 2050. Most species had a significant range contraction and lost climate space. Turnover rates were relatively high, but vary across the country. We also mapped sites retaining climatic suitability. They can be found in all Brazilian biomes, especially in the pampas region, in the southern part of the Brazilian Atlantic Forest, in the north of the Cerrado and Caatinga, and in the northwest of the Amazon. Our results provide a general overview on the likely effects of global climate change on the distribution of marsupials in the country as well as in the patterns of species richness and turnover found in regional marsupial assemblages.

Editor: David L. Roberts, University of Kent, United Kingdom

Funding: R.D.L. received a research productivity scholarship from the CNPq (grant #304703/2011-7). P.L. received a PhD scholarship from CNPq. F.V.F. and J.T.-F. received PhD scholarships from CAPES. Conservation Biogeography Lab research has been continuously supported by the CNPq, CAPES, Conservation International Brazil, and MCT-Rede CLIMA. The funders had no role in study design, data collection and analysis, decision to publish, or preparation of the manuscript.

Competing Interests: The authors have declared that no competing interests exist.

* E-mail: rdiasloyola@gmail.com

Introduction

As a result of Earth's climate warming and changes in precipitation regimes, the scientific community has a consensual agreement that conservation strategies for managing biodiversity must anticipate the impacts of climate change to be effective [1]. Most studies on climate change have been developed at local scales and use experimental, manipulative schemes, despite the much broader geographical scales at which these changes are expected to affect biodiversity patterns [2]. On the other hand, studies addressing the effects of climate change on biodiversity at continental scales are based on how species' distribution will be potentially driven by such changes, usually inferred through species distribution models [3]. These models are based on different mathematical functions that establish correlations between species' occurrences and environmental variables and, once these correlations were established, make it possible to project the model into future climates to predict species responses (assuming species' niche itself will not respond to these changes) [4].

Species distribution models have been used to predict the current and future species' distributions [5]. However, different methods for modeling species distribution and different climate models (i.e. the coupled Atmosphere-Ocean General Circulation Models, AOGCMs) may produce very distinct results increasing the uncertainties among predictions and their applicability to conservation planning [6,7]. Consequently, measuring and mapping uncertainties are necessary to increase the quality of conservation plans [8,9].

Brazil corresponds to half of South America, and concentrates more than 13% of the world's biota – in particular, *ca.* 11% of the world's mammals [10]. The country holds at least 55 marsupial species ranging from small (*ca.* 10 g) to large species (*ca.* 4 kg) distributed mostly in forest areas such as the Amazon and the Atlantic Forest [11]. However, we still know little about the distribution of marsupials in the countryside, especially in the Brazilian Cerrado, and in the Brazilian Pantanal [11]. This lack of knowledge reinforces the importance of generating species distribution models for this group. Further, marsupials are highly threatened by forest fragmentation, although we also still lack detailed information about marsupial responses to this process [12]. Such vulnerability highlights the need for studies about the effects of global changes (e.g. climate and land use changes) on the group to develop strategies for climate change adaptation related to mammal conservation in Brazil.

Here we present a comprehensive overview on the likely effects of climate change on the distribution of marsupial species inhabiting Brazil and on the patterns of marsupial species richness and turnover. We also highlight sites in which the retention of suitable climatic conditions could minimize climate-driven extinction risk for marsupials.

Materials and Methods

Taxonomy of Brazilian marsupials is still incipient, and some species have been changing their taxonomy given the increasing volume of studies on this mammal order in Brazil and its neighbor countries. Here we followed Rossi *et al.* [13] and Gardner [14].

We downloaded extent of occurrence maps of all the 55 marsupial species that occur in Brazil from the International Union for Conservation of Nature and Natural Resources (IUCN) database (www.iucnredlist.org). We overlapped these maps for each species into an equal-area grid (0.25×0.25 degrees of latitude/longitude) that covered the full extent of the country [15]. Then, we built a species by grid cell matrix, considering presences and absences of species inside grid cells. All 55 species had at least ten occurrences, which reduces model bias.

We obtained current climatic data from the WorldClim database (www.worldclim.org/current) and future climatic scenarios from CIAT (ccafs-climate.org) through WorldClim website. The Intergovernmental Panel on Climate Change (IPCC)'s Fourth Assessment Report (AR4) developed these future scenarios [16]. For each species we modeled distribution as a function of four climatic variables: annual mean temperature, temperature seasonality (standard deviation * 100), annual precipitation, and precipitation seasonality (coefficient of variation). These current climatic data were generated by interpolated climate data from 1950–2000 periods. For future climatic conditions, we used climate variables (year 2050) from four Atmosphere-Ocean General Circulation Models (AOGCMs) of the A2a and B2a green house gases emission scenarios (CCCMA-CGCM2, CSIRO-MK2.0, UKMO-HADCM3, and NIESS99) that were generated by the application of delta downscaling method on the original data from the IPCC Fourth Assessment Report (provided by International Centre for Tropical Agriculture at ccafs-climate.org). We re-scaled both current and future climate variables to our grid resolution.

We used presence and absence derived from species occurrences and climatic variables to model species distributions. We fitted six modeling methods, which differ both conceptually and statistically [4], and applied the ensemble forecasting approach within each set (see text below). We used Generalized Linear Models – GLM [17], Generalized Additive Models – GAM [18], Multivariate Adaptive Regression Splines – MARS [19], Random Forest [20], Artificial Neural Networks – ANN [21], and Generalized Boosting Regression Models – GBM [22].

We partitioned randomly presence and absence data of each species in 75% to calibration (or train) and 25% to validation (or test) and repeated this process 10 times (i.e. a cross-validation) maintaining the observed prevalence of each species. We converted continuous predictions in presence and absences finding the threshold with maximum sensitivity and specificity values in the receiver operating characteristic (or simply ROC curve). After this, we calculated the True Skill Statistics (TSS) to evaluate model performance [23]. The TSS range from −1 to +1, where values equal +1 is a perfect prediction and values equal or less of zero is a prediction no better than random [23].

We did the ensembles of forecasts to produce more robust predictions and reduce the model variability owing to the modeling methods applied and climate models used [7,9,24,25]. We projected distributions into future climate and obtained 240 projections per species within each set of methods (6 modeling

methods×4 climate models×10 randomly partitioned data) and 60 projections per species for current climatic conditions (6 modeling methods×10 randomly partitioned data) – this allowed us to generate a frequency of projections in the ensemble. We then generated the frequency of projections weighted by the TSS statistics for each species and timeframe within each set of methods. We considered the presence of a species only in cells with 50% or more of frequency of projections, but we hold a continuous value when this occurred.

Finally, we calculated species turnover between current and future species distributions in each cell as (G+L)/(SR+G), where "G" was the number of species gained, "L" the number of species lost and "SR" is the current species richness found in the cell. Then we used the total sum of squares from a two-way Analysis of Variance (ANOVA) without replication to quantify the uncertainty associated to each cell following the protocol recently proposed by [9]. We did the ANOVA using species richness as the response variable, and modeling methods and climate models as factors. Finally, we calculated the percent of variation found in each cell relative to the total uncertainty found in all cells to generate a measure of model uncertainty.

To evaluate in which sites a strategic investment in research and conservation of marsupials would be more adequate, we calculated the percent of species retaining suitable climatic conditions in each grid cell. This value was obtained by adding the number of species that occur in the cell in the present plus the number of species that were predicted to remain in that cell in the future, divided by the species richness found in the cell in the present [26].

Finally, to assess how much species richness and turnover geographic patterns would change if deforested areas were removed from the analysis both now and in 2050 we developed a spatial model of land conversion, using the Cerrado Biodiversity Hotspot as a case study. We modeled land conversion with variables from different sources. We compared the Cerrado land use between 2002 and 2008 (siscom.ibama.gov.br/monitorabiomas/index.htm) to generate a matrix of transition probability between native areas to anthropic areas. Then, we modeled the land conversion with the module Land Change Modeler - LCM, available in Idrisi Taiga Version [27], using these explanatory variables: digital elevation model and annual accumulated precipitation (data available at www.worldclim.org), proximity to roads, proximity to recent deforested areas and proximity to cities (data available at mapas.mma.gov.br/i3geo/datadownload.htm). LCM is a machine learning procedure that uses Markov Chains to project future land-use conditions. To evaluate model precision, we inverted the maps from 2002 and 2008 and the expected land-use was projected back into 1990. After this, we generated a total of 458 control points to cover the entire Cerrado by doing a visual inspection of MrSID images from 1990 (see zulu.ssc.nasa.gov/mrsid). Finally, we predicted land conversion in 2050 with a spatial resolution close to 500×500 m.

Results

For most species, TSS value was relatively high (TSS ± SD = 0.77±0.14), indicating good model fit. Patterns of marsupial species richness varied depending on the methods employed to model species distributions and the climate models used to project future climatic conditions (Fig. 1). Modeling methods accounted for 62.6% of the variation among projections, whereas climate models explained 10.3% of such variation. Greenhouse gases emission scenarios contributed little to variation among projections (2.7%). Uncertainty arising from modeling methods was high in the northeast, south and in central Brazil (regions with lower

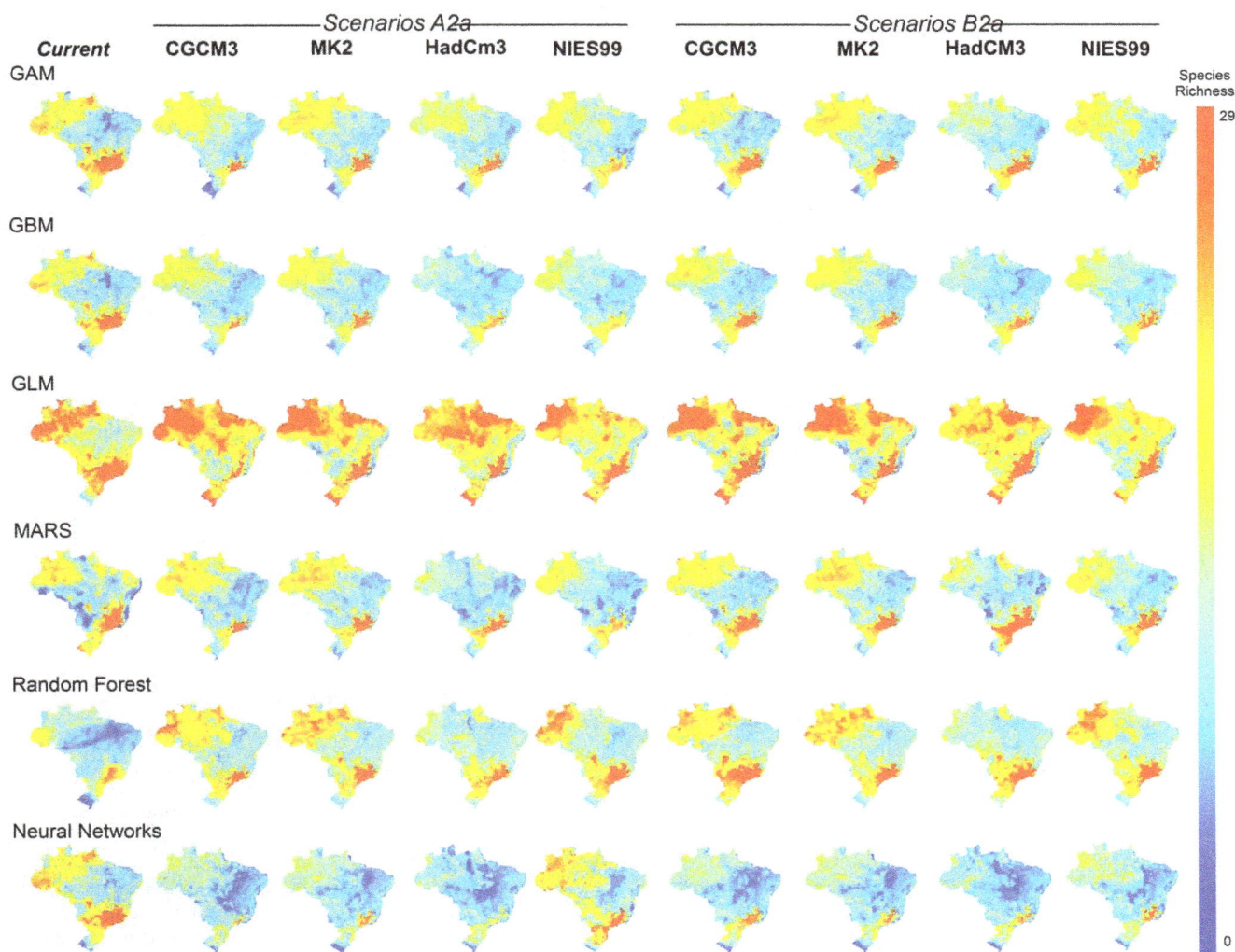

Figure 1. Marsupial species richness patterns in Brazil (current and future, 2050) forecasted by species distribution models generated by different modeling methods (Generalized Additive Models, GAM; Generalized Boosting Regression Models, GBM; Generalized Linear Models, GLM; Multivariate Adaptive Regression Splines, MARS; Artificial Neural Networks, ANN; and Random Forest), climate models (CGCM3, MK2, HadCm3, NIES99), and green house gases emission scenarios (optimist, B2a, and pessimist, A2a). See text for further details.

species richness). Uncertainties linked to climate models were higher in the north (Fig. 2).

For current time, all models indicated high species richness in the southeast of Brazil, and low richness in the south and northeast of the country, despite variation among projections (Fig. 1). GLM were an exception, indicating high richness in the northern region of Brazil. As for current time, all models forecasted species' range contraction, regardless the emission scenario (Fig. 1). Our consensual model projections (Fig. 2) forecasted species range shifts that culminate also with high richness in the southeast both for current time and for 2050. Species' range contraction was high (67% of contraction on average), although our models did not forecast species extinction until 2050 (Table 1). Although we did not observe a dramatic change in the pattern of species richness, turnover was high across the country, varying from 0% up to 95% of change in species composition. The western portion of the Brazilian Amazon, central Brazil, and the Brazilian Atlantic Forest should expect high species turnover (Table 1, Fig. 2).

Marsupials should loose much more climatic space (sites with suitable climate) than gain it, and this result is consistent even under different combinations of modeling methods and climate models (Table 1, Fig. 3). Nevertheless, there was variation in the magnitude of the loss/gain of climate space among modeling methods and climate models (Fig. 3). Random forest projected higher gains in climate space, whereas other methods showed similar projections. Similarly, climate model generated by the Hadley Centre UK (HadCm3) projected the higher losses of climate space (Fig. 3).

Regions with the highest retention of suitable climate space in Brazil overlap with those regions with low species turnover (compare Figs. 3 & 4). All Brazilian biomes had regions with high retention of adequate climate space. These regions are located in the southern part of the Atlantic Forest, southeast Pantanal, northern Cerrado and Caatinga, and eastern Amazon (Fig. 4).

Figure 5 shows the effect of habitat filtering (i.e. of including current and future land conversion) in our model predictions in the Cerrado region. Whereas the pattern in species richness and

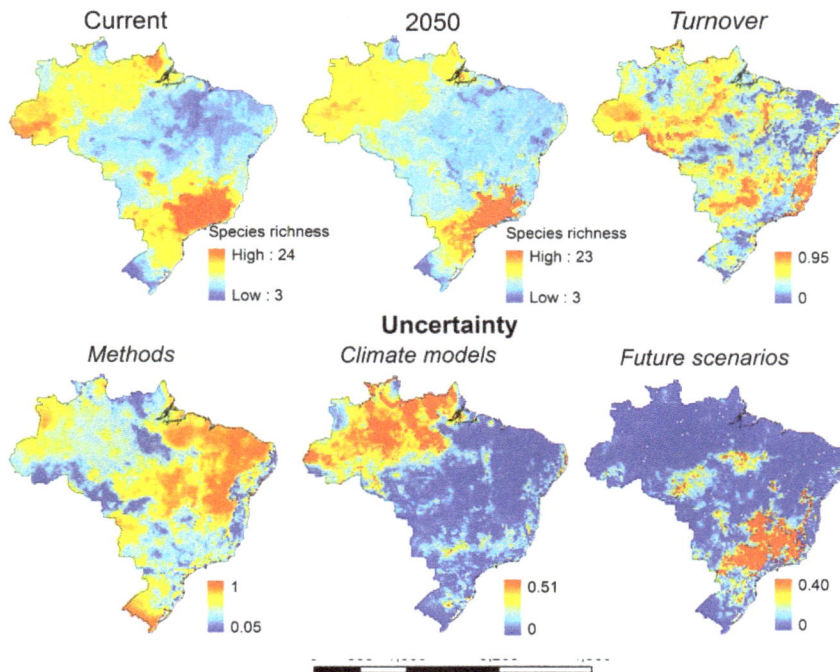

Current   2050   Turnover

Uncertainty

Methods   Climate models   Future scenarios

Figure 2. Consensus map of marsupial species richness in Brazil for current and future climatic conditions, mean turnover forecasted by model projections, and geographic patterns of model uncertainty arising from different sources: modeling methods, climate models, and green house gases emission scenarios. See text for further details.

turnover remains essentially the same, the absolute number of species found in a given cell tends to reduce as a consequence of habitat loss (Fig. 5c–e). This is clearly depicted from Fig. 5d, in which future land conversion greatly reduces the amount of available habitats in the region.

Discussion

We showed that marsupials in Brazil might loose considerable climatically suitable area within their geographic range, loosing climate space towards the year 2050. This projection holds even under different combinations of modeling methods, climate models, and green house gases emission scenarios used to generate species distribution models. Our results have important implication to mammal conservation, and for the conservation of marsupial species, in particular.

First, the use of ensemble of forecasts is preferred as oppose to model species distribution based on only one modeling method (e.g. MaxEnt) and climate model. This is because there is high variation (uncertainty) around model projections, as show here for marsupials, and elsewhere for other taxonomic groups [9,25,26,28]. Ensembles of model projections keep only the consensus-projected areas, minimizing variation among models [7]. This is especially important for conservation purposes given that model uncertainty may mislead conservation efforts ending up being cost-ineffective.

Second, conservation actions based on our marsupial species distribution models must be taken with prudence especially in the central and northeast Brazil as well as in the Amazon, because model uncertainty is higher in these regions. Uncertainties arising from distinct green house gases emission scenarios may be neglected because their level is fairly low, as suggested in other papers [9]. Nevertheless, most Brazilian biomes might hold

climatically suitable sites in which conservation action would succeed. Implementing new protected areas in these sites are highly recommended instead of implementing them in climatically unstable sites. Regions with high climate anomalies could be tracked to indicate where we should focus our attention for species extinction risk [29], but should be avoided in decision making processes as there are no guarantee on the maintenance of viable populations there. Policy maker should therefore focus on sites retaining suitable climate [26].

For marsupial species, in particular, attention should be directed to the Cerrado (central Brazil), Pantanal (southwest), Atlantic Forest (east) and the Pampas (south). These areas hold considerable extensions of suitable climates combined with low model uncertainty. We are not saying the Amazon, for instance, is not important, but action in this region require more extensive studies, based on solid field samples, habitat models and landscape assessment. Sites with high turnover rates could be also important targets for conservation action. However, protecting these sites would imply in allocating conservation resources to a constantly changing community structure, precluding an objective conservation goal such as safeguarding viable populations. In this case, habitat and population monitoring would work better, especially to indicate future habitat corridors or stepping-stones for building a regional conservation strategy. Lastly, high turnover rates might imply in unstable provisioning of ecosystem services provided by marsupial such as nutrient cycling and seed dispersal. Managers should take this into consideration when developing action plans for the group.

Even more important is to consider land conversion when planning for on-the-ground conservation actions. Our example with the Cerrado showed that land use changes might reduce dramatically the amount of available habitats for marsupial species. Therefore, under such scenarios of land conversion, loss

Table 1. Mean species richness of marsupials in Brazil (S) projected for current and future climatic conditions, different between future & current species richness (Δ), mean turnover, and percent variation (median) of species range size and its interquartile deviation obtained in each Green house gases emission scenario, modeling method, and climate model.

Emission scenario	Modeling method	Climate Model	Species richness (current climate)	Species richness (future climate)	Δ species richness	Turnover	% Range size variation (interquartile deviation)
A2a	GAM	CCCMA-CGCM2	11.28	10.79	0.49	0.50	−44.35 (51.98)
		CSIRO-MK2	11.28	10.63	0.65	0.37	−26.75 (30.59)
		HCCPR-HadCM3	11.28	9.88	1.40	0.55	−52.76 (40.44)
		NIES99	11.28	10.01	1.27	0.48	−30.77 (41.22)
	GBM	CCCMA-CGCM2	10.85	9.46	1.39	0.45	−44.97 (44.41)
		CSIRO-MK2	10.85	9.85	1	0.38	−26.93 (30.14)
		HCCPR-HadCM3	10.85	8.27	2.57	0.54	−49.66 (28.88)
		NIES99	10.85	8.97	1.87	0.48	−39.03 (30.43)
	GLM	CCCMA-CGCM2	14.2	15.4	−1.2	0.49	−33.19 (41.91)
		CSIRO-MK2	14.2	14.78	−0.58	0.35	−7.76 (26.26)
		HCCPR-HadCM3	14.2	15.05	−0.85	0.53	−34.12 (60.03)
		NIES99	14.2	14.21	−0.01	0.49	−21.18 (42.1)
	MARS	CCCMA-CGCM2	11.65	10.34	1.31	0.52	−52.04 (41.18)
		CSIRO-MK2	11.65	10.83	0.82	0.40	−28.72 (26.03)
		HCCPR-HadCM3	11.65	8.86	2.79	0.57	−46.32 (29.14)
		NIES99	11.65	9.21	2.44	0.50	−34.75 (40.05)
	RF	CCCMA-CGCM2	8.53	8.22	0.31	0.43	−21.17 (38.52)
		CSIRO-MK2	8.53	8.36	0.16	0.36	−7.83 (33.13)
		HCCPR-HadCM3	8.53	7.34	1.19	0.54	−35.58 (49.38)
		NIES99	8.53	8.27	0.25	0.45	−0.21 (0.39)
	ANN	CCCMA-CGCM2	14.42	11.83	2.59	0.45	−21.17 (38.52)
		CSIRO-MK2	14.42	12.15	2.27	0.37	−7.83 (33.13)
		HCCPR-HadCM3	14.42	10.16	4.26	0.51	−35.58 (49.38)
		NIES99	14.42	11.66	2.76	0.47	−21.04 (38.79)
B2a	GAM	CCCMA-CGCM2	11.28	10.77	0.51	0.32	−4.21(23.94)
		CSIRO-MK2	11.28	10.86	0.42	0.35	−20.24 (30.20)
		HCCPR-HadCM3	11.28	9.64	1.64	0.52	−38.78 (39.34)
		NIES99	11.28	10.53	0.75	0.42	−30.70 (32.26)
	GBM	CCCMA-CGCM2	10.85	10	0.85	0.38	−22.36 (23.24)
		CSIRO-MK2	10.85	9.97	0.88	0.37	−29.68 (27.66)
		HCCPR-HadCM3	10.85	8.47	2.37	0.53	−42.72 (35.93)
		NIES99	10.85	9.38	1.47	0.45	−34.01 (26.55)

Table 1. Cont.

Emission scenario	Modeling method	Climate Model	Species richness (current climate)	Species richness (future climate)	Δ species richness	Turnover	% Range size variation (interquartile deviation)
	GLM	CCCMA-CGCM2	14.2	15.05	-0.85	0.31	1.49 (19.73)
		CSIRO-MK2	14.2	14.44	-0.24	0.35	-13.91 (26.46)
		HCCPR-HadCM3	14.2	14.85	-0.65	0.52	-25.3 (56.3)
		NIES99	14.2	13.59	0.61	0.44	-22.71 (34.9)
	MARS	CCCMA-CGCM2	11.65	10.98	0.67	0.37	-10.82 (20.23)
		CSIRO-MK2	11.65	10.97	0.68	0.38	-21.31 (23.62)
		HCCPR-HadCM3	11.65	8.9	2.76	0.55	-36.47 (27.53)
		NIES99	11.65	10	1.65	0.45	-32.35 (25.84)
	RF	CCCMA-CGCM2	8.53	8.49	0.04	0.38	-0.03 (0.23)
		CSIRO-MK2	8.53	8.42	0.11	0.37	-10.63 (31.19)
		HCCPR-HadCM3	8.53	7.49	1.04	0.54	-34.87 (41.99)
		NIES99	8.53	8.63	-0.1	0.42	-12.87 (38.57)
	ANN	CCCMA-CGCM2	14.42	12.60	1.82	0.39	-3.34 (23.3)
		CSIRO-MK2	14.42	12.12	2.30	0.37	-10.63 (31.19)
		HCCPR-HadCM3	14.42	10.62	3.80	0.52	-34.87 (41.99)
		NIES99	14.42	11.99	2.43	0.45	-12.77 (38.57)

Generalized Additive Models, GAM; Generalized Boosting Regression Models, GBM; Generalized Linear Models, GLM; Multivariate Adaptive Regression Splines, MARS; Artificial Neural Networks, ANN; and Random Forest, RF.

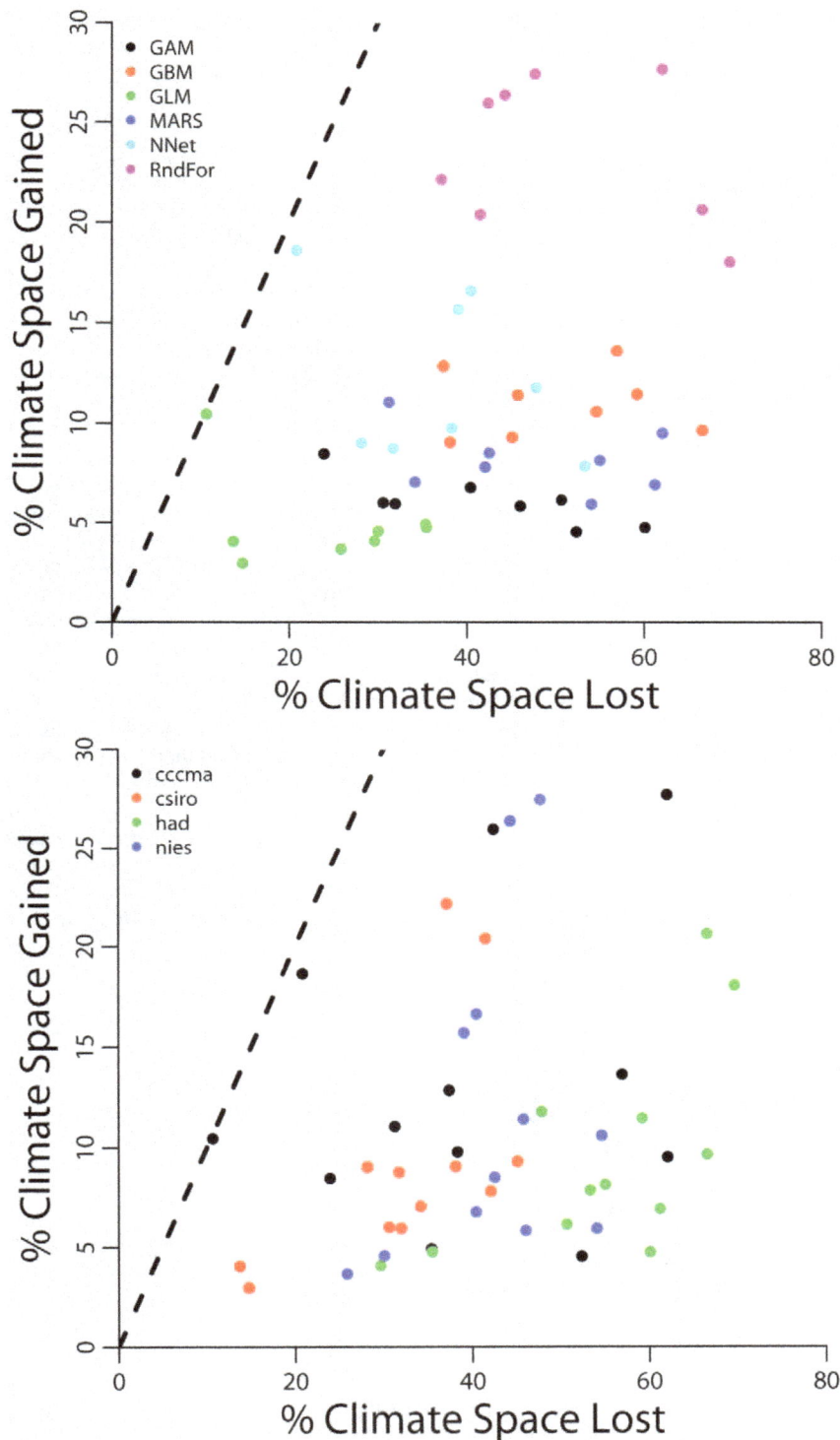

Figure 3. Proportion of climatically suitable sites (grid cells) that may be lost or gained by marsupial species in Brazil, according to six modeling methods (Generalized Additive Models, GAM; Generalized Boosting Regression Models, GBM; Generalized Linear Models, GLM; Multivariate Adaptive Regression Splines, MARS; Artificial Neural Networks, ANN; and Random Forest) and four climate models (CGCM3, MK2, HadCm3, NIES99). Values are median percentages of grid cells lost or gained for all marsupial species in a consensus of green house gases emissions scenarios (B2a and A2a). The line indicates what is expected if gains or losses of climatically suitable sites were proportional and happen by chance.

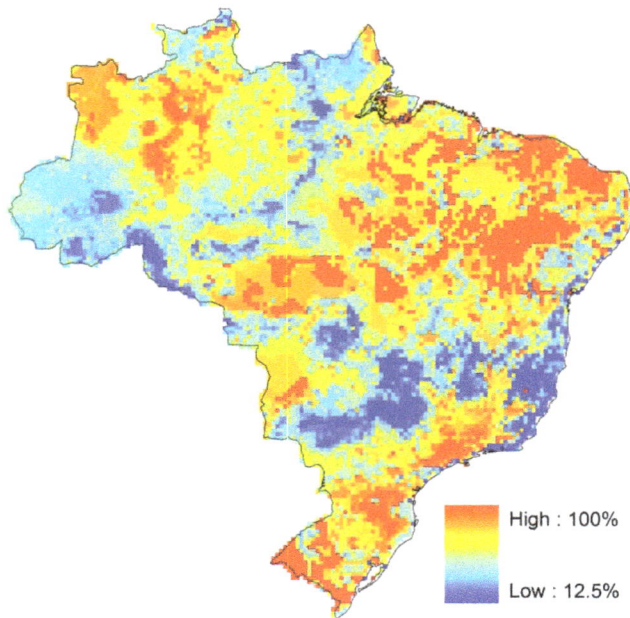

Figure 4. Percentage of species predicted to retain climatic suitability under a consensus of modeling methods, climate models, and green house gases emissions scenarios. Values are median percentages of species in each grid cell as forecasted by a consensus of all modeling methods (Generalized Additive Models, GAM; Generalized Boosting Regression Models, GBM; Generalized Linear Models, GLM; Multivariate Adaptive Regression Splines, MARS; Artificial Neural Networks, ANN; and Random Forest), climate models (CGCM3, MK2, HadCm3, NIES99), and green house gases emissions scenarios (B2a and A2a). See text for further details.

of climatically suitable areas will be even higher than our initial projections. This would also have a significant effect on resource allocation for marsupial conservation. Ultimately, land use changes are a pivotal source of uncertainty driving species distribution towards worse scenarios in the future. Here, we forecasted high species richness in southeast Brazil, which is the region with greater economic development and human population density in the country. Future analysis on the effects of species distribution under climate change should also include land use change, especially in regions like these, as a major source of uncertainty in the modeling framework. As for our projections, the southeast of Brazil, although climatically adequate, may have already lost some of marsupial species (or at least viable or large populations of these species) because of the more intense human occupation (see also Diniz-Filho *et al.* [30]).

As in any study attempting to model species distributions, this one has its own caveats. First our models assume marsupial species are in equilibrium with current climate and have unlimited dispersal to tackle suitable climates as they move in the geographic space. These are simple assumptions allowing us to model all species distribution at a time. But marsupial dispersal is clearly limited by the composition of the matrix in a fragmented context found in most regions of Brazil [31,32]. Second, we predicted future species distribution assuming that the vegetation types in Brazil will remain in the same regions of the current distribution, but such changes can occur in South America [33]. This assumption can affect our species distribution model predictions. Species' range shift outside of the current limitation of Brazil or their preferential habitat cannot be measure by our methods. Yet, we believe that this assumption will have little effect in our predictions because these changes would be a real problem only for narrow ranged species that are habitat specialists which is not the case of Brazilian marsupials, in general [11], but see [34]. Third, land use changes were not fully integrated in the analyses –

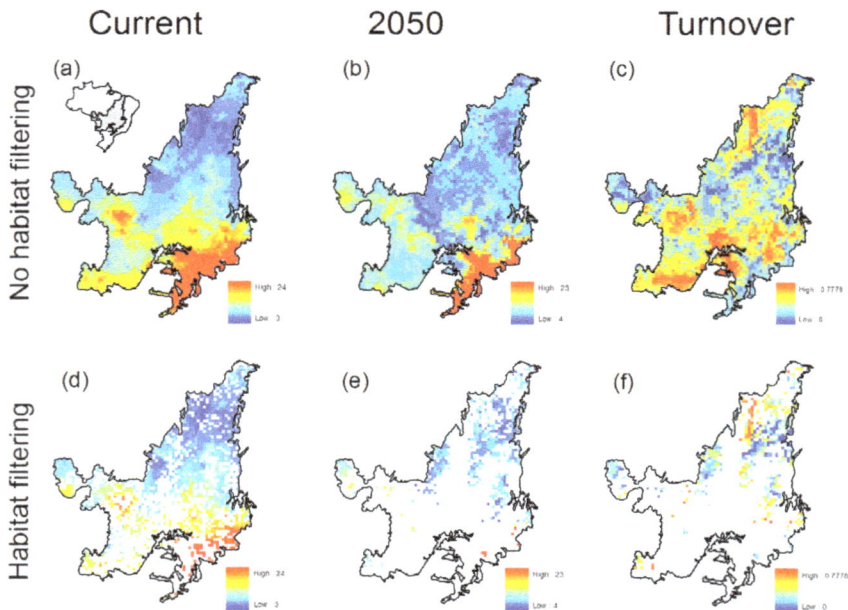

Figure 5. Consensus map of marsupial species richness in the Cerrado Biodiversity Hotspot for current and future climatic conditions, and mean turnover forecasted by model projections. Maps show patterns expected for the whole domain, no habitat filtering (a–c), and with habitat filtering (d–e). Habitat filtering followed the predictions of a spatial model of land conversion for the Cerrado, developed in this study. See text for further details.

except for the Cerrado region. As discussed above, the dynamic nature of land conversion is a key factor driving species' distribution in space and time. Lastly, in our study we implicitly assume that the marsupial fauna is Brazil is relatively well known. However, it is likely that many other unknown species remain [35]. For mammals, in particular, current estimates suggest an increase of up to 6% in species description, most being small-ranged species [35]. While creative solutions could be applied to decide where to allocate conservation efforts in the face of such lack of knowledge [36], uncertainty about the existence of species, the so-called Linnaean shortfall, remains as a difficult-to-tackle problem in species distribution modeling and conservation assessment.

To sum up, our results provide a general overview on the likely effects of global climate change on the distribution of marsupials in Brazil as well as in the patterns of species richness and turnover found in regional marsupial assemblages. Forecasts are not good - especially if future projections of land conversion are integrated to our results - but we do have time and building capacity to think hard about the problem and find solution for climate change adaptation concerning the fauna of a megadiverse country.

Acknowledgments

We thank Stuart Pimm, José Alexandre F. Diniz Filho, and an anonymous reviewer for their suggestions in an early version of this paper.

Author Contributions

Conceived and designed the experiments: RDL. Analyzed the data: PL FVF JT-F RBM. Wrote the paper: RDL.

References

1. Araújo MB, Rahbek C (2006) Ecology. How does climate change affect biodiversity? Science 313: 1396–1397. doi:10.1126/science.1131758.
2. Kerr JT, Kharouba HM, Currie DJ (2007) The macroecological contribution to global change solutions. Science 316: 1581–1584. doi:10.1126/science.1133267.
3. Pearson RG, Dawson TP (2003) Predicting the impacts of climate change on the distribution of species: are bioclimate envelope models useful? Global Ecology and Biogeography 12: 361–371. doi:10.1046/j.1466-822X.2003.00042.x.
4. Franklin J (2009) Mapping species distributions: spatial inference and prediction. Cambridge: Cambridge University Press.
5. Lawler JJ, Wiersma YF, Huettmann F (2011) Designing predictive models for increased utility: using species distribution models for conservation planning and ecological forecasting. In: Drew A, Wiersma YF, Huettmann F, editors. Predictive Modeling in Landscape Ecology. New York: Springer. pp. 271–290.
6. Pearson RG, Thuiller W, Araújo MB, Martinez-Meyer E, Brotons L, et al. (2006) Model-based uncertainty in species range prediction. Journal of Biogeography 33: 1704–1711. doi:10.1111/j.1365-2699.2006.01460.x.
7. Araújo MB, New M (2007) Ensemble forecasting of species distributions. Trends in Ecology and Evolution 22: 42–47.
8. Thuiller W, Lafourcade B, Engler R, Araújo MB (2009) BIOMOD - a platform for ensemble forecasting of species distributions. Ecography 32: 369–373. doi:10.1111/j.1600-0587.2008.05742.x.
9. Diniz-Filho JAF, Bini LM, Fernando Rangel T, Loyola RD, Hof C, et al. (2009) Partitioning and mapping uncertainties in ensembles of forecasts of species turnover under climate change. Ecography 32: 897–906. doi:10.1111/j.1600-0587.2009.06196.x.
10. Lewinsohn TM, Prado PI (2005) How Many Species Are There in Brazil? Conservation Biology 19: 619–624. doi:10.1111/j.1523-1739.2005.00680.x.
11. Cáceres NC, Monteiro-Filho ELA (2006) Os marsupiais do Brasil: biologia, ecologia e evolução. 1a ed. Campo Grande: Editora da UFSM.
12. Fernandez FAS, Pires AS (2006) Perspectivas para a sobrevivência dos marsupiais brasileiros em fragmentos florestais: o que sabemos e o que ainda precisamos aprender? In: Cáceres NC, Monteiro-Filho ELA, editors. Os marsupiais do Brasil: biologia, ecologia e evolução. Campo Grande: Editora UFMS. pp. 191–201.
13. Rossi RV, Bianconi GV, Pedro WA (2006) Ordem Didelphimorphia. In: Reis NR, Peracchi AL, Pedro WA, Lima IP, editors. Mamíferos do Brasil. Londrina: Edifurb. pp. 27–66.
14. Gardner AL (2008) Mammals of South America: marsupials, xenarthrans, shrews, and bats. V. 1. Chicago: University of Chicago Press.
15. Loyola RD, Carvalho RA, Faleiro FV, Trindade-Filho J, Lemes P, et al. (2012) Conservação da diversidade filogenética e funcional de mamíferos do Brasi. In: Freitas TRO, Vieira EM, editors. Mamíferos do Brasil: Genética, Sistemática, Ecologia e Conservação - vol II. Sociedade de Mastozoologia do Brasil. p. in press.
16. IPCC (2007) Climate Change 2007 - Impacts, Adaptation and Vulnerability: Working Group II contribution to the Fourth Assessment Report of the IPCC. Parry ML, Canziani OF, Palutikof JP, Van Der Linden PJ, Hanson CE, editors Cambridge University Press.
17. Guisan A, Edwards TC, Hastie T (2002) Generalized linear and generalized additive models in studies of species distributions: setting the scene. Ecological Modelling 157: 89–100. doi:10.1016/S0304-3800(02)00204-1.
18. Hastie T, Tibshirani R (1986) Generalized Additive Models. Statistical Science 1: 297–310.
19. Friedman JH (1991) Multivariate adaptive regression splines. Annals of Statistics 19: 1–141.
20. Breiman L (2001) Random forest. Machine Learning 45: 5–32. doi:10.1023/A.
21. Manel S, Dias JM, Buckton ST, Ormerod SJ (1999) Alternative methods for predicting species distribution: an illustration with Himalayan river birds. Journal of Applied Ecology 36: 734–747. doi:10.1046/j.1365-2664.1999.00440.x.
22. Friedman JH (2001) Greedy function approximation: a gradient boosting machine. Annals of Statistics 29: 1189–1232.
23. Allouche O, Tsoar A, Kadmon R (2006) Assessing the accuracy of species distribution models: prevalence, kappa and the true skill statistic (TSS). Journal of Applied Ecology 43: 1223–1232. doi:10.1111/j.1365-2664.2006.01214.x.
24. Marmion M, Parviainen M, Luoto M, Heikkinen RK, Thuiller W (2009) Evaluation of consensus methods in predictive species distribution modelling. Diversity and Distributions 15: 59–69. doi:10.1111/j.1472-4642.2008.00491.x.
25. Diniz-Filho JAF, Nabout JC, Bini LM, Loyola RD, Rangel TF, et al. (2010) Ensemble forecasting shifts in climatically suitable areas for Tropidacris cristata (Orthoptera: Acridoidea: Romaleidae). Insect Conservation and Diversity. doi:10.1111/j.1752-4598.2010.00090.x.
26. Garcia RA, Burgess ND, Cabeza M, Rahbek C, Araújo MB (2011) Exploring consensus in 21st century projections of climatically suitable areas for African vertebrates. Global Change Biology: n/a–n/a. doi:10.1111/j.1365-2486.2011.02605.x.
27. Eastman JR (2009) IDRISI Kilimanjaro Guide to GIS and Image Processing. Worcester, Massachusetts: Clark University.
28. Lawler JJ, Shafer SL, White D, Kareiva P, Maurer EP, et al. (2009) Projected climate-induced faunal change in the Western Hemisphere. Ecology 90: 588–597.
29. Beaumont LJ, Pitman A, Perkins S, Zimmermann NE, Yoccoz NG, et al. (2011) Impacts of climate change on the world's most exceptional ecoregions. Proceedings of the National Academy of Sciences of the United States of America 108: 2306–2311. doi:10.1073/pnas.1007217108.
30. Diniz-filho JAF, de Oliveira G, Bini LM, Loyola RD, Nabout JC, et al. (2009) Conservation biogeography and climate change in the brazilian cerrado. Natureza & Conservação 7: 100–112.
31. Prevedello JA, Vieira MV (2009) Does the type of matrix matter? A quantitative review of the evidence. Biodiversity and Conservation 19: 1205–1223. doi:10.1007/s10531-009-9750-z.
32. Prevedello JA, Forero-Medina G, Vieira MV (2011) Does land use affect perceptual range? Evidence from two marsupials of the Atlantic Forest. Journal of Zoology 284: 53–59. doi:10.1111/j.1469-7998.2010.00783.x.
33. Salazar LF, Nobre CA, Oyama MD (2007) Climate change consequences on the biome distribution in tropical South America. Geophysical Research Letters 34: L09708. doi:10.1029/2007GL029695.
34. Püttker T, Pardini R, Meyer-Lucht Y, Sommer S (2008) Responses of five small mammal species to micro-scale variations in vegetation structure in secondary Atlantic Forest remnants, Brazil. BMC ecology 8: 9. doi:10.1186/1472-6785-8-9.
35. Pimm SL, Jenkins CN, Joppa LN, Roberts DL, Russell GJ (2010) How Many Endangered Species Remain to be Discovered in Brazil? Natureza & Conservação 08: 71–77. doi:10.4322/natcon.00801011.
36. Bini LM, Diniz-Filho JAF, Rangel TFLVB, Bastos RP, Pinto MP (2006) Challenging Wallacean and Linnean shortfalls: knowledge gradients and conservation planning in a biodiversity hotspot. Diversity and Distributions 12: 475–482. doi:10.1111/j.1366-9516.2006.00286.x.

13

Climate Change Impacts on the Future Distribution of Date Palms: A Modeling Exercise Using CLIMEX

Farzin Shabani*, Lalit Kumar, Subhashni Taylor

Ecosystem Management, School of Environmental and Rural Science, University of New England, Armidale, Australia

Abstract

Climate is changing and, as a consequence, some areas that are climatically suitable for date palm (*Phoenix dactylifera* L.) cultivation at the present time will become unsuitable in the future. In contrast, some areas that are unsuitable under the current climate will become suitable in the future. Consequently, countries that are dependent on date fruit export will experience economic decline, while other countries' economies could improve. Knowledge of the likely potential distribution of this economically important crop under current and future climate scenarios will be useful in planning better strategies to manage such issues. This study used CLIMEX to estimate potential date palm distribution under current and future climate models by using one emission scenario (A2) with two different global climate models (GCMs), CSIRO-Mk3.0 (CS) and MIROC-H (MR). The results indicate that in North Africa, many areas with a suitable climate for this species are projected to become climatically unsuitable by 2100. In North and South America, locations such as south-eastern Bolivia and northern Venezuela will become climatically more suitable. By 2070, Saudi Arabia, Iraq and western Iran are projected to have a reduction in climate suitability. The results indicate that cold and dry stresses will play an important role in date palm distribution in the future. These results can inform strategic planning by government and agricultural organizations by identifying new areas in which to cultivate this economically important crop in the future and those areas that will need greater attention due to becoming marginal regions for continued date palm cultivation.

Editor: Vanesa Magar, Plymouth University, United Kingdom

Funding: The university of New England have financially supported the Work. The funders had no role in study design, data collection and analysis, decision to publish, or preparation of the manuscript.

Competing Interests: The authors have declared that no competing interests exist.

* E-mail: Fshabani@une.edu.au

Introduction

Climate is one of the principal aspects defining the potential range of plants and climate change directly affects the distribution of species [1]. Much evidence exists that the climate is changing globally, and land surface temperatures are expected to increase by 4°C between the present and 2100 [2]. Moreover, worldwide seasonal rainfall patterns are changing [2]. As a consequence, a number of serious issues arise. For example, the extent of pollution and aeroallergens will change [3]. Changes in the expansion and transmission of some infectious diseases, famine, crop failure, water shortages and population displacement are some of the other issues involved with climate change. Climate change clearly threatens different areas, such as biodiversity, agricultural production, and human health. For example, it is expected that by 2030, the risk of diarrhea will increase by 10% in some specific regions due to climate change [3]. Climate change can also have an impact on agricultural production by affecting the distribution of economically important crops due to changes in their physiology [4]. The annual income from date palms in the Middle Eastern countries decreased between 1990 and 2000 [5]. A number of factors could be involved in this reduction, and climate change could be one of them because significant losses in yield of some economically important crops have been attributed to plant diseases resulting from climate change [5]. It has been reported that climate change has caused a $438 million loss in wheat, $116 million in grapes and $67 million in sugar production in Australia and North America [6].

Date palm (*Phoenix dactylifera* L.) is a valuable plant that provides a significant source of income for both local farmers and governments in arid and semi-arid regions of the world [7]. A number of reports document the cultivation of date palms back to the 5th millennium BC. Since ancient times, the majority of date palms have continued to be grown in the hot deserts of North Africa and the Middle East, including Syria, the Persian Gulf region and north Yemen [8]. The native range of this species is from the south-eastern Azores to Pakistan, and its cultivation stems from the 4th millennium BC in Mesopotamia and Palestine [9]. The genus *Phoenix* includes up to 400 species [10–12] within the Arecaceae family. To mature, the fruit requires prolonged summer heat. Rain or high humidity during fruiting increases the risk of the fruit cracking and the onset of fungal diseases [13]. Long summers with high day and night temperatures, and mild, sunny, dry winters without prolonged frost are the ideal climatic conditions for this species [14].

Long-term management strategies to sustain economically important crops require information about the expected potential distribution and relative abundance of this plant under current and future climate scenarios. There are several distribution models that can provide information in this area, including species distribution models (SDMs), bioclimatic models and ecological niche models (ENMs). However, it has been reported that niche

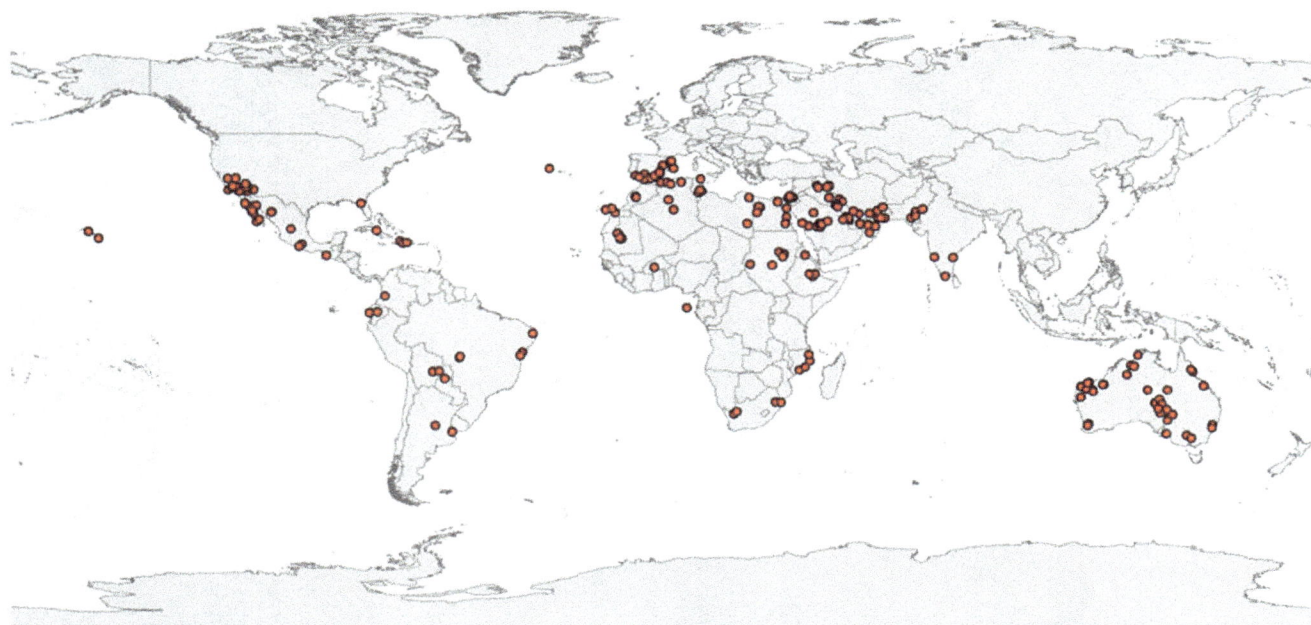

Figure 1. The current global distribution of *P. dactylifera*.

models only enable estimates of a species' fundamental niche [15] while other reports show that it provides a spatial image of the realized niche [16,17].

CLIMEX has been widely used in many different applications [18]. Taylor [19] used CLIMEX for illustrating the potential distribution of *Lantana camara L.* by 2070. Yonow [20] employed CLIMEX for mapping the distribution of the Queensland fruit fly. Sutherst [21] applied the same software for modular modeling of pests. The susceptibility of both animal and human health to parasites under future climates has also been studied using CLIMEX [22].

Table 1. CLIMEX parameter values used for *L. dactylifera* modeling.

Parameter	Mnemonic	Values
Limiting low temperature	DV0	14°C
Lower optimal temperature	DV1	20°C
Upper optimal temperature	DV2	39°C
Limiting high temperature	DV3	46°C
Limiting low soil moisture	SM0	0.007
Lower optimal soil moisture	SM1	0.013
Upper optimal soil moisture	SM2	0.81
Limiting high soil moisture	SM3	0.9
Cold stress temperature threshold	TTCS	4°C
Cold stress temperature rate	THCS	-0.01 week^{-1}
Heat stress temperature threshold	TTHS	46°C
Wet stress threshold	SMWS	0.9
Wet stress rate	HWS	0.022 week^{-1}
Heat stress accumulation rate	THHS	0.9 week^{-1}

As a consequence of climate change, the distribution of species like date palm will change [3]. It is essential to identify which regions will benefit by having the potential opportunity of cultivating date palms in the future and which may be adversely affected. Governments and agricultural organizations can prepare for this situation in advance and thereby gain significant economic advantages which can enable them to improve their economies. Alternately, regions that could be adversely affected can become aware of the situation and transition their economies. This awareness provides an opportunity to plan for alternative sources of income. With this aim, this study made use of the CLIMEX software package in developing a global model of the climate response of *P. dactylifera* based on its native and cultivated distribution. This model was then used to illustrate date palm potential distribution using two global climate models (GCM) including CSIRO-Mk3.0 and MIROC-H. These were run with the A2 SRES (Special Report on Emissions Scenarios) emission scenarios for 2030, 2050, 2070 and 2100. The A2 SRES was chosen with the assumption that, in the future, there would be high population growth coupled with slow economic growth and extensive technological change.

Methodology

CLIMEX Software

CLIMEX is a modeling software package that basically operates on an eco-physiological growth model that assumes that species encounter favorable and unfavorable seasons. Growth is maximized during favorable seasons and minimized during unfavorable seasons [23–25]. A major criticism of CLIMEX is that it does not include biotic interactions and dispersal in the modeling process. However, other factors may be incorporated after the CLIMEX modeling has been performed using GIS and RS software [26]. The key assumption behind CLIMEX is that climate is the main determinant of the distribution of plants and poikilothermal animals [27]. CLIMEX enables the user to infer parameters that describe the species' response to climate based on its geographic

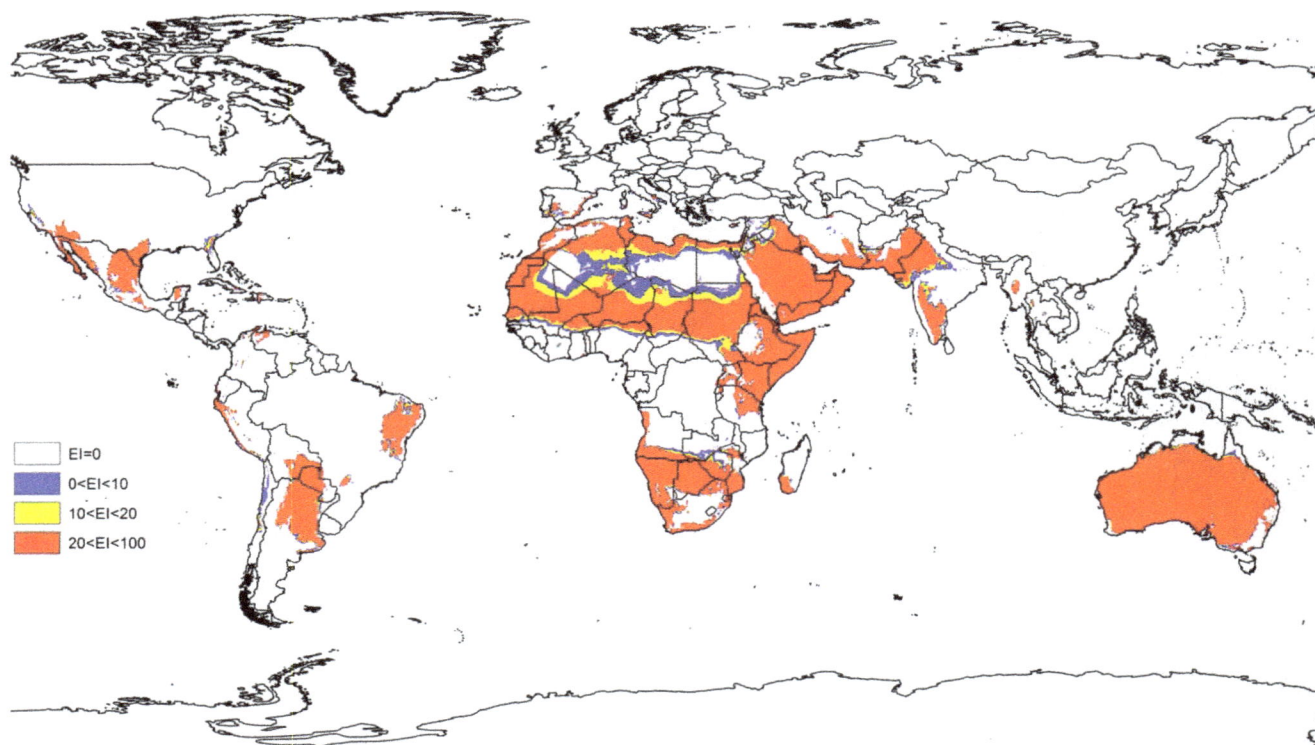

Figure 2. The Ecoclimatic Index for *P. Dactylifera,* **modeled using CLIMEX for current climate.**

range or phenological observations [23]. The Ecoclimatic Index (EI) is a general annual index of climatic suitability based on weekly calculations of growth and stress indexes. It is scaled from 0 to 100, and theoretically, species can establish if EI >0. In CLIMEX, the annual growth index (GI_A) describes the potential for population growth during favorable climate conditions. The GI_A index is determined from the temperature index (TI) and moisture index (MI) which represent the species' temperature and moisture requirements for growth. The user can describe the probability of survival of the species during unfavorable conditions using four stresses: cold, heat, dry and wet. Therefore, based on available distribution data, this software was used to develop a model of the potential distribution of *P. dactylifera* under current and future climate scenarios.

Distribution of Date Palms (*P. dactylifera*)

The Global Biodiversity Information Facility (GBIF) [28] was used to gather information on *P. dactylifera* distribution and this information was supplemented by other date palm literature [8,12,14,28–41] (Figure 1). The GBIF database contained 583 records for *P. Dactylifera*; however, 342 records did not have geographic coordinates and were removed, leaving 241 records. Duplicate records were also removed. Thus, 163 records from the GBIF database and 49 records obtained from the literature review were used in parameter fitting. These 163 records were geographically representative of the known distribution of date palms as shown in Figure 1.

Climate Data, Climate Models and Climate Scenarios

In this study, the CliMond 10′ gridded climate data were used for modeling [42]. Five climatic variables were utilized to represent historical climate (averaging period 1950–2000). These

were average minimum monthly temperature (Tmin), average maximum monthly temperature (Tmax), average monthly precipitation (Ptotal) and relative humidity at 09:00 h (RH09:00) and 15:00 h (RH15:00). These variables were also used to typify potential future climate in 2030, 2050, 2070 and 2100. The potential distribution of date palms under future climate was modeled using two Global Climate Models (GCMs), CSIRO-Mk3.0 [42] and MIROC-H (Center for Climate Research, Japan), with the A2 SRES scenario [42–44]. These two GCMs were part of the CliMond dataset and were selected from 23 GCMs based on the following criteria:

- All required variables, including temperature, precipitation, sea level pressure and humidity for CLIMEX were available

- Small horizontal grid spacing in both GCMs

- Better representation of observed climate at local scales, compared to the other GCMs [45].

In the remainder of this paper, MR and CS are used as the abbreviation of MIROC-H and CSIRO-Mk3.0, respectively.

The MR model predicts that temperature will increase by approximately 4.31°C, while the CS model predicts a rise of 2.11°C by 2100. There are also differences in rainfall patterns for CS and MR models. For example, the CS model predicts a 14% decrease in future mean annual rainfall, whereas the MR model predicts a 1% decrease [46,47].

The A2 scenario was selected to characterize one of the possible climate scenarios during 2030, 2050, 2070 and 2100. The A2 scenario covers different factors including demographic, economic and technological forces driving GHG emissions; this scenario assumes neither very high nor low global GHG emissions

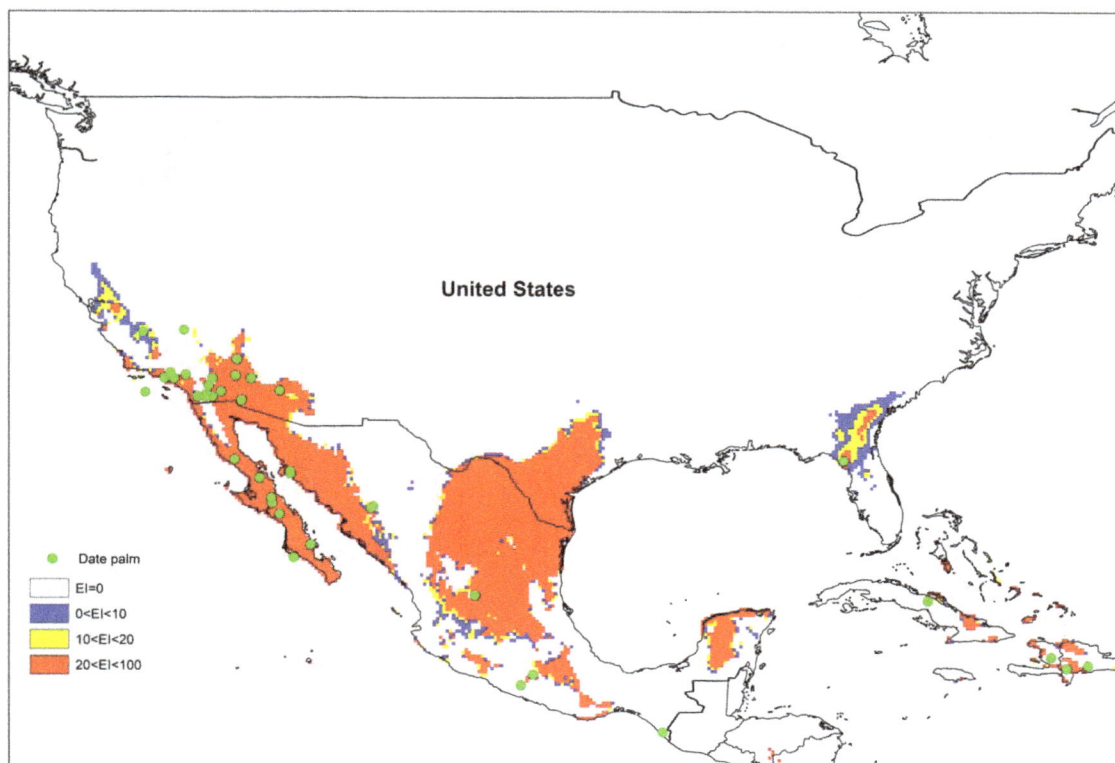

Figure 3. Current and potential distribution of *P. dactylifera* **in validation region based on EI index.**

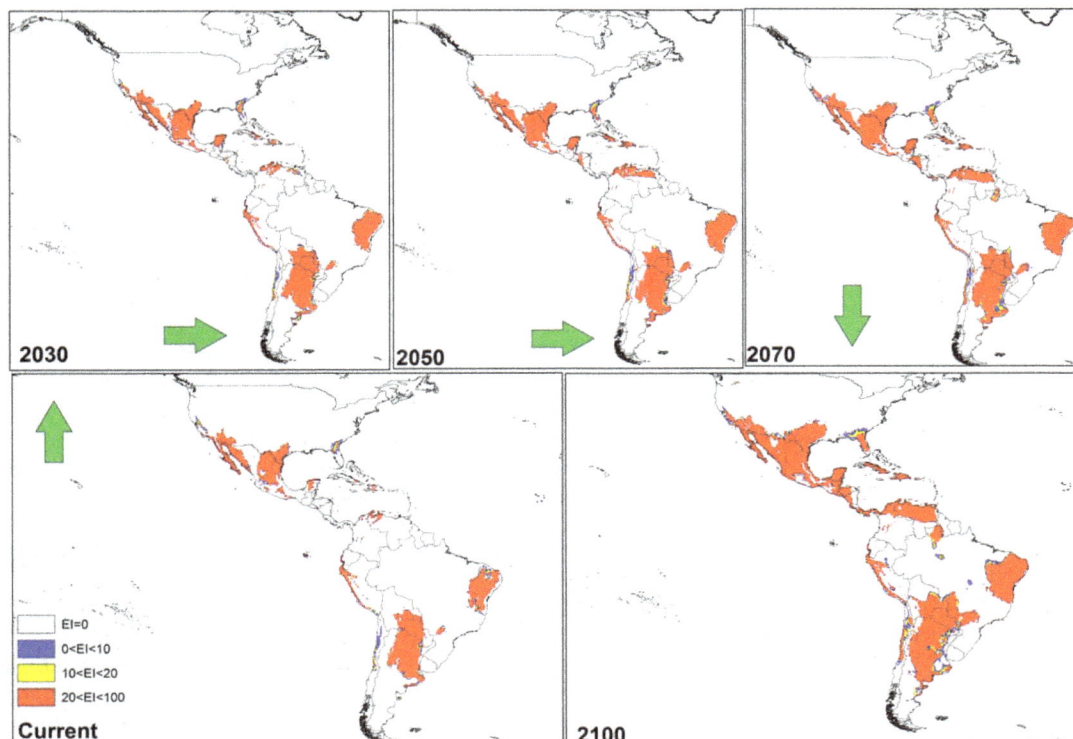

Figure 4. The climate (EI) for *P. dactylifera* **in current time and projected using CLIMEX under the CSIRO-Mk3.0 GCM running the SRES A2 scenario and for 2030, 2050, 2070 and 2100 for the North and South America continent.**

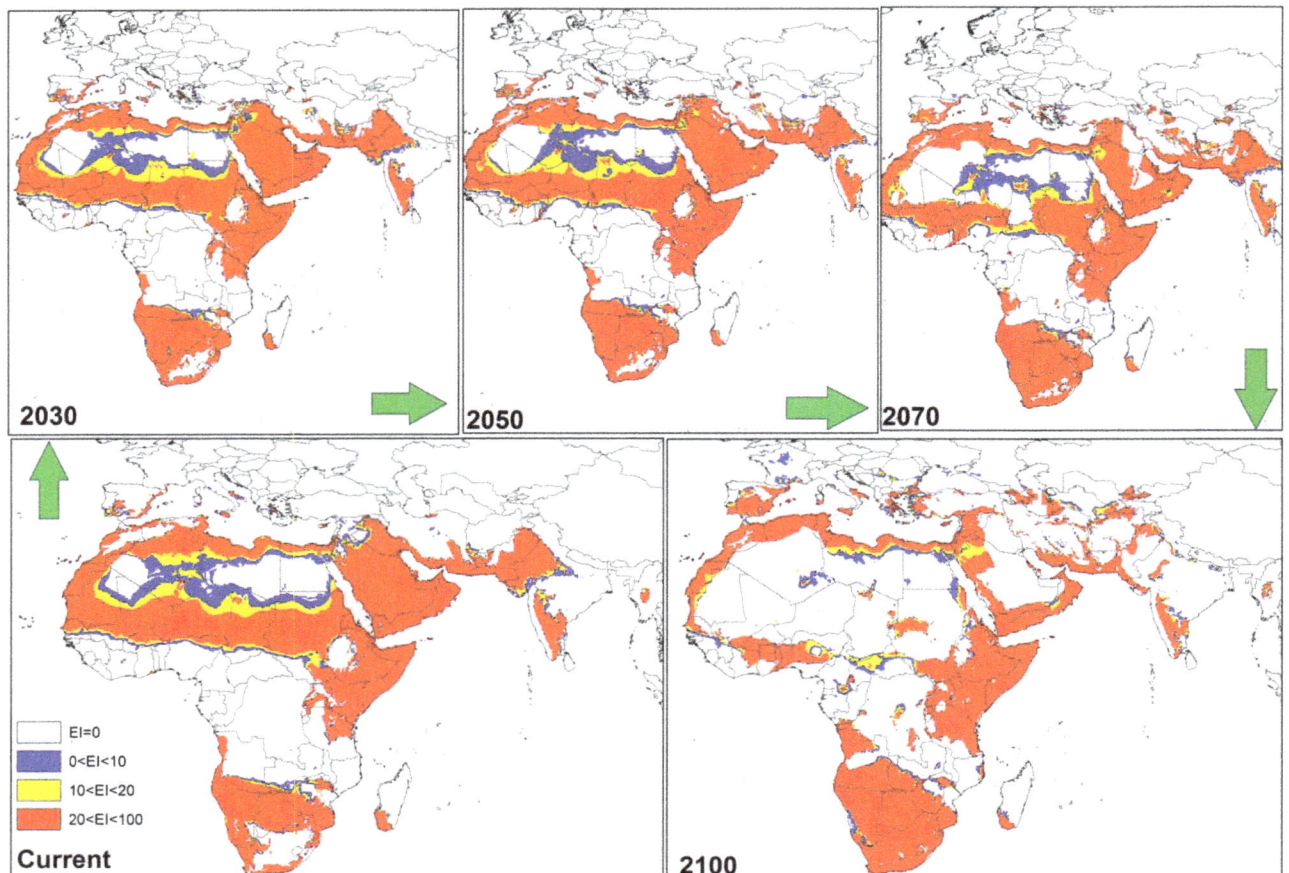

Figure 5. The climate (EI) for *P. dactylifera* in current time and projected using CLIMEX under the CSIRO-Mk3.0 GCM running the SRES A2 scenario and for 2030, 2050, 2070 and 2100 for the north and south of Africa and the Middle East.

compared to the other scenarios, such as A1F1, A1B, B2, A1T, B1 by 2100 [47].

No scenarios from the B family of SRES scenarios were included in this paper, mainly because of the observation that some parameters such as global temperature and sea level rise are presently increasing at a much greater rate than predicted by the hottest SRES scenarios [48].

Fitting CLIMEX Parameters

Using both native habitat range and agricultural distribution data in parameter fitting is highly recommended because it produces a model that approximates the potential distribution of the taxa being modeled [49]. This is because the limitations imposed by biotic influences in the species' native range may be absent in non-native locations, thus allowing it to expand its range beyond its realized Hutchinsonian niche [49,50]. In this study, parameters were fitted using the native range and the global agricultural distribution of date palms. However, the distribution data of *P. dactylifera* from North America, Mexico, and the Caribbean were not used in parameter fitting as this was set aside for model validation. The parameters were iteratively adjusted depending on satisfactory agreement between the potential and known worldwide distribution of *P. dactylifera*. The parameters were subsequently verified to ensure that they were biologically reasonable. Model validation was conducted using North American, Mexican, and Caribbean distribution data. It should be

highlighted that the wet stress threshold parameter does not have a unit, while the stress accumulation rate uses the week^{-1} unit. The heat and cold stress thresholds use degrees Celsius (°C) unit.

Cold Stress

The cold stress temperature threshold (TTCS) mechanism was used to describe the species' response to frost. Generally, the minimum winter temperature that can be tolerated by *P. dactylifera* is 10°C [11]. However, date palms have been recorded in locations as low as 4°C [28]. Therefore, intolerance to frost was incorporated by accumulating stress when the average monthly minimum temperature fell below 4°C, with the frost stress accumulation rate (THCS) set at −0.01 week^{-1}. This cold-stress mechanism allowed the species to survive in Spain (39° 635′ N and 2° 523′ W) [28]. Additionally, this value provided an appropriate fit to the observed distribution in South America, South Africa and Asia.

Heat Stress

The heat stress parameter (TTHS) was set at 46°C because it was reported that *P. dactylifera* is able to persist up to this temperature in eastern Pakistan [28]. The heat stress accumulation rate (THHS) was set at 0.9 week^{-1}, which allowed *P. dactylifera* to persist along eastern Pakistan [37,38,42] and southern Iran [8,28,31].

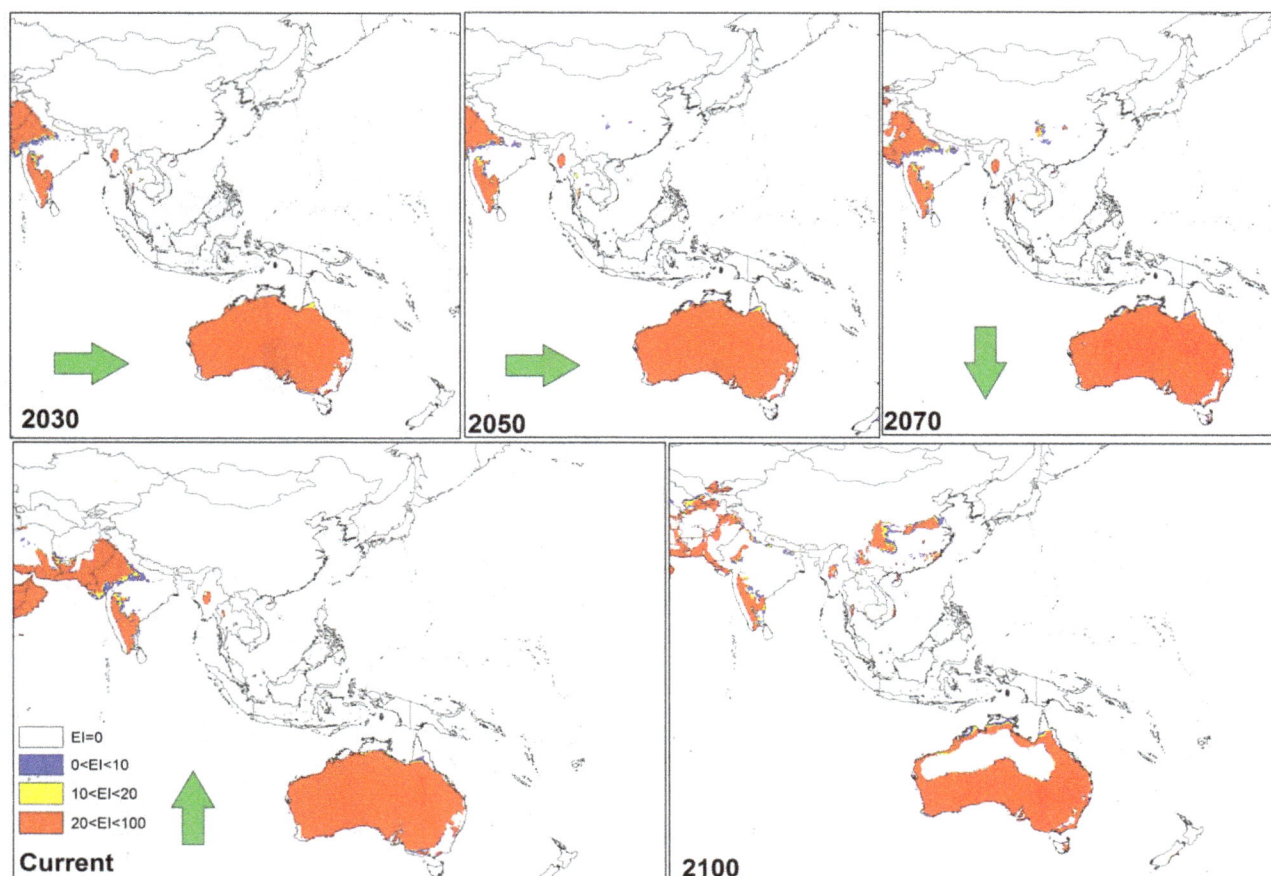

Figure 6. The climate (EI) for *P. dactylifera* in current time and projected using CLIMEX under the CSIRO-Mk3.0 GCM running the SRES A2 scenario and for 2030, 2050, 2070 and 2100 for Australia, and southern Asia.

Dry Stress

The term 'drought' refers to a period of time without significant rainfall [14]. Water stress occurs as a consequence of water loss through transpiration or evaporation during a period of time when there is a lack of available water in the soil [14]. Different degrees of water stress can be seen in a plant. When water loss is prolonged, a significant disruption in the metabolism of the plant occurs [14]. However, the date palm has developed a number of strategies to prevent dry stress. These include maintaining a high level of hydration, the ability to function while dehydrated, increasing the amount of water absorption (i.e., keeping a high level of osmotic pressure) by using abscisic acid, and by the development of an extensive root system [14]. Dry stress was not used in this study for the above reasons.

Wet Stress

August to October are the critical months when rain damage can inflict serious economic damage to the date crop [11]. A recent study observed that a total of 78.74 mm of rainfall during an 8-day period caused a greater than 50% loss in date palm yields while 86.36 mm of rainfall in 10 days led to 15% losses in date palm farms in some countries [11]. Date palms are known to suffer wet stress easily. The wet stress threshold (SMWS) was set to 0.9 and the accumulation rate (HWS) set at 0.022 week^{-1} to allow the species to grow well in arid and semi-arid regions such as Algeria, Morocco, and southern Iran.

Temperature Index

Phoenix dactylifera has been cultivated in areas with a mean annual surface temperature greater than 16°C, such as southern Iran [8,28,29], south-eastern Iraq, eastern Pakistan [11,38], and northern and central Algeria [11,39]. Western Pakistan's climatic parameters are comparable to other places suitable for date palm cultivation with the exception of its annual surface temperature, which is 13°C. Thus, the limiting low temperature (DV0) should be between 13°C and 16°C. Fourteen degrees Celsius was selected due to providing the best fit to the observed distribution of date palms in North Africa and Asia. Summer temperatures in locations which are highly climatically suitable for this species rarely exceed 46°C, thus the limiting high temperature DV3 was set at 46°C [13]. The lower (DV1) and upper (DV2) optimal temperatures were set at 20°C and 39°C, respectively, because temperatures between 20°C and 39°C are cited as favorable temperatures for date palm, depending on the varieties [14]. These numbers also provided the best fit to the observed distribution in South America, Asia, South Africa and Australia [11].

Moisture Index

In terms of soil moisture, the lower moisture threshold (SM0) was set at 0.007, to represent the permanent wilting point [27]. Furthermore, this number provided a good fit to the observed distribution of date palms in South America, Asia and the Middle

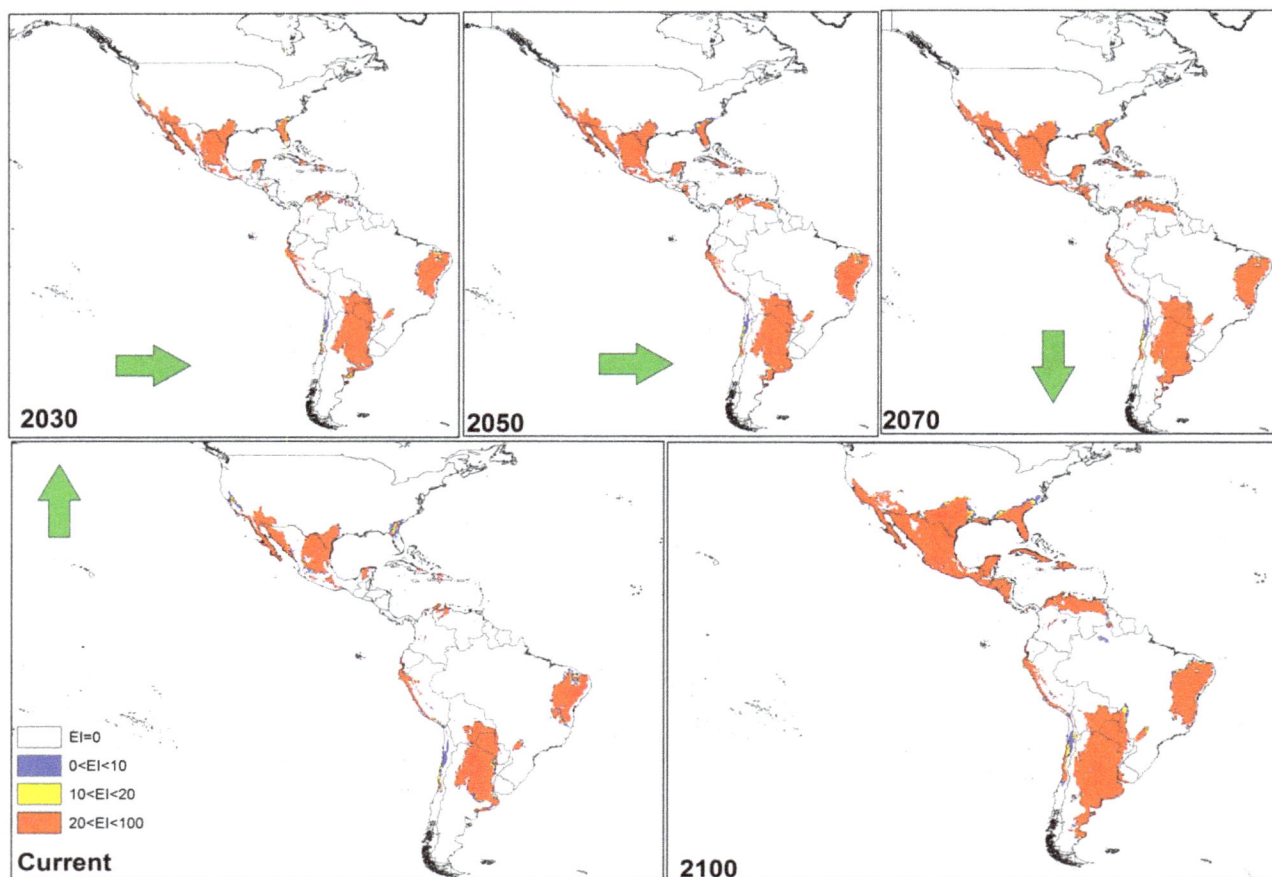

Figure 7. The climate (EI) for *P. dactylifera* **in current time and projected using CLIMEX under the MIROC-H GCM GCM running the SRES A2 scenario and for 2030, 2050, 2070 and 2100 for the North and South America continent.**

East. The lower (SM1) and upper (SM2) optimum moisture thresholds were set at 0.013 and 0.81, respectively, to improve species growth in Egypt, Saudi Arabia, Iran, India, and some countries in Africa [11]. The upper soil moisture threshold (SM3) was set at 0.9 because this species and its fruit can be negatively affected by high soil moisture [11]. Additionally, this value provided an appropriate fit to the observed distribution. All CLIMEX parameters are summarized in Table 1.

Results

Current Climate

The present distribution of native and cultivated *P. dactylifera* is illustrated in Figure 1. A comparison between the modeled global climate appropriateness (Figure 2) with the recognized distribution of this species showed that there was a good match between the Ecoclimatic Index resulting from the CLIMEX model and the current distribution of *P. dactylifera*. The modeled results indicated that the western areas of the United States, western Mexico, southeast Spain, Morocco, Portugal, central Sudan, Egypt, eastern Mozambique, central and western United Arab Emirates, southern Iran, eastern Pakistan and large parts of Australia have suitable climatic conditions for *P. dactylifera*.

Although large parts of central southern Africa were modeled to have suitable climatic conditions for *P. dactylifera* in its current known distribution, limited data were available from these regions. This could be due to a shortage of reports from these areas. Biotic

factors such as competition or lack of dispersal opportunities could preclude this species from occurring in these areas. There is also a possibility that date palm has not been grown as an economically important crop in those regions.

The current and potential distribution of *P. dactylifera* in North America, Mexico, and the Caribbean was used for model validation as shown in Figure 3. These regions were not used for model fitting. In Mexico and North America, the model projects much of the southern and south-western coast to be climatically suitable. There was a reasonably good fit between the model predictions and the actual recorded distribution data.

Future Climate

The results of the two global climate change models (GCMs) including CSIRO-Mk3.0 (CS) and MIROC-H (MR) with the A2 emission scenarios for the potential distribution of *P. dactylifera* for 2030, 2050, 2070 and 2100 are illustrated in Figures 4, 5, 6, 7, 8 and 9. For ease of discussion, the global distribution is subdivided into three regions: North and South America, Africa and Middle East, and Australia and South Asia.

a) Results from CS model. In North and South America (Figure 4), the CS GCM projected much of the south-western coast of Mexico and North America, eastern Brazil, south-eastern Bolivia, northern Venezuela, Cuba, northern Colombia, and Paraguay to be more climatically suitable for *P. dactylifera* by 2030; this expansion steadily increased by 2050, 2070 and 2100.

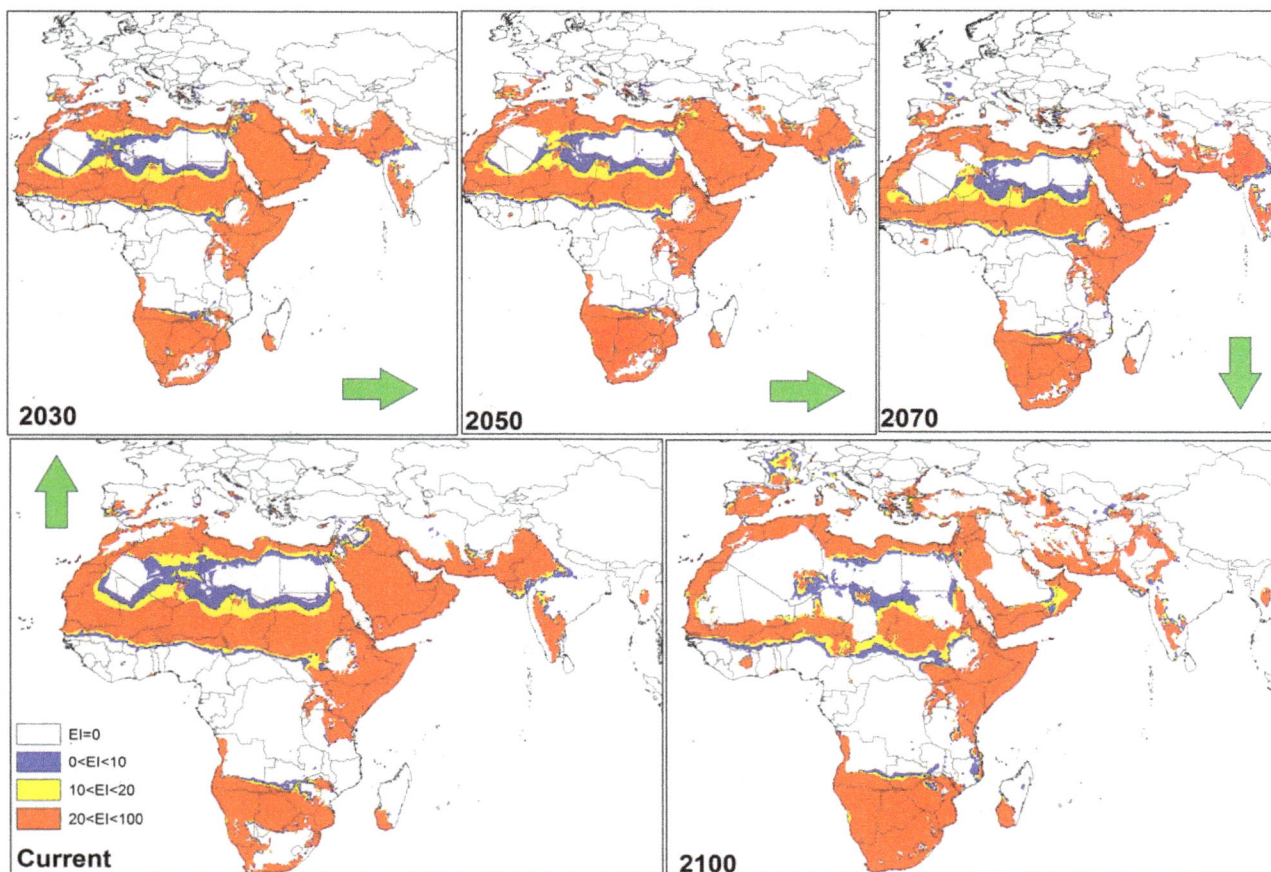

Figure 8. The climate (EI) for _P. dactylifera_ in current time and projected using CLIMEX under the MIROC-H GCM running the SRES A2 scenario and for 2030, 2050, 2070 and 2100 for the northern and southern Africa and the Middle East.

Interestingly, from northern Venezuela to the central regions, the climate was predicted to be highly suitable for date palms by 2100.

In southern Africa (Figure 5), the CS model predicted an expansion of the range of _P. dactylifera_ further inland from now to 2100. In North Africa, particularly in central and southern Algeria, Mauritania, Mali, Niger, all of the Sudan excluding the western side, and southern Tunisia were projected to become progressively less suitable (EI = 0 or EI<1–10) by 2070 and totally unsuitable by 2100 (Figure 5).

The CS GCM for the Middle East indicated that by 2030, Saudi Arabia, Iraq and western Iran would remain climatically suitable (Figure 5). However, by 2050 a reduction in climate suitability for _P. dactylifera_ was predicted for all three countries; this trend was particularly accentuated in Saudi Arabia and Iraq by 2100 (Figure 5). In Asia, especially in northern India, eastern Pakistan and southern Afghanistan (Figure 5), and in north-western Australia (Figure 6), there was a considerable reduction in climate suitability for date palms between 2050 and 2100.

b) Results from MR model. From the MR GCM, it can be seen that in the Americas, much of the south-western coast of Mexico, North America, eastern Brazil, south-eastern Bolivia, northern Venezuela, Cuba, northern Colombia, and Paraguay are projected to become climatically suitable for date palms between 2030 and 2100 (Figure 7). Moreover, the MR GCM predicted that more areas around Florida may become suitable for this species' growth by 2100 (Figure 7). The MR GCM projected that by 2100,

western Argentina would be more climatically suitable than it is currently.

The MR GCM predicted that almost all of the southern regions on the African continent may become suitable for _P. dactylifera_ in the future (Figure 8). In contrast, some countries in North Africa such as Algeria, Mali, Niger, Mauritania and Sudan are projected to become progressively less suitable, with date production becoming completely unviable by 2100 (Figure 8). However, this model projected that some countries such as Namibia, Botswana and parts of southern Zambia are likely to become highly suitable, particularly from 2070 onwards (Figure 8).

The MR GCM for the Middle East projected that Saudi Arabia, Iraq and western Iran may remain climatically suitable for date palms until 2050 (Figure 8). However, the model projected that by 2070 the climate of Saudi Arabia, Iraq and western Iran would be significantly less suitable and that by 2100, the climate in large parts of Saudi Arabia and Iraq would be unsuitable for date palm cultivation. Moreover, a considerable reduction in suitability of climate for date palms was found from 2050 to 2100 in Asia, particularly in northern India, eastern Pakistan and southern Afghanistan (Figure 8).

The results indicated that there were some differences in the projection of date palm distribution between the CSIRO-Mk3.0 and MIROC-H GCMs. These differing results were due to the varying predictions of future climate by the two GCMs.

Based on the two models, cold and wet stresses appear to be the major factors restricting date palm distribution. For example, cold

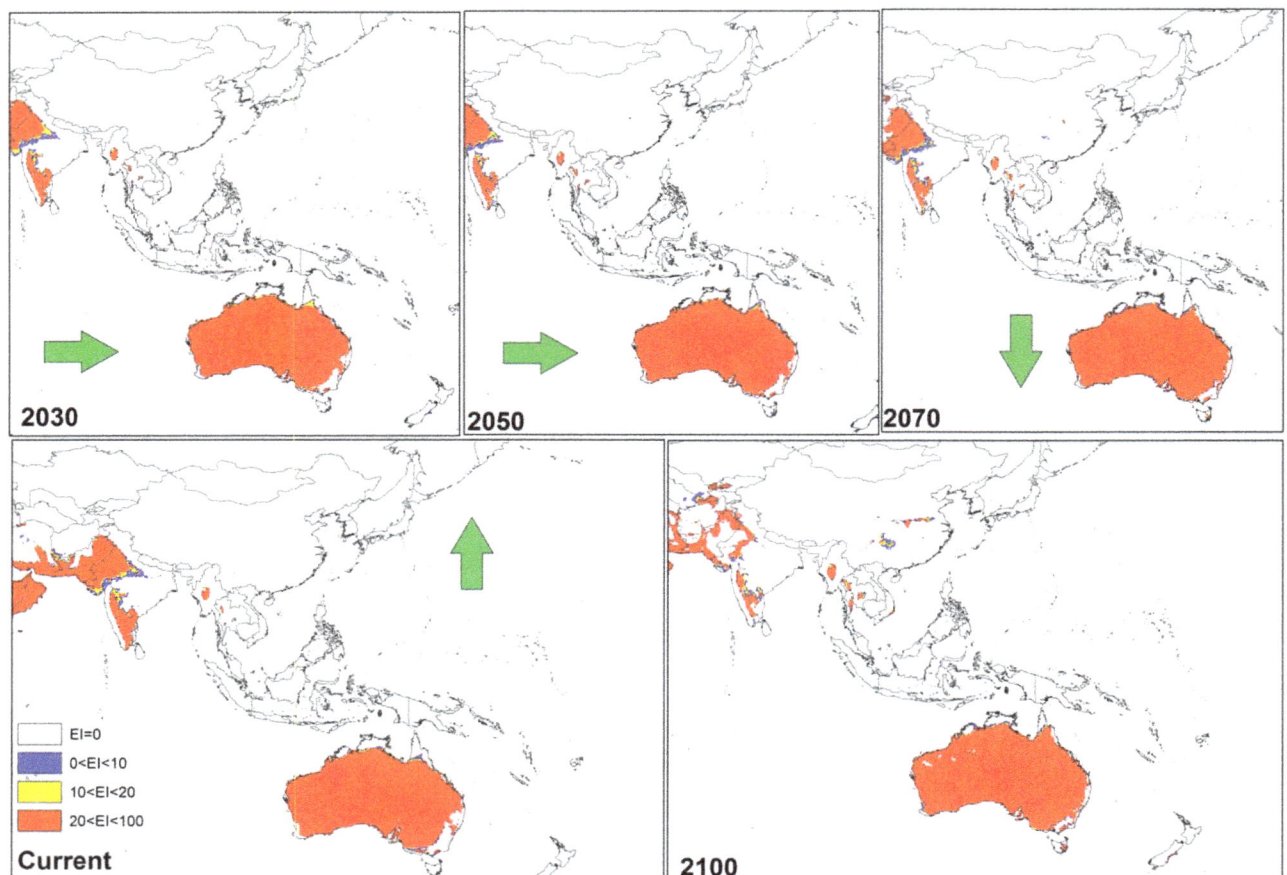

Figure 9. The climate (EI) for *P. dactylifera* in current time and projected using CLIMEX under the MIROC-H GCM running the SRES A2 scenario and for 2030, 2050, 2070 and 2100 for Australia and China.

stress is currently the main limiting factor in Canada, most parts of the United States (excluding Florida and California), Peru, Chile, and Ecuador, south-eastern Australia and most areas of China (Figures 7, 8 and 9). The same results were found for central to western Mali as a consequence of heat stress, which imposes a significant limitation for date palm establishment. Additionally, due to wet stress, *P. dactylifera* cannot be successfully grown in areas of eastern South America, such as central Guatemala, Colombia, Uruguay, and southern Chile, nor in parts of southern Africa including Angola, Zambia, Zimbabwe, and northern Madagascar. Wet stress also causes Germany, Poland, Ireland, northern Portugal, Azerbaijan, Georgia, southern India, Thailand, Burma, Bhutan, eastern Nepal, Spain and southern eastern Australia to be unsuitable for the establishment of this species. Thus, cold and wet stresses impose significant limitations for expanding the global distribution of date palm in 2030 and beyond. The current and projected distribution of cold and wet stresses can be seen for selected regions in Figures 10 and 11, respectively. In the United States, the cold and wet stresses shift northward, meaning there may be larger areas available that are not affected by the aforementioned stresses and therefore more are conducive to date palm cultivation. Our modeling showed that cold and wet stresses will no longer be the limiting factors in large parts of the United States.

Discussion

Suitable climatic areas for *P. dactylifera* under present and future climate scenarios using CLIMEX were modeled in this study. The differences in the outcomes from the two GCMs emphasize the uncertainties associated with the state of climate modeling associated with greenhouse emission patterns [14]. It is clear that different models may produce different results. It should also be highlighted that suitability projections are only potential distributions based on climatic factors and not predicted future distributions [14]. Thus, it is highly recommended that any projection of future suitable areas based on CLIMEX should also incorporate non-climatic factors such as land-use type, soil type, soil drainage and soil-nutrients [11].

Here, our model provided a good fit to the present global distribution records of date palm on the southern coast of Mexico and south-western North America, regions that were used to validate the model.

In this study, both CS and MR GCMs projected that in the Americas, some regions including the south-western coast of Mexico and North America, eastern Brazil, south-eastern Bolivia, northern Venezuela, Cuba, northern Colombia and Paraguay will become more climatically suitable towards 2100. However, the MR GCM projected Florida becoming more climatically suitable than the CS GCM between 2030 and 2100 due the projection of a greater increase in temperature and smaller decrease in the amount of rainfall in the MR GCM [46,47]. Thus, date palms

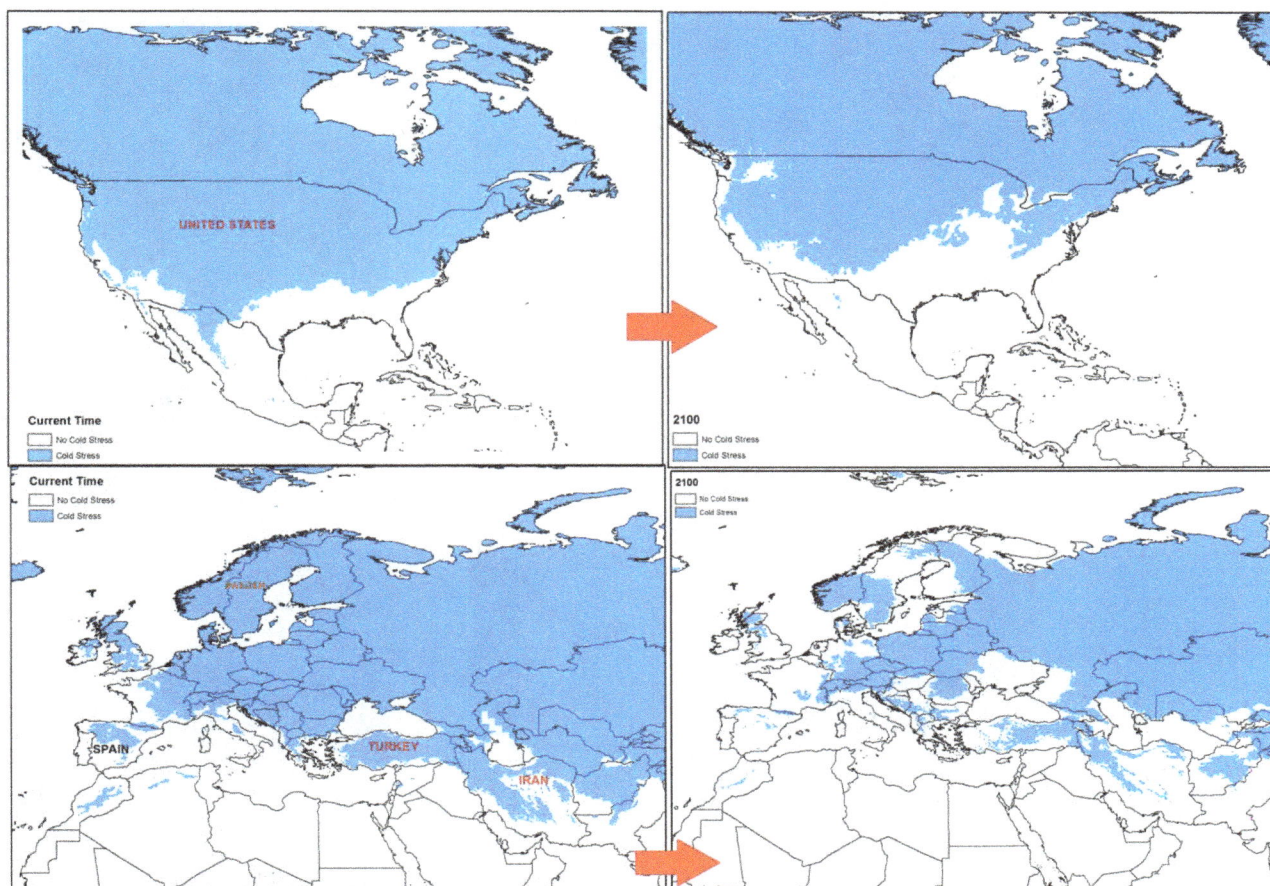

Figure 10. Comparison of the location of cold stress in some selected regions for date palm growth between current time and 2100. These areas were selected on the basis of large changes in cold stress in the future.

would not suffer any wet or cold stress in Florida. A comparison between these two models also indicated that, based on the MR GCM, more regions in western Argentina may be suitable for date palm growth compared to the CS projection (Figures 4 and 7). A comparison between the results of CS and MR GCM for southern Africa indicated that *P. dactylifera* ranges appeared to shift further inland in the future. However, the CS GCM projected that most regions in Angola may be climatically suitable by 2100, but, based on MR GCM, this suitability may be limited to the southern and coastal regions due to an increase in the wet stress in northern and eastern Angola (Figures 5 and 8).

There were some divergent results in the projection of suitable areas for date palms in North Africa and Middle Eastern countries between the CS and MR GCMs. For example, both models projected that northern Algeria, Morocco, Western Sahara, Tunisia, northern Egypt, Somalia and Kenya may become climatically suitable for *P. dactylifera*. Furthermore, both models projected that southern Algeria, eastern Mauritania, northern Mali and western Chad may become unsuitable for this species. In contrast, the CS model projected that by 2100, Mali, Niger, Chad and most parts of Sudan may not be suitable for date palm, while the MR model projected that southern Mali and Niger, eastern Chad and western Sudan may remain climatically suitable for date palms by 2100 (Figures 5 and 8).

There were some agreements and disagreements in projection of suitable areas for date palm growth in North Africa and Middle

Eastern countries between CS and MR GCMs. For example, both models projected that northern Algeria, Morocco, western Sahara, Tunisia, northern Egypt, Somalia and Kenya may become climatically suitable for *P. dactylifera* growth. Furthermore, both models projected that southern Algeria, eastern Mauritania, northern Mali and western Chad may be unsuitable for this species. On the other hand, the CS model projected that by 2100, Mali, Niger, Chad and most parts of Sudan may become unsuitable for date palm growth since the MR model projected that southern Mali and Niger, eastern Chad and western Sudan may remain climatically suitable for date palms growth by 2100 (Figures 5 and 8).

From the results (Figures 10 and 11), it is evident that currently unsuitable areas such as the western United States, southern Mexico, northern and southern Africa, may become suitable climatically by 2100 through decreasing cold and wet stresses. Iran, Turkey, and Spain are some examples where cold stress may decrease by 2100 (Figure 10). Figure 11 illustrates that wet stress in northern Gabon and eastern Quebec may decrease over the next few decades.

The results of the climate change modeling provide an indication of the possible change in the potential future distribution of *P. dactylifera*. As the climate changes, some areas where *P. dactylifera* currently occurs may become climatically unsuitable, and as a consequence, the economies in those areas may decline. For example, it was reported that Algeria and Saudi

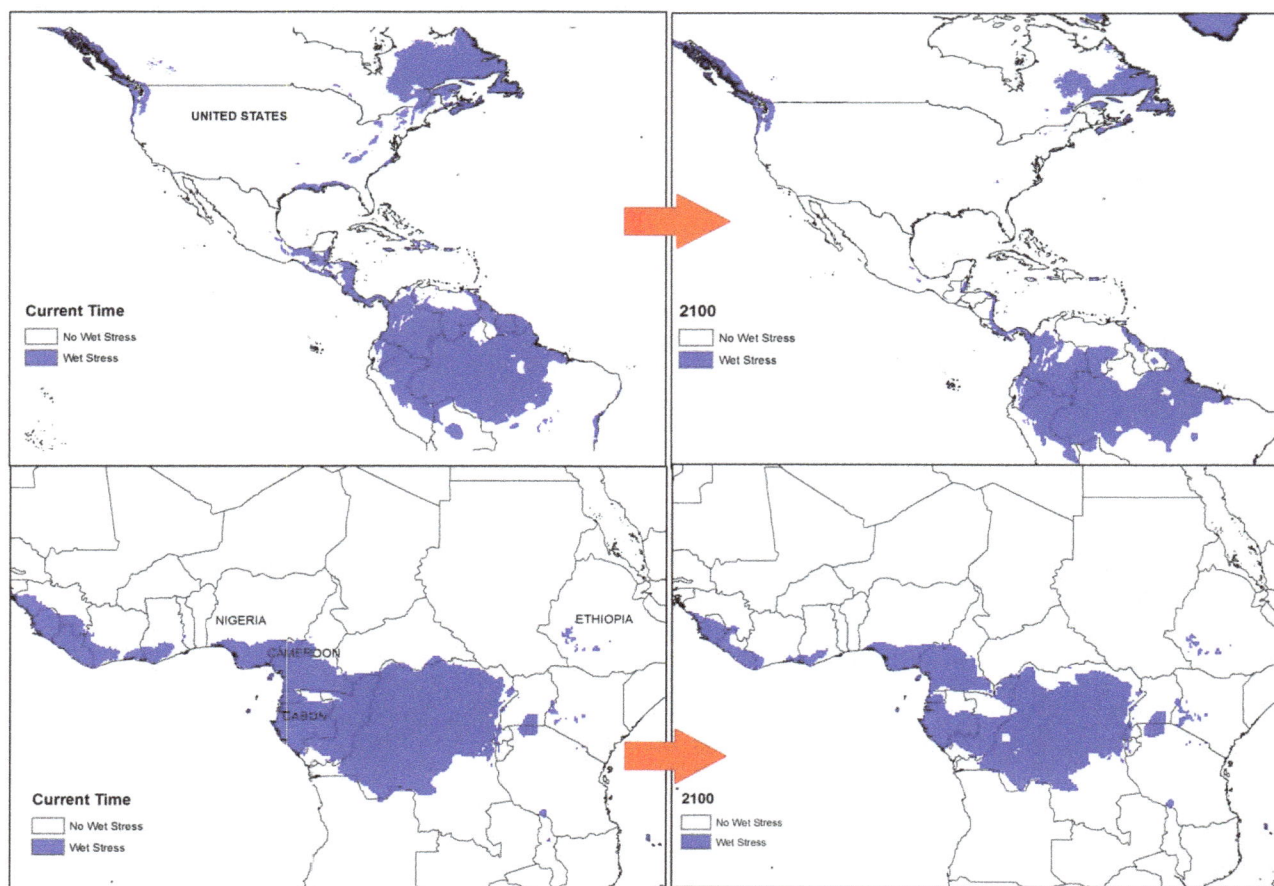

Figure 11. Comparison of the location of wet stress in some selected regions for date palm growth between current time and 2100. These areas were selected on the basis of large changes in wet stress in the future.

Arabia earned 3621 and 1378 U.S. dollars/tonne, respectively, in 1995 from exporting dates. The large disparity in price was due to their strategies in targeting different countries and the differences in date quality [11]. However, this study indicates that large areas of Algeria and Saudi Arabia may become climatically unsuitable and may not be able to cultivate this profitable crop to the same extent in the future. The results are in line with current observations of a decline in date palm production in Middle Eastern countries from 1990 to 2000 [4,5].

Consequently, the results of this paper provide some advance awareness for countries which rely on income from exporting dates. Furthermore, by making some strategic plans, many economic disadvantages can be prevented. This information is particularly important for some countries in northwestern Africa and the Middle East.

Conversely, the results indicate that some areas which are climatically unsuitable at present may become suitable for date palm cultivation in the future. These outcomes may well be useful in making informed choices about the location of date palm farms and associated industries in advance. Benin, Ghana, Cameroon, Nigeria, Venezuela and China may have the opportunity to cultivate date palms and export its produce in the future. Under future climate, *P. dactylifera* may be able to be cultivated in areas that are currently too cold or wet; this can be seen in the improved climatic suitability for Florida, Mexico, northern Venezuela, and eastern Brazil in the Americas; South Sudan and Guinea in Africa;

and Spain and France in Europe. These countries should be prepared to make use of these opportunities since, climate-wise, these areas may become highly suitable for this plant. Specifically, these maps could be used by agricultural organizations in various countries to make strategic, long-term plans. This may include research into alternative crops in areas where climate will become unfavorable for date palms.

In interpreting these results, the following should be considered:

i.   The modeling was performed based only on climate; it does not take into consideration other factors such as land uses, soil types, biotic interactions, diseases and competition.

ii.   This research was based on currently available broad-scale climate data; therefore, it only shows broad-scale shifts.

iii.   It is indicative because a certain level of uncertainty is associated with future levels of greenhouse gas emissions.

In conclusion, this research has demonstrated broad-scale shifts in areas conducive to date palm cultivation and how different areas of the world may be affected due to climate change based on broad regional-scale changes over the next hundred years using coarse scale climate data. Some regions were projected to be climatically unsuitable as a consequence of only one stress for date palm growth, such as wet stress in northern Angola. However, some regions were projected to be unsuitable as a consequence of a combination of multiple stresses; for example, the combination of wet and cold stresses imposed negative effects on date palms growth in the northern United States, meaning that the effects of stresses differ regionally. Such modeling is useful in planning future strategies and minimizing economic impacts in areas that may be adversely impacted, while preparing to take advantage of new opportunities in regions that may be positively impacted.

Author Contributions

Conceived and designed the experiments: FS LK ST. Performed the experiments: FS. Analyzed the data: FS ST. Contributed reagents/materials/analysis tools: FS LK ST. Wrote the paper: FS.

References

1. Andrewartha HG, Birch LC (1954) The distribution and abundance of animals. Chicago: University of Chicago Press. 782 p.
2. Jeffrey S, Harold A (1999) Does global change increase the success of biological invaders? Trends in Ecology and Evolution 14: 135–139.
3. McMichael A, Lendrum D, Corvalán C, Ebi K, Githeko A (2003) Climate change and human health. Available at: http://www.who.int/globalchange/publications/climchange.pdf. World Health Organization. 145–186 p. Accessed 2012 January 9.
4. Jain S (2011) Prospects of in vitro conservation of date palm genetic diversity for sustainable production. Emirates Journal of Food and Agriculture 23: 110–119.
5. Zaid A (2012) Date palm cultivation, Available: http://www.fao.org/DOCREP/006/Y4360E/y4360e07.htm#bm07.2.Accessed 2012 Mar 15.
6. Chakraborty S, Murray GM, Magarey PA, Yonow T, Sivasithamparam K, et al. (1998) Potential impact of climate change on plant diseases of economic significance to Australia. Australasian Plant Pathology 27: 15–35.
7. Jain S, Al-Khayri J, Johnson D (2011) Date palm biotechnology Springer Dordrecht Heidelberg London New York.
8. Tengberg M (2011) Beginnings and early history of date palm garden cultivation in the Middle East. Journal of Arid Environments 5: 1–9.
9. Agroforestry Tree Database. A tree species reference and selection guide, Agroforestry Tree Database. Available: http://www.worldagroforestrycentre.org/sea/Products/AFDbases/af/index.asp. Accessed 2012 Apr 21.
10. Ahmed M, Bouna Z, Lemine F, Djeh T, Mokhtar T, et al. (2011) Use of multivariate analysis to assess phenotypic diversity of date palm (*Phoenix dactylifera* L.) cultivars. Scientia Horticulturae 127: 367–371.
11. Elshibli S, Elshibli E, Korpelainen H (2009) Date Palm (*Phoenix dactylifera* L.) Plants under Water Stress: Maximisation of Photosynthetic CO_2 Supply Function and Ecotypespecific Response. "Biophysical and Socio-economic Frame Conditions for the Sustainable Management of Natural Resources" Tropentag, Hamburg. Available: http://www.tropentag.de/2009/abstracts/links/Elshibli_FGCITsVL.pdf. Accessed 2012 May 4.
12. Bokhary H (2010) Seed-borne fungi of date-palm, *Phoenix dactylifera* L. from Saudi Arabia. Saudi Journal of Biological Sciences 17: 327–329.
13. Burt J (2005) Growing date palms in Western Australia. Available: http://www.agric.wa.gov.au/objtwr/imported_assets/content/hort/fn/cp/strawberries/f05599.pdf. 2–4 p. Accessed 2012 Feb 19.
14. Jain S, Al-Khayri J, Dennis V, Jameel M (2011) Date Palm Biotechnology: Springer. 743 p.
15. Soberon J, Peterson A (2005) Interpretation of models of fundamental ecological niches and species distributional areas. Biodiversity Informatics 2: 1–10.
16. Guisan A, Zimmerman NE (2000) Predictive habitat distribution models in ecology. Ecological Modelling 135: 147–186.
17. Pearson RG, Dawson TP (2003) Predicting the impacts of climate change on the distribution of species: Are bioclimate envelope models useful? Global Ecology and Biogeography 12: 361–371.
18. Kriticos DJ, Randall RP (2001) A comparison of systems to analyze potential weed distributions. In: Groves RH, Panetta FD, Virtue JG, editors. Weed Risk Assessment. Collingwood: CSIRO Publishing. 61–79.
19. Taylor S, Kumar L, Reid N, Kriticos DJ (2012) Climate Change and the Potential Distribution of an Invasive Shrub, *Lantana camara*.L. PLoS ONE 7: e35565.
20. Yonow T, Sutherst RW (1998) The geographical distribution of the Queensland fruit fly, Bactrocera (Dacus) tryoni, in relation to climate. Australian Journal of Agricultural Research 49: 935–953.
21. Sutherst R, Floyd RB (1999) Impacts of global change on pests, diseases and weeds in Australian temperate forests. Available at: http://www.cse.csiro.au/publications/1999/temperateforests-99-08.pdf. Accessed 2012 March 21.
22. Sutherst RW (2001) The vulnerability of animal and human health to parasites under global change. International Journal for Parasitology 31: 933–948.
23. Sutherst RW, Maywald G, Kriticos DJ (2007) CLIMEX Version 3: User's Guide. In: Ltd HSSP, editor. Melbourne.
24. Sutherst RW, Maywald G (1985) A computerized system for matching climates in ecology. Agriculture Ecosystems & Environment 13: 281–299.
25. Sutherst RW, Maywald G (2005) A climate model of the red imported fire ant, Solenopsis invicta Buren (Hymenoptera : Formicidae): Implications for invasion of new regions, particularly Oceania. Environmental Entomology 34: 317–335.
26. Davis AJ, Jenkinson LS, Lawton JH, Shorrocks B, Wood S (1998) Making mistakes when predicting shifts in species range in response to global warming. Nature 391: 783–786.
27. Kriticos D, Potter K, Alexander N, Gibb A, Suckling D (2007) Using a pheromone lure survey to establish the native and potential distribution of an invasive Lepidopteran. Journal of Applied Ecology 44: 853–863.
28. Global Biodiversity Information Facility (GBIF) website. Available: http://www.gbif.org. Accessed 2012 Feb 2.
29. Woodcock l, Diana L (2010) Date Palm Plantation, Iran. Available: http://go.galegroup.com/ps/i.do?id=GALE%7CA240488438&v=2.1&u=dixson&it=r&p=LitRC&sw=w. 16: 26–29.
30. Eshraghi P, Zarghami R, Mirabdulbaghi M (2005) Somatic embryogenesis in two Iranian date palm. African Journal of Biotechnology 4: 1309–1312.
31. Shayesteh N, Marouf A (2010) Some biological characteristics of the Batrachedra amydraula Meyrick (Lepidoptera: Batrachedridae) on main varieties of dry and semi-dry date palm of Iran. 10th International Working Conference on Stored Product Protection. Portugal. 151–155.
32. Mahmoudi H, Hosseininia G (2008) Enhancing date palm processing, marketing and pest control through organic culture. Journal of Organic Systems 3: 30–39.
33. Abbas I, Mouhi M, Al-Roubaie J, Hama N, El-Bahadli A (1991) *Phomopsis phoenicola* and *Fusarium equiseti*, new pathogens on date palm in Iraq. Mycological Research 95: 509.
34. Auda H, Khalaf Z (1979) Studies on sprout inhibition of potatoes and onions and shelf-life extension of dates in Iraq. Journal of Radiation Physics and Chemistry 14: 775–781.
35. Heakal MS, Al-Awajy MH (1989) Long-term effects of irrigation and date-palm production on Torripsamments, Saudi Arabia. Geoderma 44: 261–273.
36. Al-Senaidy M, Abdurrahman M, Mohammad A (2011) Purification and characterization of membrane-bound peroxidase from date palm leaves (*Phoenix dactylifera* L.). Saudi Journal of Biological Sciences 18: 293–298.
37. Markhand G (2000) Fruit characterization of Pakistani dates. Available: http://www.salu.edu.pk/research/dpri/docs/b-003.pdf. Date Palm Research Institute. Accessed 2012 Mar 17.
38. Hasan S, Baksh K, Ahmad Z, Maqbool A, Ahmed W (2006) Economics of Growing Date Palm in Punjab, Pakistan. International Journal Of Agriculture and Biology 8: 1–5.
39. Saadi I, Namsi A, Mahamoud OB, Takrouni ML, Zouba A, et al. (2006) First report of 'maladie des feuilles cassantes' (brittle leaf disease) of date palm in Algeria. Plant Pathology 55: 572–572.
40. Elhoumaizi M, Saaidi M, Oihabi A, Cilas C (2001) Phenotypic diversity of date-palm cultivars (*Phoenix dactylifera* L.) from Morocco. Genetic Resources and Crop Evolution 49: 483–490.
41. Marqués J, Duran-Vila N, Daròs J-A (2011) The Mn-binding proteins of the photosystem II oxygen-evolving complex are decreased in date palms affected by brittle leaf disease. Plant Physiology and Biochemistry 49: 388–394.
42. Kriticos D, Webber B, Leriche A, Ota N, Macadam I, et al. (2011) Global high-resolution historical and future scenario climate surfaces for bioclimatic modelling. Methods in Ecology and Evolution 3: 53–64.
43. Gordon H, Rotstayn L, McGregor J, Dix M, Kowalczyk E, et al. (2002) The CSIRO Mk3 Climate System Model. Available: http://www.cawcr.gov.au/publications/technicalreports/CTR_021.pdf. Accessed 2012 Mar 11.
44. Intergovernmental Panel on Climate Change (2012) Special Report on Emissions Scenarios: A Special Report of Working Group III of the Intergovernmental Panel on Climate Change. Available: http://www.ipcc.ch/news_and_events/press_information.shtml#.T_mNx5HdVLo. Accessed 2012 Apr 8.
45. Hennessy K, Colman R (2007) Global Climate Change Projections. Available: http://www.ipcc.ch/pdf/assessment-report/ar4/wg1/ar4-wg1-chapter10-supp-material.pdf. Accessed 2012 Apr 22.
46. Chiew F, Kirono D, Kent D, Vaze J (2009) Assessment of rainfall simulations from global climate models and implications for climate change impact on runoff studies. 18th World IMACS Australia 3907–3914.
47. Suppiah R, Hennessy K (2007) Australian climate change projections derived from simulations performed for the IPCC 4th Assessment Report. 131–152.
48. Rahmstorf S, Cazenave A, Church JA, Hansen JE, Keeling RF, et al. (2007) Recent climate observations compared to projections. Science 316: 700–709.
49. Kriticos DJ, Leriche A (2010) The effects of climate data precision on fitting and projecting species niche models. Ecography 33: 115–127.
50. Sutherst RW (2003) Prediction of species geographical ranges. Journal of Biogeography 30: 805–816.

14

Effects of Climate Change on Range Forage Production in the San Francisco Bay Area

Rebecca Chaplin-Kramer[1], Melvin R. George[2]*

1 Natural Capital Project, Stanford University, Stanford, California, United States of America, 2 Plant Sciences Department, University of California Davis, Davis, California, United States of America

Abstract

The San Francisco Bay Area in California, USA is a highly heterogeneous region in climate, topography, and habitats, as well as in its political and economic interests. Successful conservation strategies must consider various current and future competing demands for the land, and should pay special attention to livestock grazing, the dominant non-urban land-use. The main objective of this study was to predict changes in rangeland forage production in response to changes in temperature and precipitation projected by downscaled output from global climate models. Daily temperature and precipitation data generated by four climate models were used as input variables for an existing rangeland forage production model (linear regression) for California's annual rangelands and projected on 244 12 km x 12 km grid cells for eight Bay Area counties. Climate model projections suggest that forage production in Bay Area rangelands may be enhanced by future conditions in most years, at least in terms of peak standing crop. However, the timing of production is as important as its peak, and altered precipitation patterns could mean delayed germination, resulting in shorter growing seasons and longer periods of inadequate forage quality. An increase in the frequency of extremely dry years also increases the uncertainty of forage availability. These shifts in forage production will affect the economic viability and conservation strategies for rangelands in the San Francisco Bay Area.

Editor: Ben Bond-Lamberty, DOE Pacific Northwest National Laboratory, United States of America

Funding: The California Energy Commission's Public Interest Energy Research (PIER) Program provided support for this research under the California Climate Change Center. The funders had no role in study design, data collection and analysis, decision to publish, or preparation of the manuscript.

Competing Interests: The authors have declared that no competing interests exist.

* E-mail: mrgeorge@ucdavis.edu

Introduction

California's San Francisco Bay Area is a mosaic of urban and natural lands, and tension exists between the two with scenic beauty attracting development that threatens these prized open spaces [1]. About half of the area of the eight San Francisco Bay counties is classified as rangeland, and these areas account for most of the region's open space. The private ranches on Bay Area rangelands provide a livelihood and a way of life that helps to limit urban sprawl [2]. With a growing population placing pressure on these areas for development and a changing climate posing new threats to rangeland ecosystems, understanding the degree to which their value as working landscapes will be maintained in the future is important to their conservation.

Over the next century California temperatures are projected to rise between 1.7° and 3.0°C for a lower emissions scenario, and 4.4° to 5.8°C for a higher emissions scenario [3]. Downscaled results from global climate models for the San Francisco Bay Area show a lower rise in temperatures, from 1.5° to 3.0°C by 2100 for the lower emissions scenario and 2.5° to 4.4°C for the higher emissions scenario, though considerable variation exists within the region (Fig. 1). Changes in precipitation are more uncertain, with a high degree of variability between different climate models, and even from year to year within the same model, suggesting that the region will remain vulnerable to drought. Overall, the majority of simulations indicate that total annual precipitation will decline,

mostly in the spring months, while winter precipitation will remain relatively stable [4].

The full extent of climate change impacts on rangeland forage production in California and the Bay Area in particular is uncertain. One study [5] modeled the impact of forecasted changes in precipitation patterns on California rangeland production and concluded that areas of the state suitable for cattle grazing would shift, as some areas become wetter and others become drier, depending on the climate model. Statewide, they predicted range forage production would decline between 14 and 58 percent, corresponding to a reduction in annual profits from cattle ranching of between $22 million and $92 million by 2070. Despite this statewide trend, their results for the Bay Area suggested the impacts would be more positive, with production increases projected for Santa Clara, Alameda, Contra Costa, Solano, and Napa Counties. Marin and Sonoma Counties were not included in the model. However, this precipitation-based model did not incorporate rangeland response to warming, and the authors acknowledged that their model may have overestimated the effects of precipitation. While precipitation has been shown to be an important variable for predicting annual rangeland productivity, temperature within the growing season is also important [6]. Precipitation, temperature and forage production data collected since 1935 at the San Joaquin Experimental Range in Madera County have shown that near average production can occur in low rainfall years if precipitation is well distributed and

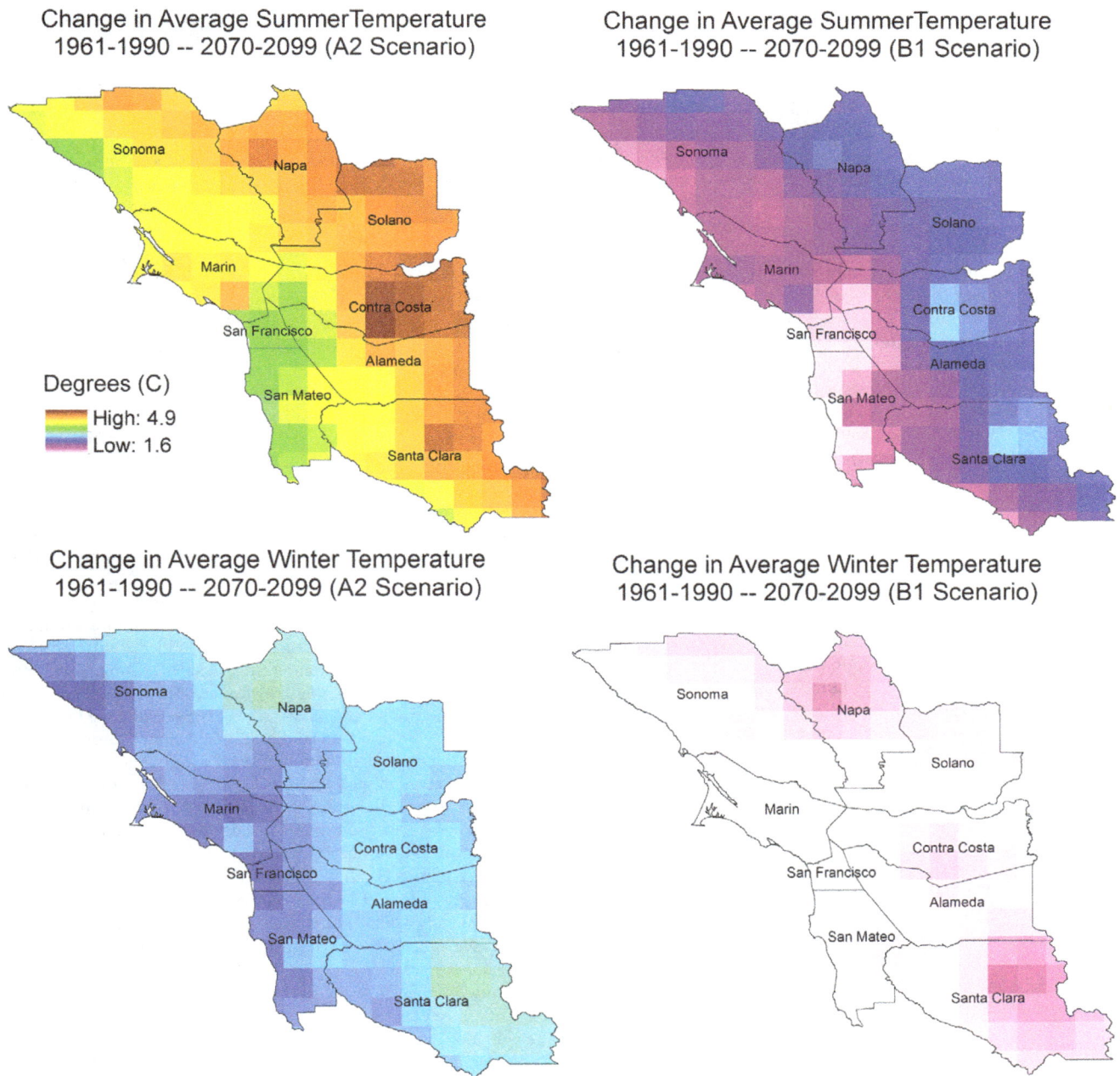

Figure 1. Historical (1961–1990) and projected (2070–2099) average temperatures for summer (June, July, August) and winter (January, February) months in the Bay Area. Temperatures reflect means of four global climate models (downscaled output from CNRM CM3, GFDL CM2.1, NCAR CCSM3.0 and NCAR PCM1).

low annual production can occur in wet years if precipitation is poorly distributed or if temperatures are below normal, as is often associated with wet weather [7]. Thus, integrating changes in both temperature and precipitation could improve forecasts of rangeland forage production.

The combined effects of seasonal temperature and precipitation patterns will influence not only productivity, but also growing season length and plant phenology in rangelands [8]. Warming has been shown to increase soil water content by accelerating plant senescence [9], which may interact with changes in precipitation to further affect water availability in rangelands. In fact, grassland ecophysiology may be less responsive to changes in total quantity of rainfall than to shifts in seasonal patterns of rainfall [10]. Early-season precipitation alone explained 49 percent of the variability in shoot-growth at the University of California Hopland Research and Extension Center (UC HREC), just north of the Bay Area [11], although additional data reduced this explanatory power to 34 percent [6]. Late-season precipitation also has a pronounced impact on Bay Area and North Coast grassland production, shown by increased shoot growth resulting from experimental water additions in the late spring [12,13].

Despite these known links between the effects of warming and precipitation in grasslands, no models incorporating the impact of warming on California rangeland production yet exist. Increases in spring plant production and an extension of the growing season has been predicted for the Great Plains [14]. However, these authors warned that increases in variance may be more important than the mean effect, because uncertainty in predicting plant growth results in suboptimal stocking decisions. They suggested that the increased variance found in their simulations would require carrying capacities to decrease from about 6.5 to 9.0 ha per animal, in order to maintain a 90 percent confidence of not overstocking. Further, more intense management would increase operating costs, and therefore may negate any benefits in forage production. In contrast to the Great Plains, where the growing season begins in the spring months following winter dormancy, the Bay Area rangeland growing season begins with the first fall rains and ends with soil moisture depletion in the spring months. Climate change in the Bay Area may be more comparable to that found in the Mediterranean climate of southern Australia. For this region, lower pasture production has been projected for future climates with lower precipitation and higher temperatures [15].

Incorporating the effects of warming into models of rangeland production in California, and the Bay Area in particular, is an important step in understanding how climate change will affect range livestock production in this region. The economic viability of rangelands is essential to maintaining the natural aesthetic that contributes to the quality of life of the residents of this unique urban-natural interface. This paper reports the projected changes in forage production in response to simulated future temperature and precipitation in the San Francisco Bay Area to better understand how climate change will impact these working landscapes so important to local conservation.

Methods

Study Area

The geographic diversity of the San Francisco Bay Area (hills, mountains, and large water bodies) produces a wide variety of microclimates. Coastal areas are generally characterized by relatively small temperature variations during the year, with cool, foggy summers and mild, rainy winters. Inland areas, especially those separated from the ocean by hills or mountains, have hotter summers and colder overnight temperatures during the winter. The rangelands in the North Bay (with its northwestern most point at 38.9375N, 123.6875W; encompassing Sonoma, Marin, Napa, and Solano Counties) are characterized by higher rainfall, a longer rainy season and cooler temperatures than those in the South Bay (southeastern most point at 36.9375N, 121.1875W; encompassing Santa Clara, Alameda, and Contra Costa Counties). San Jose, at the south end of the Bay averages fewer than 380 mm of rain annually, while Napa, in the North Bay area, can exceed 750 mm. Because range forage production is strongly influenced by temperature and precipitation, there are significant differences in growing season length and productive potential between the North and South Bay areas.

Climate Models

Climate data were acquired from Cayan et al., downscaled from global climate models using a Bias Corrected Constructed Analogues (BCCA) technique to produce two climate scenarios: the lower-emissions B1 scenario and the higher-emissions A2 scenario [4]. Four climate models produce daily temperature and precipitation projections: Centre National Recherché Meteorologique (CNRM) CM3, Geophysical Fluid Dynamics Laboratory

(GFDL) CM2.1, National Center for Atmospheric Research (NCAR) CCSM3.0, and NCAR PCM1. Each climate model was back-cast to simulate historical climate conditions (1961–1990) and represented historical climate data with accuracy [4].

Forage Production Model

Throughout California's 14.5 million acres of annual rangelands, which include the Bay Area grasslands and oak woodlands of this study, temperature is the main constraint to productivity during the growing season. Therefore, precipitation and evapotranspiration drove a simple model to determine growing season length, and temperature and growing season length drove the model for annual forage production.

Daily climate data from the four climate models were input variables for a forage production model reported by Californian researchers [16] who found that growing degree days accounted for 75 to 95 percent of the variation in growing season production (Table 1). This degree-day forage production model was run for all eight model/scenario combinations (described in Climate Models, above), for each of the 244 12 km x 12 km grid cells that comprise the Bay Area region. The mean of the output from the four climate models was taken for the A2 and B1 scenarios, and compared to output for a simulated historical period (1961–1990). All calculations and simulations were produced in the R software package [17].

Modeling the bounds of the growing season was necessary to convert a regression model based on field data into a predictive model to simulate forage production under future climate scenarios. A germinating rain that exceeds 25 mm within one week marks the start of the growing season [16]. There is no similarly well-established climatic phenomenon marking the end of the growing season; in field studies it is determined empirically, by measuring biomass until annual grasses are between the soft and hard dough stage of seed maturity [16]. Therefore, the end of the growing season in this study was simulated using a simple water balance model that was trained using CIMIS weather data for precipitation and evapotranspiration from the University of California Sierra Foothill Research and Extension Center (UC SFREC), 17 miles northeast of Marysville, California. Calculating the point at which cumulative evapotranspiration exceeded cumulative precipitation over a moving window of 60 days best predicted the peak forage date. This simple model generally came within two weeks of actual peak forage date measured at the UC SFREC, rarely extending beyond the end of May. For each year, germination and season end dates were computed according to these methods, and set the seasonal bounds within which forage production was modeled, capturing inter-annual variability in season length.

To simulate forage production, degree-days were first calculated from model-generated minimum and maximum daily temperatures above a base temperature of 5°C using the sine function method [16,18]. Accumulated degree-days (ADD), the sum of all previous degree-days from a given date, were calculated at monthly intervals from germination until the end of the growing season. Monthly standing biomass or total forage production was estimated from ADD using the regression equations from several annual rangeland sites in Table 1 [16]. Absolute forage production varied depending on the chosen equation, but as the relationship is linear, relative measures such as the change in forage production over time were very consistent, differing by only 2 to 3 percent. For this reason, future values for peak forage production are presented in terms of change from historic values.

Growth curves were constructed using the monthly forage production estimates, averaged over the window of historic (1961–

Table 1. Relationship between forage production (y, kg ha^{-1}) and accumulated degree days (x) from 10 sites in four annual rangeland counties [16].

Sample Area	County	Latitude (N)	Longitude (W)	Regression Equation	R^2
1	Yuba	39.330361	121.3476	$y = -120 + 5.2x$	0.95
2	Butte	39.384564	121.59456	$y = 14 + 4.4x$	0.91
3	Madera	37.088669	119.73461	$y = -90 + 3.8x$	0.85
4	Madera	37.089609	119.713978	$y = -141 + 3.1x$	0.82
5	Madera	37.089811	119.73698	$y = -54 + 3.9x$	0.77
6	Madera	37.095235	119.736424	$y = -280 + 4.9x$	0.88
7	Mendocino	38.997164	123.092492	$y = 77 + 2.2x$	0.74
8	Mendocino	38.986653	123.08472	$y = 138 + 2.8x$	0.76
9	Mendocino	39.006414	123.085347	$y = 96 + 4.1x$	0.91
10	Mendocino	39.003462	123.077236	$y = 82 + 2.7x$	0.74

1999) and future (2070–2099) time periods for both A2 and B1 emissions scenarios, and then averaged over all rangeland area in a county. These growth curves help to determine which parts of the season have the greatest differences between historical and future conditions, and thus hint at the mechanisms behind the difference. Differences in the first month may indicate that germination date is an important factor. Steeper slopes throughout the middle of the curve would point to the role played by warmer winter and/or spring temperatures. Differences in the slope leading up to the final time step could be at least partially explained by differences in season end date and the length of time for degree days to accumulate in that final period. Months are taken as calendar months, such that if germination occurred on September 23, the month of September would only have one week of forage production. Likewise, if the end of season date is calculated as June 2, the month of June would only have two days of additional forage production added to the overall total.

All model calculations were repeated for every year of simulated climate data from 1961 to 2099, and the data were summarized by taking the means in four windows of time: historic (1961–1999), early century (2005–2034), mid-century (2035–2064), and late century (2070–2099). Changes in peak production and season length from historic to future conditions are presented on maps of rangeland biomes (savannah, grasslands and shrublands) selected from the Existing Vegetation Types layer of the national LANDFIRE dataset [19] and overlaid on the model outputs.

Results

Forage Production

Forage production by the end of the century (2070–2099) increases in each month of the growing season relative to simulated historical forage production (1961–1990), resulting in increased total forage production under future climate conditions throughout the San Francisco Bay Area (Fig. 2). There is little difference between projections for historic vs. early-century (2005–2034) and mid-century (2035–2064) forage production for either scenario. Projected mid-century peak forage production increased only 10 to 13 percent from the historical (1961–1990) conditions for the higher-emissions A2 scenario (Table 2). However, average production for each of the eight Bay Area Counties is projected to increase 24 to 31 percent by the late century (2070–2099) for the A2 scenario, with increases of up to 40 percent in much of Northern Napa and Sonoma Counties (Fig. 3). Projected increases

for these counties in the B1 scenario are slightly higher by mid-century and more modest at the end of the century. However, late-century northeastern Santa Clara County shows an increase above 30 percent for both emission scenarios.

Season Length

The models also predicted changes in growing season length, due to changes in the simulated timing of germination (precipitation exceeding 25 mm in one week) in the fall and soil moisture depletion (evapotranspiration exceeding precipitation over a 60 day period) in the spring. The length of the growing season is projected to markedly decrease under the A2 scenario; two-week shorter seasons can be expected by late-century for much of the Bay Area (Table 2), with parts of Santa Clara showing seasons shrinking by more than three-weeks (Fig. 4). The B1 scenario shows more modest decreases in season length of a few days to a week. With decreasing rainfall, future forage season lengths in eastern Santa Clara and Alameda Counties could drop to as low as 100 days in length, a full 50 days shorter than the shortest season found in Marin or Sonoma Counties.

Inter-annual Variability

The standard errors reported in Table 2 indicate that inter-annual variability in peak forage production and season length is fairly low in the Bay Area, under both historic and future conditions. These standard errors range from 1 to 3 percent of the mean for either of these variables in any county. On the other hand, the maximums and minimums for each 30-year time period range from 38 percent below to 53 percent above the mean forage production and 29 percent above to 46 percent below mean season length (Table 3). More information can be gleaned by considering the variability as the spread of the distribution, the number of data points falling outside a range of 2 standard errors around the mean (Table 4). By this definition, inter-annual variability in peak production declines fairly dramatically throughout most of the Bay Area for both scenarios, due mainly to the number of years that production is above average in the North Bay, while Alameda and Contra Costa shows declines in variability around both sides of the mean for the B1 scenario and in the number of years production is below average in the A2 scenario. Overall inter-annual variability in peak production does not change in Santa Clara, but is skewed more negatively in future climate conditions under both scenarios (more years in which peak

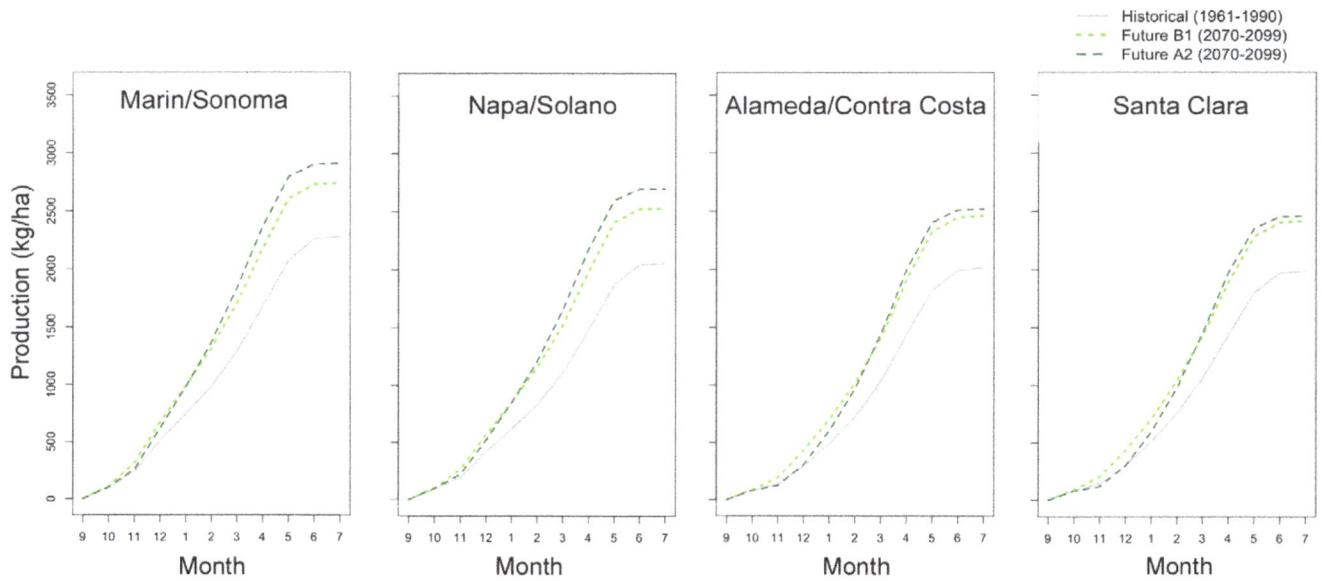

Figure 2. Seasonal growth curves for forage production in different regions under historical (1961–1990) and future (2070–2099, for A2 and B1 emissions scenarios) climate conditions. Growth curves represent the accumulated forage produced on a daily time-step, summarized at monthly intervals as the amount of total forage produced over the season by that date. Each line shows the mean production for all years within each 30-year period and for all cells within each county. Only cells containing rangelands were used (see Figure 3).

production falls below 2 standard errors below the mean occur in the future than historic conditions). Inter-annual variability in season length remains the same or even declines for most counties for the B1 scenario. In the A2 scenario variability increases slightly, with more years falling farther below average season length than historically. The variability is more centered around the mean in all future scenarios than in historical climate conditions (which were skewed slightly positively; more longer than average seasons than shorter than average seasons).

Drought Years

The climate models predict that some regions in the Bay Area would see some years with no germinating rains and therefore no growing season. Historically this would generally occur in any given location in the South Bay once over a 30 year period. While

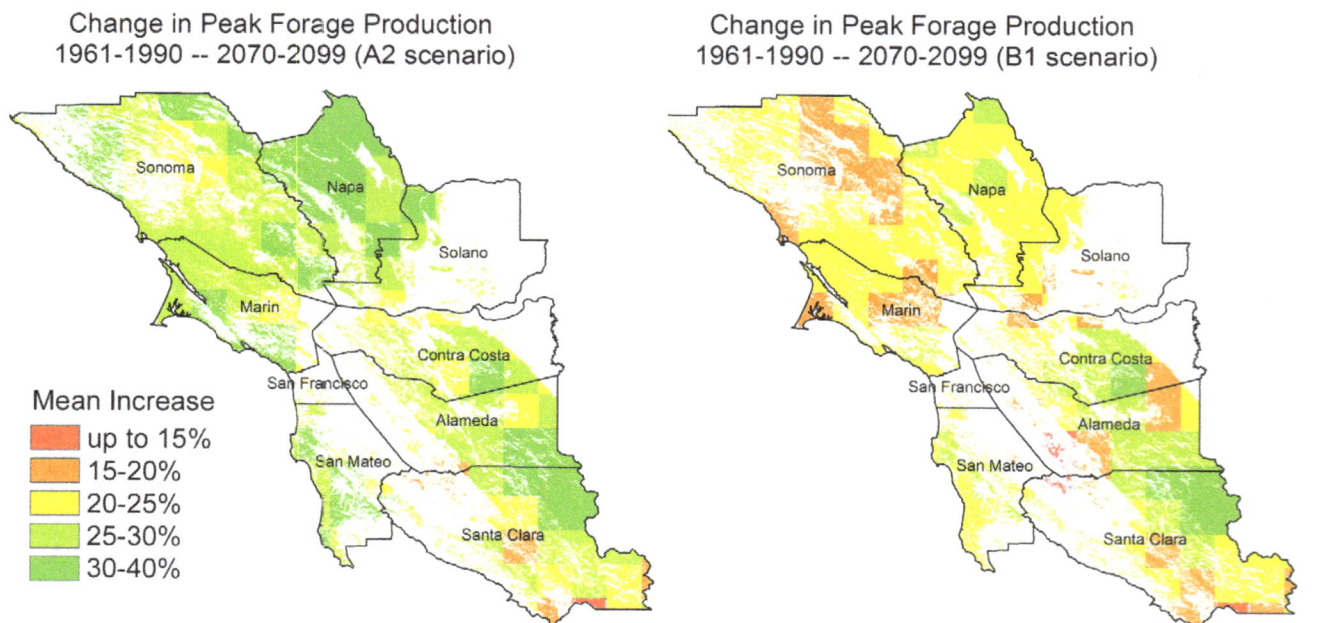

Figure 3. Change (%) in peak forage production by late-century (2070–2099), relative to historical conditions (1961–1990), shown for current rangelands (grassland, savannah, and shrubland) in the Bay Area.

Table 2. Change in peak forage production and season length compared to historic (1961–1990) conditions (mean +/− standard error).

Time Period	Marin-Sonoma		Napa-Solano		Alameda-Contra Costa		Santa Clara	
	Mean	+/−	Mean	+/−	Mean	+/−	Mean	+/−
Change in Forage Production (%): A2 Model								
2005–2034	10.1	2.2	10.4	2.5	13.1	2.8	11.8	2.5
2035–2064	11.9	2.5	13.1	2.9	13.5	3.0	13.0	3.0
2070–2099	27.5	2.8	31.1	3.0	24.9	3.1	23.6	2.9
Change in Forage Production (%): B1 Model								
2005–2034	9.4	2.9	10.1	3.1	9.5	2.9	10.8	3.1
2035–2064	16.2	2.9	19.2	3.2	17.8	3.5	17.1	3.6
2070–2099	20.0	2.6	23.0	2.8	22.1	3.1	21.5	3.1
Reference Historic Season Length (Total Days)								
1961–1990	181.1	2.5	171.8	2.5	152.8	2.78	155.3	2.9
2005–2034	−1.7	2.3	−2.4	2.5	0.3	2.9	−1.2	2.8
2035–2064	−9.7	2.4	−9.2	2.5	−9.6	3.0	−10.9	3.0
2070–2099	−15.5	2.6	−13.7	2.6	−16.3	2.8	−19.4	2.5
Change in Season Length (Days): A2 Model								
2005–2034	−2.9	3.1	−3.1	3.2	−3.4	3.6	−2.5	3.8
2035–2064	−3.3	3.0	−1.9	3.1	−3.6	3.4	−4.6	3.3
2070–2099	−5.1	2.8	−4.1	2.7	−3.3	3.4	−6.2	3.4
Change in Season Length (%): A2 Model								
2005–2034	−0.9	1.3	−1.4	1.5	0.2	1.9	−0.7	1.8
2035–2064	−5.4	1.3	−5.3	1.5	−6.3	2.0	−7.0	1.9
2070–2099	−8.6	1.4	−8.0	1.5	−10.7	1.8	−12.5	1.6
Change in Season Length (Days): B1 Model								
2005–2034	−1.6	1.7	−1.8	1.8	−2.2	2.4	−1.6	2.4
2035–2064	−1.8	1.6	−1.1	1.8	−2.3	2.2	−3.0	2.1
2070–2099	−2.8	1.5	−2.3	1.6	−2.1	2.2	−4.0	2.2

comparisons of model projections revealed that there is a high degree of variation among the four climate models, the mean shows a lower frequency of non-germination years in the southern counties for the B1 scenario compared to historic conditions, and a higher frequency for the A2 scenario. Specific outcomes supported by all models and for both scenarios are that the North Bay is almost entirely unaffected in all time periods and southeastern Santa Clara County experiences more extreme dry years under future conditions.

Discussion

Forage Production

Our model supports the results of earlier efforts incorporating only precipitation into a model for forage production in the Bay Area [5], though the increases in forage production seen here were in spite of, not because of, a shift in precipitation patterns. The agreement between these two models stands in contrast to the projections of a similar Mediterranean climate in Southern Australia [15], where lower forage production was expected under warmer and drier climate. This difference may be due to the fact that precipitation is not as limiting in the Bay Area system as winter temperatures. Projected warming for A2 and B1 scenarios result in higher standing crop throughout the growing season and

at the end of the growing season for the 2070–2099 period compared to the historical period (Fig. 2).

Comparisons of historical monthly production to projected production for the two scenarios reveal that production increases in the A2 scenario reach a maximum (greatest difference from historical conditions) in most parts of the Bay Area by the beginning of March or April (Fig. 5). In the B1 scenario, the maximum change from historical conditions occurs much earlier, by the end of October or November, but the magnitude of the maximum differences between historic and future production are much more variable than in the A2 scenario (25 to 80 percent increases for B1, 30 to 55 percent for A2). This difference between the two scenarios is likely at least partially due to earlier onset of germinating rains in the B1 scenario, as discussed below in Season Length. In the early season, when total forage production is very low, even a small absolute change in production made by a few days or a week of extra production time can make a large difference proportionally.

Season Length

In this model, season length changes with the timing of germination and plant senescence; delayed onset of germinating rains and/or earlier depletion of soil moisture will result in shorter growing seasons. Because forage quality is greatest during the

Change in Growing Season Length
1961-1990 -- 2070-2099 (A2 scenario)

Change in Growing Season Length
1961-1990 -- 2070-2099 (B1 scenario)

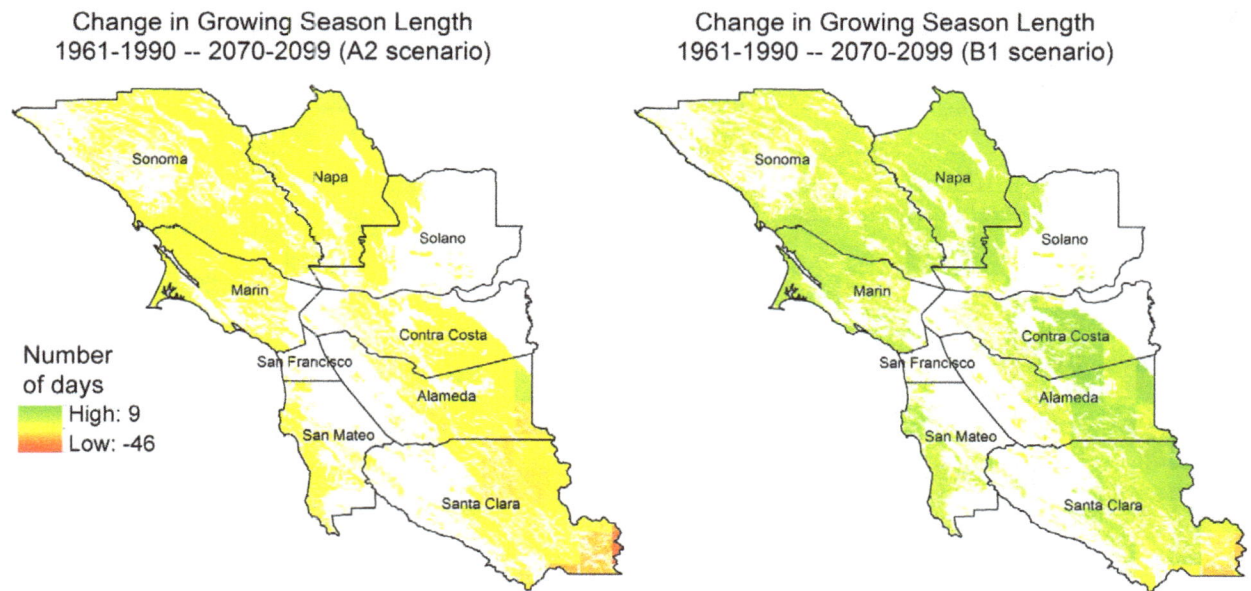

Figure 4. Change in rangeland season length by end-century (2070–2099), relative to historical conditions (1961–1990) for current rangelands in the Bay Area.

Table 3. Maximum and minimum forage production and season length, expressed as percent above and below the mean for the historic period (1961–1990), three projected 30-year periods and all periods.

Time Period	Marin-Sonoma		Napa-Solano		Alameda-Contra Costa		Santa Clara	
	Max	Min	Max	Min	Max	Min	Max	Min
	Production A2 Model							
All periods	45	32	50	34	45	28	49	35
1961–1990	23	24	30	25	34	19	33	27
2005–2034	19	24	25	24	26	26	23	17
2035–2064	24	18	30	19	36	23	41	22
2070–2099	28	24	29	26	31	29	34	23
	Production B1 Model							
All periods	35	31	39	33	53	38	46	37
1961–1990	23	24	30	25	33	19	33	27
2005–2034	32	22	35	22	30	34	35	37
2035–2064	23	34	27	31	28	30	26	34
2070–2099	25	22	27	23	40	24	34	22
	Season Length A2 Model							
All periods	24	23	24	24	26	38	26	32
1961–1990	13	18	17	20	21	17	21	25
2005–2034	15	13	16	11	22	15	20	15
2035–2064	15	18	17	15	25	19	29	17
2070–2099	22	17	24	17	26	32	29	26
	Season Length B1 Model							
All periods	22	26	21	25	28	46	24	39
1961–1990	13	18	17	20	20	17	21	25
2005–2034	18	16	20	23	21	36	23	39
2035–2064	15	25	14	21	24	22	21	22
2070–2099	16	15	15	14	28	30	24	29

Table 4. Number of outlier years (>2 standard errors above or below the mean) per time period.

	Marin-Sonoma	Napa-Solano	Alameda-Contra Costa	Santa Clara
Peak Forage Production				
Below historic mean	12	11	13	10
Above historic mean	11	10	10	10
Below mean for 2070–2099 (A2)	10	11	10	11
Above mean for 2070–2099 (A2)	7	8	10	9
Below mean for 2070–2099 (B1)	11	10	10	13
Above mean for 2070–2099 (B1)	8	8	7	8
Season Length				
Below historic mean	9	10	10	7
Above historic mean	12	12	10	11
Below mean for 2070–2099 (A2)	10	12	11	10
Above mean for 2070–2099 (A2)	10	11	10	9
Below mean for 2070–2099 (B1)	11	11	7	7
Above mean for 2070–2099 (B1)	11	10	8	8

growing season [20], periods of adequate forage quality for animal production will be shortened under future climate conditions, despite increases in forage production. The main differences between scenarios and across different regions in this study are due to the timing of germination. In the A2 scenario, the growing seasons in Santa Clara and Alameda are delayed by a week to 12 days compared to historical conditions, whereas the season in the northern Bay Area starts only slightly (2–3 days) later than historically. This intensifies the historical differences between germination dates of the North and South Bay. In contrast, earlier rains mean earlier germination (2–3 days on average, up to a week) throughout much of the Bay Area for the B1 scenario, which almost compensates for the earlier end in the B1 scenario, such that the impact of climate change on season length is more subtle than for the A2 scenario. This earlier start to the growing season also contributes to greater production in the late-century B1 scenario compared to the A2 scenario during the first few months of the season (Fig. 2).

These changes in season length and timing could have major implications for the range livestock industry in the San Francisco Bay Area. For example, delaying the start of the growing season and associated improvement in forage quality could impact the traditional fall calving season, and earlier onset of the dry season could require early weaning of calves in cow calf operations that dominate Bay Area livestock production. If the timing of these key events changes, then breeding and marketing dates may also shift by a few days or weeks. With later germination and earlier end to the growing season, managers will seek to place stock on summer pasture, including public lands, sooner and keep them there longer.

Inter-annual Variability and Drought

Inter-annual variation appears to decline with climate change throughout much of the Bay Area. This suggests that the uncertainty in stocking decisions that were cited as a major concern for climate change in the Great Plains [14] will be less of an issue here, and ranchers will likely not have to change their management response from what they do now to respond to low production years and good production years (see Conclusions). However, this low variability masks outlier years; the minimums of

38% below mean forage production do not include the years in which no forage is produced, due to drought. In fact, the occurrence of these "skipped" forage seasons can be considered an extreme case of inter-annual variability. While there exists a high degree of uncertainty over the Bay Area as a whole, extreme events in future precipitation forecast by the four different climate models are likely to increase in regions already most vulnerable to such events. Droughts so severe that forage is not produced during the growing season can be expected to increase in frequency in parts of the South Bay, which will have serious consequences for stocking decisions and the overall reliability of forage in that area.

Model Limitations and Other Climate Impacts

This simple forage production model was developed from data taken across rangelands that represent much of the Bay Area; UC HREC is ten miles north of interior Sonoma County (east of the Coast Range), and the other two research stations (San Joaquin Experimental Range in Madera County and UC SFREC in Yuba County) are inland sites that are more similar to the eastern portions of the Bay Area counties. However, the coastal regions may not be accurately represented. The curves projected for the historical period in the coastal counties of Marin and Sonoma are steeper than for the inland counties of Alameda and Santa Clara and than those from earlier empirical studies [8], especially during winter months. This difference may be explained by the generally warmer winter temperatures forecast for Marin and Sonoma County rangelands than for Alameda and Santa Clara County rangelands. Therefore, while more coastal data would improve the accuracy of model, most of the coastal effects on productivity are likely temperature effects that should be adequately modeled by accumulated degree days.

One major simplification of the processes involved with forage production in this model is that precipitation is included only to define the bounds of the growing season [16]. Precipitation effects within the growing season are not considered in the model, and it therefore does not respond to midwinter droughts, which may substantially reduce forage production in some years. At most annual rangeland locations in the Bay Area, moisture is seldom limiting during the growing season [7]. The degree day model used in this study was developed using production and weather

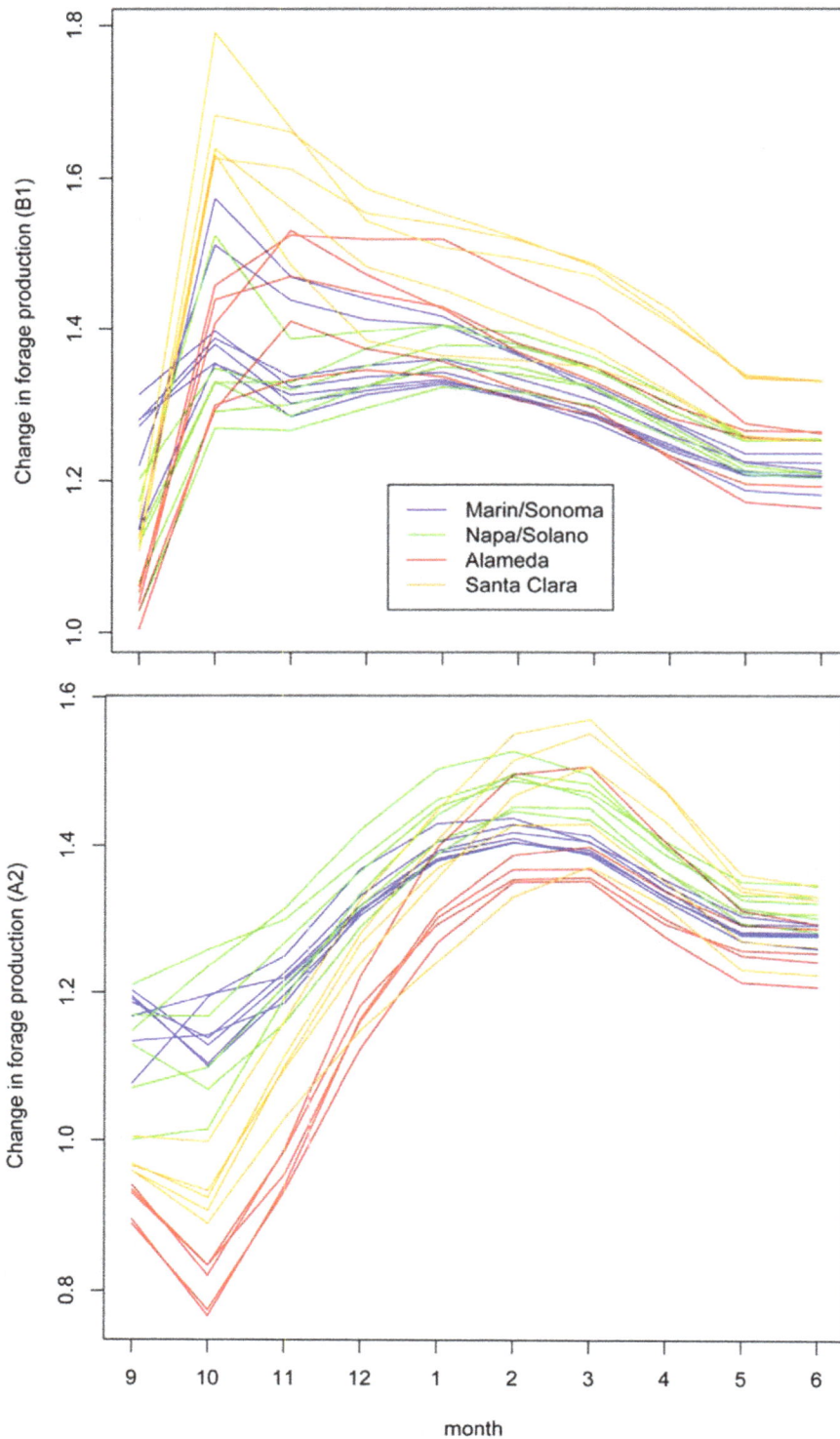

Figure 5. Change (%) in forage production from historical (1961–1990) to future (2070–2099) climate scenarios. Multiple lines of the same color represent different 12 x 12 km grid cells of rangeland in that area.

data from areas that vary quite dramatically in their precipitation regimes, with annual rainfall ranging from 13 to 53 inches [16] Modeled annual precipitation for Marin, Sonoma, Napa, and Solano Counties did not fall outside of this range in any year. Alameda, Contra Costa and Santa Clara Counties did show some

years that fell below 13 inches (330 mm), but these generally accounted for <15% of the total years. The number of years falling below that range did not increase from historical to future conditions in the B1 scenario, which means the model can be applied with confidence to all regions of the Bay Area for this scenario. The number of years precipitation in the A2 scenario fell outside that range increased in Alameda/Contra Costa Counties from 1961–1990 to 2070–2099. Results should therefore be interpreted more cautiously for these counties in the A2 scenario.

Finally, the model leaves out a number of important processes determining forage production that may be significantly altered under future conditions. Elevated atmospheric carbon-dioxide could have fertilization effects that increase the quantity of forage, while simultaneously reducing the quality by diluting the protein content [21]. Future climate could alter evapotranspiration rates, resulting in decreased soil moisture and increased water stress beyond the effects of precipitation [22], further reducing season length and potentially increasing drought frequency. Potential shifts in vegetation states resulting from projected changes in temperature and precipitation can also impact forage production at the landscape level, through shrubland expansion into existing grasslands and long-term conversion of oak woodlands to grasslands [23]. Finally, animal metabolic performance, grazing behavior and availability of stock water can all be expected to change with climate [24,25,26], and while not modeled here, their general decline with temperature will impact the overall viability of the livestock industry in the Bay Area.

Incorporating these additional factors into a model of forage production would result in a more sensitive and nuanced forecast of this important ecosystem service under climate change conditions. Our goal in this research was to apply to future climate scenarios a very simple, empirical model that despite its simplicity explains 75% of the variation in forage production under current conditions in the Bay Area. Future research should compare a simple model such as that presented here with more complex approaches, to determine how well main effects are characterized by the most basic processes. Simple models can be useful for supporting land use decisions in areas where data are limited and/or more advanced processes are poorly understood.

Conclusions

Climate change has the potential to impact the quantity and reliability of forage production, forage quality, thermal stress on livestock, water demands for both animal needs and growing forage, and large-scale rangeland vegetation patterns. This study projects increases in forage production within the growing season counterbalanced by shorter growing seasons. Increased production may result in increased carrying capacity on Bay Area rangelands. However, shorter growing seasons and increased potential for drought will increase risk. One of the primary tools for reducing drought risk is to maintain stocking rate below the carrying capacity of the land, which means the 10 to 25 percent increases in forage production forecast here may not result in substantial increases in stocking rate. Because drought is a regular occurrence

on Bay Area rangelands, especially the south eastern portion that lies in the rain shadow of the Coast Range, ranchers in these areas are already accustomed to coping with periodic drought. For the climate scenarios discussed above, grazing managers will need to strengthen their contingency planning for drought.

Overall, this model has demonstrated that shifting temperature and precipitation patterns must be considered together in order to understand the potential impacts of climate change on rangeland forage production. In a future with higher temperatures and a shorter rainy season, ranchers will need to consider management options for grazing shorter growing seasons and therefore longer dry seasons. Most of the standing biomass remaining during the dry season has senesced and is of poor nutritive quality; an extension of this period means a reduction in the availability of forage that can meet the nutrition requirements of beef cattle. These vulnerabilities to climate change are not as easily translated to economic impacts as total forage production, as each ranch has a unique set of forage sources and operational conditions. Some ranches have the flexibility to transport livestock to forage sources of higher quality during the Bay Area dry season (e.g., irrigated pastures, high elevation meadows, or wetter coastal regions), while others will graze the dry forage remaining in the Bay Area and therefore need to provide supplemental feeds including hay, protein and mineral supplements. Both of these options will increase production costs, reducing already thin profit margins, but how these additional costs will weigh against the projected gains in forage production is not well understood. However, the main message for the effects of climate change on Bay Area ranching is that it will present some opportunities as well as some challenges. The prospect of paying ranchers to graze in order to provide certain ecosystem services such as control of invasive species [27], fire hazard reduction [28], and pollination to nearby farms [29] may be an increasingly important tool to help offset the increased costs of grazing under climate change and to maintain the viability of ranching operations in the Bay Area – and the precious open spaces they support.

Acknowledgments

We thank James Bartolome, Valerie Eviner, Sheila Barry, and Sasha Gennet for their invaluable comments and constructive advice, Brian Galey and the Berkeley Geospatial Innovation Facility for technical assistance, and Mary Tyree and Dan Cayan for their help assembling the climate data.

Author Contributions

Conceived and designed the experiments: RCK. Performed the experiments: RCK. Analyzed the data: RCK. Contributed reagents/materials/analysis tools: RCK MRG. Wrote the paper: RCK MRG.

References

1. Huntsinger L, Hopkinson P (1996) Sustaining rangeland landscapes: a social and ecological process. J Range Manage 49: 167–173.
2. Forero L, Huntsinger L, Clawson WJ (1992) Land use change in three San Francisco Bay area counties: Implications for ranching at the urban fringe. J Soil and Water Conserv 47: 475–480.
3. Cayan D, Maurer EP, Dettinger MD, Tyree M, Hayhoe K, et al. (2006) Climate scenarios for California. Sacramento, CA: California Energy Commission. California Climate Change Center White Paper.
4. Cayan D, Tyree M, Pierce D, Das T (2012) Climate and Sea Level Change Scenarios for California Vulnerability and Adaptation Assessment. Sacramento, CA: California Energy Commission. Publication number: CEC-500-2012-008. Available: http://uc-ciee.org/climate-change/california-vulnerability-and-adaptation-study.
5. Shaw MR, Pendleton L, Cameron DR, Morris B, Bachelet D, et al. (2011) The impact of climate change on California's ecosystem services. Climatic Change 109: 465–484.

6. George MR, Williams WA, McDougald NK, Clawson WJ, Murphy AH (1989) Predicting Peak Standing Crop on Annual Range Using Weather Variables. J Range Manage 42:509–513.
7. George MR, Larsen RE, McDougald NM, Vaughn CE, Flavell DK, et al. (2010) Determining Drought on California's Mediterranean-Type Rangelands: The Noninsured Crop Disaster Assistance Program. Rangelands 32:16–20.
8. George M, Bartolome J, McDougald N, Connor M, Vaughn C, et al. (2001) Annual Range Forage Production. Oakland, CA: University of California. Division of Agriculture and Natural Resources Publ 8018. 9 p.
9. Zavaleta ES, Thomas BD, Chiariello NR, Asner GP, Shaw MR, et al. (2003) Plants reverse warming effect on ecosystem water balance. Proc Nat Acad of Sci 100:9892–9893.
10. Chou WW, Silver WL, Jackson RD, Thompson AW, Allen-Diaz B, et al. (2008) The sensitivity of annual grassland carbon cycling to the quantity and timing of rainfall. Global Change Biology 14: 1382–1394.

11. Murphy AH (1970) Predicted forage yield based on fall precipitation in California annual grasslands. J Range Manage 23: 363–365.

12. Suttle KB, Thomsen MA, Power ME (2007) Species interactions reverse grassland responses to changing climate. Science 315: 640–642.

13. Zavaleta ES, Shaw MR, Chiariello NR, Thomas BD, Cleland EE, et al. (2003) Grassland responses to three years of elevated temperature, CO_2, precipitation, and N deposition. Ecol Monog 73:585–604.

14. Hanson J, Baker B, Bourdon R (1993) Comparison of the effects of different climate change scenarios on rangeland livestock production. Agric Systems 41:487–502.

15. Howden SM, Crimp SJ, Stokes CJ (2008) Climate change and Australian livestock systems: impacts, research and policy issues. Austr J of Exper Agric 48:780.

16. George MR, Raguse CA, Clawson WJ, Wilson CB, Willoughby RL, et al. (1988) Correlation of degree-days with annual herbage yields and livestock gains. J Range Manage 41:193–197.

17. R Development Core Team (2011) R: A language and environment for statistical computing. R Foundation for Statistical Computing, Vienna, Austria. ISBN 3-900051-07-0. Available: http://www.R-project.org/.

18. Logan SH, Boyland PB (1983) Calculating heat units via a sine function. J Amer Soc Hort Sci 108:977–980.

19. USGS (U.S. Department of Interior, Geological Survey) (2006) The national map LANDFIRE national existing vegetation type layer. (Last updated September 2006.) Available: http://www.landfire.gov/NationalProductDescriptions21.php

20. George M, Nader G, McDougald N, Connor M, Frost B, (2001). Annual Rangeland Forage Quality. Oakland, CA: University of California. Division of Agriculture and Natural Resources Publ 8022. 13 pgs.

21. Milchunas DG, Mosier AR, Morgan JA, LeCain DR, King JY, et al. (2005) Elevated CO2 and defoliation effects on a shortgrass steppe: Forage quality versus quantity for ruminants. Agric Ecosystems & Environ 111:166–184.

22. Keshta N, Elshorbagy A, Carey S (2012) Impacts of climate change on soil moisture and evapotranspiration in reconstructed watersheds in northern Alberta, Canada. Hydrol Process 26: 1321–1331.

23. Cornwell WK, Stuart SA, Ramirez A, Dolanc CR, Thorne JH, et al. (2012) Climate Change Impacts on California Vegetation: Physiology, Life History, and Ecosystem Change. Sacramento, CA: California Energy Commission. Publication number: CEC-500-2012-023. Available: http://uc-ciee.org/climate-change/california-vulnerability-and-adaptation-study.

24. Hahn GL (1985) Management and housing of farm animals in hot environment. In: Yousef MK. editor. Stress Physiology of Livestock. Ungulates, Vol. 2. Boca Raton, FL: CRC Press. 151–176.

25. Harris NR (2001) Cattle behavior and distribution on the San Joaquin Experimental Range in the foothills of Central California (dissertation). Corvallis, OR: Oregon State University. 199 p.

26. HRC-GWRI (2011) Climate Change Implications for Managing Northern California Water Resources in the Latter 21st Century. Sacramento, CA: California Energy Commission, PIER Energy-Related Environmental Research. CEC-500-2010-051.

27. Huntsinger L, Bartolome JW, D'Antonio CM (2007). Grazing Management on California's Mediterranean Grasslands. In: Stromberg MR, Corbin JD, D'Antonio CM. editors. California Grasslands: Ecology and Management. Berkeley, CA: University of California Press. 233–253.

28. Nader G, Henkin Z, Smith E, Ingram R, Narvaez N (2007) Planned herbivory in the management of wildfire fuels. Rangelands 29:18–24.

29. Chaplin-Kramer R, Tuxen-Bettman K, Kremen C (2011) Value of wildland habitat for supplying pollination services to Californian agriculture. Rangelands 33: 33–41.

15

Designing Optimized Multi-Species Monitoring Networks to Detect Range Shifts Driven by Climate Change: A Case Study with Bats in the North of Portugal

Francisco Amorim[1]*, Sílvia B. Carvalho[1], João Honrado[1,2], Hugo Rebelo[1,3]

1 CIBIO/InBIO, Research Center in Biodiversity and Genetic Resources, University of Porto, Vairão, Portugal, **2** Department of Biology, Faculty of Sciences of the University of Porto, Porto, Portugal, **3** School of Biological Sciences, University of Bristol, Bristol, United Kingdom

Abstract

Here we develop a framework to design multi-species monitoring networks using species distribution models and conservation planning tools to optimize the location of monitoring stations to detect potential range shifts driven by climate change. For this study, we focused on seven bat species in Northern Portugal (Western Europe). Maximum entropy modelling was used to predict the likely occurrence of those species under present and future climatic conditions. By comparing present and future predicted distributions, we identified areas where each species is likely to gain, lose or maintain suitable climatic space. We then used a decision support tool (the Marxan software) to design three optimized monitoring networks considering: a) changes in species likely occurrence, b) species conservation status, and c) level of volunteer commitment. For present climatic conditions, species distribution models revealed that areas suitable for most species occur in the north-eastern part of the region. However, areas predicted to become climatically suitable in the future shifted towards west. The three simulated monitoring networks, adaptable for an unpredictable volunteer commitment, included 28, 54 and 110 sampling locations respectively, distributed across the study area and covering the potential full range of conditions where species range shifts may occur. Our results show that our framework outperforms the traditional approach that only considers current species ranges, in allocating monitoring stations distributed across different categories of predicted shifts in species distributions. This study presents a straightforward framework to design monitoring schemes aimed specifically at testing hypotheses about where and when species ranges may shift with climatic changes, while also ensuring surveillance of general population trends.

Editor: Brock Fenton, University of Western Ontario, Canada

Funding: This research was funded by Fundação para a Ciência e Tecnologia through FEDER/COMPETE, under project "EcoSensing" (PTDC/AGR-AAM/104819/2008), and by QREN/ON2 in the context of project "SIMBioN-Biodiversity Information and Monitoring System for Northern Portugal" (ON2: 3-2-14-1-1192). Hugo Rebelo and Silvia B. Carvalho were funded by Funda ão para a Ciência e Tecnologia (grants SFRH/BPD/65418/2009 and SFRH/BPD/74423/2010 respectively). The funders had no role in study design, data collection and analysis, decision to publish, or preparation of the manuscript.

Competing Interests: The authors have declared that no competing interests exist.

* E-mail: famorim@cibio.up.pt

Introduction

Ecosystems and global biodiversity are facing a decline as a direct and indirect consequence of human actions [1,2], and we are yet to experience the full impacts of anthropogenic climate change [3–5]. Effective conservation depends on our ability to define, measure, and monitor biodiversity change. Biodiversity monitoring programs usually aim at determining population trends and changes in the structure of biotic communities, often in response to environmental change, anthropogenic disturbance, or targeted management actions [6,7]. The importance of monitoring biodiversity is becoming increasingly recognized, nevertheless many ongoing monitoring schemes have been criticized for not being underpinned by clear objectives, designed to test specific scientific hypotheses or to evaluate the success of conservation actions (e.g. [7,8]), and also for not addressing the relation between cost and benefit [9]. One of the problems often identified in monitoring arises from data being collected in an *ad hoc* and fragmented way that lacks statistical and/or methodological consistency and therefore does not allow an effective

evaluation of relevant conservation questions [10]. Untargeted monitoring can result in years of wasted effort and money [11] and may even fail in acquiring the critical information to improve management options, one of the major purposes of monitoring [9].

In recent years scientists and practitioners have drawn their attention to the importance of improving methods to design monitoring schemes [7,8,12–14], and it is widely recognized that monitoring programs need to address well-defined and testable questions, understand how the focal ecosystem might work or how the monitoring targets might function, and be of management relevance (e.g. effects of a pollutant or changes in climate on features of a given ecosystem). However, effective multi-species monitoring is still lacking and, despite its value for describing conditions and detecting undesirable changes, traditionally it is not designed to determine the causes of such changes nor to track specific expected changes [15].

Criticisms on current monitoring programs are particularly evident in the case of monitoring biodiversity responses to climate change. Species may respond to human induced climate change in a variety of ways [16]. Changes in phenology, such as timing of flowering or breeding which may also lead to mismatches between the successive trophic levels [17], have been linked to climate change [18,19]. An impressive number of studies have also focused on the impact of climate change on species range shifts, from polar latitudes to tropical regions and even marine ecosystems (for an extensive review see [16]).

Species range shifts are a major challenge to conservation planning because the spatial patterns of biodiversity are expected to change, with species of high conservation interest moving from current protected areas (e.g. [20,21]), while areas that are without legal protection may become more relevant for species' conservation in the future [22,23]. Unfortunately, monitoring programs normally disregard where (and when) species ranges are predicted to shift, and thus they are not necessarily designed for detecting shifts due to climate change, although the importance of doing so has been recently acknowledged [24]. Because many species are predicted to be affected by climate change, these programs should also be optimized to monitor the highest number of species for the least cost, something which is rarely considered in the design of the monitoring schemes (but see [10,15,25]).

Due to their high mobility, bats are able to respond rapidly to environmental changes [26] rendering them as a good model to detect changes in species distributions due to climate change. In fact changes in the distribution and abundance of bats in response to climate change are already emerging. In Costa Rican cloud forest, bat species have shown an altitudinal shift from lowland areas to higher altitudes in response to climate change [27]. In Europe, *Pipistrellus kuhlii* has expanded its range northwards since the 1990s, in response to the increasing temperatures [28]; and *Pipistrellus nathusii* has also expanded its range towards higher latitudes in the U.K. [29].

Apart from their rapid response to environmental changes, there is high potential for developing bat monitoring programs, as shown by the high number of monitoring programs running worldwide [30,31]. Currently, the monitoring of cave-dwelling bat species is implemented in several countries (e.g. [32]). However, because surveying tree and crevice-dwelling species can be challenging, time consuming and expensive, monitoring of this group is seldom established. For these species, acoustic sampling is the most widespread method whenever a monitoring program is developed (e.g. [33–37]), although some limitations exist e.g. the level of activity is not necessarily proportional to abundance. Moreover factors such as differences in detectability and even temporal variations may lead to differences in species activity [38]. We are fully aware of the inherent limitations of this method, therefore we focused our monitoring networks on species with a higher detection probability.

A good example of acoustic monitoring programs is The Indicator Bats Program (http://www.ibats.org.uk/), with projects running in the UK, Eastern Europe, Ukraine, Russia and Japan. Conversely, other ongoing acoustic monitoring programs have methodological problems, mainly related to sampling design. For example, the Irish Bat Monitoring Program includes a car-based scheme that reveals information on bat populations and distributions [39]. While recognizing the importance of such information, one must be aware that road-based surveys have the potential to provide biased results, since their placement is non-random [15]. Bat monitoring programs have detected population fluctuations (see for example, the 2011 report of the UK National Bat Monitoring Program; available at http://www.bats.org.uk/), but even the best examples are subject to the most frequent problem of monitoring programs (see above): they are not designed to test which environmental changes (e.g. climate change) are leading to such fluctuations.

Predictive models, and particularly species distribution models, allow extrapolating species distribution data in space and time, based on a statistical model [40]. Such extrapolation is possible by combining observations of species occurrences with environmental variables known to influence habitat suitability and therefore species distribution. Combined with stratifications and scenarios for the relevant environmental factors, species distribution models thus have the potential to improve the spatial design and cost-efficiency of ecological monitoring networks (e.g. [41]).

The main goal of this study is to develop and test a framework to design optimized multi-species monitoring networks, able to test hypotheses about how species ranges will shift with climate change. To that end, we developed a case study based on seven bat species in the North of Portugal. We first increased the existing data about bat distribution in the study area with field work targeted at the main gap areas. Next, we used these data to predict likely suitable areas for each species under current and future climatic conditions, and identified, for each species, which areas are more likely to lose, gain or maintain climatic suitability in the future. Then, we used computational tools to optimize the allocation of monitoring stations in space. Finally, we tested whether monitoring networks designed when accounting for predicted species shifts have increased performance relatively to the ones where only the present distributions are accounted.

This study was developed within the scope of SIMBioN – Biodiversity Information and Monitoring System for the North of Portugal, a joint venture between governmental nature conservation agencies and research centers. SIMBioN was designed with the general purpose of regularly providing information on the status of regional biodiversity to support management actions, technical and political decision-making, regarding biodiversity management and conservation. The developed monitoring network should be especially sensitive to biodiversity changes occurring in native woodland areas, for that reason our research focused on the development of a monitoring program for bat species somehow associated with this native habitat type. Because the implementation and running of this network is expected to be fully executed by volunteers, several alternative networks were designed to offer different scenarios considering an unpredictable volunteer commitment. Thus, the main aim of this study is to develop the aforementioned monitoring networks always considering the logistic limitations of volunteer surveyors.

Materials and Methods

Ethics Statement

Data on species location used in this research are part of the database from Instituto da Conservação da Natureza e Florestas (ICNF) (the national authority for nature conservation and wildlife protection), and was collected following all legal requirements. Additional data collection within the scope of this research was accomplished using non-invasive methods that do not require legal permits. Sampling in Natural Protected Areas was done with the authorization of ICNF, and samples within private land were performed with the authorization of the land owners.

Overview of the analytical framework

Because our framework has many different steps we present here a brief overview of methods, which can also be found in Fig. 1. In a nutshell, using Species Distribution Models we modelled the current distributions of seven bat species and then projected the results to future conditions according to two different climatic scenarios. The difference between current and future predicted distribution allowed us to identify areas where loss, gain

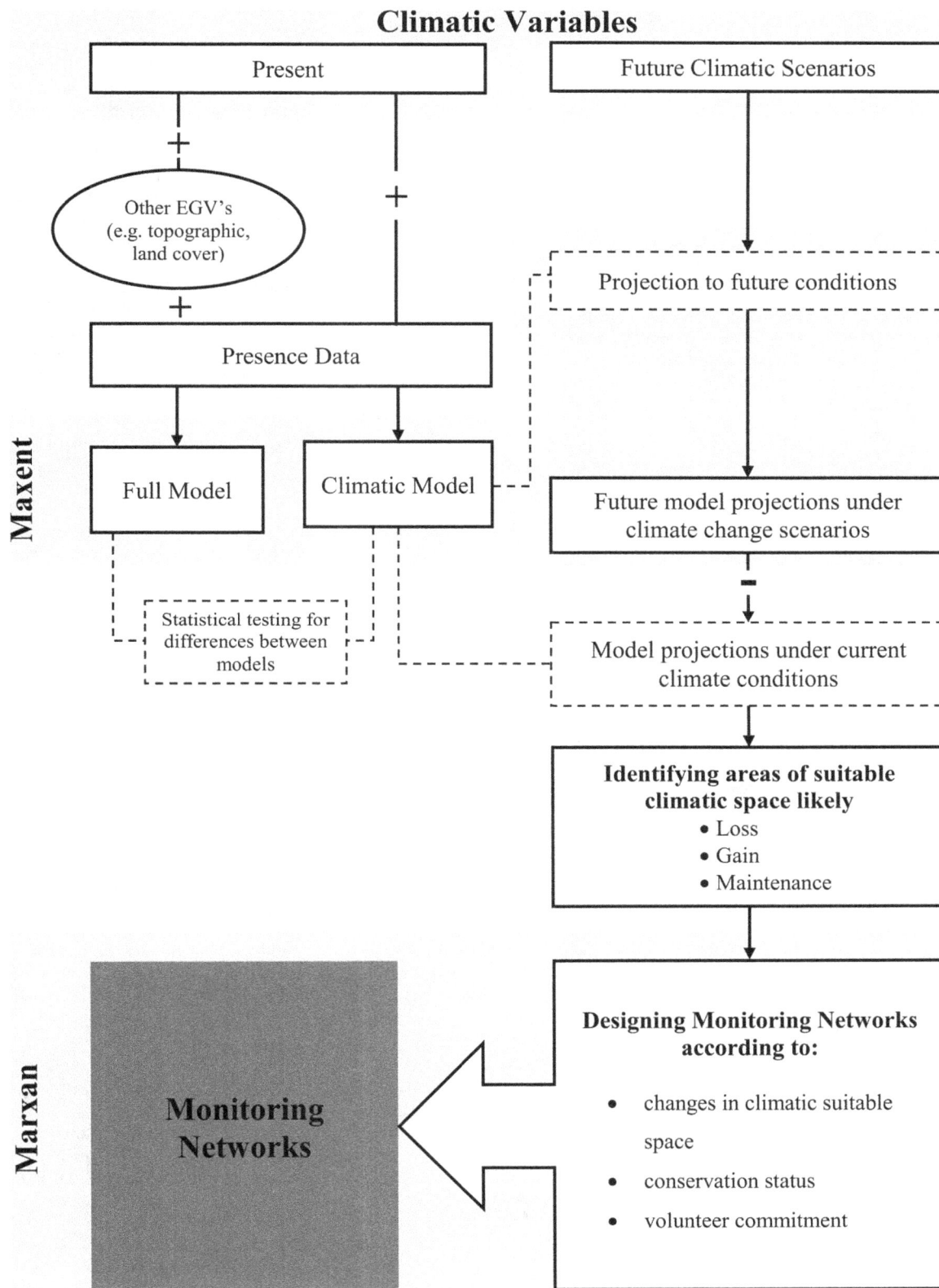

Figure 1. Framework for designing monitoring networks sensitive to climate changes. Proposed framework scheme for designing adaptive monitoring networks sensitive to climate changes. Full lines indicate data and outcomes; dash lines indicate intermediate steps. Data, variables and model addition (+) and subtraction (–) is identified.

or maintenance of suitable climatic space is more likely to occur. These results were then used in combination with conservation planning tools to optimize the location of stations for multi-species monitoring, considering areas where the three different results (loss, gain, maintenance) are more likely to occur.

Study area

The study area is located in northern Portugal (Western Europe), approximately between coordinates 40°N–42°N and 6°W–9°W (Fig. 2A). In the northwest and in high elevation areas of the northeast, the Atlantic temperate climate dominates, with mild summers and cold, rainy winters. The landscape is mountainous with native forests mainly composed of deciduous oaks (*Quercus robur*, *Q. pyrenaica*), chestnut (*Castanea sativa*), birch (*Betula celtiberica*), and ash (*Fraxinus angustifolia*). Conversely, in the northeast valleys and lowlands the climate is typically Mediterranean sub-continental, with perennial oaks (*Quercus suber*, *Q. ilex*) dominating the native woodlands.

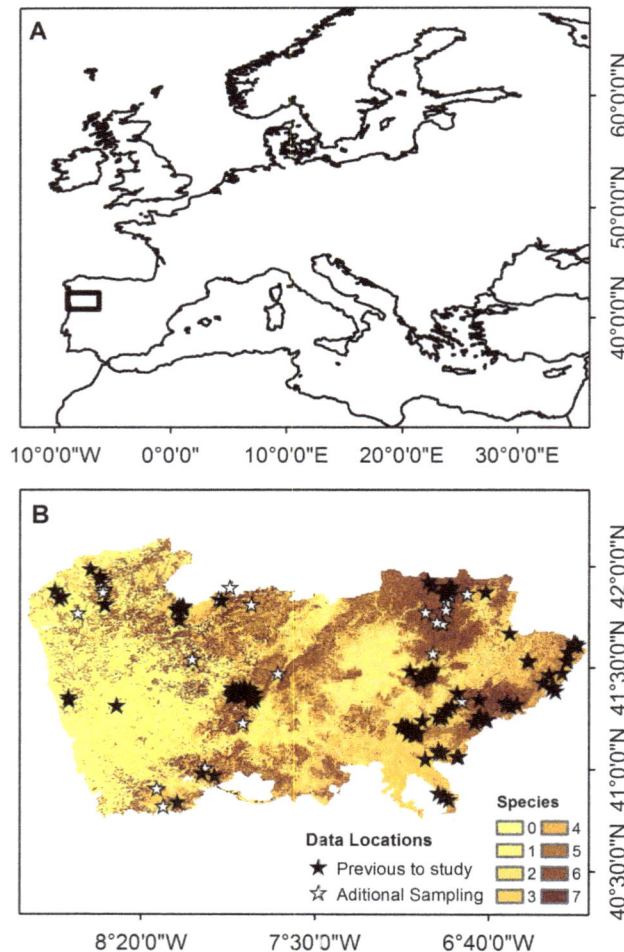

Figure 2. Study area and data locations. Location of the Study area (A) and of the available data before this study (source: Instituto da Conservação da Natureza e Florestas) and additional sampling determined by species richness as predicted by preliminary SMD for the present (B) (see methods).

Modelling Procedure

For the calculations of species distribution models we chose a presence-only technique, based on the principle of maximum entropy (implemented in Maxent, [42]). The choice of using Maxent over other modelling techniques was based on its very good predictive ability when compared with other methods [43–45]. Also, the use of presence-only data is an advantage when reliable absence data are not available or are difficult to assess. Such is the case of bats because of their elusive and nocturnal behaviour [44]. Maxent estimates the range of a species with the constraint that the expected value of each ecogeographical variable (EGV; see below) (or its transform and/or interactions) should match its empirical average (i.e., the average value for a set of sample points taken from the target species distribution [46,47]). Models were run with 80% of the presence data while the remaining 20% were used for model testing. Because Maxent randomly chooses which presence data to include in the training or test models, we ran 100 model replications and averaged them into a single model. Model calculations were done in the *autofeatures* mode with a maximum of 1000 iterations and the regularization multiplier set to 0.2. To check which variables were the most important to build the model, a Jacknife analysis of the gain was made with the presence data. Jacknife analysis measures how well an EGV distinguishes localities where the species occurs from the total area under study. All calculations were made in Maxent v.3.3.3k.

To forecast the effect of climate change, models were computed with climatic variables only. This may cause overestimations of species occurrence because the distribution of bats may be particularly constrained by land cover. Water habitats such as rivers and ponds can support high levels of bat activity [35,37] while native woodlands (i.e. oak forests) can support a high bat diversity [36,37]. To test the effect of this overestimation, we ran two models for the present conditions, one using all the EGVs including land cover, slope and altitude (hereafter "full model") and one using only the climatic variables (hereafter "climatic model"). To compare these models' predictions, we used ENMTools v1.3 [48] to measure niche breadth [49] and niche similarity to determine niche overlap [50]. We also checked the percentage of cells in which predictions from both models were in accordance by comparing the cells where presence or absence was predicted. The climatic models calibrated with present climatic variables were projected for future (2080) climatic conditions.

To determine the spatial patterns of current suitable areas for each species, as well as the likely gain and loss of suitable climatic area, all model projections were reclassified into binary presence/absence maps.

Presence Data and Environmental Variables

The studied bat species were selected according to the availability of occurrence data for the study area and their level of association to native woodland. To achieve reliable models we only considered species with more than 15 records for the study area [51]. Therefore, for model calculations we used as response variables the known locations within the study area for seven bat species: *Myotis daubentonii*, *Pipistrellus kuhlii*, *Hypsugo savii*, *Eptesicus serotinus/isabellinus*, *Nyctalus leisleri*, *Barbastella barbastellus* and *Tadarida teniotis* (source: Instituto da Conservação da Natureza e Florestas; Fig. 2B). Data were collected by several surveyors using different methods (i.e. mist-netting, acoustic sampling and direct observation). Species locations are available in Fig. S1.

In order to increase knowledge about bat distribution in the study area, additional data were collected through acoustic transects. These new surveys were performed in areas of high species richness as predicted by preliminary species distribution models. With this approach some omission errors (false absences) may occur but it allows finding bat species in areas where they were previously unknown while also increasing the geographical coverage of the presence data to be used in subsequent models. Details on the acoustic sampling methods can be found below. A set of independent EGVs, selected as environmental predictors, was considered for model calibration: annual mean temperature (°C), mean diurnal range (°C), mean temperature of warmest quarter (°C), mean temperature of coldest quarter (°C), annual precipitation (mm) (WORLDCLIM; http://www.worldclim.org), altitude, slope (source SRTM; http://www2.jpl.nasa.gov/srtm/), and land cover (Global Land Cover 2005–2006; http://postel.mediasfrance.org/ and Instituto Geográfico Português). Habitat composition and structure is known to influence bat activity, therefore land cover was reclassified into seven ecologically meaningful classes for bats [35–37,52]: urban, agriculture, production forests (mainly pines and eucalypts), scrub and regenerating forest, native woodland, water bodies, and bare ground [44].

Climatic variables (which include all the mentioned EGVs with the exception of altitude, slope and land cover) were chosen according to their reported relevance for bat physiology and survival [53–55]. After preliminary tests for the selection of the most informative variables (i.e. the ones achieving higher percentage contribution and gain in model calculations), some of the variables were excluded (max temperature of warmest month, min temperature of coldest month, and precipitation of driest month).

To forecast the effect of climate change on predicted distributions, two contrasting IPCC scenarios (A2a and B2a; http://www.worldclim.org) based on the Global Circulation Model HadCM3 were used. Scenario A2a is driven by economic growth at a regional scale, while B2a considers a regional steady growth and social awareness of environmental sustainability [56]. We used monthly averages of maximum and minimum temperatures and total precipitation, for the period of 2070–2099 (hereafter 2080), and then we calculated the bioclimatic variables according to Hijmans et al. (2005) [57] using the DivaGIS software version 7.5 (www.diva-gis.org).

Altitude, land cover and slope had a spatial resolution of approximately 280x280m. Since climatic variables had a resolution of approximately 1x1km, we downscaled these data to match the cell size of the previous EGVs following the methodological approach of Waltari et al. [58]. The study area thus included 494 700 cells, for a total extent of 21 940 km^2.

Additional sampling for presence data

In order to increase the quality of data on bat distribution within the study area and add more occurrence data for the target species to our models, additional acoustic transects were carried out between March and August 2010. Transects started one hour after sunset and lasted for three hours [37]. Each transect was walked at low speed (ca. 2 km/h) during 30 minutes using a bat detector (D240X, Pettersson Elektronik AB, Uppsala, Sweden) connected to a digital recorder (Zoom H2, Samson Technologies Inc. USA, New York). Files were saved in WAV format; sampling rate 44.1 kHz and 16 bits/sample. Bat vocalizations were analysed using sound-analysis software (BatSound Pro 3.31, Pettersson Elektronik AB, Uppsala, Sweden) with a 1024 pt FFT and Hamming window for spectrogram analysis [36,37]. Acoustic identification of bat calls was made through comparison with literature on the theme [59–61].

Monitoring networks

The main goal of our monitoring networks is to support the collection of data on presence or absence of the selected bat species throughout the 21st century. By comparing species distributions from different time intervals (i.e., between current and future sampling) it will be possible to test if observed range shifts occur due to gain, loss or maintenance of climatic suitability. Thus, to be able to test such hypothesis, monitoring stations have to be allocated under a stratified design across these different classes of predicted climatic suitability change.

To calculate the class of suitability change (referred to here as areas where loss, gain or maintenance of species climatic suitability may be observed) in each grid cell, current and future model predictions were reclassified into binary presence/absence maps. For that purpose cells with values above the 10th percentile of training presence were considered suitable for the species [44,62,63]. The 10th percentile presence value assumes that 10% of presence data may suffer from errors or lack of spatial resolution [63]. This is especially relevant when dealing with datasets gathered by several researchers (or volunteer surveyors) over large time-spans where reliability and precision has probably varied. Subsequently, each grid cell was classified into one of the three climatic suitability change classes in the following way: cells where climate is suitable in the present (1) but not in the future (0) were classified as "likely lose"; cells where climate is suitable in the present (1) and also in the future (1), were classified as "likely maintain"; cells where climate is currently unsuitable (0) but predicted to become suitable in the future (1) were classified as "likely gain". This classification was done separately for the two different climatic scenarios (A2a and B2a). Because we did not include EGVs such as land cover, slope and altitude to predict future suitability, and acknowledging their major importance for bats (particularly land cover [36,37,52]), we ensured that at least one quarter of the monitoring stations fell within areas predicted as currently suitable in the full model. By doing so we enforced that some sampling locations of the Monitoring Networks were set for areas where the values of those EGVs are currently suitable for species occurrence.

The next step consisted of determining the number of monitoring stations to be included at each suitability class for each species. Because the viability of the monitoring program is dependent on an unpredictable volunteer commitment, multiple designs were developed covering different citizen engagement scenarios, with increased number of monitoring stations for each species (MN1, MN2 and MN3). To make it possible to gradually expand the monitoring network without losing any information from previous campaigns, stations from MN1 were included in MN2 and stations from the latter were included in MN3. The number of sampling stations was set based on the level of expected commitment depicted from the levels of participation in bat detector workshops and environmental actions in the study area. Additionally, because different species have different conservation concerns, the monitoring effort allocated to each species was also set as a function of the species conservation status at the National level (for species with higher conservation status, a higher number of monitoring stations was set) [64]. Table 1 shows the minimum number of monitoring stations targeted in each network (MN1, MN2 and MN3) for each species, likely occurrence class, and full model.

We used the software Marxan [65] to identify optimized sets of monitoring stations to track range shifts in multiple bat species. Marxan is a decision-support tool which uses a simulating annealing algorithm to minimize the amount of selected sampling units whilst ensuring the representation of a set of features (species,

Table 1. Minimum number of locations targeted in each monitoring network (MN1, MN2 and MN3) for each species suitability class (G – Likely Gain; M – Likely Maintain; L – Likely Loss) and for areas predicted as currently suitable in the full model (Full).

	Conservation	MN1				MN2				MN3			
	Statuts	G	M	L	Full	G	M	L	Full	G	M	L	Full
Mdau	LC	3	3	3	3	6	6	6	6	12	12	12	12
Pkuh	LC	3	3	3	3	6	6	6	6	12	12	12	12
Hsav	DD	5	5	5	5	10	10	10	10	20	20	20	20
Nlei	DD	5	5	5	5	10	10	10	10	20	20	20	20
Eser/isa	LC	3	3	3	3	6	6	6	6	12	12	12	12
Bbar	DD	5	5	5	5	10	10	10	10	20	20	20	20
Tten	DD	5	5	5	5	10	10	10	10	20	20	20	20

Species code as follow: Myotis daubentonii (Mdau); Pipistrellus kuhlii (Pkuh); Hypsugo savii (Hsav); Nyctalus leisleri (Nlei); Eptesicus serotinus/isabellinus (Eser/isa); Barbastella barbastellus (Bbar) and Tadarida teniotis (Tten).

habitats, or other features) with a given minimum number of occurrences (occurrence target) [66]. Marxan was conceived to assist decisions about the location and design of protected area systems, but the mathematical problem underlying the optimization of monitoring network is very similar. In our case, the conservation features that we want to represent are the three classes of suitability shift for each species plus the suitable areas predicted under the full model. Reserve selection problems can also incorporate aggregation and connectivity rules which are not desirable when designing monitoring networks, because the further apart monitoring stations are, the more independent the monitoring data will be. Marxan was configured with the following parameters: algorithm – simulated annealing; number of runs – 100; penalty cost for not achieving the occurrence target – 100; iterations per simulation – 1,000,000; temperature decreases per simulation – 10,000; initial temperature and cooling factor – adaptive. For MN2, the status of the grid cells selected in the best solution of MN1 was set to 2 in order to force MN2 solution to include monitoring stations selected in MN1. This would simulate an expansion of the monitoring network as citizen engagement increases. We followed the same procedure for MN3 with the grid cells selected in the best solution of MN2. No boundary length modifier was used. Targets were set as defined in Table 1.

Performance of optimized vs. non-optimized networks

Three additional networks were designed to test whether our proposed framework has potentially increased performance in detecting species range shifts derived by climate change than the commonly used approach which only considers the current distribution of the species. For this purpose, we have rerun Marxan setting targets only for current predicted distribution of each of the seven bat species for climatic and full model. Target values were set according to the number of sample stations in each of our three monitoring networks that fall within current predicted suitability considering climatic and full model.

We used Cost Threshold function in Marxan to limit the overall number of sampling stations so that it would be equal to the number of stations selected in each of our three monitoring networks, respectively 28, 54 and 110. Differences between the performance of our framework and the testing networks were accessed by checking the proportion of the targets set in Table 1 that were not met in each of the 100 runs of the testing networks.

Results

Additional sampling for presence data

A preliminary set of species distribution models calibrated using the presence data for the targeted species available prior to this study allowed us to set 35 additional sampling sites in areas predicted to have high species richness (Fig. 2B). As a result of in-field campaigns, 418 new bat passes were recorded, adding 21 new locations for the targeted species. Species locations and presence/absence maps predicted by preliminary models are available in Fig. S1.

Current bat diversity patterns and ecological predictors

Regarding the predictive ability of the full model, test data showed only slightly lower values for AUC (area under the receiver operating characteristic curve, which ranks all locations according to their suitability [42,67]) than training data (AUC for test data ranged 0.78–0.86, AUC for training data ranged 0.86–0.93), which is an indication that no over-fitting occurred in the models. The high values of test AUC also indicate a good transferability power of the model. Likewise, climatic models showed good predictive power, with AUC ranging from 0.82–0.91 in the training data and 0.76–0.83 in the test data. For more details on the AUC values for each species see Table S1.

In the full model, land cover, altitude, annual mean temperature and temperature of coldest quarter were the most relevant EGVs for the majority of species (Fig. S2). When model fitting was based on climatic variables only, temperature of coldest quarter, temperature range, annual precipitation and annual mean temperature were the most relevant for most species (Fig. S3). A more comprehensive list of relevant predictor variables for each model and species (Table S2), as well as the corresponding response curves (Fig. S4-Fig. S10), can be found in Supporting Information.

Predicted species richness showed similar spatial patterns in the full models and in the climatic models (Fig. 3). In fact, binary predictions from both models spatially overlapped in more than 70% of the total cells for all species. Climatic models alone predicted suitability in at least 12% more cells, whereas the full model predicted unique suitable sites in less than 8% of the area (Table S3). Results from the niche overlap and D statistics ranged 0.95–0.99 and 0.78–0.88 respectively, thus confirming similar predictions between full and climatic models (Table S3). Regarding niche breadth metrics, results showed that the full model yielded a narrower niche (Table S3).

Predicted species richness showed some level of spatial structure (Fig. 3), with the majority of the highest values of species richness located in the northeast of the study area where high species richness has a more continuous distribution.

Future projections

Overall, bats in the study area have a high sensitivity to climatic changes, resulting in large extents of potential loss of climatic suitability (Fig. 4). When comparing current predicted distribution of species for both climatic scenarios, the area which could lose climatic suitability in the future for at least one species represents more than 60% of the study area (A2a: 62.5%; B2a: 64.2%) and in more than 30% likely loss was predicted for three or more species (A2a: 34.6%; B2a: 37.7%). In contrast, climatic suitability gain for at least one species is only predicted in 10.8% of the area in scenario A2a and 15.4% in scenario B2, while for scenarios A2a and B2a respectively 26.8% and 20.4% may maintain climatic suitability for the target species.

Areas where gain may occur were limited to a narrow fringe along the coastline in the western part of the study area and small isolated strongholds most of which located in the northwest of the study area (Fig. 4). Individual species maps may be found in Fig. S11 and Fig S12.

Monitoring networks

The targets set for MN1 were achieved with a total of 28 sites, while for MN2 and MN3 targets were accomplished with 54 and 110 sites, respectively (Fig. 5). In MN2 and MN3 it is possible to observe a concentration of sites in some areas. This happens mainly because there are very few areas where likely suitability gain was predicted in the future scenarios. The general distribution pattern of sites is similar for the three Monitoring Networks, nevertheless, as expected, the higher number of locations in MN2 and MN3 results in an increase of the spatial coverage of the resulting monitoring networks.

Performance of optimized vs. non-optimized networks

None of the 100 Marxan runs for each of the non-optimized networks was able to meet all the targets set in Table 1. Between 30% and 50% of the total targets were not met in either of the networks (Fig. 6A). Failure in achieving the targets was more pronounced in the climatic suitability classes of "likely gain" and "likely maintain" (Fig. 6A). Considering the achievement of targets set for individual species, we observed a large range of results, nonetheless "likely gain" and "likely maintain" were also the most problematic classes, while targets set for "likely loss" of climatic suitability were met by five out of the seven species (Fig. 6B).

Discussion

Designing and implementing optimal monitoring schemes under climate change

We presented an innovative approach representing an improvement in the design of monitoring networks that goes beyond the conventional surveillance schemes in the sense that it allows testing hypotheses about how environmental change (in our case, climate change) will drive species distributions. At the same time it produces fundamental data for the surveillance of species and communities at regional scales, which is one of the fundamental goals of biodiversity monitoring [68]. We have shown that, when compared with conventional network designs, the proposed framework has increased performance in allocating monitoring stations distributed across different categories of predicted shifts in species distributions, which is crucial to test hypotheses about the effects of climate change on species ranges. In conventional, non-optimized networks, because only the current distribution of the species was considered, we expected that selected monitoring stations were allocated more or less randomly across the species ranges, only depending on the level of species co-occurrence. Thus, we expected that for species likely to lose a great proportion of their current distribution due to climate change, it would be more difficult to allocate monitoring stations to areas that fall under the "likely maintain" class and vice versa. Our results confirmed this expectation, as most species likely to lose a great proportion of their climatic suitability (Fig. S13) were the ones with lower target achievement in the "likely maintain" class (e.g. Hsav, Nlei and Mdau in Figure 6B). We also expected that the "likely gain" class was the one less represented in the conventional networks, because these areas fall outside of the current distribution of the species. Our results also confirmed this expectation (Fig. 6), although some stations were indeed selected in areas of likely gain of climatic suitability for some species. This fact can be explained by the co-occurrence of two or more species. For instance, a monitoring station may be allocated to represent the current distribution of species A and B, and this location may

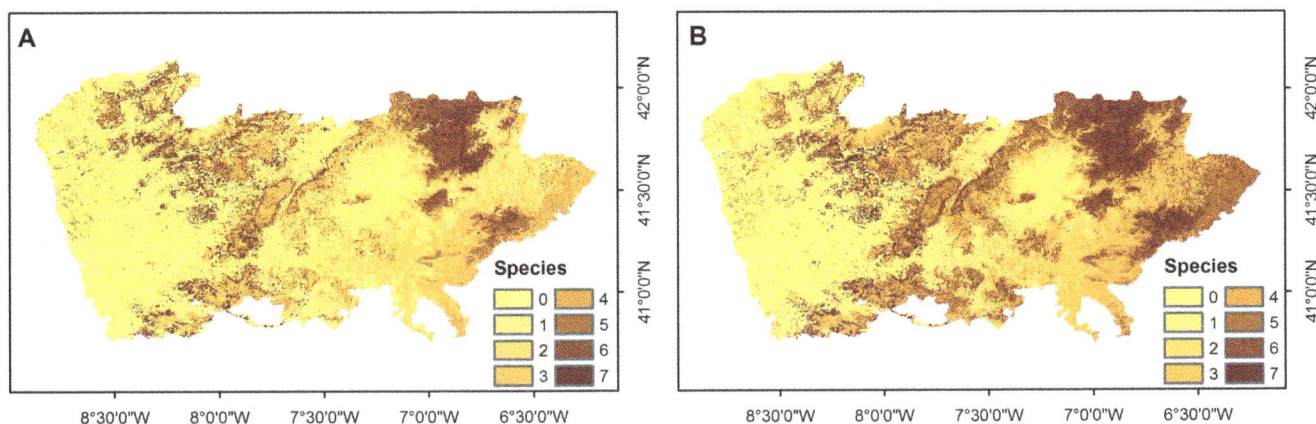

Figure 3. Current predicted species richness for full model (A) and climatic model (B). Current predicted species richness for full model (A) and climatic model (B).

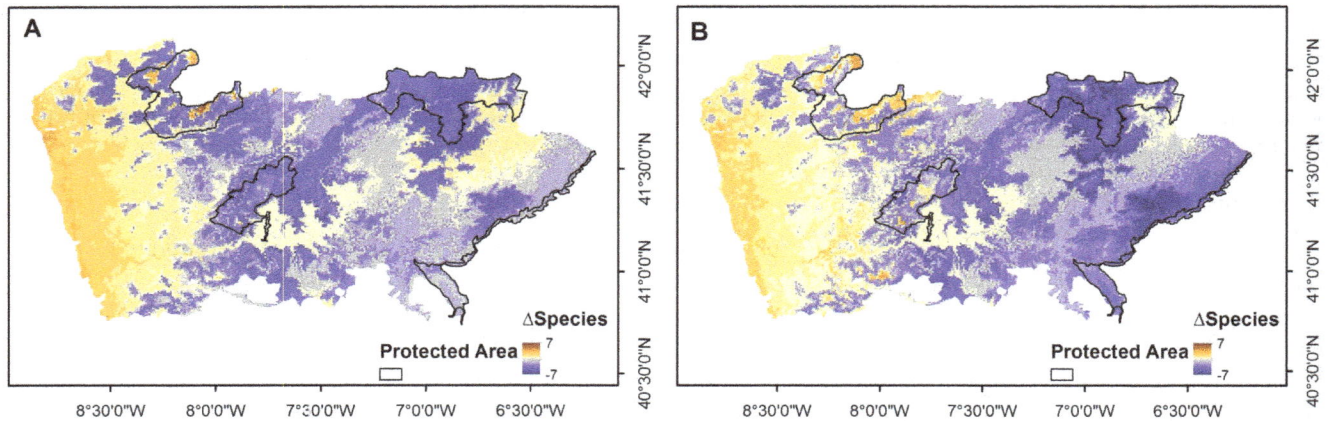

Figure 4. Predicted variation in climatic space between the present and the future. Predicted variation in climatic space for target species between the present and the future under climatic scenarios A2a (A) and B2a (B) for the year 2080. Negative △Species values indicate loss of climatic suitable space while positive indicate species gain of climatic suitable space. Most important protected areas are also represented.

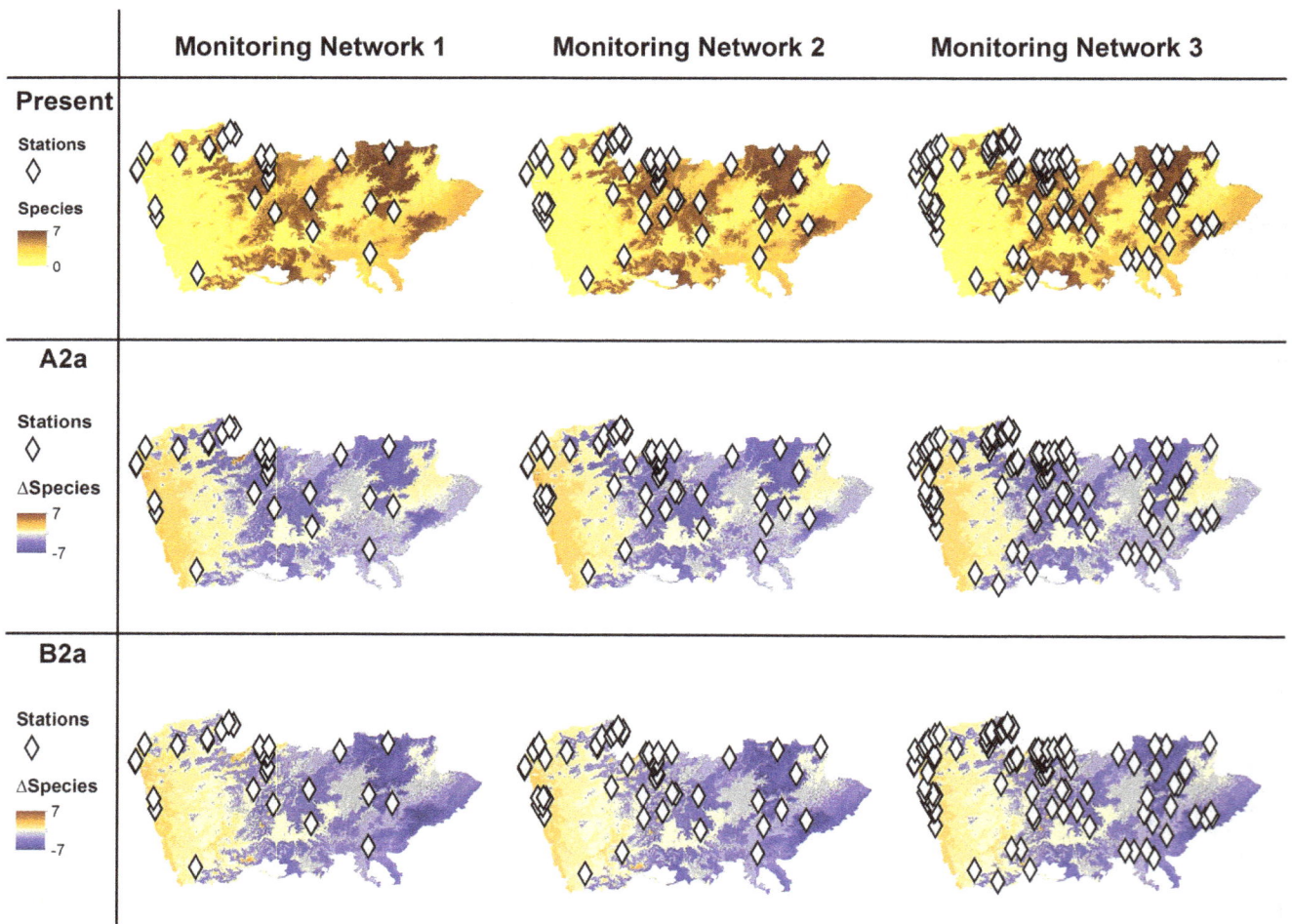

Figure 5. Sampling stations for the proposed monitoring networks (MN1, MN2 and MN3). Sampling stations for the proposed monitoring networks (MN1, MN2 and MN3) showing the present predicted species richness according to climatic model and predicted variation in climatic space between the present and the future under two climatic scenarios (A2a and B2a) for the year 2080. Negative △Species values indicate loss of climatic suitable space while positive indicate species gain of climatic suitable space.

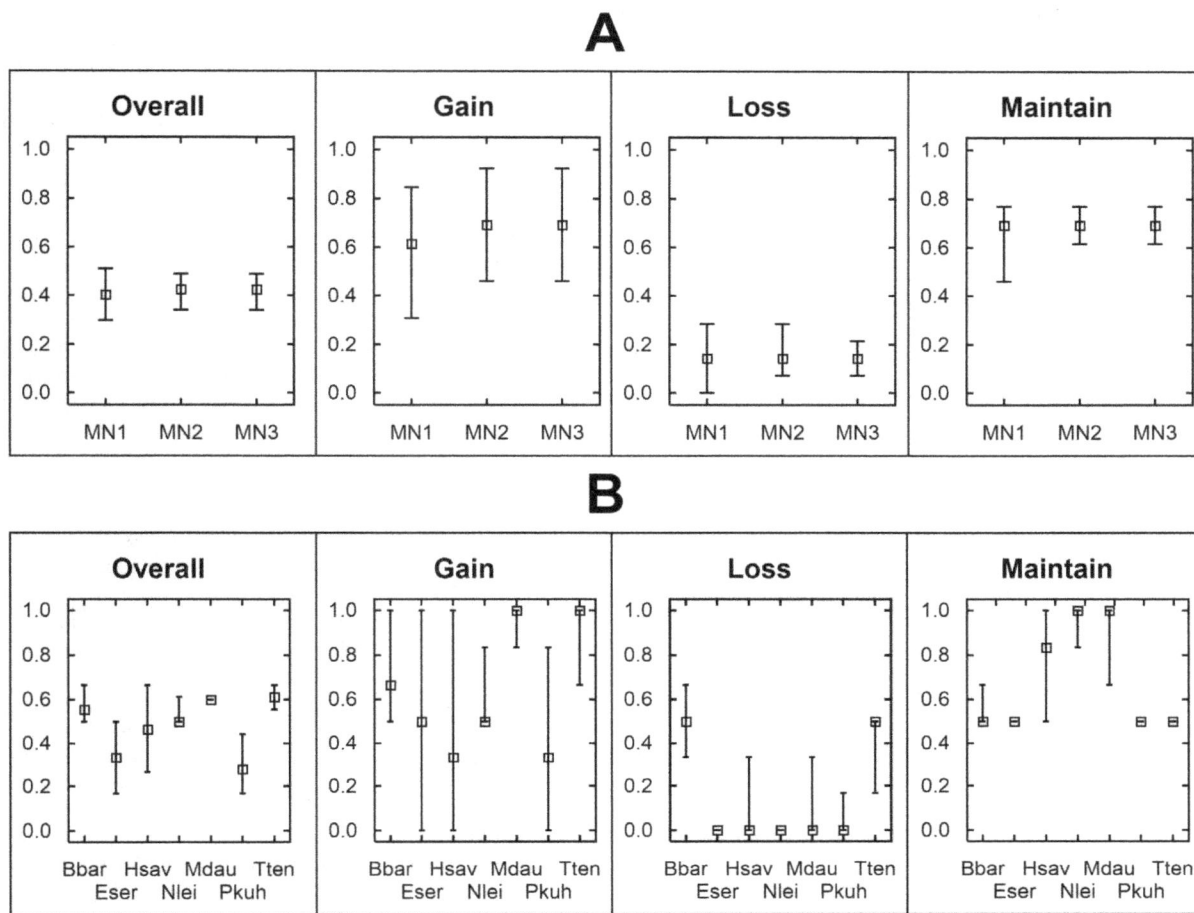

Figure 6. Proportion of targets that were not met in each of the testing networks (A) and by species (B). Proportion of targets that were not met by the 100 runs in each of the testing networks (A) and by species (B) considering overall targets and targets set for each suitability class. Median, maximum and minimum values are presented. Species code as follow: *Barbastella barbastellus* (Bbar); *Eptesicus serotinus/isabellinus* (Eser); *Hypsugo savii* (Hsav); *Nyctalus leisleri* (Nlei); *Myotis daubentonii* (Mdau); *Pipistrellus kuhlii* (Pkuh); and *Tadarida teniotis* (Tten).

be, by chance, within the area of likely gain of climatic suitability of species C. Co-occurrence may also explain why targets for the "likely lose" class were achieved for species such as *P. kuhlii* and *E. serotinus*, which have a low proportion of likely loss of climatic suitability.

Our framework also allows adjusting the sampling effort according to a frequently unpredictable volunteer commitment and to prioritize monitoring effort according to each species conservation status. Though we choose not to include the costs of implementing the sampling stations in the suggested locations, we point out that such cost can be considered when using Marxan [69]. The measures of cost can be based on any relative social, economic or ecological cost, or combinations thereof [69]. A critical example when designing monitoring networks based on volunteer effort is the costs of accessibility of different areas, which could be easily incorporated in Marxan by applying a planning unit cost factor.

Although it is highly likely that our simplest monitoring network (MN1) will not allow gathering enough information for a robust analysis, it will accomplish the fundamental goal of having sampling stations in areas where species and community structure are predicted to be sensitive to future climate changes. Moreover, it will work as a pilot survey to evaluate the statistical power of the monitoring network to detect population changes, as it is often

suggested in adaptive monitoring [7,12]. To maximize accuracy and minimize the possibility of biased conclusions being drawn about trends, a power analysis should subsequently be performed, e.g. following Walsh et al. [70]. The evaluation of volunteer efforts and the improvement of the network effectiveness could be fine-tuned following Tulloch *et al.* [71].

Operationally, the implementation of MN1 could allow attracting progressively more volunteers for the program – in fact it should be considered as the first step in the establishment of a more robust monitoring network. By working on fund-raising along with state agencies and NGO's, and depending on volunteer commitment, the underlying expectation is that the sampling effort will increase in the near future. The experience gathered with the ongoing Portuguese Bat Atlas (http://anodomorcego.wix.com/icnb), and also the example of the Portuguese Breeding Bird Atlas [72], allow some confidence on the growing commitment of citizens. Moreover, the increasing number of advanced courses and free workshops on ultra-sound recording techniques and species identification in Portugal has shown that non-specialists have strongly embraced this type of citizen science activities and transference of skills [73]. Overall these experiences strongly suggest that it is possible to successfully implement a long-term bat monitoring network in Portugal.

Model predictions and model-based simulations under climate change

Our results show that bats in the study area are highly sensitive to climate change, and though our main goal was not to determine the spatial patterns of future species richness, the findings presented here are in line with studies on the subject. A study of 28 European bat species hypothesized that a major range shift towards northern latitudes (U.K. and Fenno-Scandinavia) will occur until the end of the century, showing a significant loss of species richness in the Iberian Peninsula [74] and particularly in our study area. Other studies including amphibians, reptiles or trees [75,76] also predict a major loss of species ranges in the Atlantic climatic regions, mostly located along the north and northwest of the Iberian Peninsula. Also, the Scenarios, Impacts and Adaptation Measures (SIAM) report [77] identified the northeastern and eastern parts of our study area as highly sensitive to climate changes, which is consistent with our results, while Thuiller *et al.* [78] included the northwest Iberian Peninsula among those areas in Europe where climate change would cause highest levels of plant species loss and turnover.

Nonetheless, we should be aware that future species richness is most likely underestimated since our models only used partial information about the environmental niche for the target species. This can be critical when suitable climatic space is projected for future climate scenarios because the truncated niche can be responsible for an over-prediction of local extinctions at southern distribution edges in the northern hemisphere [79,80]. It is highly probable that southern Iberian populations of the target species could colonize the study area when ecological conditions become unsuitable at their southernmost ranges, compensating for otherwise forecasted local extinctions. Although the approach used here may reduce the models' applicability for extrapolation purposes, e.g. for predicting species–habitat interactions for other areas, times or climates [81], we stress that our main goal was to determine areas in the region that are sensitive to climate change regarding bat species, and not necessarily to predict the future distributions of the species in the region. In other words, we identified and will monitor areas where changes in species occurrence driven by future climatic shifts are more likely to occur.

In the present study, the simulations did not account for land cover changes in future projections although habitat variables are recognized to be relevant in predicting bat species distribution [36,37], which is also supported by the results obtained with our full model. Long-term changes in land cover are difficult to predict, especially in highly humanized areas because of the dependence upon economic interests and policy guidelines (among other drivers; e.g. [82]), and the inclusion of such variables would bring a higher degree of uncertainty to future projections. Although we recognize the importance of land cover, we should emphasize that the high agreement in niche overlap statistics show that full models and climatic models have a high similarity in their projections under current conditions. This result, together with the broader niche breadth obtained for the climatic model, means that the latter captures almost all niche conditions of the full model. For that reason, the use of the climatic model for future projections did not compromise our results. Nonetheless, to overcome this potential source of uncertainty, we opted to include a quarter of the sampling stations of the monitoring networks in areas where species occurrence was predicted when modelling with land cover. This approach may allow understanding whether future changes might be due to climate changes and/or to changes in land cover.

By combining different model techniques and circulation models we can achieve more robust projections, significantly reducing prediction uncertainties [83]. Although future predictions of species distributions were not the main scope of this study, these outputs can then be easily incorporated in the proposed framework. This innovative methodology can be used on any taxon or spatial scale, allowing a statistical optimization of the allocation of sampling effort in areas with high biodiversity that are also predicted to be prone to environmental changes.

Implications for conservation

To our best knowledge, the vast majority of current monitoring networks do not take into account in their design the potential changes in species distributions that may result from future climate change. By producing reliable data to detect population trends and range shifts due to climate change, the monitoring networks proposed here will provide stakeholders with important outcomes for conservation. MN1 will accomplish the immediate goal of attracting volunteers and to set a pilot survey [12]. If needed, future monitoring networks will then be fine-tuned in response to new information or new questions [7]. Power analysis will be performed during the lifetime of the project, a fundamental step because the consequences of inadequate design may not be obvious until the end of the programme, when it may be too late for amendments.

By monitoring climate change sensitive areas it will be possible to identify the most resilient populations which will be paramount for the future conservation of biodiversity. These populations will harbour unique gene pools while being source populations to colonise the new suitable areas. Moreover, it may be possible to understand species movements and consequently design corridors that promote their dispersion (e.g. [84]), focusing conservation and management efforts where they can produce the best results under environmental change scenarios. The implementation of this framework could provide an example for the development of climate change sensitive monitoring networks for other taxa and geographic contexts.

Supporting Information

Figure S1 Presence data for each target species and binarized maps of predicted occurrence according to preliminary Species Distribution Model.

Figure S2 Full model Percentage contribution of predictor variables and regularized gain With Only and Without predictor variables.

Figure S3 Climatic model Percentage contribution of predictor variables and regularized gain With Only and Without predictor variables.

Figure S4 Response curves for the EGVs most related to the predicted distribution of *Myotis daubentonii*.

Figure S5 Response curves for the EGVs most related to the predicted distribution of *Pipistrellus kuhlii*.

Figure S6 Response curves for the EGVs most related to the predicted distribution of *Hypsugo savii*.

Figure S7 Response curves for the EGVs most related to the predicted distribution of *Eptesicus serotinus/isabellinus*.

Figure S8   Response curves for the EGVs most related to the predicted distribution of *Nyctalus leisleri*.

Figure S9   Response curves for the EGVs most related to the predicted distribution of *Barbastella barbastellus*.

Figure S10   Response curves for the EGVs most related to the predicted distribution *Tadarida teniotis*.

Figure S11   Areas where each target species is likely to gain, lose or maintain suitable climatic space under scenario A2a.

Figure S12   Areas where each target species is likely to gain, lose or maintain suitable climatic space under scenario B2a.

Figure S13   Proportion of area occupied by suitability class for each species.

Table S1   AUC values for Training and Test data for both Full and Climatic models.

Table S2   Predictor variables by order of relevance for each species for both Full and Climatic models.

Table S3   Percentage of the total number of cells that overlap is observed in both Full and Climatic and ENMtools results.

Acknowledgments

We thank three anonymous reviewers for their valuable suggestions and comments.

Author Contributions

Conceived and designed the experiments: FA SBC JH HR. Performed the experiments: FA HR. Analyzed the data: FA SBC HR. Wrote the paper: FA SBC JH HR.

References

1. Macdonald DW, Service K (2007) Key Topics in Conservation Biology. Macdonald DW, Service K, editors Blackwell Publishing Ltd.
2. Rands MRW, Adams WM, Bennun L, Butchart SHM, Clements A, et al. (2010) Biodiversity conservation: challenges beyond 2010. Science 329: 1298–1303.
3. Butchart SHM, Walpole M, Collen B, van Strien A, Scharlemann JPW, et al. (2010) Global biodiversity: indicators of recent declines. Science 328: 1164–1168.
4. Dawson TP, Jackson ST, House JI, Prentice IC, Mace GM (2011) Beyond predictions: biodiversity conservation in a changing climate. Science 332: 53–58.
5. IPCC (2007) Climate Change 2007: the physical science basis. Contribution of Working Group I to the Fourth Assessment. Solomon S, Qin D, Manning M, Chen Z, Marquis M, et al., editors Cambridge: Cambridge University Press.
6. Legg CJ, Nagy L (2006) Why most conservation monitoring is, but need not be, a waste of time. J Environ Manage 78: 194–199.
7. Lindenmayer DB, Likens GE (2009) Adaptive monitoring: a new paradigm for long-term research and monitoring. Trends Ecol Evol 24: 482–486.
8. Nichols JD, Williams BK (2006) Monitoring for conservation. Trends Ecol Evol 21: 668–673.
9. McDonald-Madden E, Baxter PWJ, Fuller RA, Martin TG, Game ET, et al. (2010) Monitoring does not always count. Trends Ecol Evol 25: 547–550.
10. Franklin J, Regan HM, Hierl LA, Deutschman DH, Johnson BS, et al. (2011) Planning, implementing, and monitoring multiple-species habitat conservation plans. Am J Bot 98: 559–571.
11. McDonald-Madden E, Baxter PWJ, Fuller RA, Martin TG, Game ET, et al. (2011) Should we implement monitoring or research for conservation? Trends Ecol Evol 26: 108–109.
12. Green RE, Balmford A, Crane PR, Mace GM, Reynolds JD, et al. (2005) A Framework for Improved Monitoring of Biodiversity: Responses to the World Summit on Sustainable Development. Conserv Biol 19: 56–65.
13. Lovett GM, Burns DA, Driscoll CT, Jenkins JC, Mitchell MJ, et al. (2007) Who needs environmental monitoring? Front Ecol Environ 5: 253–260.
14. Pereira HM, David Cooper H (2006) Towards the global monitoring of biodiversity change. Trends Ecol Evol 21: 123–129.
15. Manley PN, Zielinski WJ, Schlesinger MD, Mori SR (2004) Evaluation of a multiple-species approach to monitoring species at the ecoregional scale. Ecol Appl 14: 296–310.
16. Parmesan C (2006) Ecological and Evolutionary Responses to Recent Climate Change. Annu Rev Ecol Evol Syst 37: 637–669.
17. Edwards M, Richardson AJ (2004) Impact of climate change on marine pelagic phenology and trophic mismatch. Nature 430: 881–884.
18. Visser ME, Both C (2005) Shifts in phenology due to global climate change: the need for a yardstick. Proc Biol Sci 272: 2561–2569.
19. Huntley B, Collingham YC, Willis SG, Green RE (2008) Potential impacts of climatic change on European breeding birds. PLoS One 3: e1439.
20. Araujo MB, Cabeza M, Thuiller W, Hannah L, Williams PH (2004) Would climate change drive species out of reserves? An assessment of existing reserve-selection methods. Glob Chang Biol 10: 1618–1626.
21. Hannah L, Midgley G, Andelman S, Araújo M, Hughes G, et al. (2007) Protected area needs in a changing climate. Front Ecol Environ 5: 131–138.
22. Barry S, Elith J (2006) Error and uncertainty in habitat models. J Appl Ecol 43: 413–423.
23. Buisson L, Thuiller W, Casajus N, Lek S, Grenouillet G (2010) Uncertainty in ensemble forecasting of species distribution. Glob Chang Biol 16: 1145–1157.
24. Conroy MJ, Runge MC, Nichols JD, Stodola KW, Cooper RJ (2011) Conservation in the face of climate change: The roles of alternative models, monitoring, and adaptation in confronting and reducing uncertainty. Biol Conserv 144: 1204–1213.
25. Noon BR, Bailey LL, Sisk TD, McKelvey KS (2012) Efficient species-level monitoring at the landscape scale. Conserv Biol 26: 432–441.
26. Jones G, Jacobs D, Kunz T, Willig M, Racey PA (2009) Carpe noctem: the importance of bats as bioindicators. Endanger Species Res 8: 93–115.
27. LaVal RK (2004) Impact of global warming and locally changing climate on tropical cloud forest bats. J Mammal 85: 237–244.
28. Sachanowicz K, Wower A, Bashta A-T (2006) Further range extension of Pipistrellus kuhlii (Kuhl, 1817) in central and eastern Europe. Acta Chiropterologica 8: 543–548.
29. Lundy M, Montgomery I, Russ J (2010) Climate change-linked range expansion of Nathusius' pipistrelle bat, Pipistrellus nathusii (Keyserling & Blasius, 1839). J Biogeogr 37: 2232–2242.
30. Battersby J (2010) Guidelines for Surveillance and Monitoring of European Bats. Bonn, Germany: UNEP/EUROBATS Secretariat.
31. O'Shea TJ, Bogan MA (2003) Monitoring Trends in Bat Populations of the Unitade States and Territories: Problems and Prospects.
32. Mitchell-Jones T, Bihari Z, Masing M, Rodrigues L (2007) Protecting and managing underground sites for bats. Bonn: Eurobats.
33. Kunz TH, Brock CE (1975) A comparison of mist nets and ultrasonic detectors for monitoring flight activity of bats. J Mammal 56: 907–911.
34. Walsh AL, Harris S (1996) Factors determining the abundance of vespertilionid bats in Britain: geographical, land class and local habitat relationships. J Appl Ecol: 519–529.
35. Vaughan N, Jones G, Harris S (1997) Habitat use by bats (Chiroptera) assessed by means of a broad-band acoustic method. J Appl Ecol 34: 716–730.
36. Russo D, Jones G (2003) Use of foraging habitats by bats in a Mediterranean area determined by acoustic surveys: conservation implications. Ecography (Cop) 26: 197–209.
37. Rainho A (2007) Summer foraging habitats of bats in a Mediterranean region of the Iberian Peninsula. Acta Chiropterologica 9: 171–181.
38. Kunz TH, Arnett EB, Cooper BM, Erickson WP, Larkin RP, et al. (2007) Assessing Impacts of Wind-Energy Development on Nocturnally Active Birds and Bats: A Guidance Document. J Wildl Manage 71: 2449–2486.
39. Roche N, Langton S, Aughney T, Russ JM, Marnell F, et al. (2011) A car-based monitoring method reveals new information on bat populations and distributions in Ireland. Anim Conserv 14: 642–651.
40. Franklin J (2010) Mapping species distributions: spatial inference and prediction. Usher M, Saunders D, Peet R, Dobson A, editors New York: Cambridge University Press.
41. Metzger MJ, Brus DJ, Bunce RGH, Carey PD, Gonçalves J, et al. (2013) Environmental stratifications as the basis for national, European and global ecological monitoring. Ecol Indic 33: 26–35.
42. Phillips SJ, Dudík M, Schapire RE (2004) A Maximum Entropy Approach to Species Distribution Modeling. Proceedings of the Twenty-First International Conference on Machine Learning. Banff, Canada. pp. 655–662.

43. Brotons L, Thuiller W, Araújo MB, Hirzel AH (2004) Presence-absence versus presence-only modelling methods for predicting bird habitat suitability. Ecography (Cop) 27: 437–448.

44. Rebelo H, Jones G (2010) Ground validation of presence-only modelling with rare species: a case study on barbastelles Barbastella barbastellus (Chiroptera: Vespertilionidae). J Appl Ecol 47: 410–420.

45. Elith J, Graham C, Anderson R, Dudík M, Ferrier S, et al. (2006) Novel methods improve prediction of species' distributions from occurrence data. Ecography (Cop) 29: 129–151.

46. Phillips SJ, Anderson RP, Schapire RE (2006) Maximum entropy modeling of species geographic distributions. Ecol Modell 190: 231–259.

47. Elith J, Phillips SJ, Hastie T, Dudík M, Chee YE, et al. (2011) A statistical explanation of MaxEnt for ecologists. Divers Distrib 17: 43–57.

48. Warren DL, Glor RE, Turelli M (2010) ENMTools: a toolbox for comparative studies of environmental niche models. Ecography (Cop) 01: 607–611.

49. Nakazato T, Warren DL, Moyle LC (2010) Ecological and geographic modes of species divergence in wild tomatoes. Am J Bot 97: 680–693.

50. Warren DL, Glor RE, Turelli M (2008) Environmental niche equivalency versus conservatism: quantitative approaches to niche evolution. Evolution 62: 2868–2883.

51. Wisz MS, Hijmans RJ, Li J, Peterson T, Graham CH, et al. (2008) Effects of sample size on the performance of species distribution models. Divers Distrib 14: 763–773.

52. Rainho A, Palmeirim JM (2011) The Importance of Distance to Resources in the Spatial Modelling of Bat Foraging Habitat. PLoS One 6: e19227.

53. Kunz TH, Fenton MB (2003) Bat Ecology. Chicago: University of Chicago Press.

54. Webb P, Speakman JR, Racey PA (1995) Evaporative water loss in two sympatric species of vespertilionid bat, Plecotus auritus and Myotis daubentoni: relation to foraging mode and implications for roost site selection. J Zool 235: 269–278.

55. Adams RA, Hayes MA (2008) Water availability and successful lactation by bats as related to climate change in arid regions of western North America. J Anim Ecol 77: 1115–1121.

56. Nakicenovic N, Swart R, editors (2000) Emissions Scenarios: A Special Report of Working Group III of the Intergovernmental Panel on Climate Change. Cambridge: Cambridge University Press.

57. Hijmans RJ, Cameron SE, Parra JL, Jones PG, Jarvis A (2005) Very high resolution interpolated climate surfaces for global land areas. Int J Climatol 25: 1965–1978.

58. Waltari E, Hijmans RJ, Peterson a T, Nyári AS, Perkins SL, et al. (2007) Locating pleistocene refugia: comparing phylogeographic and ecological niche model predictions. PLoS One 2: e563.

59. Russo D, Jones G (2002) Identification of twenty-two bat species (Mammalia: Chiroptera) from Italy by analysis of time-expanded recordings of echolocation calls. J Zool 258: 91–103.

60. Rainho A, Amorim F, Marques JT, Alves P, Rebelo H (2011) Chave de identificação de vocalizações dos morcegos de Portugal continental: 16.

61. Walters CL, Freeman R, Collen A, Dietz C, Brock Fenton M, et al. (2012) A continental-scale tool for acoustic identification of European bats. J Appl Ecol 49: 1064–1074.

62. Suárez-Seoane S, García de la Morena EL, Morales Prieto MB, Osborne PE, de Juana E (2008) Maximum entropy niche-based modelling of seasonal changes in little bustard (Tetrax tetrax) distribution. Ecol Modell 219: 17–29.

63. Raes N, Roos MC, Slik JWF, Van Loon EE, Steege H ter (2009) Botanical richness and endemicity patterns of Borneo derived from species distribution models. Ecography (Cop) 32: 180–192.

64. Cabral MJ, Almeida J, Almeida PR, Dellinger T, Ferrand de Almeida N, et al. (2005) Livro vermelho dos vertebrados de Portugal. 2ª ed. Lisboa: Instituto de Conservação da Natureza/Assírio & Alvim.

65. Ball IR, Possingham HP (2000) MARXAN (V1.8.2): Marine Reserve Design Using Spatially Explicit Annealing, a Manual. Differences: 67.

66. Ball IR, Possingham HP, Watts M (2009) Marxan and relatives: software for spatial conservation prioritisation. In: Moilanen A, Wilson KA, Possingham HP, editors. Spatial conservation prioritisation: quantitative methods and computational tools. Oxford, UK: Oxford University Press. pp. 185–195.

67. Zweig MH, Campbell G (1993) Receiver-Operating Clinical Medicine (ROC) Plots: A Fundamental Evaluation Tool in. Clin Chem 39: 561–577.

68. McComb B, Zuckerberg B, Vesely D, Jordan C (2010) Monitoring animal populations and their habitats: a practitioner's guide. Taylor & Francis Group.

69. Game ET, Grantham HS (2008) Marxan User Manual: For Marxan version 1.8. 10. University of Queensland, St. Lucia, Queensland, Australia, and Pacific Marine Analysis and Research Association, Vancouver. Ardron J, Klein C, Nicolson D, Possingham H, Watts M, editors University of Queensland, St. Lucia, Queensland, Australia, and Pacific Marine Analysis and Research Association, Vancouver, British Columbia, Canada.

70. Walsh A, Catto C, Hutson T, Racey P, Richardson P, et al. (2001) The UK's National Bat Monitoring Programme: Final Report 2001. London: Bat Conservation Trust.

71. Tulloch AIT, Mustin K, Possingham HP, Szabo JK, Wilson K a. (2013) To boldly go where no volunteer has gone before: predicting volunteer activity to prioritize surveys at the landscape scale. Divers Distrib 19: 465–480.

72. Equipa Atlas (2008) Atlas das aves nidificantes em Portugal (1999-2005). Lisboa: Instituto de Conservação da Natureza e da Biodiversidade, Sociedade Portuguesa para o Estudo das Aves, Parque Natural da Madeira e Secretaria Regional do Ambiente e do Mar. Assírio & Alvim.

73. ICNF (2013) Agreement on the Conservation of Populations of European bats: Report on Implementation of the Agreement in Portugal (2013/18 Advisory Committee Meeting).

74. Rebelo H, Tarroso P, Jones G (2010) Predicted impact of climate change on European bats in relation to their biogeographic patterns. Glob Chang Biol 16: 561–576.

75. Carvalho SB, Brito JC, Crespo EJ, Possingham HP (2010) From climate change predictions to actions - conserving vulnerable animal groups in hotspots at a regional scale. Glob Chang Biol 16: 3257–3270.

76. Benito Garzón M, Sánchez de Dios R, Sainz Ollero H (2008) Effects of climate change on the distribution of Iberian tree species. Appl Veg Sci 11: 169–178.

77. Santos FD, Forbes K, Moita R (2002) Climate Change in Portugal. Scenarios, Impacts and Adaptation Measures - SIAM Project. Santos FD, Forbes K, Moita R, editors Lisboa: Gradiva.

78. Thuiller W, Lavorel S, Araújo MB, Sykes MT, Prentice IC (2005) Climate change threats to plant diversity in Europe. Proc Natl Acad Sci U S A 102: 8245–8250.

79. Barbet-Massin M, Thuiller W, Jiguet F (2010) How much do we overestimate future local extinction rates when restricting the range of occurrence data in climate suitability models? Ecography (Cop) 33: 878–886.

80. Huntley B, Collingham YC, Green RE, Hilton GM, Rahbek C, et al. (2006) Potential impacts of climatic change upon geographical distributions of birds. Ibis (Lond 1859) 148: 8–28.

81. Braunisch V, Bollmann K, Graf RF, Hirzel AH (2008) Living on the edge—Modelling habitat suitability for species at the edge of their fundamental niche. Ecol Modell 214: 153–167.

82. Verburg PH, Eickhout B, Meijl H (2007) A multi-scale, multi-model approach for analyzing the future dynamics of European land use. Ann Reg Sci 42: 57–77.

83. Araújo MB, New M (2007) Ensemble forecasting of species distributions. Trends Ecol Evol 22: 42–47.

84. Williams P, Hannah L, Andelman S, Midgley G, Araújo M, et al. (2005) Planning for Climate Change: Identifying Minimum-Dispersal Corridors for the Cape Proteaceae. Conserv Biol 19: 1063–1074.

16

Estimating How Inflated or Obscured Effects of Climate Affect Forecasted Species Distribution

Raimundo Real[1], David Romero[1], Jesús Olivero[1], Alba Estrada[2], Ana L. Márquez[1]*

1 Biogeography, Diversity, and Conservation Research Team, Department of Animal Biology, Faculty of Sciences, University of Malaga, Malaga, Spain, **2** Instituto de Investigación en Recursos Cinegéticos IREC, (CSIC-UCLM), Ciudad Real, Spain

Abstract

Climate is one of the main drivers of species distribution. However, as different environmental factors tend to co-vary, the effect of climate cannot be taken at face value, as it may be either inflated or obscured by other correlated factors. We used the favourability models of four species (*Alytes dickhilleni*, *Vipera latasti*, *Aquila fasciata* and *Capra pyrenaica*) inhabiting Spanish mountains as case studies to evaluate the relative contribution of climate in their forecasted favourability by using variation partitioning and weighting the effect of climate in relation to non-climatic factors. By calculating the pure effect of the climatic factor, the pure effects of non-climatic factors, the shared climatic effect and the proportion of the pure effect of the climatic factor in relation to its apparent effect (ρ), we assessed the apparent effect and the pure independent effect of climate. We then projected both types of effects when modelling the future favourability for each species and combination of AOGCM-SRES (two Atmosphere-Ocean General Circulation Models: CGCM2 and ECHAM4, and two Special Reports on Emission Scenarios (SRES): A2 and B2). The results show that the apparent effect of climate can be either inflated (overrated) or obscured (underrated) by other correlated factors. These differences were species-specific; the sum of favourable areas forecasted according to the pure climatic effect differed from that forecasted according to the apparent climatic effect by about 61% on average for one of the species analyzed, and by about 20% on average for each of the other species. The pure effect of future climate on species distributions can only be estimated by combining climate with other factors. Transferring the pure climatic effect and the apparent climatic effect to the future delimits the maximum and minimum favourable areas forecasted for each species in each climate change scenario.

Editor: Kimberly Patraw Van Niel, University of Western Australia, Australia

Funding: The research was funded by the Ministerio de Ciencia e Innovación of Spain and FEDER (project CGL2009-11316/BOS). D. Romero is a PhD student at the University of Malaga with a grant of the Ministerio de Educación y Ciencia (AP 2007-03633). The funders had no role in study design, data collection and analysis, decision to publish, or preparation of the manuscript.

Competing Interests: The authors have declared that no competing interests exist.

* E-mail: almarquez@uma.es

Introduction

Species distribution models (SDMs) are becoming increasingly important tools for conservation biology, because determining which factors drive the distribution patterns can help to adopt more specific and appropriate strategies for the management and conservation species [1]. This knowledge is also the basis for making good forecasts on the effect of climate change on future species distributions or suitable areas, which is a new challenge for environmental managers [2, 3]. However, the estimation of impact of the climate change on future species distribution is complex and related to different kinds of uncertainties [4–9], including the inability to assess the weight of climate as a driver of species distribution.

Climate envelope models are widely used to forecast future species distributions under climate change scenarios [10–13]. Some authors argue against the validity of using SDMs based on climatic variables alone as tools to forecast future species distributions, because they consider that other factors play a role in the distributions and that these factors are not taken into account in the models [3, 5, 14–16]. Apart from climate, species distributions may be controlled by spatial trends, topography, human activity, biotic interactions, history, and population

dynamics, among others [17–20]. As species may show differential responses to these factors [3, 5, 21], their combined importance should be assessed before projecting species distribution models to the future. In addition, the effect of climate can only be assessed in the context of the other influential factors, because its pure effect on species distributions could be obscured or overrated by correlated aspects, becoming evident only when all the relevant factors are considered together [18, 19, 22, 23].

Variation partitioning techniques have been used to separate the effects of different factors on species richness [24, 25], on abundance [26], on ecological communities [27], on species assemblages [28] and also on species distributions [29, 30]. These techniques have also been used to segregate the pure effect of different factors (topography, climate, human activity, spatial situation and lithology) on species distributions [31]. However, these techniques have not been used to relate the pure climatic effect to its apparent effect, being the latter in correlation with other factors. This is of fundamental importance, because the apparent climatic effect could be misrepresenting the true role of climate on species distributions due to the effect of other correlated factors. Therefore, the potential changes in species distributions related to climate change could be distorted and lead to misleading conclusions about the species vulnerability or their risk of extinction.

Mountains are areas of interest regarding the early detection and study of the signals of climate change and its impact on ecological systems [32]. Mountain species are particularly sensitive to climate change [33–36], because mountain areas have more pristine habitats than lowland landscapes, and because these species can track climate change over shorter distances [37, 38]. Predicting the possible effects of climate change on the future distributions of these kinds of species is of fundamental importance in conservation plans.

We evaluated the relative contribution of climate in mainland Spain to the forecasted favourability of four vertebrate mountain species whose distributions are related to climate and to altitude or slope, and for which published distribution models are available both for the present and for the future according to the apparent effect of climate [8, 19]. The aim of this work is to propose a method by which to analyze the relative contribution of climate in relation to non-climatic factors (spatial, topographic, and human) and to distinguish between its apparent and its pure effect in models designed to forecast how favourable areas for species could vary because of climate change. Our results underline the possibly misleading outcome of not considering the pure climatic effect in the projections of the SDMs to the future.

Methods

The study area

Mainland Spain is located in southwestern Europe and covers an area of 493,518 km^2. Its latitude (40° N), geographical position and complex orography make its climate heterogeneous, with a precipitation gradient (100–2500 mm) decreasing mainly eastward and southward, and a temperature gradient (6°–18°C) decreasing mainly northward [39]. In Spain, five homogeneous climatic precipitation regions [39] can be distinguished: 1) the North Atlantic coast, which has abundant and regular precipitation due to the continuous arrival of Atlantic frontal systems; 2) the central area that receives wet and cold air intrusions from frontal Atlantic systems and presents low precipitations; 3) the eastern coast, which is characterized by irregular and scanty annual precipitation, with large variability due to severe rainfall events produced by wet and warm air intrusions from the Mediterranean Sea [40]; 4) the southeastern region, which is a dry desert-like area with very little rainfall; and 5) the southwestern region, which has more regular and abundant rainfall influenced by Atlantic winds.

This makes Spain particularly appropriate for analyzing the effect of different climate change scenarios on species distributions (e.g., [19]).

The species

We analysed the distribution in mainland Spain of four vertebrate species whose distributions are positively associated with altitude or slope. We chose an amphibian (Baetic midwife toad, *Alytes dickhilleni*), a reptile (Lataste's viper, *Vipera latasti*), a bird (Bonelli's eagle, *Aquila fasciata*) and a mammal (Iberian wild goat, *Capra pyrenaica*). *A. dickhilleni* is a small toad, between 32.8 and 56.5 mm in length, endemic to Spain and located exclusively in the mountainous systems of the southwestern part of the Iberian Peninsula. It lives on rough and steep terrains, in cracks and crevices next to streams, springs and pools. Species reproduction occurs in permanent water points. *V. latasti* is a venomous viper species found in southwestern Europe and northwestern Africa, which can reach 70 cm in length. Its distribution in Spain is relegated by human activity to mountainous and sparsely

populated areas. *A. fasciata* is a small to medium-size eagle, 55–65 cm in length. It is one of the rarest raptors in Europe and is a priority target species for special conservation measures in Spain (Council Directive 2009/147/EC and National Real Decreto 439/1990). It occupies mountain ranges, small hills and plains, where it breeds mainly in cliffs. *C. pyrenaica* is an endemic goat, only found in the mountainous areas of Spain. It is a species with strong sexual dimorphism, males are larger than females and their horns are three times longer and thicker than those of females. It lives in both forests and grassy expanses in mountains at altitudes between 500 and 2500 meters.

Baseline models

As starting point, we used the current favourability models obtained for *A. dickhilleni*, *V. latasti*, *A. fasciata* and *C. pyrenaica* using different climate change scenarios in mainland Spain, available in Márquez *et al.* [19]. The distributions of the four species were modelled using variables related to climate, spatial situation, topography, and human activity (see Table S1 for more specific details of the variables) and with the favourability function as the modelling technique [41–43]. Climatic variables were provided by the Agencia Estatal de Meteorología (AEMET), which regionalized the general circulation models CGCM2 (Canadian Climate Centre for Modeling) and ECHAM4 (Max Planck Institut für Meteorologie) and the A2 and B2 emission scenarios for Spain. Mean values of the climatic variables were obtained for the periods: 1961–1990, 2011–2040, 2041–2070 and 2071–2100 (Figure S1). For each species a total of four factor models were obtained, related to climate alone, spatial situation alone, topography alone, and human activity alone, respectively. A combined model was then obtained which took into account the four environmental factors (climatic and non-climatic) together (see Márquez *et al.* [19] for more details). In this way, the favourability values for each species in each cell at the present time (F_p) were obtained. Future favourability values for each species according to the apparent effect of climate (F_{fClim}) in each cell, as well as an analysis of the impact of climate model choice and scenario choice on expected favourability, were obtained by Real *et al.* [8] by replacing the current (1961–1990) climatic values in the combined favourability models with those expected according to each AOGCM and SRES for the following time periods (2011–2040, 2041–2070, 2071–2100).

Variation partitioning

We segregated the pure effect of climate from the effect of the other factors in the models using a variation partitioning procedure similar to that of Borcard *et al.* [44], Barbosa *et al.* [29] and Muñoz *et al.* [45] with some modifications. Thus, we specified how much of the variation of the combined favourability model was explained by the pure effect of climate (not affected by the covariation with other factors in the model), and which proportion of the climate effect cannot be distinguished from that of the other factors (shared effect) [45].

The portion of the variation in the model apparently explained by climate was estimated using the coefficient of determination of the linear regression of the logit function (y) of the model on the climatic variables included in it (R^2_{Clim}) for the period 1961–1990; the part apparently explained by the non-climatic factors (R^2_{NClim}) was obtained in a similar manner. The pure effect of climate (R^2_{pClim}) was then assessed by subtracting from 1 the variation of the combined model explained by the non climatic factors ($R^2_{pClim} = 1 - R^2_{NClim}$). The pure effect of the non-climatic factors was obtained by subtracting from 1 the variation explained by climate ($R^2_{pNClim} = 1 - R^2_{Clim}$). The effect shared by climate and non-climatic factors was obtained by subtracting from 1 both pure effects ($R^2_{ClimNClim} = 1 - R^2_{pClim} - R^2_{pNClim}$). We used the adjusted R^2 in all cases [46] although, given the high number of cells used (n = 5167 10x10 km^2 UTM cells), the difference between R^2 and adjusted R^2 was very small. This partitioning procedure was applied only to the portion of variation explained by the model, not over the total variation of the species distribution [44, 46], as it is the explanatory model which is calibrated to be transferred to the future. We expressed these effects as percentages and considered them to be the percentage of model variation attributable to the pure climatic effect (PCE), the pure non-climatic effect (PNCE) and the shared effect of climate and non-climatic factors (SCE).

We estimated the proportion of the apparent climatic effect represented by the pure effect of climate as $(\rho) = \dfrac{R^2_{pClim}}{R^2_{Clim}}$. We calculated the logit function expected for the future in each cell according to the pure effect of climate (y_{fClim}) by applying the expression $y_{fClim} = y_p + \rho(y_f - y_p)$, where y_p is the logit function at the present time and y_f is the logit function expected for the future according to the apparent effect of climate. We obtained the future favourability according to the pure climatic effect (F_{fPClim}) using the expression:

$$F_{fPClim} = \frac{\exp^{y_{fClim}}}{\frac{n_1}{n_0} + \exp^{y_{fClim}}}$$

where n_1 is the number of presences and n_0 the number of absences (see formula 7 in Real et al. [41]).

This way of inferring the effect of climate differs from usual projections according to climate change scenarios, which are customarily based on the apparent effect of climate. The difference between the sum of areas forecasted to be favourable according to the apparent climatic effect (F_{AC}), calculated using the expression $F_{AC} = \sum F_{fClim}$, and those forecasted according to the pure climatic effect (F_{PC}), calculated using the expression $F_{PC} = \sum F_{fPClim}$, was computed and expressed as the relative proportion of discrepancy $R = (F_{AC} - F_{PC})/F_{AC}$.

Results

Climate had a more significant effect than non-climatic factors on *A. fasciata* (Table 1). However, for the other species (*A. dickhilleni*, *V. latasti* and *C. pyrenaica*), the non-climatic effect was more important than the climatic effect (Table 1). In addition, regarding *A. dickhilleni*, the percentage of variation of the model attributable to a Shared Climatic Effect (SCE) was very important, which means that the apparent effect of climate could be due in large part to other correlated factors. Regarding *A. fasciata* and *C. pyrenaica*, the values of SCE were negative in the majority of the models. These negative values measure the amount of reciprocal obscuring caused by factors that have opposite geographic effects on the explained favourability, so that in these species the apparent climatic effect under-represents the pure climatic effect.

Figure 1 shows three examples of the differences between forecasted favourabilities taking into account the apparent climatic effect (F_{AC}) and the pure climate effect (F_{PC}), when F_{PC} is lower, similar or higher than F_{AC}. The differences between the two future forecasted favourabilities ($F_{AC} - F_{PC}$) and the relative proportion of discrepancy between both types of effects ($R = (F_{AC} - F_{PC}) / F_{AC}$) are shown in Table 2. Figure 2 shows the spatial distribution of the difference between the forecasted favourabilities taking into account the apparent climatic effect and the pure climatic effect

Table 1. Variation partitioning of the combined favourability models for the period 1961–1990.

AOGCM-SRES		*A. dickhilleni*	*V. latasti*	*A. fasciata*	*C. pyrenaica*
CGCM2-A2	PNCE	36.20	56.13	67.07	78.63
	PCE	6.80	24.19	66.73	38.40
	SCE	57.00	19.68	−33.80	−17.03
	ρ	0.107	0.551	2.026	1.797
CGCM2-B2	PNCE	37.27	59.27	57.78	77.99
	PCE	7.02	37.62	72.82	37.82
	SCE	55.71	3.10	−30.60	−15.80
	ρ	0.112	0.924	1.725	1.718
ECHAM4-A2/B2	PNCE	66.43	43.61	46.92	80.08
	PCE	12.11	19.45	79.28	6.40
	SCE	21.46	36.94	−26.19	13.53
	ρ	0.361	0.345	1.493	0.321

Values shown are the percentages of variation explained by the Pure Non-Climatic Effect (PNCE), the Pure Climatic Effect (PCE) and the Shared Climatic Effect (SCE). ρ: Proportion of pure climatic factor in relation to the whole climatic factor.

($|F_{fClim} - F_{fPClim}|$) for the species and situations described in Figure 1.

Discussion

The inclusion of climatic and non-climatic factors in SDMs is recommended not only because it can improve fit and increase their predictive accuracy [47, 48], but also because the effect of climate can only be assessed in the context of the other influential factors [18, 19, 22, 23]. Our results show that the correlation of influential non-climatic factors with temperature and precipitation could either inflate or obscure the apparent effect of climate, and that this modification of the apparent effect of climate would remain hidden if non-climatic variables were not included in the SDM. Even the use of the latitude and longitude of every cell alone may pinpoint certain areas of origin, dispersion, or past vicariance events driving current distributions, which results in a historically-caused spatial pattern that may coincide with specific climatic characteristics [49]. Consequently, the true effect of climate should be assessed in the context of spatial influences both on species distributions and on climate [50]. We used human, topographic and spatial variables as non-climatic predictors that, although correlated with climatic variables, can influence species distributions for reasons not directly linked to climate [18]. It was by taking into account these non-climatic factors and removing their effects statistically that we identified the underlying pure climate-distribution relationships, which could then be used in forecasting their distribution shifts under climate change [3].

However, the inclusion of climatic variables together with non-climatic, static variables entails other kinds of problems. Stanton et al. [51] considered that static variables such as elevation, latitude or longitude may hinder the accuracy of future predictions, as the relationships between them and climatic variables is likely to change in the future, and that including such variables in the SDM is likely to result in models which underestimate the effects of climate change. Our results confirm that this may be the case, although these effects may be under- or over-estimated. Our procedure is a way to gauge these relationships and assess the

Apparent effect   Pure effect

ECHAM4-A2 (2071-2100)

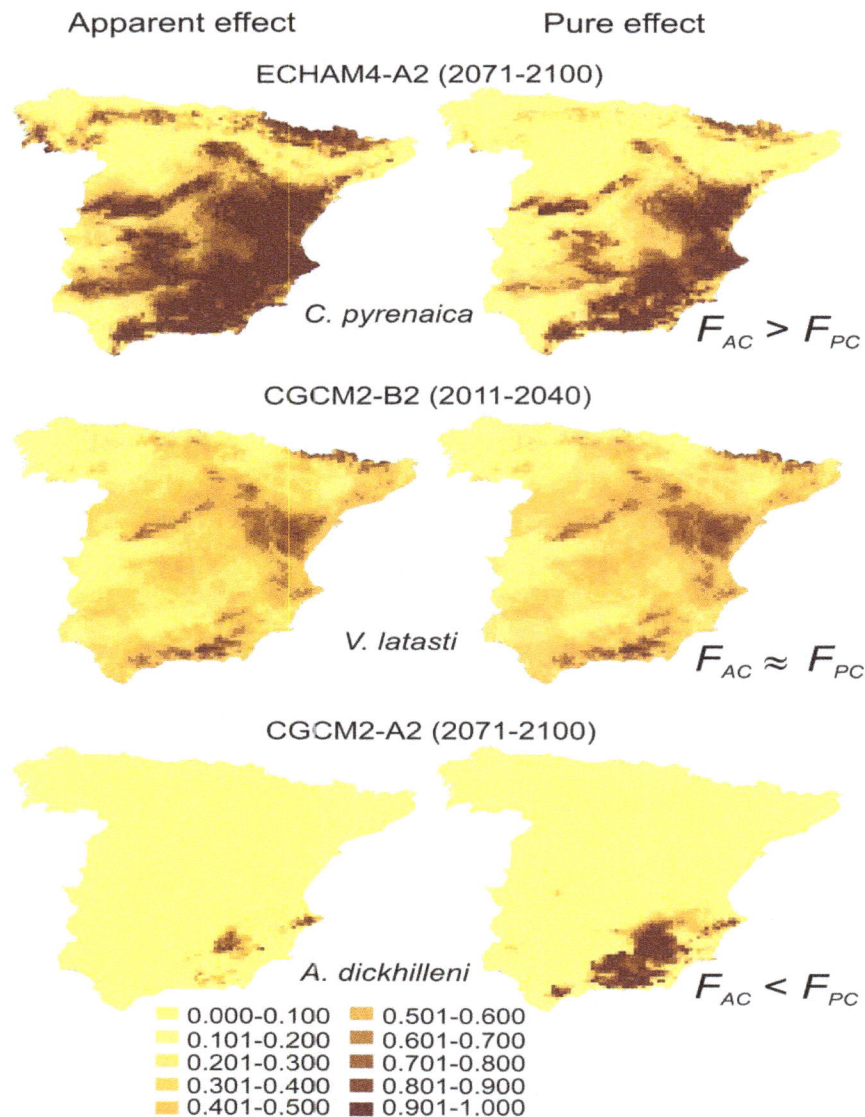

C. pyrenaica

$F_{AC} > F_{PC}$

CGCM2-B2 (2011-2040)

V. latasti

$F_{AC} \approx F_{PC}$

CGCM2-A2 (2071-2100)

A. dickhilleni

$F_{AC} < F_{PC}$

0.000-0.100	0.501-0.600
0.101-0.200	0.601-0.700
0.201-0.300	0.701-0.800
0.301-0.400	0.801-0.900
0.401-0.500	0.901-1.000

Figure 1. Forecasted favourability. Distribution of the future favourability forecasted according to the apparent climatic effect (F_{AC}) and that forecasted according to the pure climatic effect (F_{PC}). These maps represent three situations: $F_{AC} < F_{PC}$, $F_{AC} \approx F_{PC}$, $F_{AC} > F_{PC}$. The examples shown are those where the difference or similarity were most evident.

maximum extent to which the current correlation between these static variables and climate may affect the climatic parameters in the SDM.

On the other hand, the shared climatic effect is equally attributable to climate or to other correlated factors, so in our current state of knowledge the exact effect of climate cannot be determined with precision, although it lies somewhere between the apparent effect and the pure effect. The uncertainty related to the differences between both effects vary spatially and their intensity depends on the species (see Figure 2). This kind of uncertainty is added to other sources of uncertainty associated with forecasting future species distributions [8, 52, 53], among them those derived from assuming that the species' climate tolerances will remain constant through time, which is one serious limitation to the customary use of SDMs. However, despite these uncertainties, the estimation of species range shifts is the basis to predict where the species are likely to move under different future conditions [54].

More reliable predictions of species distribution responses to future climate conditions depend on developing more rigorous statistical analyses of the available data and on the combination of different factors, as well as on placing limits on the different uncertainties involved in the scientific forecasting of future events [19]. Transferring the pure climatic effect and the apparent climatic effect to the future allows us to delimit the maximum and minimum effect of climate on the species distributions.

Most of the favourability models of all the species considered in this study included three or four factors (spatial situation, topography, human activity and climate) (see Table 3 in Márquez *et al.* [19]) that we summarized into a climatic factor and a non-climatic factor. Pure climatic effect (PCE), pure non-climatic effect (PNCE) and shared climatic effect (SCE) on the future favourable areas for *A. dickhilleni*, *V. latasti*, *A. fasciata* and *C. pyrenaica* differed substantially (see Table 1). For *A. dickhilleni*, *V. latasti* and *C. pyrenaica* the PNCE was more important than the PCE, which

$$(|F_{fClim} - F_{fPClim}|)$$

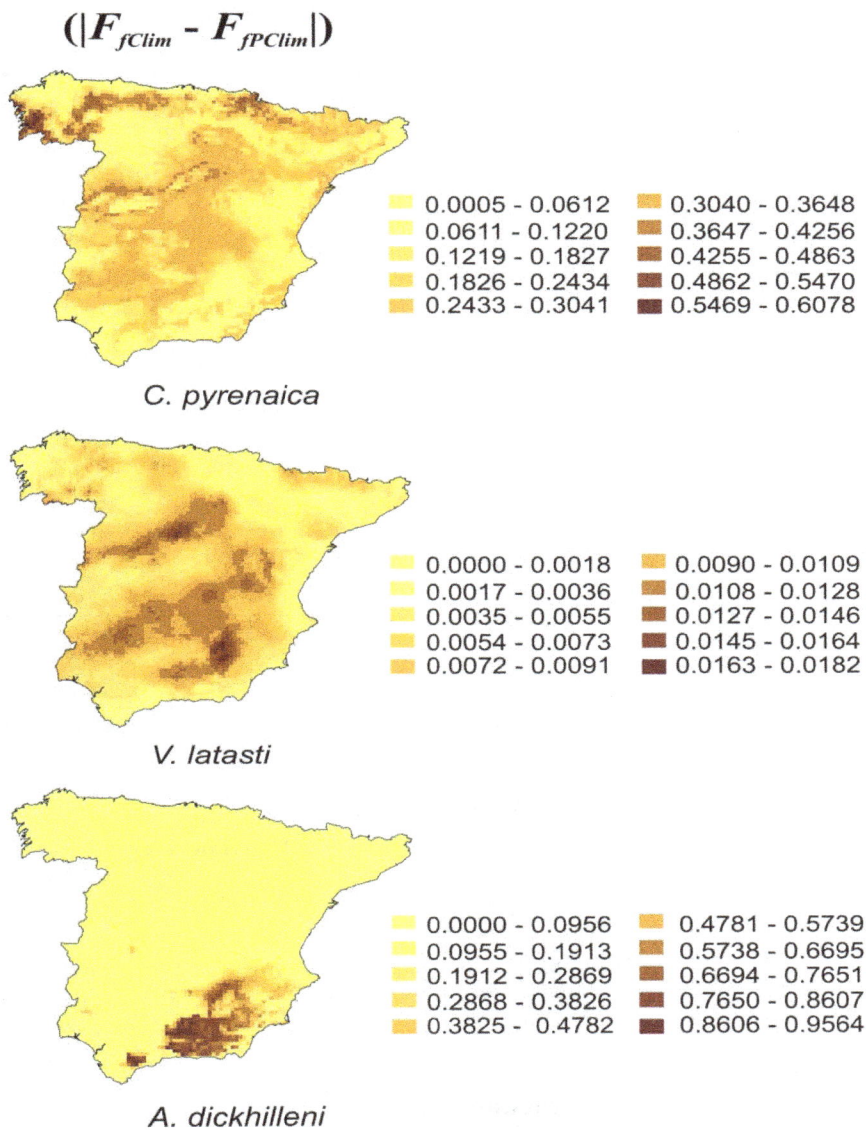

C. pyrenaica

0.0005 - 0.0612	0.3040 - 0.3648
0.0611 - 0.1220	0.3647 - 0.4256
0.1219 - 0.1827	0.4255 - 0.4863
0.1826 - 0.2434	0.4862 - 0.5470
0.2433 - 0.3041	0.5469 - 0.6078

V. latasti

0.0000 - 0.0018	0.0090 - 0.0109
0.0017 - 0.0036	0.0108 - 0.0128
0.0035 - 0.0055	0.0127 - 0.0146
0.0054 - 0.0073	0.0145 - 0.0164
0.0072 - 0.0091	0.0163 - 0.0182

A. dickhilleni

0.0000 - 0.0956	0.4781 - 0.5739
0.0955 - 0.1913	0.5738 - 0.6695
0.1912 - 0.2869	0.6694 - 0.7651
0.2868 - 0.3826	0.7650 - 0.8607
0.3825 - 0.4782	0.8606 - 0.9564

Figure 2. Differences between forecasted local favourabilities. Distribution of the uncertainty associated with differences between the favourabilities forecasted according to the apparent and the pure climatic effect ($|F_{fClim} - F_{fPClim}|$) for the three species and situations represented in Figure 1.

suggests that their future distributions will be more related to non-climatic environmental variables, such as biotic interactions, past human activities or past contingent events, than with the climatic factor [31, 55]. For *A. fasciata*, and in most scenarios for *C. pyrenaica*, the SCE was negative, that is, the climatic effect and the non-climatic effect can be reciprocally obscured by their opposite effect on the explained favourability (Table 1) [24, 46]. In these cases the apparent climatic effect under-represents the pure climatic effect. Future favourabilitiy for these cases taking into account the pure climatic effect would represent their maximum future favourable area, which is higher than that forecasted according to the apparent climatic effect (Table 2). The future areas favourable to *A. dickhilleni* forecasted according to the apparent climatic effect differed from those forecasted according to the pure climatic effect by 61% on average, which was the highest difference in the four species considered (Table 2). In this case, the apparent climatic effect was highly inflated by non-climatic factors.

The SCE could be a measure of uncertainty related to the complex interactions existing between climate and non-climatic factors. In some cases it represents the obscured climatic effect (when it has a negative value) and in other cases the inflated climatic effect (when it has a positive value). In any case, the SCE represents the uncertainty associated with the possibility of misunderstanding the effect of climate due to the effect of other correlated factors. This improves the usefulness of this kind of model for understanding species' potential responses to climate change, although possible changes in species-environment correlations through time can, nevertheless, place a limit on the predictive performance of these models [56]. According to Pearson and Dawson [10], understanding the complex interaction between the many factors affecting distributions is needed for the performance of more realistic simulations of the effect of climate change on species distributions. Models that take the effect of climate at face value yield future potential favourable areas that

Table 2. Differences between the favourability forecasted according to the apparent climatic effect (F_{AC}) and to the pure climatic effect (F_{PC}) for each species and period of time.

	AOGCM-SRES	A. dickhilleni		V. latasti		A. fasciata		C. pyrenaica	
		F_{AC}-F_{PC}	R	F_{AC}-F_{PC}	R	F_{AC}-F_{PC}	R	F_{AC}-F_{PC}	R
2011–2040	CGCM2-A2	21.874	0.043	−501.387	−0.518	−339.467	−0.151	−429.507	−0.209
	CGCM2-B2	68.640	0.122	−30.569	−0.017	−353.400	−0.146	−413.958	−0.204
	ECHAM4-A2	241.694	0.297	−385.007	−0.255	−215.186	−0.092	192.548	0.092
	ECHAM4-B2	282.001	0.303	−279.382	−0.167	−336.028	−0.125	216.448	0.121
2041–2070	CGCM2-A2	−150.240	−0.489	−416.164	−0.339	−721.372	−0.277	−753.816	−0.294
	CGCM2-B2	−28.377	−0.063	−62.350	−0.048	−517.740	−0.195	−531.307	−0.230
	ECHAM4-A2	216.533	0.270	−406.433	−0.277	−656.160	−0.158	406.952	0.197
	ECHAM4-B2	15.849	0.034	50.292	0.023	−724.439	−0.202	559.879	0.246
2071–2100	CGCM2-A2	−335.978	−4.230	−328.051	−0.225	−864.559	−0.309	724.801	0.176
	CGCM2-B2	−269.066	−1.262	512.495	0.173	−613.793	−0.217	−551.656	−0.237
	ECHAM4-A2	−9.056	−0.022	350.620	0.133	−285.609	−0.114	943.755	0.335
	ECHAM4-B2	148.574	0.227	214.544	0.089	−578.274	−0.177	813.794	0.311
Mean absolute percentage of change			61.33		18.86		18.05		22.10

R: Relative proportion of change ((F_{AC} - F_{PC}) / F_{AC}).

are, depending on the species, overestimated or underestimated. Using the method proposed in this paper, models may more realistically assess the levels of potential threat or opportunities to species of climate change.

Conclusion

In contrast to the tendency of not using correlated variables in spatial distribution models due to the possibility of the resulting coefficients can being unstable, we have to deal with the fact that in nature most factors are correlated; thus analyses that separate the pure and combined effect of the relevant factors should be performed. Given that the apparent effect of climate can be either inflated or obscured by other correlated factors, transferring both the pure climatic effect and the apparent climatic effect to the future allows us to delimit the maximum and minimum favourable areas forecasted for each species in each climate change scenario, thus permitting us to assess the uncertainty associated with the possibility of misrepresenting the effect of climate. This also allows us to detect and control the over- or under-estimation of the effect of climate change (either positive or negative) on future species distributions that is implicit in current climate envelope models. This may make models more complex and harder to perform, but

the output would be closer to what may be scientifically forecasted taking into account this kind of uncertainty.

Supporting Information

Figure S1 Precipitations and temperatures. Annual precipitations and mean annual maximum temperatures for each period and combination of circulation model and scenario used.

Table S1 Factors and variables. Explanatory factors and variables used to model the species distributions of Márquez et al 2011 [19].

Acknowledgments

We are grateful to P. Acevedo for his comments on earlier drafts of the manuscript. We thank the Agencia Estatal de Meteorología of Spain for providing the climatic data.

Author Contributions

Conceived and designed the experiments: RR DR. Analyzed the data: ALM JO AE. Wrote the paper: ALM RR.

References

1. Guisan A, Zimmermann NE (2000) Predictive habitat distribution models in ecology. Ecol Model 135: 147–186.
2. Huntley B, Collingham YC, Willis SG, Green RE (2008) Potential impacts of climatic change on European breeding birds. PLoS One 3(1): e1439. doi:10.1371/journal.pone.0001439.
3. Duncan RP, Cassey P, Blackburn TM (2009) Do climate envelope models transfer? A manipulative test using dung beetle introductions. P Roy Soc B-Biol Sci 276: 1449–1457.
4. Beaumont LJ, Hughes L, Pitman AJ (2008) Why is the choice of future climate scenarios for species distribution modelling important? Ecol Lett 11: 1135–1146.
5. Dormann CF, Schweiger O, Arens P, Augenstein I, Aviron ST, et al. (2008) Prediction uncertainty of environmental change effects on temperate European biodiversity. Ecol Lett 11: 235–244.
6. Baer P, Risbey JS (2009) Uncertainty and assessment of the issues posed by urgent climate change. An editorial comment. Climatic Change 92:31–36

7. Diniz-Filho JAF, Bini LM, Rangel TF, Loyola RD, Hof C, et al. (2009) Partitioning and mapping uncertainties in ensembles of forecasts of species turnover under climate change. Ecography 32: 897–906.
8. Real R, Márquez AL, Olivero J, Estrada A (2010) Species distribution models in climate change scenarios are not useful yet for informing emission policy planning: an uncertainty assessment using fuzzy logic. Ecography 33: 304–314.
9. Engler R, Randin CF, Thuiller W, Dullinger S, Zimmermannk NE, et al. (2011) 21st century climate change threatens mountain flora unequally across Europe. Global Change Biol 17: 2330–2341.
10. Pearson RG, Dawson TP (2003) Predicting the impacts of climate change on the distribution of species: are bioclimate envelope models useful? Global Ecol Biogeogr 12: 361–371.
11. Araujo MB, Thuiller W, Pearson RG (2006) Climate warming and the decline of amphibians and reptiles in Europe. J Biogeogr 33: 1712–1728.
12. Thuiller W, Broennimann O, Hughes G, Alkemade JRM, Midgley GF, et al. (2006) Vulnerability of African mammals to anthropogenic climate change

under conservative land transformation assumptions. Global Change Biol 12: 424–440.

13. Green RE, Collingham YC, Willis SG, Gregory RD, Smith KW, et al. (2008) Performance of climate envelope models in retrodicting recent changes in bird population size from observed climatic change. Biol Lett 4: 599–602.

14. Stefanescu C, Herrando S, Páramo F (2004) Butterfly species richness in the north-west Mediterranean Basin: the role of natural and human-induced factors. J Biogeogr 31: 905–916.

15. Ritchie EG, Martin JK, Johnson CN, Foz BJ (2009) Separating the influences of environment and species interactions on patterns of distribution and abundance: competition between large herbivores. J Anim Ecol 78: 724–731.

16. WallisDeVries MF, Baxter W, Van Vliet AJH (2011) Beyond climate envelopes: effect of weather on regional population trends in butterflies. Oecologia 167: 559–571.

17. Real R, Márquez AL, Estrada A, Muñoz AR, Vargas JM (2008) Modelling chorotypes of invasive vertebrates in Mainland Spain. Divers Distrib 14: 364–373.

18. Aragón P, Lobo JM, Olalla-Tárrega MA, Rodríguez MA (2010) The contribution of contemporary climate to ectothermic and endothermic vertebrate distributions in glacial refuge. Global Ecol Biogeogr 19: 40–49.

19. Márquez AL, Real R, Olivero J, Estrada A (2011) Combining climate with other influential factors for modelling climate change impact on species distribution. Climatic Change 108: 135–157.

20. Acevedo P, Jiménez-Valverde A, Melo-Ferreira J, Real R, Alves PC (2012) Parapatric species and the implications for climate change studies: a case study on hares in Europe. Global Change Biol 18: 1509–1519.

21. Willis SG, Thomas CD, Hill JK, Collingham YC, Telfer MG, et al. (2009) Dynamic distribution modelling: predicting the present from the past. Ecography 32: 5–12.

22. Lavergne S, Thuiller W, Molina J, Debussche M (2005) Environmental and human factors influencing rare plant local occurrence, extinction and persistence in the Mediterranean region. J Biogeogr 32: 799–811.

23. Dubey S, Shine R (2011) Predicting the effects of climate change on reproductive fitness of an endangered montane lizard, Eulamprus leuraensis (Scincidae). Climatic Change 107: 531–547.

24. Jiménez-Valverde A, Lobo JM (2007) Determinants of local spider (Araneidae and Thomisidae) species richness on a regional scale: climate and altitude vs. habitat structure. Ecol Entomol 32: 113–122.

25. Hortal J, Rodríguez J, Nieto-Díaz M, Lobo JM (2008) Regional and environmental effects on the species richness of mammal assemblages. J Biogeogr 35: 1202–1214.

26. Randin CF, Jaccard H, Vittoz P, Yoccoz NG, Guisan A (2009) Land use improves spatial predictions of mountain plant abundance but not presence-absence. J Veg Sci 20: 996–1008.

27. Peres-Neto PR, Legendre P (2010) Estimating and controlling for spatial structure in the study of ecological communities. Global Ecol Biogeogr 19: 174–184.

28. Godinho C, Rabaça JE, Segurado P (2010) Breeding bird assemblages in riparian galleries of the Guadiana River basin (Portugal): the effect of spatial structure and habitat variables. Ecol Res 25: 283–294.

29. Barbosa AM, Real R, Márquez AL, Rendón MA (2001) Spatial, environmental and human influences on the distribution of otter (Lutra lutra) in the Spanish provinces. Divers Distrib 7: 137–144.

30. Galantinho A, Mira A (2009) The influence of human, livestock, and ecological features on the occurrence of genet (Genetta genetta): a case study on Mediterranean farmland. Ecol Res 24: 671–685.

31. Acevedo P, Real R (2011) Biogeographical differences between the two Capra pyrenaica subspecies, C. p. victoriae and C. p. hispanica, inhabiting the Iberian Peninsula: Implications for conservation. Ecol Model 222: 814–823.

32. Beniston M (2003) Climatic change in mountain regions: a review of possible impacts. Climatic Change 59: 5–31.

33. Klanderud K, Birks HJB (2003) Recent increases in species richness and shifts in altitudinal distributions of Norwegian mountain plants. Holocene 13: 1–6.

34. Peñuelas J, Boada MA (2003) Global change-induced biome shift in the Montseny mountains (NE Spain). Global Change Biol 9: 131–140.

35. Wilson RJ, Gutiérrez D, Gutiérrez J, Martínez D, Agudo R, et al. (2005) Changes to the elevational limits and extent of species ranges associated with climate change. Ecol Lett 8: 1138–1146.

36. Pauli H, Gottfried M, Reiter K, Klettner C, Grabherr G (2007) Signals of range expansions and contractions of vascular plants in the high Alps: observations (1994–2004) at the GLORIA* master site Schrankogel, Tyrol, Austria. Global Change Biol 13: 147–156.

37. Wilson RJ, Gutiérrez D, Gutiérrez J, Monserrat VJ (2007) An elevational shift in butterfly species richness and composition accompanying recent climate change. Global Change Biol, 13: 1873–1887.

38. Gasner MR, Jankowski JE, Ciecka AL, Kyle KO, Rabenold KN (2010) Projecting the local impacts of climate change on a Central American montane avian community. Biol Conser 143: 1250–1258.

39. Muñoz-Díaz D, Rodrigo F (2004) Spatio-temporal patterns of seasonal rainfall in Spain (1912–2000) using cluster and principal component analysis: comparison. Ann. Geophys 22: 1435–1448.

40. García-Ortega E, Fita L, Romero R, López L, Ramis C, et al. (2007) Numerical simulation and sensitivity study of a severe hail-storm in northeast Spain. Atmos Res 83: 225–241.

41. Real R, Barbosa AM, Vargas JM (2006) Obtaining environmental favourability functions from logistic regression. Environ Ecol Stat 13: 237–245.

42. Nielsen C, Hartvig P, Kollmann J (2008) Predicting the distribution of the invasive alien Heracleum mantegazzianum at two different spatial scales. Diver Distrib 14: 307–317.

43. Barbosa AM, Real R, Vargas JM (2010) Use of coarse-resolution models of species' distributions to guide local conservation inferences. Conserv Biol 24: 1378–1387.

44. Borcard D, Legendre P, Drapeau P (1992) Partialling out the spatial component of ecological variation. Ecology 87: 2614–2625.

45. Muñoz AR, Real R, Barbosa AM, Vargas JM (2005) Modelling the distribution of Bonelli's eagle in Spain: implications for conservation planning. Divers Distrib 11: 477–486.

46. Peres-Neto PR, Legendre P, Dray S, Borcard D (2006) Variation partitioning of species data matrices: estimation and comparison of fractions. Ecology 87: 2614–2625.

47. Heikkinen R, Luoto M, Araújo MB, Virkkala R, Thuiller W, et al. (2006) Methods and uncertainties in bioclimatic envelope modelling under climate change. Prog Phys Geog 30: 751–777.

48. Triviño M, Thuiller W, Cabeza M, Hickler T, Araújo MB (2011) The contribution of vegetation and landscape configuration for predicting environmental change impacts on Iberian birds. PLoS One 6: e29373. doi:10.1371/journal.pone.0029373.

49. Pereira MC, Itami RM (1991) GIS-based habitat modeling using logistic multiple regression: A study of the Mt. Graham red squirrel. Photogramm Eng Rem S 57: 1475–1486.

50. Márquez AL, Real R, Vargas JM (2004) Dependence of broad-scale geographical variation in fleshy-fruited plant species richness on disperser bird species richness. Global Ecol Biogeogr 13: 295–304.

51. Stanton JC, Pearson RG, Horning N, Ersts P, Akçakaya HR (2012) Combining static and dynamic variables in species distribution models under climate change. Methods Ecol Evol 3: 349–357.

52. Kharouba HM, Algar AC, Kerr JT (2009) Historically calibrated predictions of butterfly species' range shift using global change as a pseudo-experiment. Ecology 90: 2213–2222.

53. Buisson L, Thuiller W, Casajus N, Lek S, Grenouillet G (2010) Uncertainty in ensemble forecasting of species distribution. Global Change Biol 16: 1145–1157.

54. Lawler JJ, Shafer SL, White D, Kareiva P, Maurer ED, et al. (2009) Projected climate impacts for the amphibians of the Western Hemisphere. Conser Biol, 24: 38–50.

55. Santos X, Brito JC, Sillero N, Pleguezuelos JM, Llorente GA, et al. (2006) Inferring habitat-suitability areas with ecological modelling techniques and GIS: A contribution to assess the conservation status of Vipera latastei. Biol Conser 130: 416–425.

56. Rubidge EM, Monahan WB, Parra JL, Camero SE, Brashares JS (2010) The role of climate, habitat, and species co-occurrence as drivers of change in small mammal distributions over the past century. Global Change Biol 17: 696–708.

17

Spatial Heterogeneity in Ecologically Important Climate Variables at Coarse and Fine Scales in a High-Snow Mountain Landscape

Kevin R. Ford[1]*, **Ailene K. Ettinger**[1], **Jessica D. Lundquist**[2], **Mark S. Raleigh**[2], **Janneke Hille Ris Lambers**[1]

1 Department of Biology, University of Washington, Seattle, Washington, United States of America, **2** Department of Civil and Environmental Engineering, University of Washington, Seattle, Washington, United States of America

Abstract

Climate plays an important role in determining the geographic ranges of species. With rapid climate change expected in the coming decades, ecologists have predicted that species ranges will shift large distances in elevation and latitude. However, most range shift assessments are based on coarse-scale climate models that ignore fine-scale heterogeneity and could fail to capture important range shift dynamics. Moreover, if climate varies dramatically over short distances, some populations of certain species may only need to migrate tens of meters between microhabitats to track their climate as opposed to hundreds of meters upward or hundreds of kilometers poleward. To address these issues, we measured climate variables that are likely important determinants of plant species distributions and abundances (snow disappearance date and soil temperature) at coarse and fine scales at Mount Rainier National Park in Washington State, USA. Coarse-scale differences across the landscape such as large changes in elevation had expected effects on climatic variables, with later snow disappearance dates and lower temperatures at higher elevations. However, locations separated by small distances (~20 m), but differing by vegetation structure or topographic position, often experienced differences in snow disappearance date and soil temperature as great as locations separated by large distances (>1 km). Tree canopy gaps and topographic depressions experienced later snow disappearance dates than corresponding locations under intact canopy and on ridges. Additionally, locations under vegetation and on topographic ridges experienced lower maximum and higher minimum soil temperatures. The large differences in climate we observed over small distances will likely lead to complex range shift dynamics and could buffer species from the negative effects of climate change.

Editor: Francesco de Bello, Institute of Botany, Czech Academy of Sciences, Czech Republic

Funding: Research was supported by the University of Washington Department of Biology (KF, AE), the American Alpine Club (KF), the Seattle chapter of the ARCS Foundation (KF), an NSF Doctoral Dissertation Improvement Grant (DEB-1010787) (AE), the University of Washington Royalty Research Foundation (JH) and the US Department of Energy (DOE#DE-FC02 06ER64159) (JH). This material is based on work supported by the NSF Graduate Research Fellowship Program (DGE-0718124– KF, AE), NSF grant no. CBET-0931780 (JL) and by NASA Headquarters under the NASA Earth and Space Science Fellowship Program (NNX09AO22H) (MR). The funders had no role in study design, data collection and analysis, decision to publish, or preparation of the manuscript.

Competing Interests: The authors have declared that no competing interests exist.

* E-mail: kford10@gmail.com

Introduction

Biologists have long recognized the fundamental role climate plays in determining the geographic distributions of species and biomes [1–3]. As a result, climate change is expected to induce shifts in the geographic ranges of species. This prediction is supported by the many observations of upward or poleward range shifts over the last 100 years consistent with observed warming [4] as well as range shifts inferred from the fossil record [5,6]. Alarmingly, models of the impacts of future anthropogenic climate change on species ranges have forecasted widespread extinction risks as the climatic niche of many species disappears or shifts faster than species can likely migrate [7,8].

However, these projections of climate change-induced range shifts (and subsequent extinction risks) are sensitive to the spatial scale at which the analyses are conducted [9]. Most range shift assessments rely on gridded maps of climate variables with grid cell sizes ranging from 800×800 m (e.g. PRISM [10] and WorldClim [11]) up to 50×50 km (e.g. [12]). The finer scale maps (800×800

m grid cells) capture a wide variety of climatic patterns, but the scales of these maps are still far coarser than the scales at which organisms experience their environment. Thus, these climate maps may hide fine-scale differences in climate that are important for organism distributions [13]. For example, north and south facing slopes separated by tens of meters may receive different amounts of solar radiation and experience very different temperature regimes [14,15], which could lead to differences in species composition within these microhabitats.

The implication of significant fine-scale climatic heterogeneity that is not captured by coarse-scale climate maps is that projections based on these maps could fail to capture important range shift dynamics. For example, cool microhabitats (such as north-facing slopes in the Northern Hemisphere) near the contracting edge or core of a species' distribution may allow populations of that species to persist if individuals can disperse to them from warmer microhabitats (such as south-facing slopes in the Northern Hemisphere), even if most of the surrounding landscape becomes unsuitably warm (as long as these microhabitat types comprise a total area large enough to support a population [16]). At the same time, warm microhabitats beyond the advancing edge of a species' range may provide the first sites of colonization that allow that species to migrate to new locations. Thus, instead of needing to move hundreds of meters upward or hundreds of kilometers poleward to track suitable climate, many species may only need to move tens of meters from one microhabitat to another and could be buffered from the negative effects of climate change [17].

For such fine-scale climatic heterogeneity to strongly influence range dynamics, however, fine-scale differences in climate must be large relative to coarse-scale differences. We addressed this issue by examining the magnitude of fine-scale heterogeneity relative to coarse-scale heterogeneity in snow disappearance date and growing season soil temperature. Specifically, we deployed 284 microclimate sensors across a ~1500 m elevation gradient spanning forest, subalpine and alpine biomes at Mount Rainier National Park. Our objectives were to 1) quantify snow disappearance date and soil temperature as a function of coarse-scale differences in elevation and exposure to storm tracks (i.e. being on the windward or leeward side of the mountain) and fine-scale differences in vegetation structure or topography, 2) compare fine-scale differences in climatic variables (that would be missed by climate models) to coarse-scale differences (that would be captured by climate models), and 3) determine whether fine-scale patterns in climatic variables related to topography (but not vegetation structure) are correlated with fine-scale patterns in vegetation characteristics. We focus on snow disappearance date and growing season soil temperature because snow disappearance date influences the length of the growing season (especially important in this region where the growing season can be very short due to the persistence of large winter snowpacks) while soil temperature strongly influences plant growth rates and other physiological processes [18]. These variables have also been shown to be strongly associated with patterns of distribution, abundance, productivity and diversity of plant species, at our sites and others [19–21]. Additionally, both variables are likely to change in the coming decades as a result of anthropogenic climate change, with rising temperatures and declining snowpacks leading to warmer and longer growing seasons [22,23].

It has long been known that climate can vary dramatically at fine spatial scales (reviewed in [14] and [24]), but these patterns have only recently begun to be studied explicitly and systematically with respect to the impacts of climate change on species distributions. Studies have found that locations separated by only tens of meters experienced mean seasonal soil temperatures that differed by 3–7°C, equivalent to the average temperature difference experienced in locations separated by hundreds of meters in elevation or hundreds of kilometers in latitude [15,21,25–28]. Moreover, such large differences in temperature are known to strongly influence organismal performance [18,29] and are greater than the expected increase in temperature due to climate change in many parts of the globe [30]. Similarly large differences were also observed in air temperature, snow cover duration or snow disappearance date over fine spatial scales in these studies. Our paper builds on these case studies and is notable for its large sample size of 284 sensors (important for assessing microclimate patterns in a statistically rigorous way), its explicit comparison of coarse- and fine-scale climatic heterogeneity (important for assessing the biases of coarse-scale models) and the broad environmental gradients covered (important for assessing how widespread these biases may be).

Methods

Ethics Statement

We obtained the necessary scientific research permits from Mount Rainier National Park, where all data collection occurred, before conducting this study. We did not sample any protected species.

Study Area

Mount Rainier National Park encompasses 95,354 ha of land in the western Cascade Mountains in Washington State, USA (Figure 1). The region experiences a temperate, maritime climate with mild, dry summers and cool, wet winters that produce large snowpacks. Elevation ranges from 518 m in the deep valley floors to 4392 m at the peak of Mount Rainier, the volcano located in the middle of the Park. The mountainous terrain produces steep climatic gradients: temperature decreases and precipitation increases with elevation, while the rainshadow effect produced by the volcano leads to lower precipitation on the east side of the Park. There are two primary climate stations in the Park. At the station located at 842 m elevation, mean annual temperature is 6.6°C and mean annual precipitation is 2030 mm; at the 1654 m station, mean annual temperature is 3.7°C and mean annual precipitation is 3005 mm (1981–2010 normals, NOAA National Climate Data Center – www.wrcc.dri.edu/Climsum.html).

The large climatic gradients create three major biomes in the Park. The forest biome extends from the lowest elevations of the Park up to about 1600–2000 m and is dominated by coniferous trees. The subalpine biome typically extends about 300 m above the upper limit of the forest and is a mosaic of conifer tree patches and subalpine meadows. The alpine biome occupies the highest elevations, stretching from 1900–2300 m to the summit of Mount Rainier, and consists of large patches of mostly continuous alpine meadows (dominated by forbs, grasses and dwarf shrubs) near the lower limit of the biome, with exposed rock, glaciers, bare soil, and cryptogams (mostly mosses, lichens, algae and cryptobiotic soil crusts) predominating above.

Study Design

From September 2009 through October 2010, we deployed 284 soil temperature sensors (HOBO Pendants made by the Onset Computer Corporation and iButtons made by Maxim Integrated Products) across Mount Rainier at elevations ranging from 638 m to 2164 m as part of two different plant ecology studies where microclimate was measured as an explanatory variable. The first study took place in the forest biome and spanned the elevational range of forests on the south side of Mount Rainier. The second study took place in the subalpine and alpine biomes, with study sites set up at the lower limit of the subalpine biome and the upper limit of alpine meadows on three sides of the mountain (Figure 1). Microclimate data from these studies were ideally suited for our questions as they covered large elevational gradients with sensors at each location stratified by vegetation or topographic features expected to influence microclimate. Due to differences in study design (described below), we analyzed the microclimate data from the two studies separately. The sensors remained in place and logged data until we collected them in September/October 2010.

For each sensor, we calculated the values of four climatic variables: snow disappearance date, and average daily mean, maximum and minimum soil temperature. We could assess snow cover from soil temperature measurements because snow insulates soil from fluctuations in air temperature so that temperatures beneath the snowpack in this region remain constant around 0°C. Thus, the soil temperature data allowed us to determine whether

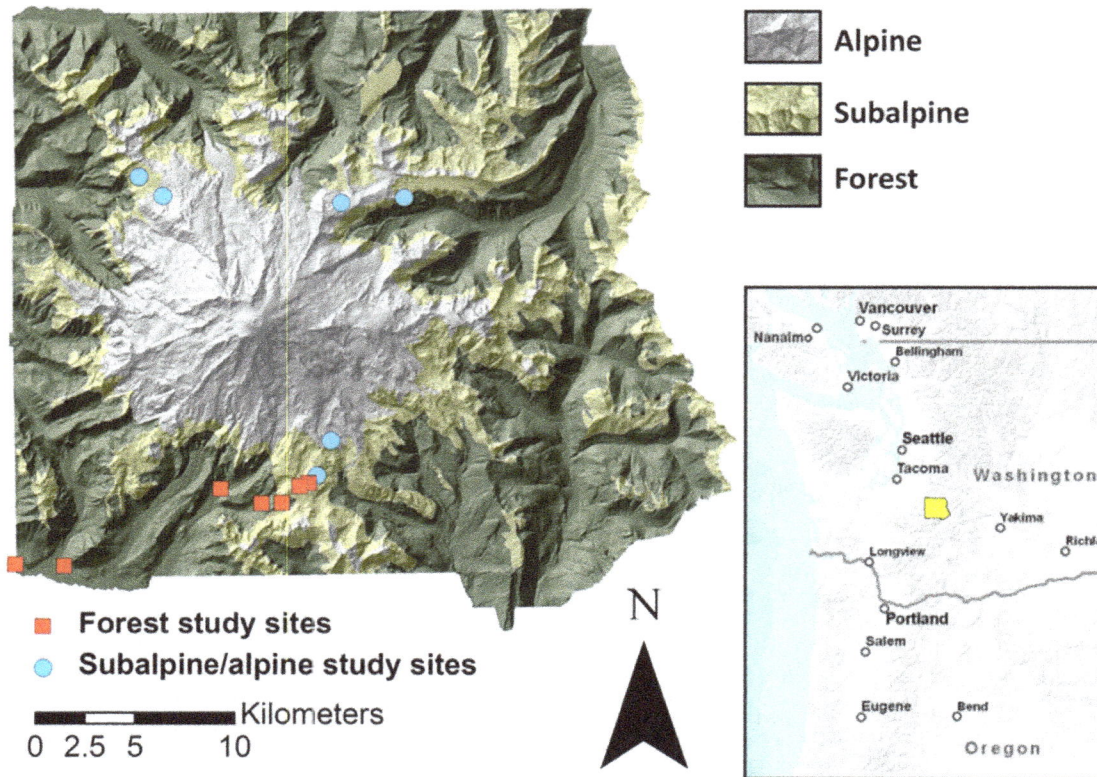

Figure 1. Study area. Mount Rainier National Park and its three major biomes, along with study site locations. Shading depicts topographic relief.

snow was covering the sensor for each day the sensor was deployed using an algorithm that considers daily temperature ranges and maxima [31,32]. We calculated average daily mean, maximum and minimum soil temperature for periods in Summer/Fall 2010 when all sensors in a study experienced snow-free conditions and reported data. This period was August 14 through October 3, 2010 for the sensors in the forest study and August 11 through August 18, 2010 for the sensors in the subalpine/alpine study. This meant we only used a portion of the snow-free temperature data for some sensors, even though temperatures outside this period are likely to also be ecologically relevant. However, it was necessary to use the same time period for all sensors in a study so that temporal differences in the snow-free period between locations did not confound our analysis of spatial differences in temperature.

Arrays of sensors were deployed at 13 sites throughout the Park. For the forest study, we established three study sites along an elevation gradient in Summer 2009, allowing us to calculate snow disappearance date in 2010. We quantified growing season soil temperature in Summer/Fall 2010 at these three sites plus an additional four sites along the same elevation gradient. Within each site, we deployed sensors under gaps in the forest canopy caused by tree falls ("gaps") and in locations under intact canopy within 20 m of one of the gaps ("non-gaps"). Each study site contained five of these gap/non-gap pairs. Gaps were ~130 m² on average. Within each gap or non-gap location, we placed one sensor in a 5.5×1.5 m area where all understory vegetation up to 2 m tall had been experimentally removed since early Summer 2009 (the "removed" plot) and one sensor at an adjacent location 2 m away where the vegetation had been left undisturbed (the "control" plot) (Figure 2A, Table 1).

For the subalpine/alpine study, we quantified microclimate at study sites on three sides of the mountain (south, northwest and northeast) which have different exposures to storm tracks and experience different precipitation regimes. On each side, we established two study sites, one close to the lower limit of the subalpine biome and one close to the upper limit of continuous alpine meadows, (below this limit, the ground is mostly vegetated while above it is almost entirely rock, glaciers and bare soil). These sites were about 200–300 m apart in elevation on each side of the mountain. Within each site, we established six linear transects that ran from a depression in the landscape up to a ridge (transects parallel to the slope) and were about 20 m in length. Within each transect, two sensors were located in the depression and two sensors were located on the ridge (Figure 2B, Table 1).

At each of the sensors in the subalpine/alpine biomes (where fine-scale sensor placement was stratified by topographic position and not vegetation structure), we measured vegetation characteristics in order to compare patterns in microclimate to ecological patterns. At the study sites near the lower limit of the subalpine biome, where closed-canopy forests transition to open meadows with increasing elevation, we measured percent cover by tree canopy above each sensor using a spherical densiometer (a gridded, hemisphere-shaped mirror used to estimate percent cover by foliage above a point on the ground), allowing us to assess the density of trees (higher values of tree canopy cover implies more or bigger trees). At the study sites near the upper limit of alpine meadows, where meadows transition to bare ground with increasing elevation, we estimated the percent of ground covered by vegetation at each sensor using a square-shaped PVC frame (1×1 m) placed on the ground adjacent to the sensor. String tied to the PVC frame created 100 evenly spaced grid points, allowing us

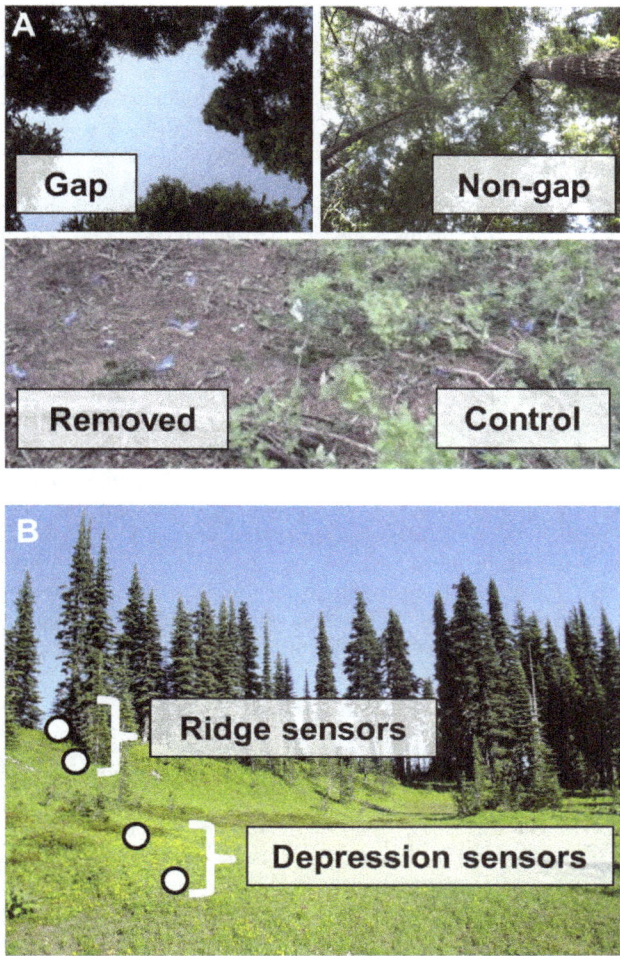

Figure 2. Temperature sensor deployment. Sensor deployment in (A) forest and (B) subalpine/alpine biomes. At each elevation in the forest biome (A), sensors were placed in gaps in the forest canopy (top left) and non-gaps with intact forest canopy (top right). Within each of these canopy types, sensors were located in plots where understory vegetation was removed (bottom left) and control plots where it was left undisturbed (bottom right). In the subalpine/alpine biomes (B), temperature sensors were located along transects running from depressions in the landscape to ridges.

to count the number of grid points overlaying vegetation in the area within the frame.

Data Analysis

We used linear mixed effects models (LMMs) to characterize the relationships between the potential drivers of climate and each of the four climatic response variables [33]. The LMMs allowed us to estimate the effects of explanatory variables and their two-way interactions on the response variable ("fixed effects") while statistically controlling for the effects of randomly selected experimental units on the response variable ("random effects"). At the forest sites, the drivers of climate were elevation, canopy structure (gap vs. non-gap) and understory structure (removed vs. control treatments), with gap/non-gap pair designated as a random effect. At the subalpine/alpine sites, the drivers of climate were side of the mountain (south vs. northwest vs. northeast), elevation (upper limit of alpine meadows vs. lower limit of the

Table 1. Details of temperature sensor deployment.

Biome(s)	Time span of deployment	# sites	# sensors per site	Total # sensors	Type of sensor	Sensor accuracy	Data logging interval	Sensor location
Forest	Summer 2009–Fall 2010	3	8	24	HOBO Pendants made by the Onset Computer Corporation	±0.53°C from 0° to 50°C	2 hours	Soil surface
Forest	Summer–Fall 2010	7*	20	140	HOBO Pendants made by the Onset Computer Corporation	±0.53°C from 0° to 50°C	1 hour	Soil surface
Subalpine/alpine	Summer 2009–Summer 2010	6	24	144	iButtons made by Maxim Integrated Products	±0.5°C from −10° to 65°C, or ±1°C from −30° to 70°C[†]	1, 2 or 4 hours[†]	3 cm below soil surface

*These sites include the three sites with sensors deployed in Summer 2009.
[†]Differences in accuracy and logging intervals were due to differences in the specific model of iButton sensor used.

subalpine biome, treated as a categorical variable since there were only two values of elevation on each side of the mountain) and topographic position (depression vs. ridge), with sensor transect designated as a random effect. We verified that the residuals of these models were normally distributed, to validate our use of linear mixed effects models (rather than generalized linear mixed effects models).

For each model, we used Akaike's information criterion (AIC) to select the most parsimonious combination of fixed and random effects to derive the "best-fit" model. Specifically, we used a three-step process following [33] where we (1) used AIC to determine the optimal random effects structure, selecting amongst several LMMs (fit with restricted maximum likelihood) that had different random effect terms (no random effects, random intercepts, random slopes or both), but the same fixed effect terms (which included all main effect and two-way interaction terms for each explanatory variable); (2) determined the optimal combination of fixed effect terms by using AIC to select amongst models (fit with maximum likelihood) with all possible combinations of fixed effect terms (but sharing the optimal random effects structure selected in the first step); and (3) fit a model with the random effects structure selected in the first step (which could be no random effects) and the fixed effects structure selected in the second step and considered this model to be our final "best-fit" model. This final model was fit with restricted maximum likelihood if it included random effects or maximum likelihood if it did not. All models were fit in R version 2.12.0 using the lme4 package for the LMMs [34,35]. See Appendix S1 for more details of the model fitting and selection procedure.

We assessed the significance of the model coefficients using Markov chain Monte Carlo sampling implemented with the MCMCglmm package in R [36] or t-tests (when no random effects were included in the best-fit model). We then used the explanatory variable coefficients of the best-fit models to calculate the magnitude of differences in microclimate response variables relative to differences in the explanatory variables. For example, if the coefficients related to topographic position in the model of snow disappearance date at the subalpine/alpine sites indicated that the difference between ridges and depressions was 20 days, controlling for differences in other variables, then the effect of topographic position on snow disappearance date would be equal to 20 days. In order to compare the effects of elevation (which we consider a coarse-scale driver of climate) to the effects of other explanatory variables, we calculated the difference in snow disappearance date or temperature between two points 100 m apart in elevation for each model (controlling for differences in other variables). Like differences in climate amongst different sides of the mountain, differences in climate caused by a 100 m difference in elevation can typically be captured by coarse-scale climate models (e.g. PRISM, while differences caused by vegetation structure and fine-scale topography cannot. If one of the explanatory variables was not included in the best-fit model, we included the main effect of that variable in the final model for comparative purposes. This happened for one explanatory variable in one model (the understory structure variable in the snow disappearance date model in the forest study).

For sites in the subalpine/alpine biomes, we also fit linear models (LMs) to characterize the relationships between each of the four microclimate variables and percent cover by tree canopy at the lower elevation sites, and the relationships between each of the four microclimate variables and percent cover by ground vegetation at the higher elevation sites, for a total of eight LMs. In these models, the response variable was the vegetation characteristic while the explanatory variables were the microcli-

mate variable, side of the mountain (included as a covariate) and their interaction. Using LMMs with sensor transect designated as a random effect did not improve model fit for any of the relationships, so we used the simpler LMs for all of these analyses. Next, we used AIC to select the best-fit LM. With one exception, this best-fit model included both the microclimate variable and side of the mountain, but not their interaction, as explanatory variables. In these cases, we used t-tests to assess the significance of the microclimate variable coefficient in the best-fit model in order to assess the significance of that microclimate variable. When modeling percent tree canopy cover at the lower elevation sites as a function of average daily minimum temperature and side of the mountain, neither explanatory variable nor their interaction was included in the best-fit model (i.e. the best-fit model was the null model with only an intercept). For this situation, we performed a t-test on the minimum temperature coefficient in a model that included both minimum temperature and side of the mountain, but not their interaction, as explanatory variables (i.e. a model with the same structure as the best-fit model for all the other vegetation-climate analyses) in order to assess the significance of minimum temperature.

We calculated the proportion of variance in the response variable explained by variance in the fixed effect explanatory variables (r^2) for all models, following [37].

Results

Variations in climate were explained by both coarse- and fine-scale drivers, with best-fit models having r^2 values ranging between 0.20 and 0.94 (Table 2). As expected, higher elevations experienced later snow disappearance dates and lower average daily mean and minimum temperatures (Figure 3A, B, D). However, the relationship between elevation and average daily maximum temperature was weak, and variability in this parameter was dominated by variability amongst locations at similar elevations (Figure 3C). At the subalpine/alpine sites, snow disappearance date and temperature varied depending on what side of the mountain sensors were on – e.g. the south side experienced later snow disappearance dates on average, probably because this side of the mountain is the most exposed to winter storms and receives the largest amount of winter precipitation. However, there were also substantial differences amongst locations at similar elevations for each of these variables that could be attributed to vegetation structure or topographic position (Figures 3, 4). We also assessed heterogeneity in growing degree days (GDD – calculated as the sum of daily mean soil temperatures for all days where the daily mean soil temperature was over 5°C), which showed patterns very similar to those of snow disappearance date (results not shown due to limitations of the data – sensors were not deployed long enough to estimate GDD for the full year or growing season, which could bias comparisons of GDD amongst locations).

Forest Biome: Stratification by Vegetation Structure at Fine Scales

Snow disappearance date was later at higher elevations and in canopy gaps, while understory vegetation structure had little effect (Figure 3A). We also found that the effect of fine-scale differences in canopy structure (gaps vs. non-gaps) on snow disappearance date was similar to the effects of coarse-scale differences in elevation (100 m elevation differences) (Figure 4A). Thus, snow disappearance date differed as much at fine scales (where locations differed in forest canopy structure) as it did over coarse spatial scales.

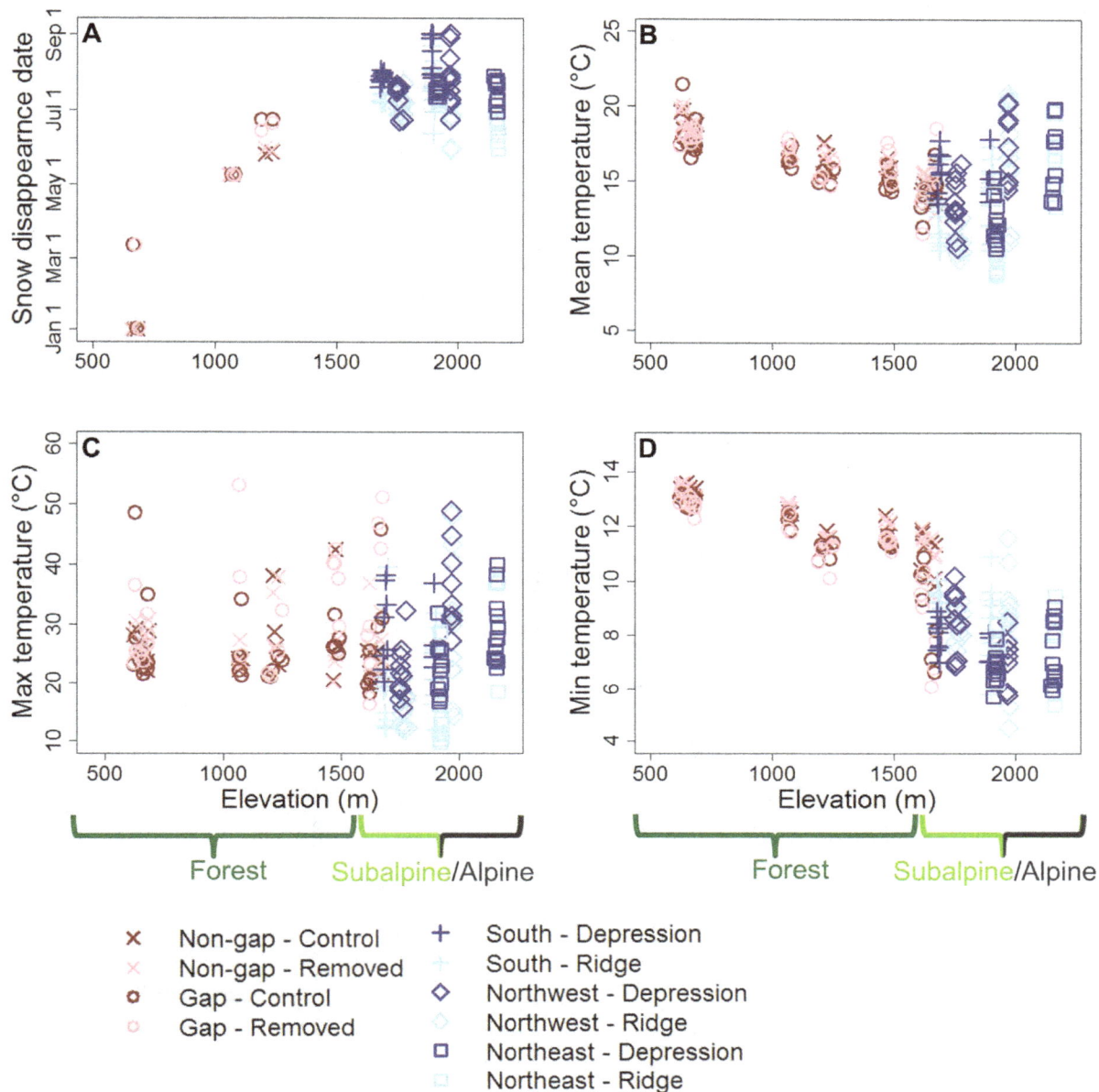

Figure 3. Patterns in climate. (A) Snow disappearance date in 2010 and average daily (B) mean, (C) maximum and (D) minimum soil temperature for a representative week during the growing season (August 11–18, 2010) plotted against elevation. Note the differences in scale on the axes showing temperature values. Points represent individual sensors with symbol type and color designating sampling stratification for forest (dark and light red) and subalpine/alpine sites (dark and light blue). "Non-gap"/"gap" refer to canopy structure categories while "control"/"removed" refer to understory structure categories (forest sites). "South"/"northwest"/"northeast" refer to sides of the mountain while "ridge"/"depression" refer to topographic positions (subalpine/alpine sites). Approximate biome ranges are shown below the elevation axes.

As expected, growing season soil temperature generally declined with increasing elevation. Canopy gaps had higher maximum temperatures, but lower minimum and mean temperatures relative to non-gaps (Figure 3B–D). Canopy structure had a similar or greater effect on temperature than a 100 m change in elevation for average daily maximum and minimum temperature (Figure 4C, D). Locations where understory vegetation was removed experienced higher maximum and mean temperatures, but lower minimum temperatures, relative to control plots where vegetation was undisturbed (Figure 3B–D). Understory structure had a greater effect on average daily maximum temperature than a 100 m change in elevation, but had weaker effects on average

daily mean and minimum temperature (Figure 4B–D). Overall, there was about as much heterogeneity in temperature at fine scales (differing vegetation structure) as there was at coarse scales (100 m differences in elevation).

Subalpine/alpine Biomes: Stratification by Topographic Position at Fine Scales

Snow disappearance date was later on the south side of the mountain, at higher elevations and in topographic depressions (Figure 3A). Furthermore, the effect of fine-scale topographic differences (depressions vs. ridges) on snow disappearance date was

Table 2. Best-fit models for the climatic response variables.

Study	Climatic response variable	Model formula*	r^2
Forest: stratification by vegetation structure	Snow disappearance date	SDD = f { **elev**+**canopy**+understory† + (1|pair) }	0.94
Forest: stratification by vegetation structure	Average daily mean temperature	Tmean = f { **elev**+canopy+understory+elev:canopy+**elev:understory**+**canopy:understory**+(0+canopy|pair) }	0.35
Forest: stratification by vegetation structure	Average daily maximum temperature	Tmax = f { elev+**canopy**+understory+**elev:canopy**+**elev:understory**+**canopy:understory**+(0+canopy|pair) }	0.40
Forest: stratification by vegetation structure	Average daily minimum temperature	Tmin = f { **elev**+canopy+**understory**+**elev:canopy**+**canopy:understory**+(0+canopy|pair) }	0.20
Subalpine/alpine: stratification by topographic position	Snow disappearance date	SDD = f { **side**+**elev**+**topo**+elev:**topo**+**side:elev**+(1|tran) }	0.60
Subalpine/alpine: stratification by topographic position	Average daily mean temperature	Tmean = f { **side**+elev+topo+**elev:topo**+**side:elev** }	0.60
Subalpine/alpine: stratification by topographic position	Average daily maximum temperature	Tmax = f { **side**+elev+topo+**elev:topo**+**side:elev** }	0.52
Subalpine/alpine: stratification by topographic position	Average daily minimum temperature	Tmin = f { side+elev+**topo**+side:elev }	0.30

*Parameters in bold have significant coefficients ($p < 0.05$). For the forest biome study, elev = elevation; canopy = forest canopy. structure, gap or non-gap; understory = understory vegetation structure, removed or control; pair = gap/non-gap pairings. For the subalpine/alpine biomes study, side = side of the mountain, south or northwest or northeast; elev = elevation; topo = topographic position, ridge or depression; tran = sensor deployment transect. The colon indicates an interaction effect between two explanatory variables. The parentheses indicate the term is a random effect – all other terms are fixed effects. If fe is a particular fixed effect and re is a particular random effect, then (1\|re) indicates the intercept was allowed to vary randomly with respect to re while (0+fe\|re) indicates the interaction of fe and re was allowed to vary randomly. The case of both the intercept and interaction being allowed to vary randomly was not included in any of the best-fit models.

†Understory was not included in the best-fit model, based on AIC, but was retained for comparative purposes.

Forest biome

A Snow disappearance date

$r^2 = 0.94$

Effect (days)

Understory | Canopy | Elevation

B Mean temperature

$r^2 = 0.35$

Effect (°C)

Understory | Canopy | Elevation

C Max temperature

$r^2 = 0.40$

Effect (°C)

Understory | Canopy | Elevation

D Min temperature

$r^2 = 0.20$

Effect (°C)

Understory | Canopy | Elevation

Driver of climate

fine-scale ← → coarse-scale

Subalpine/alpine biomes

E Snow disappearance date

$r^2 = 0.60$

Topography | Elevation | Side

F Mean temperature

$r^2 = 0.60$

Topography | Elevation | Side

G Max temperature

$r^2 = 0.52$

Topography | Elevation | Side

H Min temperature

$r^2 = 0.30$

Topography | Elevation | Side

Driver of climate

fine-scale ← → coarse-scale

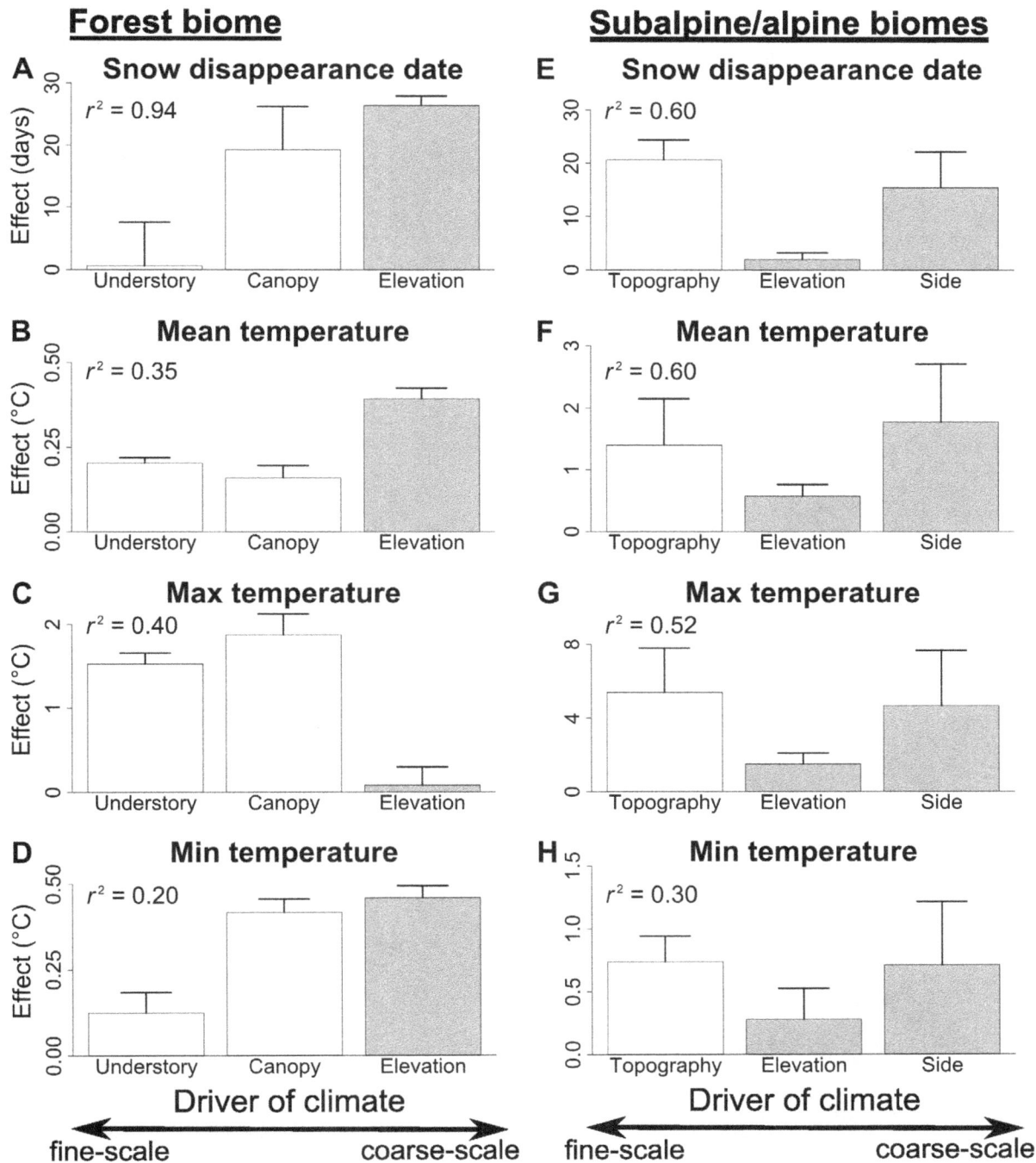

Figure 4. Effects of fine- and coarse-scale drivers of climate. The effects of fine- and coarse-scale drivers of climate on snow disappearance date and the average daily values of mean, maximum and minimum growing season soil temperature. Bars show differences in snow disappearance date or temperature attributed to the effect of different drivers of climate by the best-fit model, with standard error. The effect of elevation was standardized to the effect of a 100 m difference in elevation. Bars filled with gray represent drivers that are coarse enough in scale to be captured by typical climate models (>1 km) while unfilled bars represent drivers too fine in scale to be captured by these models (≤20 m). Fine-scale drivers of climate often had a greater effect on snow or soil temperature than coarse-scale drivers.

similar to the effects of coarse-scale differences in elevation (100 m difference in elevation) and side of the mountain (Figure 4E). In other words, snow disappearance date differed as much over fine spatial scales as it did over coarse spatial scales.

Growing season soil temperatures during our sampling period were lower on the northeast side of the mountain than on the northwest and south sides, potentially because the meadows are at higher elevations on this side of the mountain. On a given side of the mountain, higher elevations (the upper limit of alpine meadows) had higher mean and maximum temperatures, but lower minimum temperatures, than lower elevations (the lower limit of the subalpine biome). Compared to ridges, depressions had

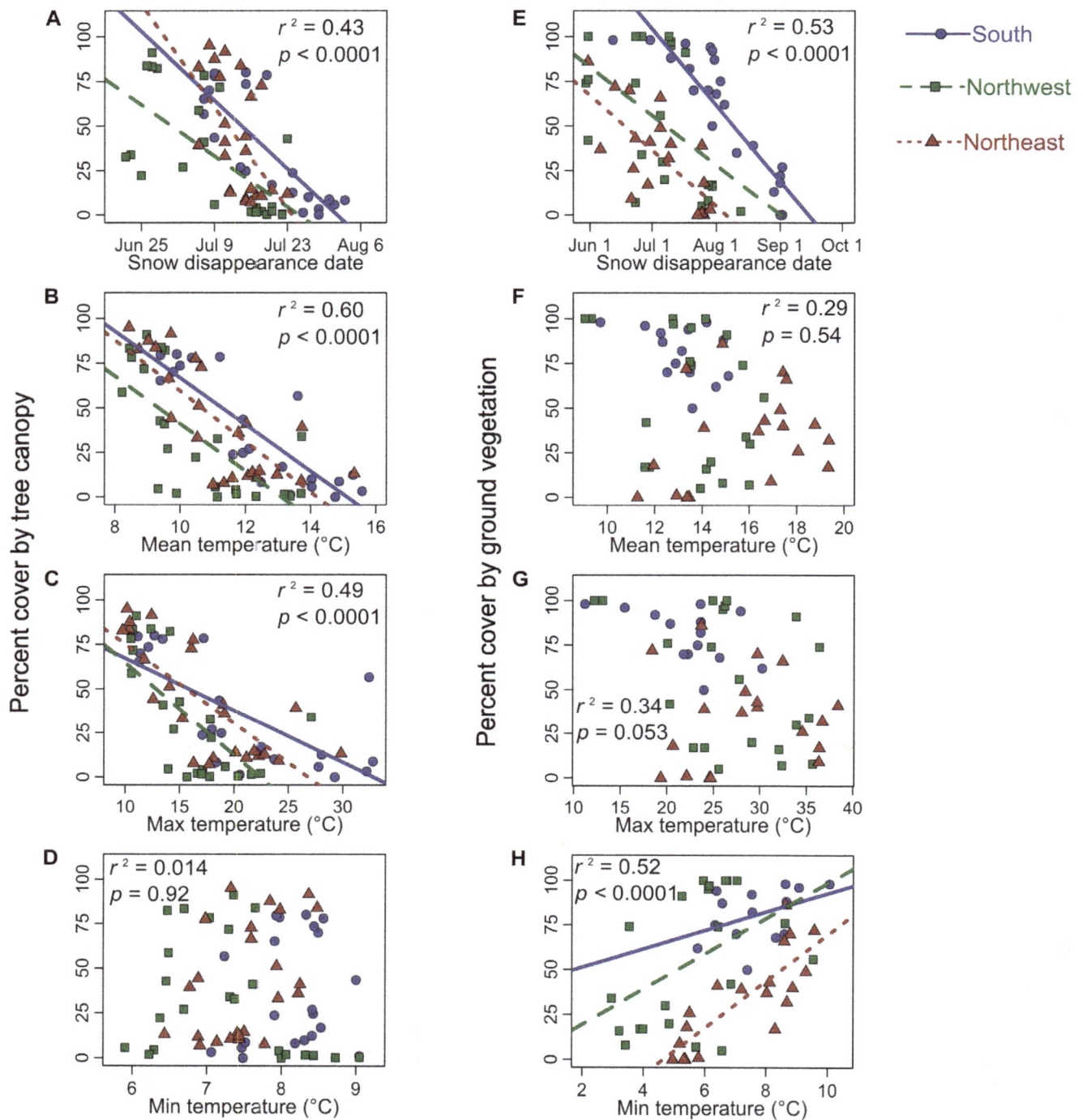

Figure 5. Relationships between vegetation characteristics and microclimate. (A–D) Percent cover by tree canopy at sites near the lower limit of the subalpine biome and (E–H) percent cover by ground vegetation at sites near the upper limit of alpine meadows plotted against the four microclimate variables (snow disappearance date and average daily mean, maximum and minimum soil temperature) on each of the three sides of the mountain. The r^2 values are for models that included the microclimate variable and side of the mountain as explanatory variables, while the p values indicate the significance of the microclimate variable in these models. Regression lines are shown for significant p values (<0.05).

higher mean and maximum temperatures but lower minimum temperatures (Figure 3B–D). We found that the effects of fine-scale topographic differences (depressions vs. ridges) were similar to the effects of coarse-scale differences in elevation and side of the mountain for average daily mean, maximum and minimum

temperature (Figure 4F–H). Overall, there was as much heterogeneity in temperature at fine scales as there was at coarse scales.

Several microclimate variables were significantly correlated with vegetation characteristics (Figure 5). At study sites near the lower limit of the subalpine biome, percent cover by tree canopy was lower where snow disappearance date was later and average daily mean and maximum soil temperatures were higher ($p<0.0001$). At study sites near the upper limit of alpine meadows, percent cover by ground vegetation was lower where snow disappearance date was later and average daily minimum soil temperature was lower ($p<0.0001$).

Discussion

Our study suggests that climatic heterogeneity at the fine spatial scales most organisms experience their environment is substantial, implying that projections based on coarse-scale climate models will not capture the full complexity of range shifts in response to climate change. Specifically, we found large differences in snow disappearance date and growing season soil temperatures over small distances (Figures 3, 4), differences that were sometimes as large as those experienced when travelling hundreds of meters upward in elevation or several kilometers to a different side of the mountain. These microclimate variables have been shown to be strongly associated with plant species distributions and abundances [20,21], suggesting that the microclimate heterogeneity we observed is important for plant communities. We also found that vegetation characteristics (canopy and ground vegetation cover) can be strongly correlated with the microclimate variables influenced by fine-scale topographic features (Figure 5), further suggesting that the fine-scale climatic heterogeneity we observed is ecologically important. However, because we did not measure species distributions or abundances in this study, we cannot conclusively state that the microclimate heterogeneity we observed is linked to species distributions or abundances at Mount Rainier. Nonetheless, understanding fine-scale climatic heterogeneity will likely be critical for management, as cool or snowy microhabitats could provide an important buffer against the negative effects of climate change on biodiversity. Thus, when assessing potential species range shifts in response to climate change, it is critical for ecologists to consider fine-scale patterns in climate in addition to other important factors such as broad-scale climate patterns, dispersal constraints, biotic interactions and evolutionary dynamics.

Explanations of Fine-scale Climatic Heterogeneity

In the forest biome, a complex interplay between elevation and vegetation structure is likely responsible for the heterogeneous patterns in snow disappearance date and soil temperature we observed. For example, locations under tree canopy gaps likely experienced later snow disappearance dates than locations under an intact canopy (Figure 3A) because tree canopies intercept snowfall where it can rapidly sublimate or melt instead of being incorporated into the snowpack on the ground [38]. Tree canopies also increase incoming longwave radiation (which increases ablation rates) and this effect can sometimes be greater than the effect of canopies decreasing incoming shortwave radiation by shading the snowpack (which reduces ablation rates), leading to a net effect of canopies increasing ablation rates [39]. Although the presence of trees has also been shown to lead to longer snow persistence by shading the snowpack and decreasing wind speeds (reducing incoming sensible and latent heat fluxes) [38], these effects appear to be relatively weak at our study sites. Increased shading from tree canopies and understory vegetation in forest locations probably led to substantially lower maximum soil temperatures (Figure 3C). But these low sky exposure locations also experienced higher minimum soil temperatures (Figure 3D), probably due to vegetation emitting more longwave radiation (which warms the surface) than the night sky [14]. Differences in mean soil temperatures appeared to be the net effect of these two

counteracting influences of sky exposure, with mean soil temperatures being higher in the shadier non-gap locations, but lower in the shadier undisturbed understory vegetation locations (Figure 3B).

Similarly, in the subalpine/alpine biomes we found that both coarse- and fine-scale features had large effects on climate. For example, snow disappeared substantially later from depressions in the landscape than from ridges only ~20 m away, likely because snow typically collects in these depressions while it is blown off of ridges and because shading from surrounding slopes can reduce ablation rates [24]. Feedbacks between vegetation and climate are also likely to influence fine-scale climatic variability. At the lower elevation sites, for example, patches of trees with trunks sticking out above the snowpack emit substantial amounts of longwave radiation which quickens the ablation of snow next to the tree patch and can lead to earlier snow disappearance dates. Trees can also intercept snowfall, reducing snowpack accumulation under canopy and resulting in earlier snow disappearance [38]. These effects can lead to a positive feedback, where trees establish in microsites with earlier snow disappearance dates (e.g. ridges), and the established trees lead to even earlier snow disappearance dates and more tree establishment. This result is consistent with previous studies from subalpine meadows in the region that have documented increased tree establishment on ridges that tend to have earlier snow disappearance dates [40,41].

A striking pattern to emerge from our data was that mean and maximum soil temperatures were greater at higher elevations within the subalpine/alpine biomes (though minimum soil temperatures were lower). Feedbacks between climate and vegetation likely play important roles in producing this temperature pattern. First, tree cover declines with increasing elevation, leading to less shading and potentially higher daytime soil temperatures, especially during the sunny growing season when our data were collected. This explanation is supported by the negative correlation we observed between percent canopy cover (a measure of tree density) and mean/maximum soil temperature in the subalpine/alpine biomes (Figure 5B, C). Second, ground vegetation density declines with increasing elevation, which can lead to lower organic matter content in the soil and lower soil moisture levels. The lower moisture levels probably cause the soil to have a lower heat capacity, leading to greater temperature change per unit of energy input and hence higher maximum temperatures and lower minimum temperatures. This second explanation is supported by the pattern of low soil organic matter content and soil water holding capacity at high elevations in Mount Rainier's subalpine/alpine biomes (Appendix S2). Soil characteristics also have important effects on vegetation in subalpine/alpine environments [42], creating the possibility for complex feedbacks amongst soil, vegetation and climate. These two hypotheses are not mutually exclusive, and further study is needed to assess the importance of each. Regardless, our results suggest that even if patterns in climate are ultimately responsible for patterns in vegetation, the feedback effect of vegetation on soil temperature (either directly, or indirectly through the effects of vegetation on soil characteristics which then affect temperature) appears to at times be stronger than the original forcing of physiographic effects on soil temperature.

Implications of Fine-scale Climatic Heterogeneity for Species Distributions in a Warming World

Since snow disappearance date and growing season temperature vary dramatically over short distances, species whose distributions are primarily constrained by these climate variables may not need to migrate long distances to remain in suitable habitat even when there are large changes in climate. For example, in the subalpine/ alpine biomes we found that the average difference in snow disappearance date between depressions and ridges separated by only ~20 m was often one month or more. This is an especially large difference considering the ground is typically only free of snow for 3–5 months out of the year at these elevations. Snow manipulation experiments have shown that differences of this magnitude have large impacts on the phenology, species composition, diversity and productivity of plant communities [20]. Thus, these snowy microhabitats have the potential to serve as refugia for species in a warmer world and provide linkages to new areas of suitable climate, implying that fine-scale climatic heterogeneity could buffer species from climate change [27,43,44], as it may have done during past periods of rapid climate change [45]. Given that we did not stratify our sensors along all gradients likely to produce fine-scale differences in climate (e.g. wind direction, aspect – [14,24]), our results may even be an underestimate of the magnitude of fine-scale heterogeneity.

The importance of topographic heterogeneity for creating climatic heterogeneity shown in this study and others [15,21,25–28] also suggests that mountainous regions will be important for providing climatic refugia in a warming world. However, mountain biotas will still likely face unique challenges. For example, organisms currently living on or near summits will not be able to shift upwards to track suitable climate, and deep valleys between mountains will likely pose serious obstacles to poleward shifts [7]. Broad-scale modeling will continue to be important for addressing these problems. Furthermore, fine-scale environmental heterogeneity does not guarantee that biodiversity will be buffered from climate change. It is possible for heterogeneity to produce small, isolated patches of habitat that cannot support many species, producing a negative effect on diversity [16]. Thus, whether the net effect of heterogeneity on maintaining diversity will be positive during a period of rapid climate change remains an open question.

Different kinds of cool or snowy microhabitats will likely differ in their effectiveness as refugia in a warming world. First, the abundance of microhabitat types will strongly influence how effective they can be as refugia. For example, depressions in the landscape in the subalpine/alpine biomes may have a high likelihood of serving as refugia because they are a common topographic feature. Second, the longevity of microhabitat types will affect their ability to act as refugia. For example, canopy gaps may disappear relatively quickly as trees establish in them, forcing species that might use gaps as refugia to migrate amongst gaps, which may not be possible for some species (though others may be adapted to this migration). In contrast, depressions in the landscape could provide more long-term refugia. Third, the temporal climatic heterogeneity experienced in microhabitat types may affect how well they can serve as refugia. For example, gaps had lower daily minimum and higher daily maximum temperatures compared to non-gaps, showing that these microhabitats experience a wide variety of temperatures. This heterogeneity may favor some species but not others. A final complicating factor influencing how and whether microhabitat types can serve as climatic refugia are the non-climatic conditions associated with them. For example, depressions may differ from other topographic positions in soil characteristics, which could prevent some species from using them as snowy microrefugia.

An important caveat to these findings is that they are based on one year of data. Spatial patterns in climate can change from year to year due to differences in prevailing synoptic weather patterns [46–48], so the patterns we observed in the year we conducted this study may not represent typical spatial patterns. However, the year

of our study was a fairly typical year in terms of snow disappearance date and in terms of growing season air temperature for the past few decades (Appendix S3). And although spatial patterns in climate can vary year to year, the patterns are generally constant from one year to the next, especially in terms of snow [49–52]. For example, locations with later snow disappearance dates in one year tend to have later snow disappearance dates in other years, even though the spatially averaged snow disappearance date varies from year to year.

Challenges and Opportunities for Management

To best protect biodiversity in a period of rapid climate change, conservation biologists and resource managers will require realistic assessments of future species distributions [53]. Thus, incorporating fine-scale climatic heterogeneity is essential for improving projections of species range shifts and extinction risks. Current coarse-scale models of the relationships between climate and species distributions ignore fine-scale heterogeneity and may therefore overestimate the distance species must migrate to track suitable climate (because forecasted range shifts are necessarily at the resolution of the model), and overpredict habitat loss and extinction risks ([54–57], but see [58]). Ecologists have previously criticized these bioclimate envelope models for only predicting where the climate that is currently associated with a species distribution will shift to and failing to account for biotic factors that could affect a species' ability to track these climate shifts (dispersal limitations, biotic interactions and evolutionary changes) [59]. However, the limitation of model spatial resolution could undermine predictions not only of species' abilities to track shifts in climate but also of the climate shifts themselves.

In addition to more realistically forecasting range shifts, knowledge of fine-scale climatic heterogeneity may also allow managers to increase species and ecosystem resilience to climate change. For example, protecting microhabitats with cooler temperatures or later snow disappearance dates could become increasingly important as climate change occurs because these microhabitats may provide critical refugia for species. Additionally, our results suggest that planting seedlings and sowing seeds at a variety of microhabitats when restoring degraded sites is an important bet-hedging strategy because it could increase the probability that species establish in microsites that remain suitable as climate change progresses, even if those microsites are only marginally suitable now. Thus, information on fine-scale climate heterogeneity has the potential to be useful for natural area protection and restoration when taken together with other important factors such as edaphic constraints, biotic interactions, genetic diversity and financial costs [60]. More detailed and longer term studies are needed to assess whether microclimate heterogeneity can contribute substantially to plant establishment and restoration efforts in a warming world.

Conclusions

We have shown that snow disappearance date and growing season soil temperature vary dramatically over small distances due to differences in vegetation structure and topography. In fact, fine-scale features such as gaps in the forest canopy or small depressions in the landscape can produce differences in snow disappearance date and soil temperature as large as those produced by shifting hundreds of meters up a mountain slope. This large degree of fine-scale spatial heterogeneity may provide an important buffer against the negative effects of rapid climate change, as many species may only need to migrate tens of meters from one microhabitat to another in order to track suitable climate, as opposed to shifting hundreds of meters upward in elevation or hundreds of kilometers poleward. Climate change will undoubtedly pose serious threats to biodiversity, but knowledge of fine-scale climatic heterogeneity may allow managers to better assess and potentially alleviate some of these threats.

Acknowledgments

Thanks to Erin Curtis, Rachel Konrady, Tony Krueger, Jack Lee, Gerald Lisi, Anna O'Brien, Mitch Piper, Elli Jenkins Theobald and Courtney Wenneborg for field assistance. We also want to thank Lou Whiteaker and Bret Christoe at Mount Rainier National Park for advice and help with permitting as well as Josh Lawler, Jeremy Littell, Regina Rochefort, Catharine Copass Thompson, Pete Wyckoff and two anonymous reviewers for providing helpful feedback on the manuscript.

Author Contributions

Conceived and designed the experiments: KF AE JHRL. Performed the experiments: KF AE. Analyzed the data: KF. Contributed reagents/materials/analysis tools: KF AE MR JHRL. Wrote the paper: KF AE JL MR JHRL.

References

1. Humboldt Av, Bonpland A (1807) Essay on the geography of plants. Paris: Levrault Schoell. 155 p.
2. Merriam CH, Steineger L (1890) Results of a biological survey of the San Francisco Mountain region and the desert of the Little Colorado, Arizona. North American Fauna Report 3. Washington, DC: US Department of Agriculture, Division of Ornithology and Mammalogy. 136 p.
3. Whittaker RH (1975) Communities and ecosystems. New York: Macmillan. 385 p.
4. Chen I, Hill JK, Ohlemueller R, Roy DB, Thomas CD (2011) Rapid range shifts of species associated with high levels of climate warming. Science 333: 1024–1026. doi: 10.1126/science.1206432.
5. Davis M, Shaw R (2001) Range shifts and adaptive responses to Quaternary climate change. Science 292: 673–679. doi: 10.1126/science.292.5517.673.
6. Huntley B (2005) North temperate responses. In: Lovejoy TE, Hannah L, editors. Climate change and biodiversity. New Haven, Connecticut: Yale University Press. pp. 109–124.
7. Fischlin A, Midgley GF, Price JT, Leemans R, Gopal B, et al. (2007) Ecosystems, their properties, goods, and services. In: Parry ML, Canziani OF, Palutikof JP, Linden PJvd, Hanson CE, editors. Climate change 2007: Impacts, adaptation and vulnerability. Contribution of Working Group II to the Fourth Assessment Report of the Intergovernmental Panel on Climate Change. Cambridge, UK and New York, NY, USA: Cambridge University Press. pp. 211–272.
8. Loarie SR, Duffy PB, Hamilton H, Asner GP, Field CB, et al. (2009) The velocity of climate change. Nature 462: 1052–1057. doi: 10.1038/nature08649.
9. Levin SA (1992) The problem of pattern and scale in ecology. Ecology 73: 1943–1967.
10. Daly C, Halbleib M, Smith JI, Gibson WP, Doggett MK, et al. (2008) Physiographically sensitive mapping of climatological temperature and precipitation across the conterminous United States. Int J Climatol 28: 2031–2064. doi: 10.1002/joc.1688.
11. Hijmans RJ, Cameron SE, Parra JL, Jones PG, Jarvis A (2005) Very high resolution interpolated climate surfaces for global land areas. Int J Climatol 25: 1965–1978. doi: 10.1002/joc.1276.
12. Thuiller W, Lavorel S, Araujo MB, Sykes MT, Prentice IC (2005) Climate change threats to plant diversity in Europe. Proc Natl Acad Sci U S A 102: 8245–8250. doi: 10.1073/pnas.0409902102.
13. Guisan A, Zimmermann NE (2000) Predictive habitat distribution models in ecology. Ecol Model 135: 147–186.
14. Geiger R, Aron RH, Todhunter P (2009) The climate near the ground. Lanham, MD: Rowman & Littlefield Publishers, Inc. 623 p.
15. Ackerly DD, Loarie SR, Cornwell WK, Weiss SB, Hamilton H, et al. (2010) The geography of climate change: Implications for conservation biogeography. Divers Distrib 16: 476–487. doi: 10.1111/j.1472-4642.2010.00654.x.
16. Tamme R, Hiiesalu I, Laanisto L, Szava-Kovats R, Partel M (2010) Environmental heterogeneity, species diversity and co-existence at different spatial scales. Journal of Vegetation Science 21: 796–801. doi: 10.1111/j.1654-1103.2010.01185.x.
17. Willis KJ, Bhagwat SA (2009) Biodiversity and climate change. Science 326: 806–807. doi: 10.1126/science.1178838.
18. Larcher W (2003) Physiological plant ecology: Ecophysiology and stress physiology of functional groups. Berlin: Springer-Verlag. 513 p.
19. Franklin JF, Moir WH, Hemstrom MA, Greene SE, Smith BG (1988) The forest communities of Mount Rainier National Park. Scientific Monograph Series No. 19. Washington, DC: US Department of the Interior, National Park Service. 194 p.
20. Wipf S, Rixen C (2010) A review of snow manipulation experiments in Arctic and alpine tundra ecosystems. Polar Res 29: 95–109. doi: 10.1111/j.1751-8369.2010.00153.x.
21. Scherrer D, Körner C (2011) Topographically controlled thermal-habitat differentiation buffers alpine plant diversity against climate warming. J Biogeogr 38: 406–416. doi: 10.1111/j.1365-2699.2010.02407.x.
22. Mote P, Salathé EP (2009) Future climate in the Pacific Northwest. In: Elsner MM, Littell J, Binder LW, editors. The Washington climate change impacts assessment. Seattle, WA: Climate Impacts Group, Center for Science in the Earth System, Joint Institute for the Study of the Atmosphere and Oceans, University of Washington. pp. 21–43.
23. Elsner MM, Cuo L, Voisin N, Deems JS, Hamlet AF, et al. (2009) Implications of 21st century climate change for the hydrology of Washington State. In: Elsner MM, Littell J, Binder LW, editors. The Washington climate change impacts assessment. Seattle, WA: Climate Impacts Group, Center for Science in the Earth System, Joint Institute for the Study of the Atmosphere and Oceans, University of Washington. pp. 69–106.
24. Clark MP, Hendrikx J, Slater AG, Kavetski D, Anderson B, et al. (2011) Representing spatial variability of snow water equivalent in hydrologic and land-surface models: A review. Water Resources Research 47: doi:10.1029/2011WR010745.
25. Fridley JD (2009) Downscaling climate over complex terrain: High finescale (<1000 m) spatial variation of near-ground temperatures in a montane forested landscape (Great Smoky Mountains). Journal of Applied Meteorology and Climatology 48: 1033–1049. doi: 10.1175/2008JAMC2084.1.
26. Millar CI, Westfall RD (2010) Distribution and climatic relationships of the American pika (Ochotona princeps) in the Sierra Nevada and western Great Basin, USA; periglacial landforms as refugia in warming climates. Arctic Antarctic and Alpine Research 42: 76–88. doi: 10.1657/1938-4246-42.1.76.
27. Scherrer D, Körner C (2010) Infra-red thermometry of alpine landscapes challenges climatic warming projections. Global Change Biol 16: 2602–2613. doi: 10.1111/j.1365-2486.2009.02122.x.
28. Wundram D, Pape R, Loeffler J (2010) Alpine soil temperature variability at multiple scales. Arctic Antarctic and Alpine Research 42: 117–128. doi: 10.1657/1938-4246-42.1.117.
29. Deutsch CA, Tewksbury JJ, Huey RB, Sheldon KS, Ghalambor CK, et al. (2008) Impacts of climate warming on terrestrial ectotherms across latitude. Proc Natl Acad Sci U S A 105: 6668–6672. doi: 10.1073/pnas.0709472105.
30. Meehl GA, Stocker TF, Collins WD, Friedlingstein P, Gaye AT, et al. (2007) Global climate projections. In: Solomon S, Qin D, Manning M, Chen Z, Marquis M, et al, editors. Climate change 2007: The physical science basis. Contribution of Working Group I to the Fourth Assessment Report of the Intergovernmental Panel on Climate Change. Cambridge, UK and New York, NY, USA: Cambridge University Press. pp. 747–845.
31. Lundquist JD, Lott F (2008) Using inexpensive temperature sensors to monitor the duration and heterogeneity of snow-covered areas. Water Resour Res 44: W00D16. doi: 10.1029/2008WR007035.
32. Raleigh MS, Rittger K, Moore C, E., Henn B, Lutz J, A., et al. (2013) Ground-based testing of MODIS fractional snow cover in subalpine meadows and forests of the Sierra Nevada. Remote Sensing of Environment 128: 44–57.
33. Zuur AF, Ieno EN, Walker NJ, Saveliev AA, Smith GM (2009) Mixed effects models and extensions in ecology with R. New York: Springer. 574 p.
34. R Development Core Team (2010) R: A language and environment for statistical computing.
35. Bates D, Maechler M, Bolker B (2011) lme4: Linear mixed-effects models using S4 classes. R package version 0.999375-39.
36. Hadfield JD (2010) MCMC methods for multi-response generalized linear mixed models: The MCMCglmm R package. Journal of Statistical Software 33: 1–22.
37. Xu RH (2003) Measuring explained variation in linear mixed effects models. Stat Med 22: 3527–3541. doi: 10.1002/sim.1572.
38. Varhola A, Coops NC, Weiler M, Moore RD (2010) Forest canopy effects on snow accumulation and ablation: An integrative review of empirical results. Journal of Hydrology 392: 219–233. doi: 10.1016/j.jhydrol.2010.08.009.
39. Sicart JE, Pomeroy JW, Essery RLH, Hardy J, Link T, et al. (2004) A sensitivity study of daytime net radiation during snowmelt to forest canopy and

atmospheric conditions. J Hydrometeorol 5: 774–784. doi: 10.1175/1525-7541(2004)005<0774:ASSODN>2.0.CO;2.

40. Rochefort RM, Peterson DL (1996) Temporal and spatial distribution of trees in subalpine meadows of Mount Rainier National Park, Washington, USA. Arct Alp Res 28: 52–59.

41. Zald HSJ, Spies TA, Huso M, Gatziolis D (2012) Climatic, landform, microtopographic, and overstory canopy controls of tree invasion in a subalpine meadow landscape, Oregon Cascades, USA. Landscape Ecol 27: 1197–1212. doi: 10.1007/s10980-012-9774-8.

42. Körner C (2003) Alpine plant life: Functional plant ecology of high mountain ecosystems, 2nd edition. Berlin: Springer. 344 p.

43. Dobrowski SZ (2011) A climatic basis for microrefugia: The influence of terrain on climate. Global Change Biol 17: 1022–1035. doi: 10.1111/j.1365-2486.2010.02263.x.

44. Keppel G, Van Niel KP, Wardell-Johnson GW, Yates CJ, Byrne M, et al. (2012) Refugia: Identifying and understanding safe havens for biodiversity under climate change. Global Ecol Biogeogr 21: 393–404. doi: 10.1111/j.1466-8238.2011.00686.x.

45. Hof C, Levinsky I, Araujo MB, Rahbek C (2011) Rethinking species' ability to cope with rapid climate change. Global Change Biol 17: 2987–2990. doi: 10.1111/j.1365-2486.2011.02418.x.

46. Lundquist JD, Cayan DR (2007) Surface temperature patterns in complex terrain: Daily variations and long-term change in the central Sierra Nevada, California. Journal of Geophysical Research-Atmospheres 112: D11124. doi: 10.1029/2006JD007561.

47. Bennie JJ, Wiltshire AJ, Joyce AN, Clark D, Lloyd AR, et al. (2010) Characterising inter-annual variation in the spatial pattern of thermal microclimate in a UK upland using a combined empirical-physical model. Agric For Meteorol 150: 12–19. doi: 10.1016/j.agrformet.2009.07.014.

48. Lundquist JD, Minder JR, Neiman PJ, Sukovich E (2010) Relationships between barrier jet heights, orographic precipitation gradients, and streamflow in the northern Sierra Nevada. J Hydrometeorol 11: 1141–1156. doi: 10.1175/2010JHM1264.1.

49. Erickson T, Williams M, Winstral A (2005) Persistence of topographic controls on the spatial distribution of snow in rugged mountain terrain, Colorado, United States. Water Resour Res 41: W04014. doi: 10.1029/2003WR002973.

50. Deems JS, Fassnacht SR, Elder KJ (2008) Interannual consistency in fractal snow depth patterns at two Colorado mountain sites. J Hydrometeorol 9: 977–988. doi: 10.1175/2008JHM901.1.

51. Sturm M, Wagner AM (2010) Using repeated patterns in snow distribution modeling: An Arctic example. Water Resour Res 46: W12549. doi: 10.1029/2010WR009434.

52. Egli L, Jonas T, Gruenewald T, Schirmer M, Burlando P (2012) Dynamics of snow ablation in a small alpine catchment observed by repeated terrestrial laser scans. Hydrol Process 26: 1574–1585. doi: 10.1002/hyp.8244.

53. Hannah L, Midgley G, Andelman S, Araujo M, Hughes G, et al. (2007) Protected area needs in a changing climate. Frontiers in Ecology and the Environment 5: 131–138.

54. Austin MP, Van Niel KP (2011) Impact of landscape predictors on climate change modelling of species distributions: A case study with *Eucalyptus fastigata* in southern New South Wales, Australia. J Biogeogr 38: 9–19. doi: 10.1111/j.1365-2699.2010.02415.x.

55. Barrows CW, Murphy-Mariscal ML (2012) Modeling impacts of climate change on Joshua trees at their southern boundary: How scale impacts predictions. Biol Conserv 152: 29–36. doi: 10.1016/j.biocom.2012.03.028.

56. Luoto M, Heikkinen RK (2008) Disregarding topographical heterogeneity biases species turnover assessments based on bioclimatic models. Global Change Biol 14: 483–494. doi: 10.1111/j.1365-2486.2007.01527.x.

57. Randin CF, Engler R, Normand S, Zappa M, Zimmermann NE, et al. (2009) Climate change and plant distribution: Local models predict high-elevation persistence. Global Change Biol 15: 1557–1569. doi: 10.1111/j.1365-2486.2008.01766.x.

58. Trivedi MR, Berry PM, Morecroft MD, Dawson TP (2008) Spatial scale affects bioclimate model projections of climate change impacts on mountain plants. Global Change Biol 14: 1089–1103. doi: 10.1111/j.1365-2486.2008.01553.x.

59. Pearson RG, Dawson TP (2003) Predicting the impacts of climate change on the distribution of species: Are bioclimate envelope models useful? Global Ecol Biogeogr 12: 361–371.

60. Rochefort RM, Kurth LL, Carolin TW, Riedel JL, Mierendorf RR, et al. (2006) Mountains. In: Apostol D, Sinclair M, editors. Restoring the Pacific Northwest: The art and science of ecological restoration in Cascadia. Washington, DC: Island Press. pp. 241–275.

18

On the Applicability of Temperature and Precipitation Data from CMIP3 for China

Chiyuan Miao[1]*, Qingyun Duan[1], Lin Yang[2], Alistair G. L. Borthwick[3]

1 State Key Laboratory of Earth Surface Processes and Resource Ecology, College of Global Change and Earth System Science, Beijing Normal University, Beijing, People's Republic of China, **2** State Key Laboratory of Resources and Environmental Information System, Institute of Geographical Sciences and Natural Resources Research, Chinese Academy of Sciences, Beijing, People's Republic of China, **3** Department of Civil & Environmental Engineering, University College Cork, Cork, Ireland

Abstract

Global Circulation Models (GCMs) contributed to the Intergovernmental Panel on Climate Change (IPCC) Fourth Assessment Report (AR4) and are widely used in global change research. This paper assesses the performance of the AR4 GCMs in simulating precipitation and temperature in China from 1960 to 1999 by comparison with observed data, using system bias (*B*), root-mean-square error (*RMSE*), Pearson correlation coefficient (*R*) and Nash-Sutcliffe model efficiency (*E*) metrics. Probability density functions (PDFs) are also fitted to the outputs of each model. It is shown that the performance of each GCM varies to different degrees across China. Based on the skill score derived from the four metrics, it is suggested that GCM 15 (ipsl_cm4) and GCM 3 (cccma_cgcm_t63) provide the best representations of temperature and precipitation, respectively, in terms of spatial distribution and trend over 10 years. The results also indicate that users should apply carefully the results of annual precipitation and annual temperature generated by AR4 GCMs in China due to poor performance. At a finer scale, the four metrics are also used to obtain best fit scores for ten river basins covering mainland China. Further research is proposed to improve the simulation accuracy of the AR4 GCMs regarding China.

Editor: Juan A. Añel, University of Oxford, United Kingdom

Funding: This work was financially supported by the National Natural Science Foundation of China (no. 41001153), the National Key Basic Special Foundation Project of China (no. 2010CB951604 and no. 2010CB428402), and the Fundamental Research Funds for the Central Universities. The funders had no role in study design, data collection and analysis, decision to publish, or preparation of the manuscript.

Competing Interests: The authors have declared that no competing interests exist.

* E-mail: miaocy@vip.sina.com

Introduction

The global temperature has significantly increased in recent decades, according to both direct measurements and credible proxy data. The IPCC-AR4 indicates that the rise in global-average surface temperature has been particularly pronounced since about 1950, with an updated trend of 0.74±0.18 °C during 1906–2005 [1]. IPCC-AR4 also estimates that on average a warming of about 0.2 °C/decade may occur during the next two decades. Another key variable, precipitation is expected to increase under global warming at high latitudes and in the vicinity of the equator, but decrease in the subtropics [2]. It is believed that precipitation will experience an overall increase on average due to there being greater evaporation [3]. Climate change can have major impacts on vulnerable natural systems and sensitive human systems at local, regional and national scales. Accordingly, there is an urgent need for an improved understanding of climate change, its consequences, and mitigation and adaptation strategies [4].

To provide information to support IPCC-AR4, more than 20 modeling groups around the world conducted climate change simulations using different GCMs. These IPCC-AR4 GCMs can be used to simulate present-day and projected future climate conditions under different scenarios, and hence inform decision makers regarding potential mitigation measures and adaptation strategies. However, the theoretical description of the climate remains incomplete, and simplifying assumptions are inherent when building these GCMs [5]. Epistemic and aleatory uncer-

tainties in climate models introduce biases into the simulations, and so GCMs are unable to represent fully the intensity and frequency of observed data on climate characteristics [6–8]. Several researchers have assessed the performance of GCMs from the global [5], national [9,10] and regional [11,12] scales respectively. The results have demonstrated that not all GCMs are able to provide a similarly accurate description of the present climate [13–18]. Furthermore, it should be noted that the performance of the AR4 GCMs is not uniformly consistent over large geographical areas, especially for the extreme climate variables [11,12]. Consequently, the accuracy of any GCM should be established through validation studies before using it to predict future climate scenarios [19]. Although accurate simulation of the present climate does not guarantee that forecasts of future climate will be reliable [5], it is generally accepted that the agreement of model predictions with present observations is a necessary prerequisite in order to have confidence in the quality of a model [17], and models that reproduce accurately the present climate are more likely to provide reasonably accurate predictions of future climate [20].

China has experienced gradual warming throughout the 20th Century consistent with the warming observed at global scale. It was reported that the mean annual surface air temperature in mainland China increased by about 1.3 °C from 1951 to 2004 due to the greenhouse effect and rapid urbanization [21]; the warming rate of about 0.25 °C/decade is more than twice the global warming rate. No significant trend in mean precipitation occurred during this period taking China as a whole [22], however, the North and northeast regions experienced a 12% decline in precipitation from 1960 to 2005 while the South had increasing rainfall during the summer and winter seasons [23] due to the East Asian Monsoon variability [24–26]. In short, the climate in China varies considerably in space and time due to the scale and complexity of its land topography [27].

The present paper aims to assess the performance of the IPCC-AR4 GCMs in the simulation of precipitation and temperature throughout mainland China (excluding Taiwan island) from the spatial scale (country and large river basin) and temporal scale (intra- and inter- annual) respectively.

Materials and Methods

Data

Observed data on surface air temperature and precipitation for the period from 1960 to 1999 were obtained from the National Meteorological Information Center, China Meteorological Administration. Daily measurements of daily temperature and precipitation were acquired from a total of 731 meteorological stations (Figure 1), and subjected to quality control processes including homogenization, cross-validation, and topographic correction [28]. Following Chinese Bureau of Meteorology Standards, monthly and annual climatic datasets were derived from daily data, and interpolated onto a grid at $1° \times 1°$ resolution comprising a total of 1023 cells covering mainland China. Further details of the quality control processes and the archived raw data are given by the National Meteorological Information Center (NMIC, available at http://cdc.cma.gov.cn). It should also be noted that the observed data from each meteorological station were interpolated onto the grid with topographic correction provided by a high resolution digital elevation model (DEM). However, topographic corrections were not applied during the grid interpolations of the outputs of the AR4 GCMs, which unavoidably impacts on model accuracy.

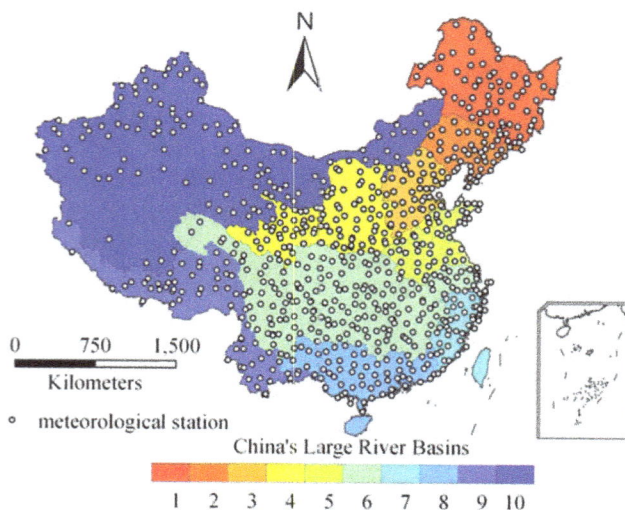

Figure 1. Locations of meteorological stations and major river basins in mainland China. The color coding relates to the following river basins: 1 Songhua River; 2 Liaohe River; 3 Haihe River; 4 Yellow River; 5 Huaihe River; 6 Yangtze River; 7 southeast drainage area rivers; 8 Pearl River; 9 southwest drainage area rivers; 10 northwest drainage area rivers.

Monthly temperature and precipitation simulation data were produced by the 24 AR4 GCMs as part of the Coupled Model Intercomparison Project Phase 3 (CMIP-3) of the World Climate Research Programme (WCRP). The data are stored in a multi-model dataset [29,30] archived by the Program for Climate Model Diagnosis and Intercomparison (PCMDI, available at https://esg. llnl.gov:8443/index.jsp). The CMIP-3 data outputs used herein are taken from the Twentieth Century (20C3M) experiment, which has the most realistic forcings. Table 1 lists the various models. Further details of model status, model documentation, related references, etc., are available from the website http://www-pcmdi.llnl.gov/ipcc/info_for_analysts.php. The spatial resolution of the models used in the present analysis varied from $1.125° \times 1.125°$ for GCM 16 (miroc3_2_hires model) to $5° \times 4°$ for GCM 10 (giss_model_e_h model) [1]. For comparison purpose, all model results were interpolated to the same resolution as that of the observed data ($1° \times 1°$ grid). All model outputs are taken from a single result (run 1).

Skill score metrics

In analyzing the simulated temperature and precipitation patterns from 1960 to 1999, four metrics were used to indicate the overall agreement between the predictions P from each AR4 GCM and the measured observations O. The first metric is the system bias (B) between the 40-year mean values of the simulated and observed data:

$$B = \sum (\bar{P} - \bar{O}) \tag{1}$$

in which the overbar indicates a time-average.

The second metric is the spatially (and temporally) aggregated root-mean-square error ($RMSE$) between the simulated and observed data [7]:

$$RMSE = \sqrt{\frac{1}{n} \sum_{i=1}^{n} (P_i - O_i)^2} \tag{2}$$

where the summation is taken over a total of n spatial grid points (or temporal units). Thus, the smaller is the value of $RMSE$, the closer are the point-wise magnitudes of the simulated and observed climate characteristics.

The third metric is the Pearson correlation coefficient that quantifies similarities between the spatial (and temporal) patterns of the predicted and the observed values:

$$R = \frac{\sum_{i=1}^{n} (P_i - \bar{P})(O_i - \bar{O})}{\sqrt{\sum_{i=1}^{n} (P_i - \bar{P})^2} \cdot \sqrt{\sum_{i=1}^{n} (O_i - \bar{O})^2}} \tag{3}$$

where, again, all summations are over a total of n spatial grid points (or temporal units). The Pearson correlation coefficient $R \in [-1, 1]$, with $R \sim 1$ implying a close match between the (spatial or temporal) patterns of simulated and observed climate characteristics, and $R \sim 0$ indicating a lack of similarity. When $R \sim -1$, the respective simulated and observed fields are similar in pattern, but their point-wise (spatial or temporal) variations are oppositely signed.

The fourth metric is the Nash-Sutcliffe model efficiency (E) [31] that assesses quantitatively the accuracy of the (spatial or temporal) patterns of the model outputs from:

Table 1. List of the global climate models used in this research.

GCM	Model	Source
1	bccr_bcm2_0	Bjerknes Centre for Climate Research, Norway
2	cccma_cgcm3_1	Canadian Centre for Climate Modelling and Analysis
3	cccma_cgcm_t63	Canadian Centre for Climate Modelling and Analysis
4	cnrm_cm3	Centre National de Recherches Meteorologiques, France
5	csiro_mk3_0	Australian Commonwealth Scientific and Research Org.
6	csiro_mk3_5	Australian Commonwealth Scientific and Research Org.
7	gfdl_cm2_0	Geophysical Fluid Dynamics Laboratory, United States
8	gfdl_cm2_1	Geophysical Fluid Dynamics Laboratory, United States
9	giss_aom	Goddard Institute of Space Studies(NASA), United States
10	giss_model_e_h	Goddard Institute of Space Studies(NASA), United States
11	giss_model_e_r	Goddard Institute of Space Studies(NASA), United States
12	iap_fgoals1_0_g	Institute of Atmospheric Physics, China
13	ingv_echam4	National Institute of Geophysics and Volcanology, Italy
14	inmcm3_0	Institute for Numerical Mathematics, Russia
15	ipsl_cm4	Institut Pierre Simon Laplace, France
16	miroc3_2_hires	Center for Climate System Research, Japan
17	miroc3_2_medres	Center for Climate System Research, Japan
18	miub_echo_g	Meteorological Institute of the University of Bonn, Germany
19	mpi_echam5	Max-Planck-Institute for Meteorology, Germany
20	mri_cgcm2_3_2a	Meteorological Research Institute, Japan
21	ncar_ccsm3_0	NCAR Community Climate System Model, USA
22	ncar_pcm1	NCAR Parallel Climate Model, USA
23	ukmo_hadcm3	Hadley Centre for Climate Prediction, UK
24	ukmo_hadgem1	Hadley Centre for Climate Prediction, UK

$$E = 1 - \frac{\sum\limits_{i=1}^{n} (O_i - P_i)^2}{\sum\limits_{i=1}^{n} (O_i - \bar{O})^2} \quad (4)$$

where $E \in [-\infty, 1]$. In interpreting the results, it should be noted that $E \sim 1$ indicates better accuracy, $E = 0$ indicates the predictions have accuracy equal to the mean of the observations, and $E < 0$ indicates that the observed mean is better than the model as a predictor [32].

Results

System bias analysis

Figure 2 shows the system bias for each of the 24 AR4 GCMs obtained by comparing the 40-year overall mean values of the observed and simulated data on annual mean precipitation and temperature. The GCMs give reasonably accurate predictions of the temperature, but are less successful at reproducing the precipitation. All GCMs overestimate precipitation throughout China, with GCM 10 (giss_model_e_h) giving the maximum system bias, which is almost double the annual mean precipitation. It should be noted that the 40-year mean of observed precipitation is 563.1 mm/yr. For temperature, 18 models underestimate the annual mean temperature, and the rest overestimate. The maximum system bias (for temperature simulation) comes from GCM 7 (gfdl_cm2_0) with a value of −4.30 K/yr. GCM 16 (miroc3_2_hires) performs best with a low system bias of 0.065 K/yr.

Spatial simulations after removal of system bias

In order to assess systematically the performance of the AR4 GCMs, the system bias is removed from each dataset by multiplying the monthly model data by the ratio of overall mean model data to overall mean observed data. Figure 3 shows that the calculated spatial correlation coefficients obtained are invariably over 0.5, except for GCM 12 (iap_fgoals1_0_g) and GCM 22 (ncar_pcm1). This indicates that the majority of the AR4 GCMs provide satisfactory simulations of the spatial distribution of annual mean precipitation during 1960–1999. GCM 10 (giss_model_e_h) and GCM 11 (giss_model_e_r) give the poorest simulations comparatively speaking, with $E < 0$ and the highest values of RMSE. GCM 23 (ukmo_hadcm3) is the best overall at simulating the spatial distribution of annual precipitation (with $R = 0.85$ and $E = 0.71$). Turning to annual mean temperature, the spatial correlation coefficients obtained for the AR4 GCMs are invariably over 0.8, and, in general, it can be seen from Figure 3 that the AR4 GCMs are better at simulating the spatial distribution of annual mean temperature than that of precipitation. Similar to the comparative performance for annual mean precipitation, GCM 10 (giss_model_e_h) and GCM 11 (giss_model_e_r) perform worst spatially for annual mean temperature in terms of E and RMSE. GCM 13 (ingv_echam4) gives the most accurate results in terms of the spatial distribution of the annual mean temperature data, with $R = 0.96$ and $E = 0.93$.

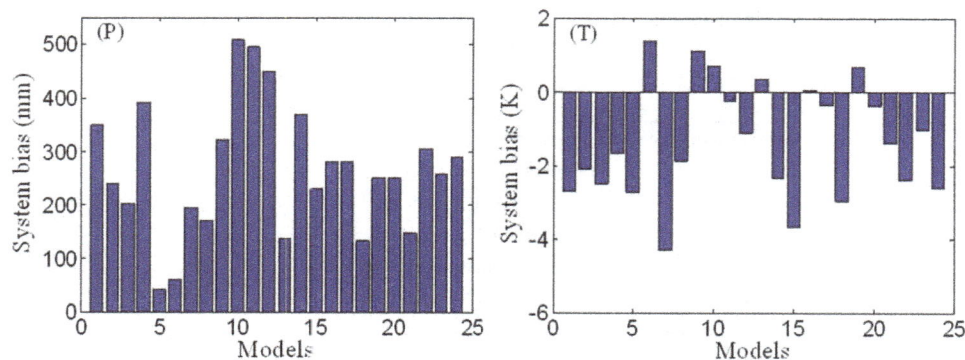

Figure 2. Bar chart indicating the system bias of different AR4 GCMs with regard to annual mean precipitation (*P*) and temperature (*T*) in mainland China during 1960–1999.

Figure 4 presents the best performance spatial distribution simulations (from GCM 23 - ukmo_hadcm3 for precipitation and GCM 13 - ingv_echam4 for temperature). The plots confirm that the AR4 GCMs reproduce the important spatial characteristics of precipitation and temperature in mainland China. The climate warm and wet in South China, but cool and dry in northwest China and northeast China. Figure 4 also provides error contours obtained using the best performing models. Regions where the precipitation error is largest are indicated by the superimposed ellipses in southeast China and West China. The maximum error in temperature simulation occurs in northeast China and West China.

Temporal simulations after removal of system bias

Figure 5 shows the temporal performance of AR4 GCMs in simulating annual mean precipitation and annual mean temperature for mainland China. By comparison with Figure 3, the results presented in Figure 5 show that the GCMs provide less accurate inter-annual temporal than spatial simulations. This is especially the case for precipitation where, for all GCMs, $R<0.4$ and $E<0$, implying that the observed mean is a better predictor than the model. For temperature simulation, R is mainly between 0.3–0.5, and E remains unacceptable. It is generally accepted that the warming in the late 20th century in AR4 GCMs was likely mainly due to increases of greenhouse gases [9,33]. Hence, the time series after linear detrending will represent the performance of GCMs in simulating inter-annual variabilities more objective. It is found that the correlation after linear detrending between the observation and the GCM simulation is weaker further ($R<0.2$, $E<0$ for precipitation, and $R<0.3$, $E<0$ for temperature in all GCMs) when comparing with the original time series. Consequently, direct use of the GCM outputs to model the inter-annual variation is not recommended in mainland China, especially for the precipitation simulation.

Figure 6 illustrates the performance of the AR4 GCMs in temporal simulation of the 10-year moving average precipitation and temperature. Although the AR4 GCMs are much less accurate at representing the 10-year moving average precipitation than the temperature, the precipitation results are nevertheless considerably improved in comparison with the temporal inter-annual variability of mean precipitation (in Figure 5); and the results from GCM 3 (cccma_cgcm_t63) appear to be relatively acceptable. Turning to 10-year moving average temperature, it can be seen that the trend is accurately simulated by all the models except GCM 7 (gfdl_cm2_0) and GCM 22 (ncar_pcm1). GCM 16 (miroc3_2_hires) performs best with $E=0.83$ and $R=0.96$. In

short, the results imply that the AR4 GCM simulations give an approximate view of the inter-decadal and long-term trends of temperature over China.

Figure 7 plots the simulated and observed forty-year average monthly precipitation and temperature in mainland China. The observed precipitation and temperature results show the steep onset of summer rainfall associated with the summer monsoon, which peaks sharply in July. Almost all the AR4 GCMs succeed in capturing the seasonal variation characteristics of a single peak, except the monthly precipitation simulations by GCMs 10 (giss_model_e_h), 11 (giss_model_e_r) and 18 (miub_echo_g). Almost all the AR4 GCMs overestimate the precipitation in winter and spring, and underestimate the precipitation in summer. The resulting gross estimation error implies that these models are unlikely to be directly useful for hydrological impact assessment. In general, the AR4 GCMs simulate the forty-year average monthly temperature more accurately than the corresponding precipitation. All models capture the bell-shape of the forty-year average monthly temperature profile. However, certain models (i.e. GCM 5 - csiro_mk3_0, GCM 14 - inmcm3_0 and GCM 24 - ukmo_hadgem1) predict temperatures that are too hot in summer and too cold in winter. Other models (such as GCM 19 - giss_aom and GCM 18 miub_echo_g) predict a climate that is too cool in summer and too warm in winter. Overall, GCM 3 (cccma_cgcm_t63) gives the most accurate forty-year average climate simulation.

Probability density functions after removal of system bias

Figure 8 shows the probability density functions (PDFs) of annual mean precipitation and annual mean temperature. The PDF for the observed annual mean precipitation covers a range from 475 mm to 678 mm, and is slightly skewed (due to the monsoon effect). In general, the simulated PDF for annual mean precipitation is similar but narrower and taller than the observed PDF, especially for GCM 9 (giss_aom) and GCM 20 (mri_cgcm2_3_2a) which give the two highest peaks in the left hand plot of Figure 8. The PDF for observed annual mean temperature ranges from 277.8 K to 280.5 K, and is again asymmetric with a steep rising limb and a broader undular tail. Typically, the AR4 GCMs simulate a wider, lower PDF profile for annual mean temperature that is symmetric and possibly Gaussian, except GCM 1 (bccr_bcm2_0) and GCM 5 (csiro_mk3_0), which give the two largest peaks in the right hand plot of Figure 8.

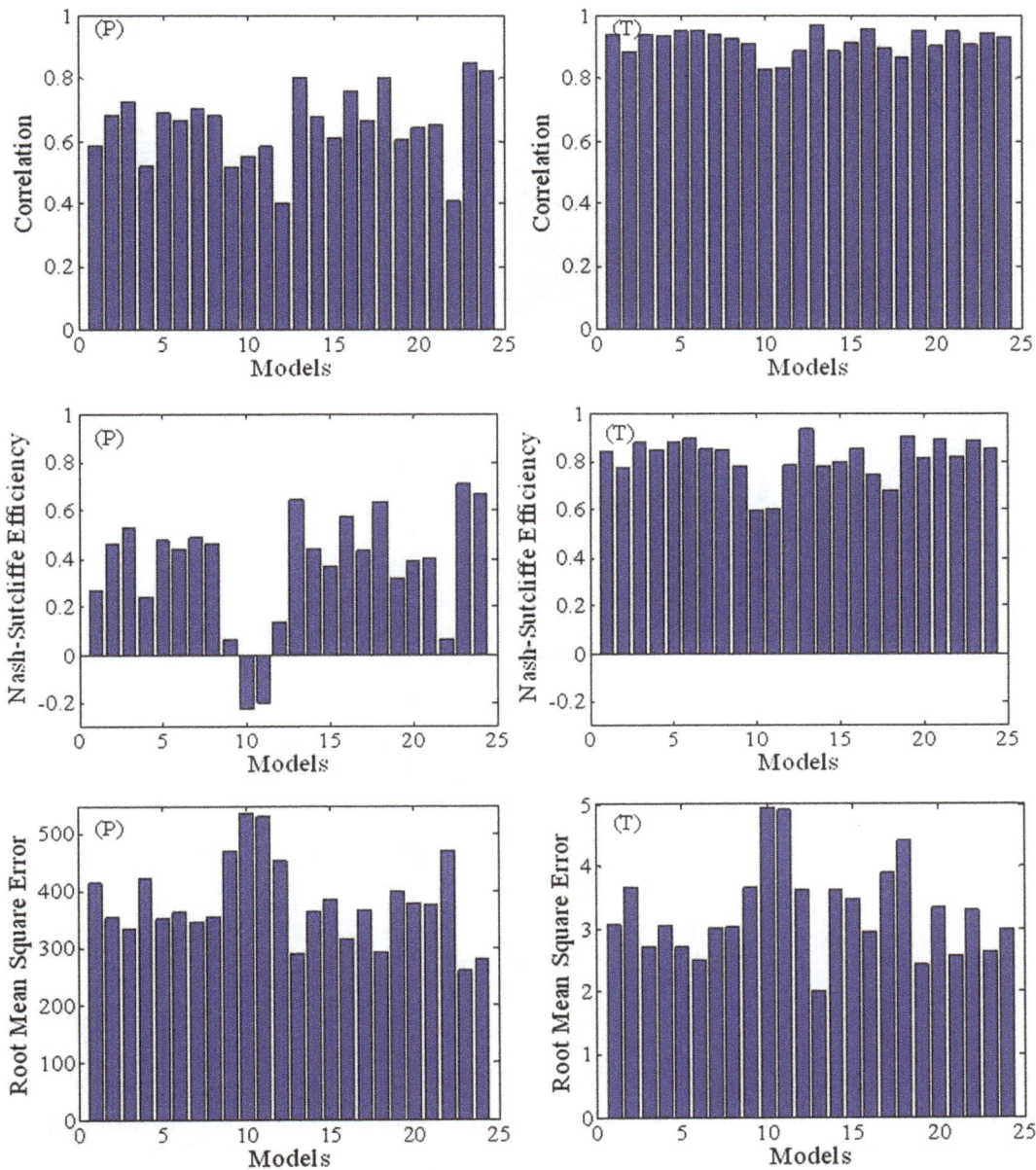

Figure 3. Bar chart indicating relative performances of the AR4 GCMs with regard to the spatial simulation of annual mean precipitation (*P*) mm and temperature (*T*) K in mainland China during 1960–1999.

Discussion

The present study has demonstrated that the AR4 GCMs exhibit a wide range of performance skills in reproducing the recent (1960–1999) observed climate throughout mainland China. Measured and simulated surface air temperatures and precipitations have been interpreted in terms of spatial distributions, inter-annual and intra-annual trends, and PDFs in order to evaluate the performance of the AR4 GCMs. The results demonstrate that certain models are unsuitable for application to China, with little capacity to simulate the spatial variations in climate across the country. It should be emphasized however that the present conclusions should not be generalized to other climate variables or to other regions of the world.

In general, the simulations are more accurate in space than time, and temperature is better simulated than precipitation. When carrying out research into continental precipitation, it is found that the AR4 GCMs exhibit systematic model bias with most models displaying aggregated precipitation variability magnitudes that are larger than observed [19]. The present study shows similar overestimation to be the case for AR4 GCM simulations of temperature and precipitation in China. The ubiquitous system bias means that caution should be applied when using outputs from the AR4 GCMs in hydrological and ecological assessments..

Several potentially disturbing factors complicate the agreement between the models and reality. The first is model resolution. It is believe that higher resolution does not automatically lead to improved model accuracy [6]. The present research confirms this

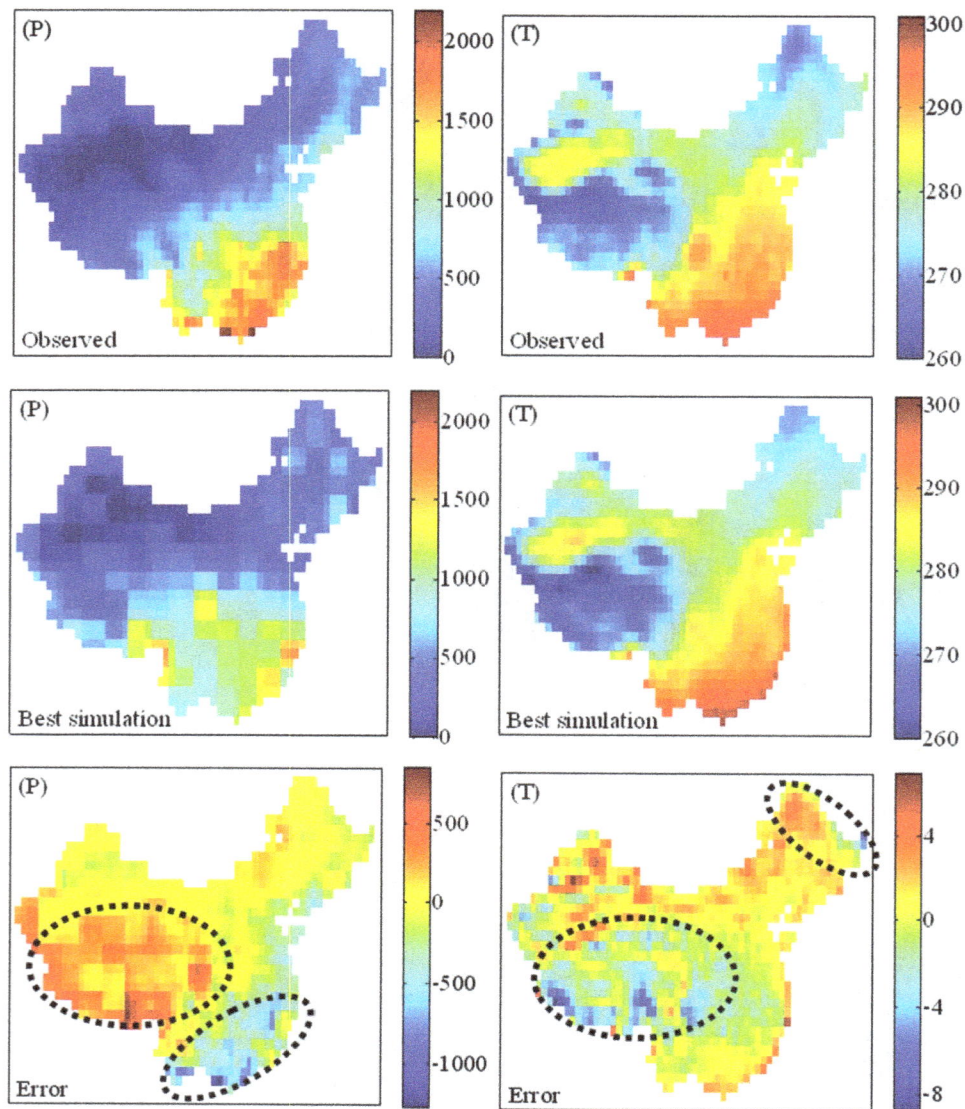

Figure 4. Spatial simulation results for annual mean precipitation (*P*) and temperature (*T*) distributions in mainland China. Best simulation means the best model's simulated results; ukmo_hadcm3 is the best performance precipitation simulation and ingv_echam4 is the best performance temperature simulation; Error is the error between the observed and the best model simulation.

view. Whereas GCM 13 (ingv_echam4) gives the best spatial temperature simulation, its resolution is only moderate. However, the degree of resolution must have an effect on the accuracy of the spatial simulation of a given model, even after the model outputs have been interpolated onto a grid of uniform resolution ($1° \times 1°$ in the present study).

The second factor relates to the quality of the observed data. Meteorological stations are non-uniformly distributed in China, with stations particularly sparse in the West and northwest of China. Although interpolation is used to deal with the scarcity of meteorological data in these regions, the results do not properly represent the actual climate due to limitations of the interpolation techniques used. The lack of meteorological stations in the West of China is therefore mainly responsible for the occurrence of the largest errors in this region (Figure 4). On the other hand, the observed climatological records often contain inhomogeneities, which is defined as a change point (a time point in a series such

that the observations have a different distribution before and after this time) in the data series [34]. The causes of inhomogeneity can be induced by several non-climatic factors: changes in measurement practices, station relocations, changes in the surroundings of a station over the years, etc. [35]. The homogeneity of observed data has not been detected in this research. If the inhomogeneity is identified and then homogenization techniques are performed to compensate for the biases produced by the inhomogeneities, it is possible potentially improve the agreement of AR4 GCMs with the observations.

The third factor is scale. The AR4 GCMs were used to simulate changes in the climate as a result of slow alterations to certain parameters (such as the greenhouse gas concentration and the solar constant), which affect the energy balance at the global scale. Previous research has shown that data from the AR4 GCMs can accurately reproduce the spatial variations in climate characteristics in Iberia [14], Australia [36], North America [37], and global

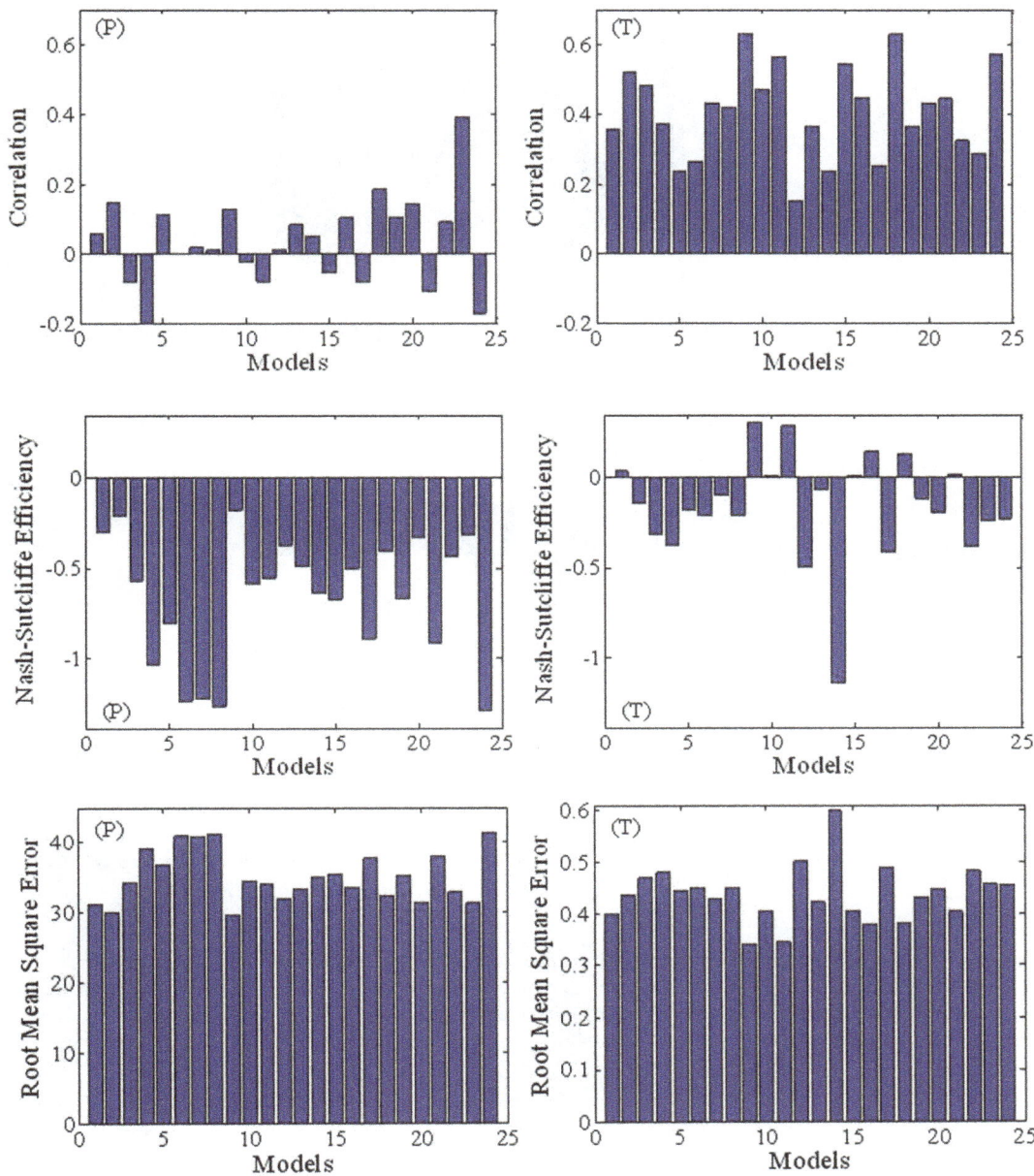

Figure 5. Bar chart indicating relative performances of the AR4 GCMs regarding the temporal simulation of inter- annual mean precipitation (P) and temperature (T) in mainland China during 1960–1999.

continents [19]. However, China is a region of particularly complicated topography, with the Tibetan Plateau to the West and various mountain chains in the northern and central regions [27]. China lies mainly in the northern temperate zone and experiences an annual monsoon season. Consequently, China's climate differs considerably from region to region, making accurate local simulation by the AR4 GCMs less likely. This scale discrepancy certainly influences AR4 GCM performance when applied to regions within China.

The fourth factor relates to forcing agents of GCMs. Three possible forcing agents have been identified to contributors to the 20th century global warming [8,38], mainly includes anthropogenic greenhouse gases forcing (CO_2, CH_4, N_2O, etc.), natural forcing (sulfate aerosols and ozone change) and the internal variability of climate system itself (North Atlantic Oscillation, NAO and El Niño – Southern Oscillation, ENSO). Although preexisting researches have suggested the late 20th warming was likely mainly due to increases of greenhouse gases [33], it was reported that the inclusion of natural forcing has improved the simulation [9]. Some GCMs have not included the time-varying natural forcings, such as bccr_bcm_2_0, csiro_mk3_0 etc.. On the other hand, the internal variability of the climatic system is still not full considered in AR4 GCMs. All of these certainly influence the performance of inter-annual simulation. Consequently, AR4 GCMs give unsatisfactory simulations of the inter-annual temporal variability but acceptable inter-decadal variability simulations.

In practice, any user of AR4 GCM data would certainly hope to choose the best model for a particular region, and skill score

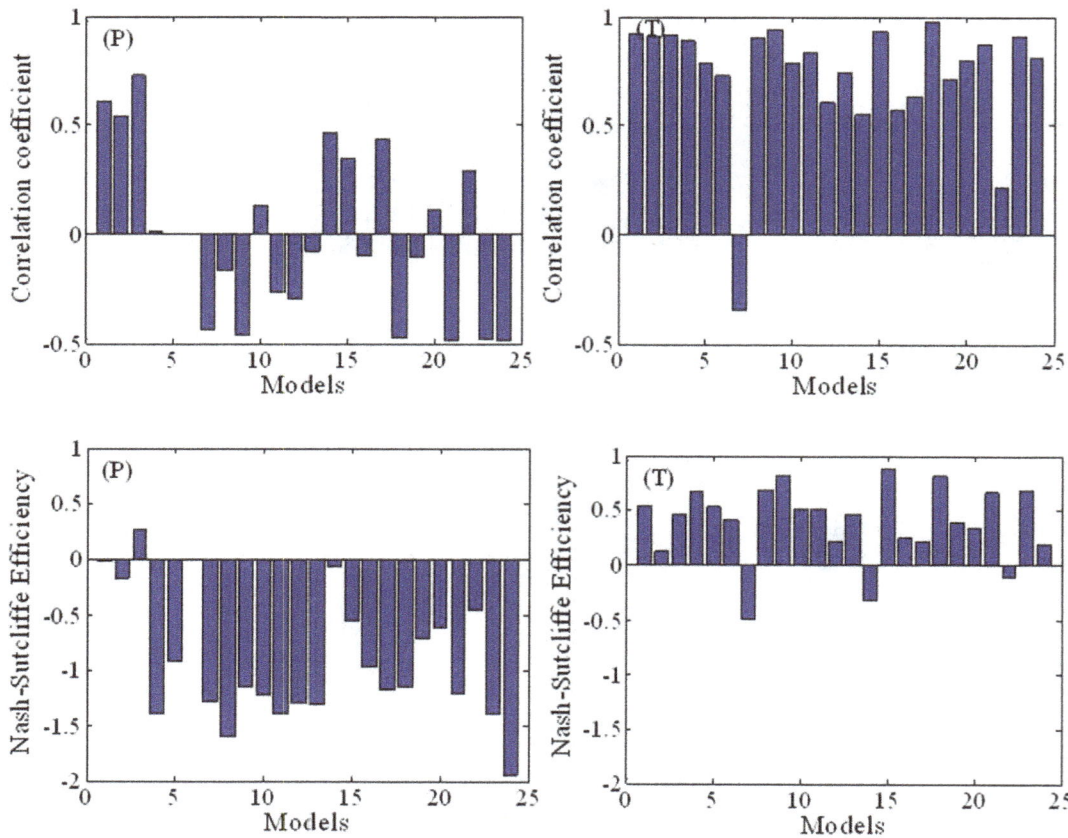

Figure 6. Bar chart indicating relative performances of the AR4 GCMs regarding the temporal simulation of the 10-year moving averages of precipitation (P) and temperature (T) in mainland China during 1960–1999.

metrics provide a way of ranking the AR4 GCMs. However, the present study shows that no one model is best at spatial, inter-annual, and intra-annual simulations of both precipitation and temperature. From the results, it is obvious that the inter-annual

simulations (temperature and precipitation) by AR4 GCMs are not suitable for direct application. It is recommended that techniques for improving annual simulation should be first investigated. Previous research has indicated that multi-model ensemble

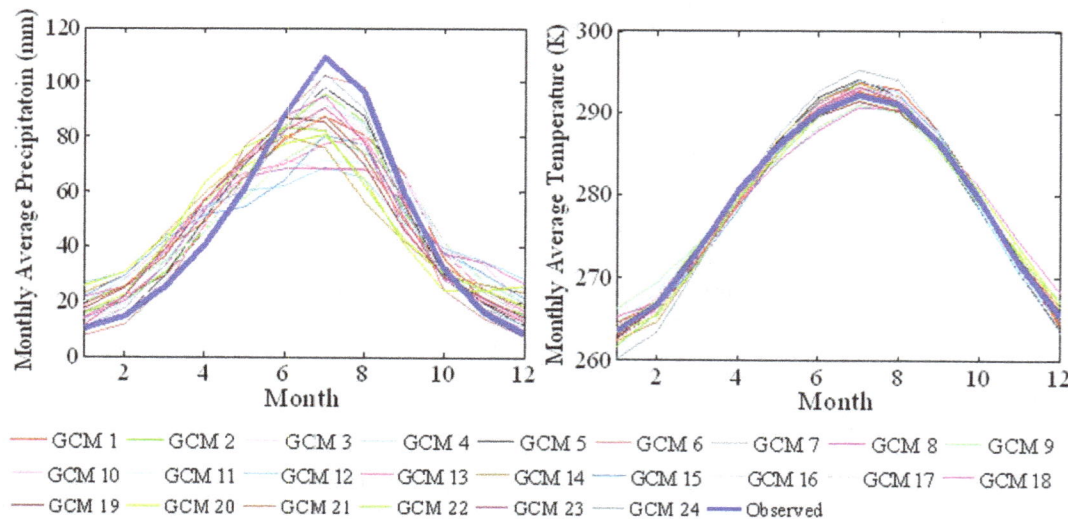

Figure 7. Observed and simulated forty-year averages of monthly precipitation and temperature throughout a calendar year in mainland China during 1960–1999.

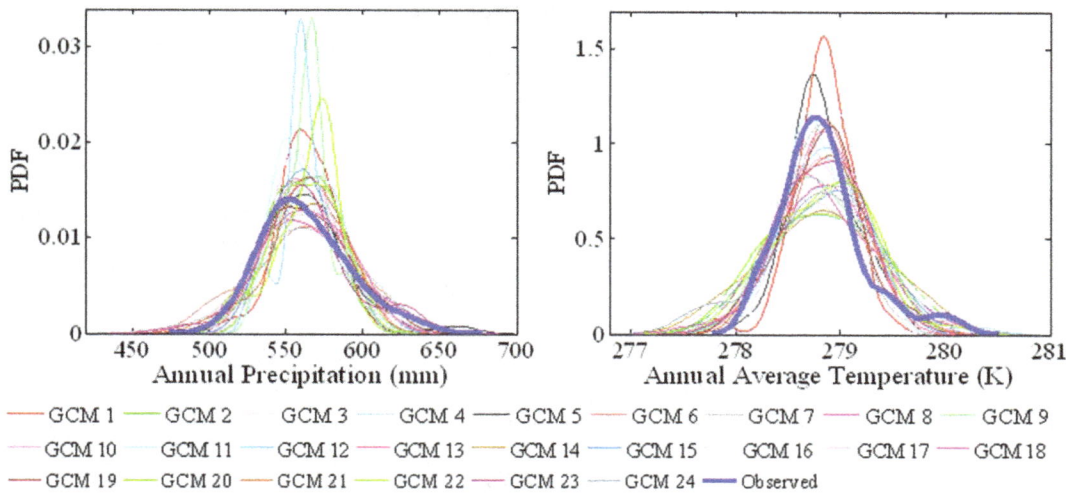

Figure 8. Probability density functions for annual mean precipitation and annual mean temperature in mainland China during 1960–1999.

simulation can produce better agreement with observed data than any single model [7,19]. Various ensemble prediction methods have been proposed, including simple model averaging [19,39], reliability ensemble averaging [40], and Bayesian model averaging [7]. The present study has focused on assessing the performance of each AR4 GCM in assessing certain climate characteristics in China. Multi-model ensemble prediction is recommended as the next step.

Based on comprehensive performance of the models simulation for spatial distribution and inter-decadal trend, we recommend that the best climate models to China are GCM 15 (ipsl_cm4) for temperature, and GCM 3 (cccma_cgcm_t63) for precipitation. As shown in Figure 1, China can be divided into ten large or aggregate river basins comprising the Songhua River, Liaohe River, Haihe River, Yellow River, Huaihe River, Yangtze River, southeast drainage area rivers, Pearl River, southwest drainage area rivers and northwest drainage area rivers. Table 2 lists the models which give the best simulations after 10-year moving average in precipitation and temperature for each basin. The results once again demonstrate that no AR4 GCM performs consistently the best throughout mainland China.

The AR4 GCMs do not perform uniformly well in simulating the characteristic spatial and temporal behaviors of temperature and precipitation in China. No one model is best at all simulations. In general, the AR4 GCMs tend to overestimate the precipitation and temperature over China. Furthermore, the model simulations are better at fitting the spatial than the annual temporal behavior,

and provide more accurate simulations of temperature than precipitation. By ranking the models according to the four skill score metrics, it has been found that the most appropriate climate models for application to mainland China are GCM 15 (ipsl_cm4) for temperature and GCM 3 (cccma_cgcm_t63) for precipitation. We recommend that AR4 GCM outputs of annual changes in temperature and precipitation are not applied directly to scenarios specific to mainland China. Instead, it is necessary that the data accuracy be improved.

Precipitation and temperature are climate parameters that directly affect hydrological processes, agricultural production, ecosystem restoration, and environmental protection in China. At river basin scale, GCM 14 (inmcm3_0), GCM 12 (iap_f-goals1_0_g), GCM 10 (giss_model_e_h), GCM 12 (iap_f-goals1_0_g), GCM 19 (mpi_echam5), GCM 18 (miub_echo_g), GCM 23 (ukmo_hadcm3), GCM 22 (ncar_pcm1), GCM 13 (ingv_echam4) and GCM 2 (cccma_cgcm3_1) provide the best results in simulating the inter-decadal precipitation trends in the Songhua River, Liaohe River, Haihe River, Yellow River, Huaihe River, Yangtze River, southeast drainage area rivers, Pearl River, southwest drainage area rivers and northwest drainage area rivers respectively. GCM 18 (miub_echo_g), GCM 24 (ukmo_hadgem1), GCM 18 (miub_echo_g), GCM 18 (miub_echo_g), GCM 18 (miub_echo_g), GCM 11 (giss_model_e_r), GCM 11 (giss_mo-del_e_r), GCM 15 (ipsl_cm4), GCM 10 (giss_model_e_h), GCM 9 (giss_aom) are recommended respectively when simulating the inter-decadal temperature trends in the same regions. The results

Table 2. Top ranked climate model for different river basins.

Precipitation simulation									
Basin	1	2	3	4	5	6	7	8	9	10
GCM	14	12	10	12	19	18	23	22	13	2
Temperature simulation									
Basin	1	2	3	4	5	6	7	8	9	10
GCM	18	24	18	18	18	11	11	15	10	9

have shown that each AR4 GCM performs differently in different regions of China, particularly with respect to precipitation.

Further research is required regarding simulation of the climate characteristics of China. This includes further assessment of the accuracy of AR4 GCMs (by considering other climate variables and skill score metrics), application of the homogenization techniques, and using uncertainty analyses and multi-model ensemble predictions to improve the reliability of the model outputs.

Acknowledgments

We are grateful to the Program for Climate Model Diagnosis and Intercomparison (PCMDI) for collecting and archiving the model data, and to the National Meteorological Information Center (China) for collecting and archiving the observed climate data.

Author Contributions

Conceived and designed the experiments: QYD CYM. Performed the experiments: CYM QYD LY AGLB. Analyzed the data: CYM LY. Contributed reagents/materials/analysis tools: QYD CYM LY. Wrote the paper: CYM QYD LY AGLB.

References

1. IPCC (2007) Climate Change 2007: The Physical Sciences Basis. In: Solomon S, Qin D, Manning M, Chen Z, Marquis M, et al. (eds). Contribution of Working Group 1 to the Fourth Assessment Report of the Intergovernmental Panel on Climate Change. Cambridge University Press, Cambridge, United Kingdom and NewYork, USA, 966 pp.

2. Watterson IG, Whetton PH (2011) Distributions of decadal means of temperature and precipitation change under global warming. J Geophys Res 116: D07101, doi:10.1029/2010JD014502.

3. Lofgren BM (2004) Global warming effects on Great Lakes water: more precipitation but less water? In: 18th Conference on Hydrology, 8th Annual Meeting of the American Meteorological Society, Seattle, WA, 2004, pp.3.

4. Black R, Bennett SRG, Thomas SM, Beddington JR (2011) Climate change: migration as adaptation. Nature 478(7370): 447–449.

5. Reichler T, Kim J (2008) How well do coupled models simulate today's climate? B Am Meteorol Soc 89(3): 303–311.

6. Kiktev D, Sexton DMH, Alexander L, Folland CK (2003) Comparison of modeled and observed trends in indices of daily climate extremes. J Climate 16(22): 3560–3571.

7. Duan Q, Phillips TJ (2010) Bayesian estimation of local signal and noise in multimodel simulations of climate change. J Geophys Res 115: D18123, doi:10.1029/2009JD013654.

8. Frame DJ, Aina T, Christensen CM, Faull NE, Knight SHE, et al. (2009) The climateprediction.net BBC climate change experiment: design of the coupled model ensemble.Philos Transact A Math Phys Eng Sci 367(1890):855–70.

9. Zhou T, Yu R (2006) Twentieth century surface air temperature over China and the globe simulated by coupled climate models. J Climate 19: 5843–5858.

10. Brunsell NA, Jones AR, Jackson TL, Feddema JJ (2011) Seasonal trends in air temperature and precipitation in IPCC AR4 GCM output for Kansas, USA: evaluation and implications. Int J Climatol 30(8): 1178–1193.

11. Li H, Feng L, Zhou T (2010) Multi-model Projection of July-August Climate Extreme Changes over China under CO_2 Doubling. Part II: Temperature. Adv Atmos Sci 28(2): 448–463.

12. Li H, Feng L, Zhou T (2011) Multi-model Projection of July-August Climate Extreme Changes over China under CO_2 Doubling. Part I: Precipitation. Adv Atmos Sci 28(2): 433–447.

13. Dutton JA (2002) Opportunities and priorities in a new era for weather and climate services. B Am Meteorol Soc 83: 1303–1311.

14. Nieto S, Rodríguez-Puebla C (2006) Comparison of precipitation from observed data and general circulation models over the Iberian Peninsula. J Climate 19(17): 4254–4275.

15. Im ES, Kwon WT, Ahn JB, Giorgi F (2007). Multi-decadal scenario simulation over Korea using a one-way double-nested regional climate model system. Part 1: recent climate simulation (1971–2000). Clim Dynam 28(7): 759–780.

16. Delire C, Ngomanda A, Jolly D (2008) Possible impacts of 21st century climate on vegetation in Central and West Africa. Global Planetary Change 64(1–2): 3–15.

17. Errasti I, Ezcurra A, Sáenz J, Ibarra-Berastegi G (2011) Validation of IPCC AR4 models over the Iberian Peninsula. Theor Appl Climatol 103(1): 61–79.

18. Masson D, Knutti R (2011) Spatial-scale dependence of climate model performance in the CMIP3 ensemble. J Climate 24(11): 2680–2692.

19. Phillips TJ, Gleckler PJ (2006) Evaluation of continental precipitation in 20th Century climate simulations: The utility of multimodel statistics. Water Resour Res 42(3): W03202, doi:10.1029/2005WR004313.

20. Coquard J, Duffy PB, Taylor KE, Iorio JP (2004) Present and future surface climate in the western USA as simulated by 15 global climate models. Clim Dynam 23(5): 455–472.

21. Ren G, Xu M, Chu Z (2005) Changes of surface air temperature in China during 1951–2004. Climatic Environ Res 10(4):711–727.

22. Li P, Guo M, Wang L, Li Q, Xu B, et al. (2011) Research of dynamics and relationship of precipitation and temperature in the recent 60 years in China. Eng Sci 04: 29–36.

23. Piao S, Ciais P, Huang Y, Shen Z, Peng S, et al. (2010) The impacts of climate change on water resources and agriculture in China. Nature 467(7311): 43–51.

24. Yu R, Wang B, Zhou T (2004) Tropospheric cooling and summer monsoon weakening trend over East Asia. Geophys Res Lett 31: L22212, doi:10.1029/2004GL021270

25. Yu R, Zhou T (2007) Seasonality and three-dimensional structure of the interdecadal change in East Asian monsoon. J Climate 20: 5344–5355.

26. Zhou T, Gong D, Li J, Li B (2009) Detecting and understanding the multi-decadal variability of the East Asian Summer Monsoon - Recent progress and state of affairs. Meteorol Z 18(4): 455–467.

27. Gao X, Shi Y, Song R, Giorgi F, Wang Y, et al. (2008) Reduction of future monsoon precipitation over China: comparison between a high resolution RCM simulation and the driving GCM. Meteorol Atmos Phys 100(1): 73–86.

28. Zhao T, Guo W, Fu C (2008) Calibrating and evaluating reanalysis surface temperature error by topographic correction. J Climate 21(6): 1440–1446.

29. Meehl GA, Arblaster JM, Tebaldi C (2007) Contributions of natural and anthropogenic forcing to changes in temperature extremes over the United States. Geophys Res Lett 34: L19709, doi: 10.1029/2007GL030948.

30. Meehl GA, Covey C, Delworth T, Latif M, McAvaney B, et al. (2007) The WCRP CMIP3 multimodel dataset: a new era in climate change research. B Am Meteorol Soc 88: 1383–1394.

31. Nash JE, Sutcliffe JV (1970) River flow forecasting through conceptual models. Part 1 – A discussion of principles. J Hydrol 10(3):282–290.

32. Warner GS, Stake JD, Guillard K, Neafsey J (1997) Evaluation of EPIC for a shallow New England soil: II. Soil nitrate. T ASAE 40(3): 585–593.

33. Stott PA, Tett SFB, Jones GS, Allen MR, Mitchell JFB, et al. (2000) External control of 20th century temperature by natural and anthropogenic forcings. Science 290: 2133–2137.

34. Beaulieu C, Seidou O, Ouarda TBMJ, Zhang X (2009) Intercomparison of homogenization techniques for precipitation data continued: Comparison of two recent Bayesian change point models. Water Resour Res 45(8): W08410, doi: 10.1029/2008WR007501.

35. Ducré-Robitaille J, Vincent LA, Boulet G (2003) Comparison of techniques for detection of discontinuities in temperature series. Int J Climatol 23(9): 1087–1101.

36. Maxino CC, McAvaney BJ, Pitman AJ, Perkins SE (2008) Ranking the AR4 climate models over the Murray-Darling Basin using simulated maximum temperature, minimum temperature and precipitation. Int J Climatol 28(8): 1097–1112.

37. Ahlfeld DP (2006) Comparison of climate model precipitation forecasts with North American observations. In: proceedings of the XVI international conference on computational methods in water resources (eds PJ . Binning, P . Engesgaard, H . Dahle, GF . Pinder and WG . Gray), Copenhagen, Denmark, June, 2006, 8 pp.

38. Li L, Wang B, Zhou T (2007) Impacts of external forcing on the 20th century global warming. Chinese Sci Bull 52(22): 3148–3154.

39. Lambert SJ, Boer GJ (2001) CMIP1 evaluation and intercomparison of coupled climate models. Clim Dynam 17(2): 83–106.

40. Giorgi F, Mearns LO (2002) Calculation of average, uncertainty range, and reliability of regional climate changes from AOGCM simulations via the "Reliability Ensemble Averaging (REA) " Method. J Climate 15(10): 1141–1158.

19

Impact of Future Climate on Radial Growth of Four Major Boreal Tree Species in the Eastern Canadian Boreal Forest

Jian-Guo Huang[1*¤a], **Yves Bergeron**[1], **Frank Berninger**[2¤b], **Lihong Zhai**[2], **Jacques C. Tardif**[3], **Bernhard Denneler**[1†]

1 Chaire industrielle CRSNG-UQAT-UQAM en Aménagement Forestier Durable, Université du Québec en Abitibi-Témiscamingue, Rouyn-Noranda, Québec, Canada, **2** Département des sciences biologiques, Université du Québec à Montréal, Montréal, Québec, Canada, **3** Centre for Forest Interdisciplinary Research (C-FIR), University of Winnipeg, Winnipeg, Manitoba, Canada

Abstract

Immediate phenotypic variation and the lagged effect of evolutionary adaptation to climate change appear to be two key processes in tree responses to climate warming. This study examines these components in two types of growth models for predicting the 2010–2099 diameter growth change of four major boreal species *Betula papyrifera*, *Pinus banksiana*, *Picea mariana*, and *Populus tremuloides* along a broad latitudinal gradient in eastern Canada under future climate projections. Climate-growth response models for 34 stands over nine latitudes were calibrated and cross-validated. An adaptive response model (A-model), in which the climate-growth relationship varies over time, and a fixed response model (F-model), in which the relationship is constant over time, were constructed to predict future growth. For the former, we examined how future growth of stands in northern latitudes could be forecasted using growth-climate equations derived from stands currently growing in southern latitudes assuming that current climate in southern locations provide an analogue for future conditions in the north. For the latter, we tested if future growth of stands would be maximally predicted using the growth-climate equation obtained from the given local stand assuming a lagged response to climate due to genetic constraints. Both models predicted a large growth increase in northern stands due to more benign temperatures, whereas there was a minimal growth change in southern stands due to potentially warm-temperature induced drought-stress. The A-model demonstrates a changing environment whereas the F-model highlights a constant growth response to future warming. As time elapses we can predict a gradual transition between a response to climate associated with the current conditions (F-model) to a more adapted response to future climate (A-model). Our modeling approach provides a template to predict tree growth response to climate warming at mid-high latitudes of the Northern Hemisphere.

Editor: Ben Bond-Lamberty, DOE Pacific Northwest National Laboratory, United States of America

Funding: This project was financially supported by the Natural Science and Engineering Research Council of Canada, Ouranos Consortium, and Canada Research Chair programs. The funders had no role in study design, data collection and analysis, decision to publish, or preparation of the manuscript.

Competing Interests: The authors have declared that no competing interests exist.

* E-mail: huang500@purdue.edu

¤a Current address: Department of Forestry and Natural Resources, Purdue University, West Lafayette, Indiana, United States of America
¤b Current address: Department of Forest Sciences, University of Helsinki, Helsinki, Finland

† Deceased.

Introduction

Modeling future growth of forests under climate change is a challenge since our understanding of tree physiology and growth, and the role and rate of genetic adaptation to a rapidly shifting climate is currently limited. Previous modeling studies have attempted to predict potential changes in tree growth, net primary productivity, and forest productivity under increased greenhouse gases emissions scenarios (generally $2 \times CO_2$) using either empirical/statistical models [1,2] or process-based models [3,4]. Tree ring based studies aimed at predicting the impact of future climate usually assume that the relationships between tree growth and climate are linear and constant through time. For example, Chhin et al. [5] using a site-specific empirical regression model, reported negative impact of future climate warming on productivity of lodgepole pine (*Pinus contorta* Dougl. ex Loud. var. *latifolia* Engelm.) in Alberta. Tree ring based studies, however, often do not address the fact that the relationships between growth and climate may change and they often extrapolate growth responses beyond the range where they have been validated.

In this study, we predict future radial growth of trees based on two theoretical assumptions. Given that tree growth conditions may be changing with a warming climate over time, the first assumption is that future response of trees to climate warming at the local scale can be best approximated by the response of trees to climate currently occurring at more southern latitudes. Here we defined this assumption based-model as "Adaptive response model (A-model)". By shifting the growth model to more southern models as climate warming occurs, the potential effects (e.g., drought effect highlighted in the southern models [6]) will be involved. For example, different climate effects that the northern models may not indicate will be involved, as e.g. trees develop shoots which may be better adapted physiologically for the warming climate. Using the correlation pattern along the gradient 48–50°N in eastern Canada, among sites, between species [black spruce (*Picea mariana* (Mill.) BSP), and jack pine (*Pinus banksiana* Lamb.)], and through time (1825–1993), Hofgaard et al. [7] found significant linear relationship between climate change over time and climate change over latitude. This study supports our first assumption.

In contrast, the second assumption states that future tree growth can be best predicted by the climate-growth equations obtained from the local stands when considering genetic constraints or the lagged effect of genetics on growth. Here we defined this assumption based-model as "Fixed response model (F-model)". This assumption is supported by Savva et al. [8] who reported in a provenance study that northern populations of jack pine transferred to a southern latitude did not benefit from the warmer climate conditions due to an inherent earlier onset of growth cessation than the local populations. This might indicate a genetic constraint or a lagged effect in genetic response to climate [9] since the onset of bud set and growth cessation, and maximum radial growth rate are genetically-constrained photoperiodic responses [10]. Growth simulations contrasting these two assumptions would provide alternative scenarios for better understanding how future climate warming will impact the growth of trees and forests in eastern Canada.

To accomplish this, we designed a dendrochronological study across a broad latitudinal gradient from 46–54°N in eastern Canada. In our first examination of this data, we systematically investigated the radial growth response of four major boreal species paper birch (*Betula papyrifera* Marsh.), jack pine, black spruce, and trembling aspen (*Populus tremuloides* Michx.) to past climate change in eastern Canada and found their growth responses to the past 50-years of climate warming in this vegetation transition zone differed among species [6]. This broad latitudinal gradient design may provide an analogue of future climate warming given that the northern growth conditions might be changing to the southern ones with warming. In this study, we further forecast the radial growth of these species over this broad spatial scale using the A- and F- models noted above. We hypothesized that 1) the climate-growth calibration models developed from the southern stands can be used to predict future growth of stands growing in the north; 2) the northern trees may be able to benefit from future warming to enhance growth.

The specific objectives of the current paper were fourfold. First, we updated the calibration of the climate-growth models for each species at different latitudes reported in Huang *et al.* [6] through extending the calibration period to the maximum length of chronology and climate data for each species at each site. Second, we forecast the potential mean growth change (MGC) of a species for a given latitude from 2010–2099 based on four different IPCC Emissions Scenarios generated from three General Circulation Models (GCMs) and one from the Canadian Regional Climate Model (CRCM) using the above calibration models. Third, we quantified whether the predicted MGC of a given species at different latitudes differs among different climate change projections using different calibration climate-growth models through decomposition of variance. This process can allow us to determine the most appropriate calibration model for predicting future radial growth at a given latitude over time. Last, we predict the potential MGC of each species for a given latitude from 2010–2099 using the A- and F- models constructed on the two assumptions, respectively. This study expands how we simulate future growth

rate of trees and forests to consider both environmental and genetic effects.

Materials and Methods

Study Area

The study area is located along the Quebec-Ontario border over a latitudinal gradient ranging from Petawawa (approximately 46°N) in the south to Radisson (approximately 54°N) in the north (Fig. 1). The topography along the gradient is generally flat and uniform with low-elevation hills and rock outcrops. The climate of the region is dominated by dry polar and moderate polar air masses in winter, and by moist maritime and moist tropical air masses in summer [11]. A climate gradient followed the latitudinal gradient, as described in Huang et al. [6]. A vegetation transition zone between the mixedwood and the coniferous-dominated boreal forest occurs at approximately 49°N [12]. The tree line is about 500 km north of the northernmost stands.

Tree-ring Increment Data

Ring width data sets for aspen, birch, spruce and pine used in this study were exactly the same as reported in Huang et al. [6], i.e. residual chronologies from 8 aspen stands, 8 birch stands, 9 spruce stands and 9 pine stands (2 cores per tree, 20 trees per stand). All tree-ring chronologies were developed according to standard dendrochronological techniques: crossdating, standardization (60-year or more flexible spline detrending was used), calculation of residual ring-width chronologies, as well as reduction of noise created by insect defoliations [6,13]. All residual chronologies were then used to calibrate the current climate-growth response. The expressed population signal, which is a statistic used to evaluate the reliability of a chronology, was above the generally accepted cutoff value of 0.85 for all time spans of the residual chronologies [14].

Instrumental Climate Data for Updating Calibration Models

Instrumental climate data used for updating the calibration models at each of 12 locations were interpolated from ANUSPLIN (version 4.3) [15] for the period 1901–2003 (see details in [6]). Climate variables used in the model included monthly maximum temperature (T_{max}), monthly minimum temperature (T_{min}), and monthly total precipitation (P_{total}). In addition, the monthly Canadian Drought Code (CDC) from May to October was also calculated for each climate data set as described by Girardin & Wotton [16] using monthly T_{max} and P_{total} generated by ANUSPLIN. The CDC is a numerical index representing the average moisture content of deep and compact organic layers, and was used to investigate if soil moisture variability had any impact on tree growth in the region.

Climate Change Projections for Future Growth Simulation

For each of the 12 locations, the simulations obtained from 8 climate change scenarios through three GCMs and the Canadian third-generation coupled global climate model (CRCM3) were used for growth predictions. They included three sets of data generated from CGCM3 under the A1B, A2, and B1 scenarios [17], two sets of data generated from UK Hadley Centre (HadCM3) under the A2 and B2 scenarios [18], two sets of data generated from Max Planck Institut für Meteorologie (ECHAM4) under the A2 and B2 scenarios [19], and one set of data generated from CRCM3 under the A2 scenario. They were referred to

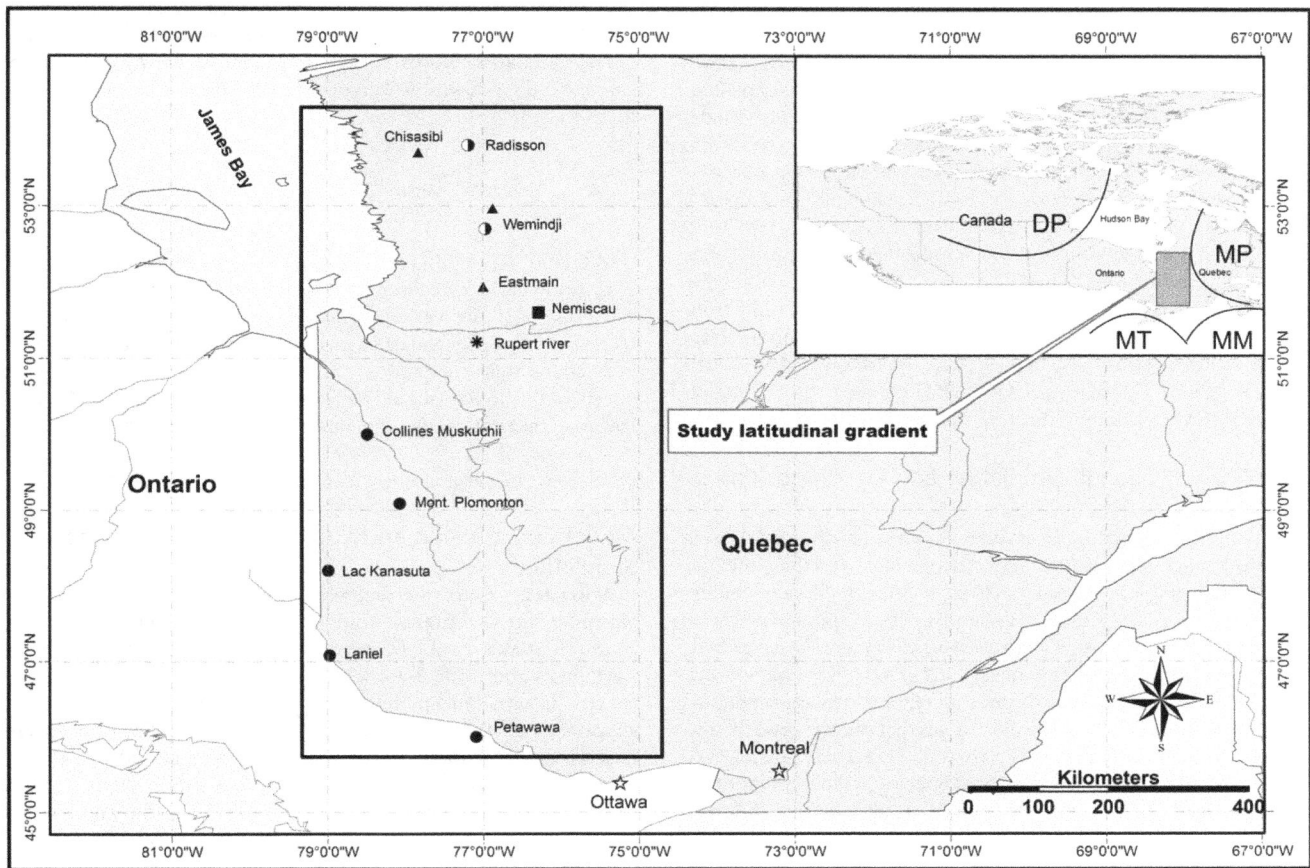

Figure 1. The sampling sites for paper birch, jack pine, black spruce, and trembling aspen in the eastern Canadian boreal forest, where all four species (•), only two conifers (▲), only two deciduous species (half solid circle), only aspen, spruce and pine (*), and only birch (■) were sampled at the site. The origins of major air mass types affecting the climate of the region are also indicated: dry polar (DP), moist polar (MP), moderate moist (MM), and moist tropical (MT) [11].

CA1B, CA2, CB1, HA2, HB2, EA2, EB2 and MA2 respectively in this paper. A brief description of the four emissions scenarios and the corresponding GCMs were given in Table S1.

The regional projections were used to estimate the climate anomalies using a delta method in which the predicted changes, or anomalies, of climate variables (their mean and variance) obtained from climate future simulations are added to the current climate normal [20]. These anomalies were applied to a series of years randomly selected from 1980 to 2000, as reported by Lapointe-Garant et al. [21]. T_{max} and P_{total} data obtained from this process were then used to calculate the CDC from May to October for each of the 12 locations. All climate change projection data during 1961–2099 were used for future growth simulations.

Updating Climate-growth Calibration Models using Instrumental Climate Data

Based on tree-ring chronologies and instrumental climate data from 1950–2003, Huang et al. [6] calibrated and reported the climate-growth calibration models for all species/sites over the gradient. In the current study, to obtain the most reliable calibration models for future simulation, it would be preferable to use the maximum length of the chronologies and instrumental climate data to update the calibration models reported in Huang et al. [6]. Least-squares stepwise multiple regression employing a backward selection was used to update these empirical climate-

growth models. The covariates included 17 T_{max}, 17 T_{min}, and 17 P_{total} monthly variables (previous May to current September), and 11 CDC at each latitude/species. This updating procedure resulted in the shortest and longest calibration period being, respectively, 1946–2003 (BS at 46°N) and 1902–2003 (10 latitudes/species, see Table S2). The common longest calibration period among all species/latitudes was 1946–2003.

For each latitude/species, the statistically important monthly climate variables were retained during modeling trials, and multi-month combinations of these climate variables were then calculated to reduce the number of predictors to establish the best calibration model (minimum tolerance $P = 0.05$). Variance inflation factors (VIF) were also calculated to detect multicollinearity among the variables [22]. VIFs were generally lower than the accepted value of 3. The minimum Akaike Information Criterion (AIC) [23] was used to choose the best calibration model. In general, less than 10 combined climate variables were retained in the final model (Table S3).

The performance of each regression model was cross-validated using a split sample calibration–verification scheme [24]. First, climate data during the full calibration period was split into two subperiods at the year of 1965, thus resulting in a calibration subperiod and a verification subperiod, respectively. For example, when the full calibration period is 1902–2003, then the calibration and verification subperiod is 1902–1965, and 1966–2003, respectively. Second, the regression coefficients obtained from the equation fit to the calibration subperiod were applied to the climate variables over the verification periods to produce a series of tree-ring increment estimates. The same process was repeated, reversing the subperiods for model calibration and model verification. Tree-ring increment estimates obtained from the above procedure were further compared with tree-ring observations to assess the strength of the model using the Pearson R^2, reduction of error (RE, its values can range from $-\infty$ to a maximum of 1 and any positive value indicates that the model has some skill), product means test (PM), and sign test (ST counts the numbers of agreements/disagreements in sign of deviation from the means in the observed and estimated series) [25]. All analyses were conducted using SAS 10 (SAS Corporation, Cary, North Carolina, USA). The program VFY [26] was used for calculation of the RE, PM and ST verification statistics.

Mean Growth Change Simulation and Decomposition of Variance

Each of the updated calibration climate-growth models was used to predict ring-width index (dependent variable) for the period 2010–2099 based on each of 8 sets of climate change scenarios data (independent variables). The simulated mean growth change (MGC) was expressed as a percentage of growth (ring-width index) anomalies over 1962–1991. Because we are unclear whether each of the calibration climate-growth models of a species has the same capability to predict future growth, the decomposition of variance [27] was thus performed to quantify whether the predicted MGC of a given species at different latitudes differed among different climate change scenarios using different

calibration models. The decomposition of variance was performed on a large dataset of MGC (8 or 9 calibration models ×8 climate change scenarios ×8 or 9 latitudes). The variance of the whole simulated dataset was broken down into different variance parts as follows:

$$\delta^2_{TOT} = \delta^2_{MOD} + \delta^2_{SCE} + \delta^2_{LAT} + \delta^2_{INT} + \delta^2_{CLI}$$

Where δ^2_{TOT} is the variance of the whole simulated growth and δ^2_{MOD} the variance caused by different calibration climate-growth models, δ^2_{SCE} the variance caused by different climate change scenarios, δ^2_{LAT} the variance caused by different latitudes at which predictions were made, δ^2_{INT} the variance caused by interactive effects of the above factors (all possible interactions), and δ^2_{CLI} the variance that cannot be explained by the above factors, and thus be considered to be caused by future climate variability. The breakdown of variances was calculated using ANOVA.

Model Construction and Future Mean Growth Change Simulation

A-model. Since the first assumption insists that with climate warming the calibration model currently developed for the southern stands could be used to predict future growth of the northern stands, the A-model was constructed as follows. First, a future 30-year mean temperature was calculated for three subperiods 2010–2039, 2040–2069, and 2070–2099, respectively, based on the 8 climate change scenarios and 9 latitudes (Table 1). Second, comparisons between the moving future 30-year mean temperature (e.g., 1981–2010, 1982–2011, and so on) and the past

Table 1. Comparisons of 30-year mean annual temperatures (±SD) between instrumental climate data for 1961–1990 and future mean GCMs simulation climate data from eight climate change scenarios for 2010–2039, 2040–2069, and 2070–2099 over latitude of 46–54°N, as well as the corresponding models identified and applied at each latitude from 48–54°N.

Latitude (°N)			Time Period		
		1961–1990	2010–2039	2040–2069	2070–2099
46	T (°C)	4.25 (0.65)	5.83 (0.81)	7.31 (0.96)	8.75 (1.14)
	Models	N/A	N/A	N/A	N/A
47	T (°C)	3.02 (0.75)	4.47 (0.84)	5.97 (0.98)	7.45 (1.20)
	Models	N/A	N/A	N/A	N/A
48	T (°C)	1.14 (0.78)	2.66 (0.84)	4.18 (0.97)	*5.67 (1.21)*
	Models		L48-L47	L48-L46	*L47-L46*
49	T (°C)	−0.63 (0.80)	0.82 (0.82)	2.36 (0.96)	3.88 (1.22)
	Models		L49-L48	L48-L47	L48-L46
50	T (°C)	−0.89 (0.85)	0.67 (0.85)	2.24 (0.97)	3.79 (1.25)
	Models		L50-L48	L49-L47	L48-L46
51	T (°C)	−1.54 (0.92)	0.13 (0.85)	1.73 (0.94)	3.32 (1.23)
	Models		L51-L48	L49-L47	L48-L46
52	T (°C)	−3.06 (1.00)	−1.28 (0.89)	0.38 (0.97)	2.02 (1.27)
	Models		L52-L50	L51-L48	L49-L47
53	T (°C)	−3.21 (1.01)	−1.42 (0.90)	0.23 (0.98)	1.89 (1.28)
	Models		L53-L50	L51-L48	L49-L47
54	T (°C)	−3.18 (1.07)	−1.35 (0.94)	0.34 (1.02)	2.03 (1.32)
	Models		L54-L50	L51-L48	L49-L47

Note: Models at 46°N and 47°N were not determined (N/A); The underscored italic text indicates the simulations with uncertainties. Standard deviation (SD) of climate projections is interannual variation of the average of the mean of the 8 climate change scenarios.

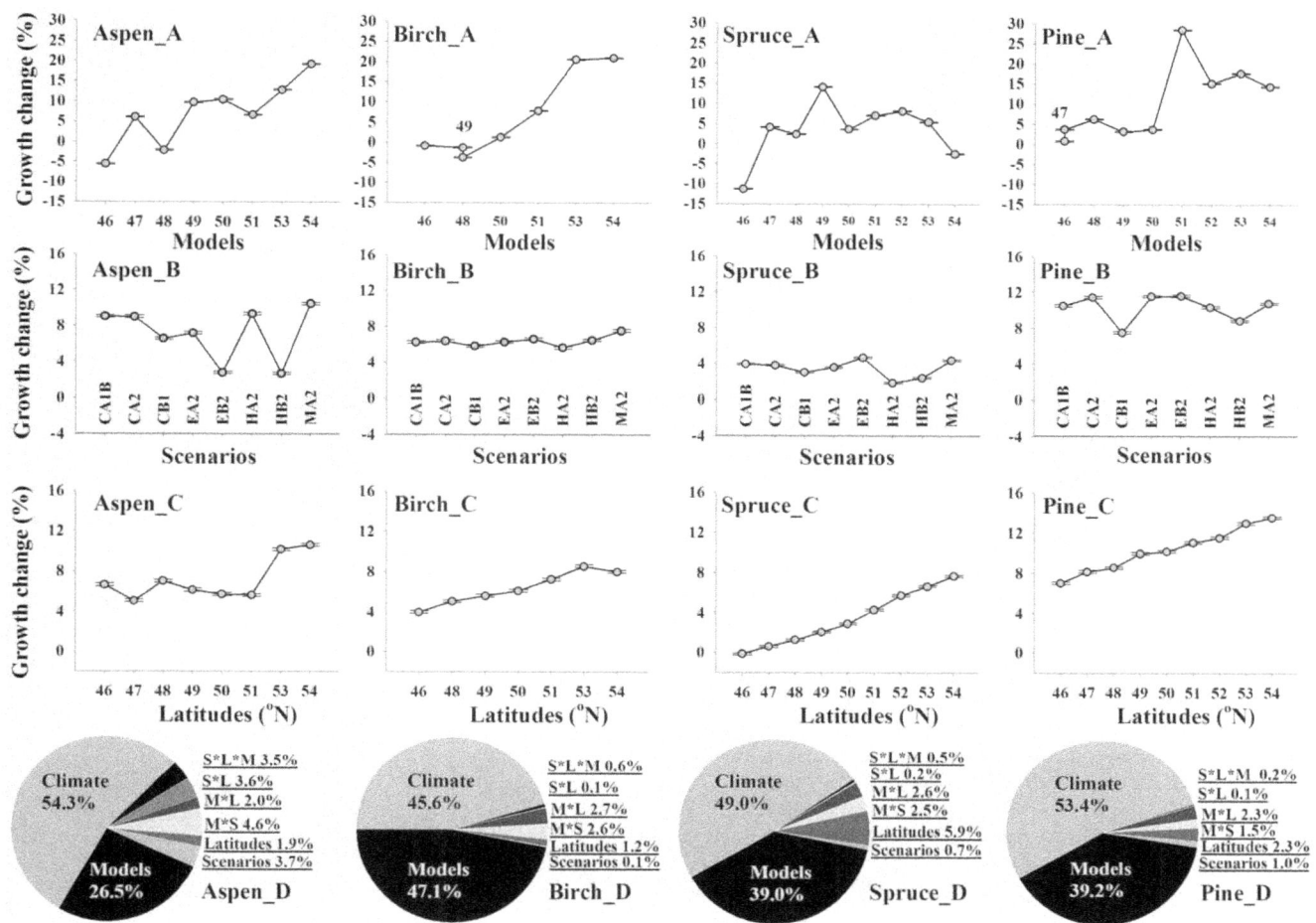

Figure 2. Partitioning of variance in mean simulated growth change of paper birch, jack pine, black spruce, and trembling aspen along the gradient. The pie plot shows that the proportions of variance in mean growth change for each species explained by the calibration models (M), scenarios (S), latitudes (L), and the interactive effects of the above three factors, as well as the climate. Grey dots represent the least square means of potential growth change of each species predicted from 2010 to 2099. Grey short lines represent the standard error of the means.

30-year (1961–1990) ANUSPLIN mean temperature at different latitudes were made to identify the latitude at which future mean temperature at a given latitude for a given subperiod approximates that of a southern stand. Third, the calibration models established from the given northern latitude to the best southern latitude identified above were involved to predict future MGC using 8 sets of climate change scenario data simulated for the given northern latitude, given that those climate conditions might be gradually changing from the given latitude to the best southern latitude with warming over time.

The calibration models identified in above procedure for each latitude/species are listed in Table 1. In many cases several models were involved for future growth simulation. For instance, the growing condition of aspen at 54°N during the future subperiod 2010–2039 is assumed to be gradually changing to that at a more southern latitude (e.g., 50°N) when the 2010–2039 mean temperature at 54°N is near the 1961–1990 mean temperature of 50°N (southern limit). In this case, the calibration models from 54, 53, 52, 51 and 50°N were all selected for the 2010–2039 growth simulations of stands at 54°N. We calculated the growth response by interpolating the growth models based on average yearly temperatures. For the next subperiod (e.g., 2040–2069), the southern limit chosen above (e.g., 50°N) was considered as the

northern limit, and a more southerly site yet (than 50°N) was chosen, and the calibrated models established between these two limits were used sequentially for future growth simulation. The same principle was applied to each of all the latitudes/subperiods. However, when future climate range was beyond the range of our data (such as for 46–47°N and during 2070–2099 at 48°N, Table 1), extrapolation was thus not performed and future growth in these latitudes were not further discussed. The 95% confidence interval was built for each prediction, which indicates the uncertainty due to different models and climate scenarios. Decomposition of variance was also calculated.

F-model. Our second assumption was that future tree growth would be best predicted by the climate-growth equations obtained from the given local stands due to their local genetic constraints. The F-model was constructed and future growth was simulated. This was a simpler process where the calibrated climate-growth model obtained for each species in each local stand was employed to simulate future MGC based on each of 8 climate change scenario data simulated for the given latitude. The MGCs calculated for the 8 climate change scenarios were further averaged for comparison with the A-model simulations. The 95% confidence interval was also built, which indicates the uncertainty due to climate scenarios.

Results

Updated Climate-growth Calibration Models

As listed in Table S2, the significant and positive RE values reveal that the models are reasonably robust over the full length of the calibration period. Both the PM and ST results suggest significant predictive skills to reproduce the magnitude and direction of year-to-year changes. Among four species, higher R^2 and RE values in most of pine stands along the gradient indicate that pine models are more robust than that of the other species, and should have high fidelity to predict radial growth. As shown in Table S3, the main climate variables selected for aspen models mostly include previous summer- and autumn temperatures, current growing season temperatures and current precipitation. The main climate variables for birch models include previous autumn temperatures and current spring and summer temperatures. The models for two conifers highlight mostly winter and the average growing season temperatures.

Future Climate Anomalies

As shown in Fig. S1 and Table S4, compared to mean climate during 1961–1990, the greatest positive temperature anomalies (+2 to +6°C) were predicted by ECHAM4 (EA2 and EB2), and the smallest positive anomalies (+1 to +4°C) were simulated by HadCM3 (HA2 and HB2). Higher precipitation anomalies (+40 to +180 mm) were predicted by CGCM (CA1B, CA2, and CB1) and CRCM3 (MA2) than that predicted by ECHAM4 and HadCM3. Severe drought anomalies (+10 to +40 units) were more frequently predicted by ECHAM4 (EA2 and EB2) than other models.

Mean Growth Change and Decomposition of Variance

As illustrated in Fig. 2 D, climate variability accounts for about half of the simulated variability (45.6%–54.3%), followed by the calibration model differences representing 26.5%–47.1% of simulated variability, scenario differences representing 0.1%–3.7%, latitude differences representing 1.2%–5.9%, and other interactive effects representing 4.2%–13.7%. As shown in Fig. 2 A, the northern models for aspen, birch and pine generally predicted a large growth increase, whereas the southern models predicted a minor growth change. The models for spruce mostly show a moderate increase in growth with increasing latitude except for the southernmost and northernmost stands. Among different latitudes (Fig. 2 C), the MGC was predicted to increase towards the north (5.1% to 10.6% from south to north for aspen, 4.0% to 8.6% for birch, −0.2% to 7.7% for spruce, and 7.0% to 13.6% for pine). Among different scenarios (Fig. 2 B), relatively less variability in simulated MGC was found. Within the GCMs, the A2 scenario generally resulted in better growth in aspen and pine than the B2 scenario, yet no such difference was found for other two species.

Mean Simulated Growth Change using the A-model

As shown in Fig. 3 and Fig. 4, the A-model simulation results indicate that aspen would have a positive MGC (10–15% growth increase) during 2010–2099 at 53–54°N, and a moderate growth increase (less than 10%) during 2010–2069 followed by a growth decrease (less than 10%) during 2070–2099 at 50–51°N. For stands at 49°N, a moderate growth increase (less than 8%) during 2010–2049 was predicted, followed by growth fluctuations (increases/decreases) during 2050–2069 and a growth decrease (less than 10%) after 2070. For stands at 48°N, a low growth increase (less than 5%) during the next one or two decades followed by a gradual growth decrease thereafter were simulated, with a large 95% confidence interval. As shown by the pie plot in Fig. 3 and Fig. 4, the results of partitioning the variance showed

that from south to north the variance in simulated MGC explained by climate variability increases (53.5% to 73.4%), whereas that explained by the calibration models decreases (40.9% to 18.4%). Both together accounted for the largest proportion of the variance, yet other factors like climate change scenarios and scenario×calibration model interactions explained only very little variance.

Birch would show moderate growth increases (less than 20%) until 2040s followed by a gradual growth decrease (less than 20%) thereafter at 53–54°N. For stands at 51°N, a weak growth increase (less than 10%) during 2010–2039 was predicted, followed by gradual growth decrease thereafter. Stands at 48–50°N would show a moderate growth decrease (less than 10%) during 2010–2099. Partitioning of variance indicates that the climate variability explained more variance for stands at 48–51°N (84.2% to 96.1%) than that for stands at 53–54°N (61.5% to 63.9%), whereas the calibration models explained less variance for stands at 48–51°N (0.0% to 5.9%) than that for stands at 53–54°N (26.3% to 27.6%). Other factors explained only very little variance.

Spruce at 52–54°N was predicted to show an obvious growth increase (up to about 20%) until 2099. For stands at 49–51°N, a growth increase (less than 15%) during 2010–2069 was simulated, followed by a moderate growth decrease during 2070–2099, with a large 95% confidence interval. Stands at 48°N would show a weak growth increase (less than 6%) during the coming decade, followed by a gradual growth decrease thereafter. The results of partitioning of variance showed that the variance in simulated MGC explained by climate variability increases from 45.2% to 74.1%, and that explained by the calibration models decreases from 47.7% to 15.5%, from south to north, respectively. Other factors explained only very little variance.

In contrast to the previous three species, pine at all latitudes would show a consistent growth increase until 2099. Up to 20% growth increase in stands at 52–54°N and up to 10% growth increase in stands at 48–51°N are expected. The results of partitioning of variance showed that, from south to north, the variance in simulated MGC explained by the climate variability decreases from 92.0% to 35.2%, and explained by the calibration models increases from 2.7% to 59.3%. Other factors also explained very little variance.

Mean Simulated Growth Change using the F-model

As illustrated in Fig. 5 and Fig. S2, the F-model simulation results showed that except for an expected growth decrease until 2099 at 46 and 48°N, aspen would have a moderate growth increase during 2010–2099 at most of the latitudes, with the highest growth increase (up to 40%) at 53–54°N. Birch would show a potential growth increase until 2099 at 51–54°N, and a growth decrease during most of the 21st century at 46–50°N. Like aspen, the fastest growth increase (up to 40–50%) is expected for stands at 53–54°N. Spruce would show a moderate growth increase north of 48°N, relatively minor growth fluctuations at 47–48°N, and a linear growth decrease at 46°N to as low as 30% in 2099. Pine would show a strong linear growth increase (up to 60%) at 51–54°N, moderate growth increases (10%) at 48–50°N, and a weak, fluctuating growth increase from 2010–2099.

Discussion

Our simulations showed that tree ring based growth simulations depend strongly on the assumptions of the growth model. There were consistent and large differences between the A- and the F-model. Further differences were, however, caused by the climate change scenarios and natural climate variability.

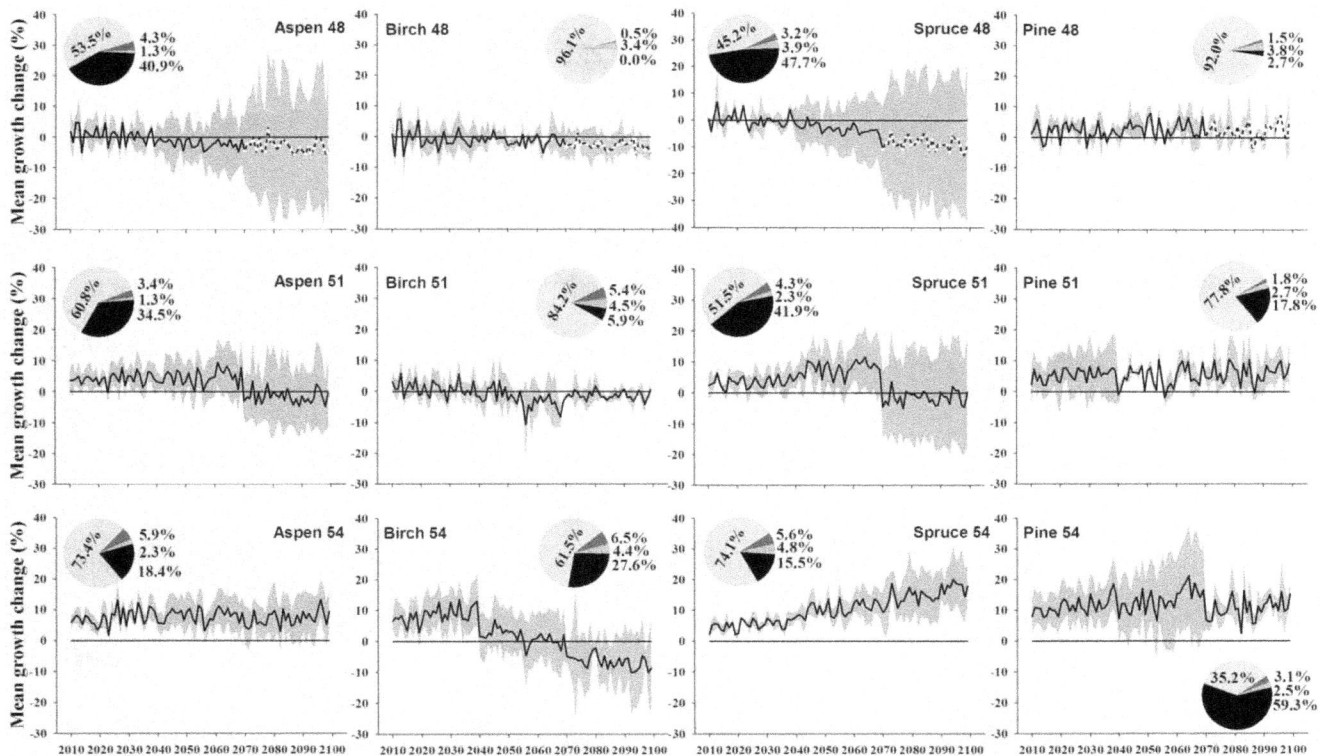

Figure 3. The predicted mean growth change of trembling aspen, paper birch, black spruce, and jack pine at 48, 51 and 54°N under the A-model. The pie plots indicate the proportion of variance in mean potential growth change explained by, counter clock wise, the calibration models (dark), scenarios (grey), calibration models *scenarios (darker grey), and climate (blue). The grey zones are 95% confidence interval. White dashed lines indicate the estimation of mean growth change with uncertainties.

Decomposition of Variance

The decomposition of variance showed that climate variability will remain a major source of growth variation. There are large differences between the various growth models, indicating that our poor understanding of the growth responses is a second important source of uncertainty. Surprisingly, the choice of the climate scenario and the latitude was not such an important effect. We should also note that the decomposition of variance deals with variations of simulated growth indices. These exclude possible interactions of the environment with site factors such as soils. Furthermore, other factors such as insect outbreaks might change the growth [28].

Comparison between the A-model and F-model

Species adjustments to environmental change occur in the short term through physiological plasticity of individuals and in the long term through the evolutionary process of selection, migration, mutation, and drift [29]. With rapid warming predicted in the current century, slow maturing organisms like trees may not be able to keep pace with this change [30]. Evolution and migration may, therefore, play a minor role in the survival of tree species, whereas phenotypic variation may be the most important key process [9]. The main feature of the A-model is that it varies over time. That is, a gradual changing environment for tree growth is highlighted because more southern models involved may highlight different growth responses to climate over spatial gradient. This modeling concept, i.e., a changing environment, is similar to the provenance experiments that often transfer a northern original population (seeds or seedlings) to a more southern location to

detect its fitness compared to the local populations [31]. Compared to the provenance experiments, however, our modeling approach has the advantages of sampling at a broad spatial scale and short duration, focusing on mature trees and forests, with much less cost. Its disadvantage is that it cannot separate environmental from genetic effects like a provenance trail does. Consequently, this model might reveal an alternative scenario of growth when tree growth conditions will be changing with warming.

In contrast, the F-model is similar to the previous empirical tree-ring modeling studies that often used the local models to simulate future tree growth under climate change scenarios [5,32]. Although this approach is identical to methods that use static and linear relations of climate and tree growth to extrapolate future growth trends, its innovative aspect is the ecological interpretation of the approach, not the approach itself. Common garden experiments showed that the null transfer for most of forest tree species studied in North America is optimal for current climates in terms of temperature, suggesting that local populations generally have higher fitness than non-local populations in local climates [33,34]. Therefore, the local model established from the current stands might be the best model to quantify the growth response to a rapidly changing climate. Growth responses of trees in the model appear to be generally larger and the model seems to emphasize strong temperature responses of northern populations and to put little weight on the effects of drought on tree growth.

A common feature of these two models is that their northern forests all experience a large growth increase, but minor growth changes were predicted for southern forests. This is in agreement

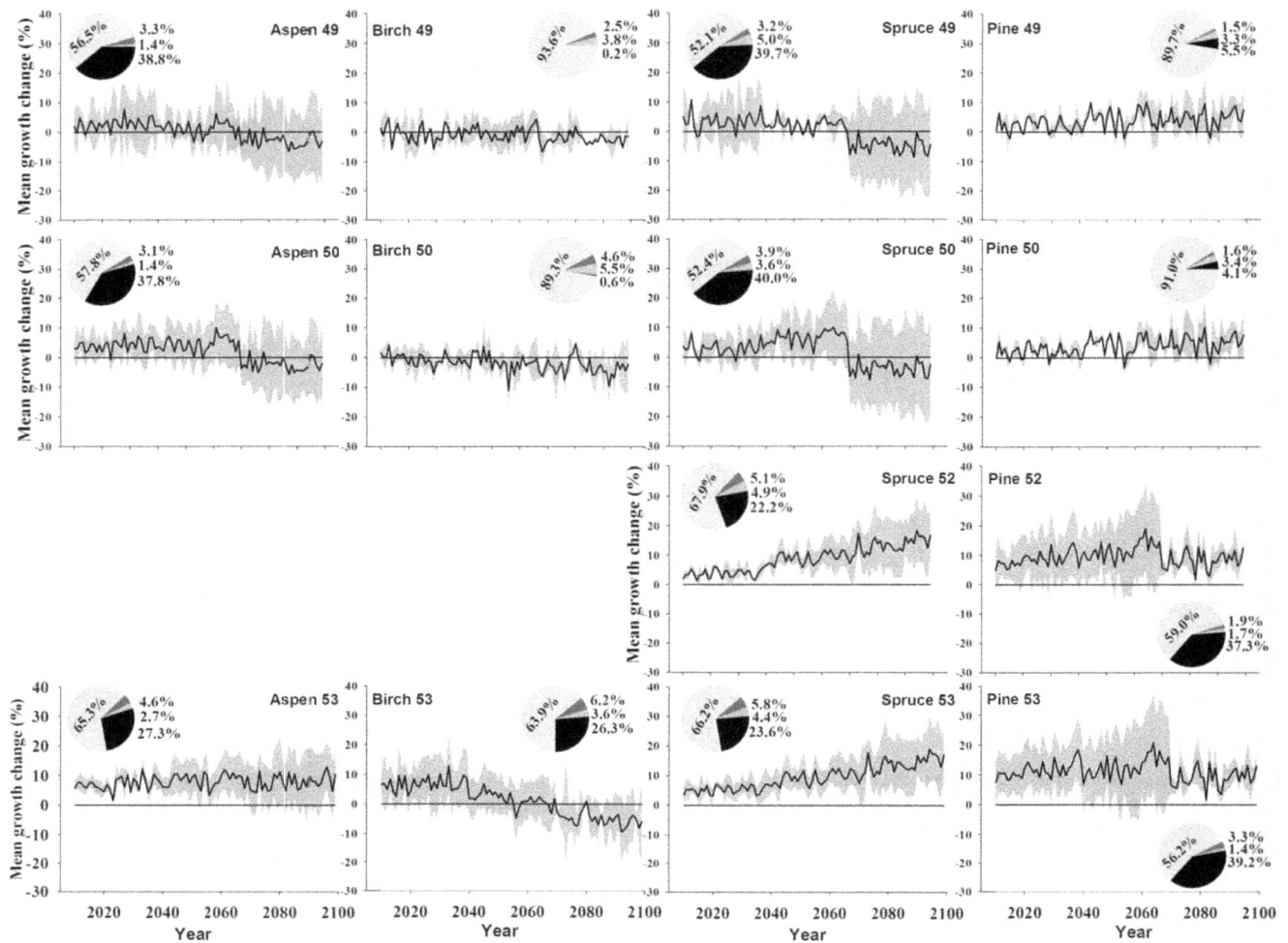

Figure 4. The predicted mean growth change of trembling aspen, paper birch, black spruce, and jack pine at 49, 50, 52 and 53°N under the A-model. The pie plots indicate the proportion of variance in mean potential growth change explained by, counter clock wise, the calibration models (dark), scenarios (grey), calibration models *scenarios (darker grey), and climate (blue). The grey zones are 95% confidence interval. White dashed lines indicate the estimation of mean growth change with uncertainties.

with the principle of the limiting factors that has been commonly used in dendrochronology [35]: with increasing latitude, temperature plays a more important role in limiting radial growth, whereas synergistic factors (temperature, precipitation, drought) are more significant in the south [36]. Hence, with climate warming, the northern stands would lose their temperature limitations, leading to a large growth increase due to extension of the growing season, earlier budbreak and growth stimulation, as well as less damage by severe cold temperatures; The southern stands would become more drought stressed, thus tending to show minor growth change [36]. Generally speaking, in the north where tree populations are occupying climates much colder than their optima, trees should have the smallest losses but more gain in growth with climate warming, whereas in the south, where the discrepancies between the inhabited and optimal environments are the least, the negative effect of a warming climate will be the greatest [33]. Rehfeldt et al. [33], for instance, predicted more gains in productivity of the *P. contorta* Doug. ex Loud. forests of British Columbia at high latitudes and elevations, which would overcompensate the losses projected for lower elevations in the south under future warming.

Simulation Output from the A-model and F-model

Future consistent growth increases in northern stands predicted by the two models are in agreement with the findings of provenance experiments [9,37] and forest growth modeling studies [21,38]. For instance, Rehfeldt *et al.* [9] reported that the immediate short-term response of *P. sylvestris* to global warming should be strongly positive for populations inhabiting severe climates. Thomson et al. [37] documented that the northern black spruce provenances in Ontario currently achieve better height growth when moved to more southern locations. Eggers et al. [38] projected the forest resources from 2000 to 2100 for 15 European countries under different climate scenarios, and observed significantly increased growth in northern Europe, but minor growth change in southern Europe.

Consistent growth decreases or minor growth change forecasted in southern stands by the two models are supported by the current provenance experiments and other modeling studies [9,39]. For example, several provenance experiments on black spruce and jack pine in Ontario revealed that their current maximum growth is achieved in the central latitudes (approximately between 45 and 48°N) [37,40]. Potential drought or heat stress -induced growth

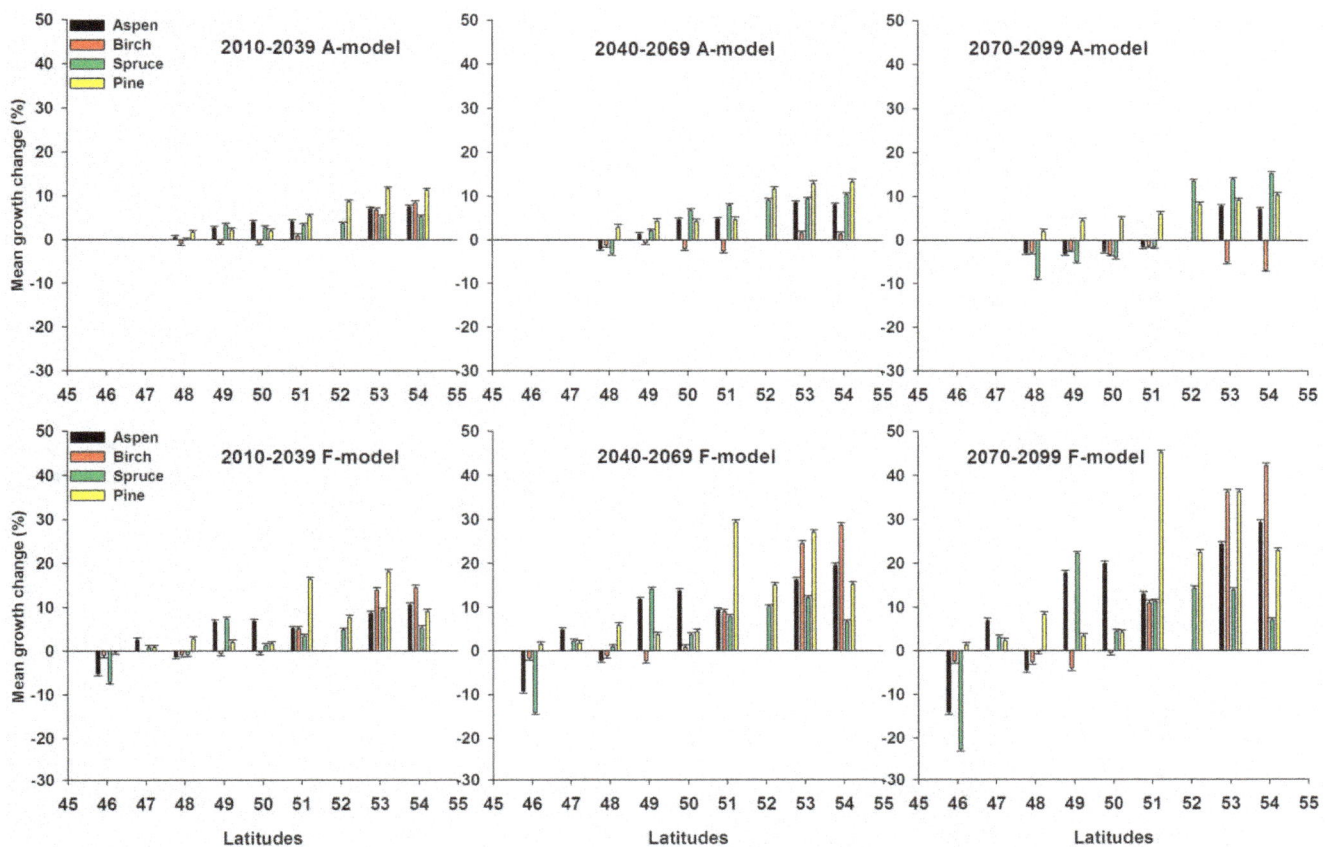

Figure 5. Comparison between average growth change of paper birch, jack pine, black spruce, and trembling aspen predicted by the A-model, and that predicted by the F-model at each latitude from 48 to 54°N over time slices 2010–2039, 2040–2069, and 2070–2099 in the eastern Canadian boreal forest. The error bars were shown by the short dashed lines.

decline will be expected with future warming [41]. Rehfeldt et al. [9] pointed out that climate warming would negatively influence the southern populations of *P. sylvestris* that currently inhabit mild climates. Our results are also supported by other studies in Europe [42] and North America [36,43] documenting drought-caused growth decline or loss in several boreal tree species in southern part of their range. Compared to the other species, consistent growth increase predicted for the southern pine stands during 2010–2099 might be due to its better drought-tolerance than others or more variance explained in the model.

Rationale of the Predicted Future MGC Under the Two Models

Under the A-model, all stands from 48 to 54°N were predicted to show moderate growth increase, then a decrease with time rather than a steep linear growth change over time. These results corroborate tree physiological studies which show that the optimum growth of trees often occurs in an intermediate climate conditions (moderately warm and moist) [44]. Therefore trees might not be able to continue to enhance growth after reaching the optimum during the latter part of the current century. Furthermore, future climate-induced changes in disturbance regimes such as insect outbreaks might also have a negative effect on growth [45]. Last, but not least, this moderate growth change calculated through means of growth estimates obtained by several linear models from different latitudes is equivalent to that estimated on the same aspen dataset by a

nonlinear model, which integrated both climate and non-climate factors [21]. Overall the A-model might indicate a lower boundary for future growth.

In contrast, simulated growth change under the F-model showed high variability and large range of growth responses across species/stands. It assumes that actually, the current populations at mid to high latitudes established after a long evolutionary process and thus are close to their optimal growth conditions [33]. This assumption has also been supported by the provenance trial studies that observed growth decrease when moving the populations growing at mid latitude to more southern latitudes [8,37,40]. The conditions for the near optimal growth would persist only in the next decades; thereafter tree growth would decrease with climate warming. The climate-growth response model established at the local stands would thus be almost the best model and may give best growth prediction during the current century. In any case, the simulated large growth change under the F-model is based on the assumption that future climate-growth relationships would be constant over time. In fact, this would be possible only during the coming decades and the models will be very likely to shift from the northern ones to more southern ones with warming, i.e., growth simulation under the A-model. Hence the F-model prediction might indicate an upper boundary for future growth. Altogether, we infer that under climate warming, potential MGC of these four species in eastern Canada might be somewhere between the two simulation results obtained by the A-model and F-model, respectively.

Mechanism of Adaptation, Acclimation and Plasticity

Aspen, birch, spruce, and pine are typically fire-adapted, wind-pollinated species, and bear high genetic variability [34,46]. The trees currently growing at the sites have genetically adapted to the local environmental and climate conditions, and would not be able to fully adapt to future climatic conditions considering the rate of changes and the pace of genetic selection [34,47]. However, they can demonstrate some growth plasticity to respond to climate warming such as early onset of xylem cell production [48,49], spring early flowering, budburst, and shoot extension [49,50]. As the next generation develops acclimation of seeds to new climatic environment and selection of best adapted genotypes will contribute to a better response to the changing climate. Common garden experiments have shown high among-population levels of genetic variation for quantitative traits associated with adaptation, geographic climatic gradients, and genotype-by-environment interactions [34], providing strong evidence of local adaptation of populations to climate [51]. Therefore, as time elapses we can predict a gradual transition between a response to climate associated with the current conditions (the F-model) to a more adapted response to the future climate (the A-model). It may take several generations and a constant climate for trees to reach an optimal response to future climate.

Model Limitations

Our simulation results might be biased by different periods of calibration for different tree species/latitudes or lower explained variance in the calibration model (adjR^2). Future predictions might be also influenced by other external factors that were not involved in the models, such as fire disturbances [16,], species competition [52], as well as the potential for a direct CO_2 fertilization effect [53] or possible nitrogen deposition [54]. Moreover, choosing a calibration model derived from the southern populations for predicting a northern population neglects the genetic differences (in contrast to the phenotypic differences). However, if the southernmost populations are already genetically adapted but the northernmost populations are not able to adapt to the warming by 2099, there would be genetic differences between the southernmost and northernmost populations. As a result, the southernmost models-based growth prediction for the northern population would have already included genetic differences in the simulation. Therefore the current approach does not conclusively allow for disentangling genotypic and phenotypic variation. Finally, future growth change prediction was based on the standardized tree-ring indices, not the absolute growth change in ring width. Actual growth can be calculated through multiplying mean annual ring-width growth during 1961–1990 with the percentage change predicted by the two models (Table S5).

Implications

Overall, the main trend demonstrated by both the A- and F-model simulation results is that stands to the north of the vegetation transition zone of 49°N would mostly increase growth, whereas those at south of 49°N would decrease growth in the current century. In terms of species, the results indicate that pine and birch might benefit the most and least, respectively, from climate warming, and spruce and aspen are intermediate species. Our results suggest that climate warming will not favour deciduous species in eastern Canada in the long term, though it may for the next few decades. This result is contrary to the general expectation of increased growth of boreal deciduous species [55]. Our results altogether further suggest that there might be a potential shift in forest composition and structure from south to north. That is, the southern boreal mixedwood forest might gradually develop into the coniferous-dominated boreal forest with warming over time. As a consequence, forest productivity might be increasing in the north but decreasing in the south. This will further have profound impacts on the sustainable forest management and boreal carbon cycles and equilibrium. Our modeling approach and concept in the study could be used as a template to investigate growth response of other tree species to climate warming at mid-high latitudes of the Northern Hemisphere, which may allow us to further quantify how forest growth, structure, and composition will respond to and shift in a future warming climate.

Supporting Information

Figure S1 Positive climate anomalies of future three subperiods 2010–2039, 2040–2069, and 2070–2099 (from bottom to top) from each model in relative to mean climate during the period 1961–1990 (See Table S4 for the reference values) along the latitudinal gradient in eastern Canada. Abbreviations: Tmax (anomalies for mean maximum temperature during each subperiod), Tmin (anomalies for mean minimum temperature during each subperiod), Precipitation (anomalies for mean total precipitation during each subperiod), drought code (anomalies for mean drought code during each subperiod). See definition for scenarios abbreviations in the text. The climate anomalies values were shown in the figures.

Figure S2 The predicted mean growth change of trembling aspen, paper birch, black spruce, and jack pine at 46–54°N under the F-model. The calibrated model for paper birch at 47°N was not established and thus the predicted mean growth change was not shown.

Table S1 General circulation models (GCM) including Canadian third-generation coupled global climate model (CGCM3), UK Hadley Centre HadCM3, Max Planck Institut für Meteorologie ECHAM4 and Canadian Regional Climate Model (CRCM3), their corresponding scenarios and the storylines (from the worst to the best) that describe the relationships between the forces driving greenhouse gas and aerosol emissions and their evolution during the 21st century applied in this study.

Table S2 Statistics of the model calibration and verification for trembling aspen, paper birch, black spruce and jack pine along the gradient. Significance level is at p<0.5. *Note*: r: correlation coefficient; R^2: explained variance; adjR^2: square of the multiple correlation coefficients following adjustment for loss of degrees of freedom; SE: standard error of the predictions; RE: reduction of error statistic, which is a measure of shared variance between the actual and modelled series, but is usually lower than the calibration R^2. A positive value signifies that the regression model has some skill [24]. PM: product means test [35]; ST: sign test [35]. The italic texts indicate insignificant values.

Table S3 The calibrated full-period climate-growth (Tree-Ring Index, TRI) models for trembling aspen, paper birch, black spruce, and jack pine along the latitudinal gradient from 46°N to 54°N. *Note*: The chronology full period and adj R^2 of each model were listed in Table S2. Monthly climate variables were abbreviations in the model, for example climate variables in May, p5p and p5 indicates precipitation in the previous and current May, respectively; tmax5p and tmax5 indicates maximum temperature in the previous and current May, respectively; tmin5p and tmin5 indicates minimum temperature in the previous and current May, respectively; dc5p and dc5 indicates drought code in the previous and current May.

Table S4   The mean climate values (±SD) during the reference period of 1961–1990 over the latitudinal gradient 46–54°N in eastern Canada. *Note*: Max T and Min T: Mean maximum, and minimum temperature during 1961–90; Precipitation: mean total annual precipitation during 1961–90; Drought code is mean drought code during 1961–90.

Table S5   Mean annual ring width [RW (±SD) mm] growth of the four species from 1961 to 1990 over the

latitudinal gradient 46–54°N in eastern Canada. *Note:* NA indicates that stands were not found at 52°N.

Acknowledgments

We thank field and lab assistants for help with this project. We also thank Dr. K. Stadt, Dr. F. Tremblay, the reviewers and the Academic editor for reviewing the manuscript. We appreciate all the contributions of Dr. Bernhard Denneler who was one of the main drivers for the project but died in 2007 because of cancer.

Author Contributions

Conceived and designed the experiments: JGH BD YB. Performed the experiments: JGH BD LHZ. Analyzed the data: JGH YB FB JT. Contributed reagents/materials/analysis tools: JGH YB FB LHZ JT. Wrote the paper: JGH YB FB LHZ JT.

References

1. Rathgeber C, Nicault A, Guiot J, Keller T, Guibal F, et al. (2000) Simulated response of *Pinus halepensis* forest productivity to climatic change and CO_2 increase using a statistical model. Global Planet Change 26: 405–421.
2. Girardin MP, Raulier F, Bernier PY, Tardif JC (2008) Response of tree growth to a changing climate in boreal central Canada: A comparison of empirical, process-based, and hybrid modelling approaches. Ecol Model 213: 209–228.
3. Running SW, Coughlan JC (1988) A general model of forest ecosystem processes for regional applications I. Hydrologic balance, canopy gas exchange and primary production processes. Ecol Model 42: 125–154.
4. Berninger F, Nikinmaa E (1997) Implications of varying pipe model relationships on Scots Pine growth in different climates. Funct Ecol 11: 146–156.
5. Chhin S, Hogg EH, Lieffers VJ, Huang SM (2008) Potential effects of climate change on the growth of lodgepole pine across diameter size classes and ecological regions. For Ecol Manag 256: 1692–1703.
6. Huang JG, Tardif J, Bergeron Y, Denneler B, Berninger F, et al. (2010) Radial growth response of four dominant boreal tree species to climate along a latitudinal gradient in the eastern Canadian boreal forest. Global Change Biol 16: 711–731.
7. Hofgaard A, Tardif J, Bergeron Y (1999) Dendroclimatic response of *Picea mariana* and *Pinus banksiana* along a latitudinal gradient in the eastern Canadian boreal forest. Can J For Res 29: 1333–1346.
8. Savva Y, Denneler B, Koubaa A, Tremblay F, Bergeron Y, et al. (2007) Seed transfer and climate change effects on radial growth of *P. banksiana* populations in a common garden in Petawawa, Ontario, Canada. For Ecol Manag 242: 636–647.
9. Rehfeldt G, Tchebakova NM, Parfenova YI, Wykoff WR, Kuzmina NA, et al. (2002) Intraspecific responses to climate in *Pinus sylvestris*. Global Change Biol 8: 912–929.
10. Körner C, Basler M (2010) Phenology under global warming. Science 327: 1461–1462.
11. Sheridan SC (2002) The redevelopment of a weather-type classification scheme for North America. Int J Climat 22: 51–68.
12. Gauthier S, De Grandpre L, Bergeron Y (2000) Differences in forest composition in two boreal forest ecoregions of Quebec. J vegetation Sci 11: 781–790.
13. Huang JG, Tardif J, Denneler B, Bergeron Y, Berninger F (2008) Tree-ring evidence extends the historic northern range limit of severe defoliation by insects in the aspen stands of western Quebec, Canada. Can J For Res 38: 2535–2544.
14. Wigley TML, Briffa KR, Jones PD (1984) On the average value of correlated time series, with applications in dendroclimatology and hydrometeorology. J Climate Applied Meteorology 23: 201–213.
15. Hutchinson MF (2004) ANUSPLIN Version 4.3. Center for Resource and Environmental Studies, Australian National University, [WWW document] URL Available: http://fennerschool.anu.edu.au/publications/software/anusplin.php Accessed: 12 September 2011].
16. Girardin MP, Wotton BM (2009) Summer moisture and wildfire risks across Canada. J Applied Meteorol Climatol 48: 517–533.
17. Flato GM, Hibler WD, Lee WG (2000) The Canadian center for climate modelling and analysis global coupled model and its climate. Climate Dyn 16: 451–467.
18. Collins M, Tett SFB, Cooper C (2001) The internal climate variability of HadCM3, a version of the Hadley Centre coupled model without flux adjustments. Climate Dyn 17: 61–81.
19. Roeckner E, Arpe K, Bengtsson L, Christoph M, Claussen M, et al. (1996) The atmospheric general circulation model ECHAM-4: model description and simulation of present-day climate. Report No. 218. Max Planck Institute for Meteorology, Hamburg, Germany.
20. Ramirez-Villegas J, Jarvis A (2010) Downscaling global circulation model outputs: The Delta method decision and policy analysis working paper No.1. Available: WWW document] URL http://www.ccafs-climate.org/downloads/docs/Downscaling-WP-01.pdf Accessed: 2 January 2013.
21. Lapointe-Garant MP, Huang JG, Gea Izquierdo G, Raulier F, Bernier P, et al. (2010) Use of tree rings to study the effect of climate change on trembling aspen in Quebec. Global Change Biol 16: 2039–2051.
22. Belsley DA, Kuh E, Welsch RE (1980) Regression diagnostics: Identifying influential data and sources of collinearity. John Wiley, New York, USA.
23. Akaike H (1974) A new look at the statistical model identification. IEEE Transactions on Automatic Control 19: 716–723.
24. Cook ER, Kairiukstis L (1990) Methods of Dendrochronology: applications in the environmental sciences. Kluwer Academic Publishers, Dordrecht, The Netherlands.
25. Cook ER, Briffa KR, Jones PD (1994) Spatial regression methods in dendroclimatology: a review and comparison of two techniques. Int J Climat 14: 379–402.
26. Holmes RL (1999) Dendrochronology Program Library. Laboratory of Tree-Ring Research, University of Arizona, Tucson, Arizona.
27. Lawler JJ, Edwards Jr TC (2006) A variance-decomposition approach to investigating multiscale habitat associations. Condor 108(1): 47–58.
28. Drobyshev I, Simard M, Bergeron Y, Hofgaard A (2010) Does soil organic layer thickness affect climate-growth relationships in the black spruce boreal ecosystem? Ecosystems 13(4): 556–574.
29. Futuyma DJ (1979) Evolutionary biology. Sinauer Associates, Inc, Sunderland, MA, 565p.
30. Ledig FT, Rehfeldt GE, Sáenz-Romero C, Flores-López C (2010) Projections of suitable habitat for rare species under global warming scenarios. Amer J Botany 97: 970–987.
31. Rehfeldt G, Tchebakova NM, Milyutin LI, Parfenova EI, Wykoff WR, et al. (2003) Assessing population response to climate in *Pinus sylvestris* and *Larix* spp. of Eurasia with climate-transfer models. Eurasian J For Res 6: 83–98.
32. Laroque CP, Smith DJ (2003) Radial-growth forecasts for five high-elevation conifer species on Vancouver Island, British Columbia. For Ecol Manag 183: 313–325.
33. Rehfeldt G, Ying CC, Spittlehouse DL, Hamilton Jr DA (1999) Genetic responses to climate in *Pinus contorta*: niche breadth, climate change, and reforestation. Ecol Monogr 69: 375–405.
34. Aitken SN, Yeaman S, Holliday JA, Wang T, Curtis-McLane S (2008) Adaptation, migration or extirpation: climate change outcomes for tree populations. Ecol Appl 1: 95–111.
35. Fritts HC (1976) Tree Rings and Climate. Academic Press, New York, NY, USA.
36. Ma ZH, Peng CH, Zhu Q, Chen H, Yu GR, et al. (2012) Regional drought-induced reduction in the biomass carbon sink of Canada's boreal forest. PNAS www.pnas.org/cgi/doi/10.1073/pnas.1111576109.
37. Thomson AM, Riddell CL, Parker WH (2009) Boreal forest provenance tests used to predict optimal growth and response to climate change: 2. *P. mariana*. Can J For Res 39: 143–153.
38. Eggers J, Lindner M, Zudin S, Zaehle S, Liski J (2008) Impact of changing wood demand, climate and land use on European forest resources and carbon stocks during the 21st century. Global Change Biol 14: 2288–2303.
39. Andalo C, Beaulieu J, Bousquet J (2005) The impact of climate change on growth of local white spruce populations in Quebec, Canada. For Ecol Manag 205: 169–182.
40. Thomson AM, Parker WH (2008) Boreal forest provenance tests used to predict optimal growth and response to climate change: 1. *P. banksiana*. Can J For Res 38: 157–170.

41. Peng CH, Ma ZH, Lei XD, Zhu Q, Chen H, et al. (2011) A drought-induced pervasive increase in tree mortality across Canada's boreal forest. Nature Climate Change doi:10.1038/nclimate1293.

42. Reich PB, Oleksyn J (2008) Climate warming will reduce growth and survival of Scots pine except in the far north. Ecol Letters 11: 588–597.

43. Matyas C (1994) Modeling climate change effects with provenance test data. Tree Physiol 14: 797–804.

44. Kramer P, Kozlowski T (1979) Physiology of woody plants. Academic Press, San Diego, CA, USA.

45. Kurz W, Dymond CC, Stinson G, Rampley GJ, Neilson ET, et al. (2008) Mountain pine beetle and forest carbon feedback to climate change. Nature 452: 987–990.

46. Gauthier S, Simon JP, Bergeron Y (1992) Genetic structure and variability in jack pine populations: effects of insularity. Can J For Res 22: 1958–1965.

47. Petit RJ, Aguinagalde I, de Beaulieu JL, Bittkau C, Brewer S, et al. (2003) Glacial refugia: hotspots but not melting pots of genetic diversity. Science 300: 1563–1565.

48. Huang JG, Bergeron Y, Zhai LH, Denneler B (2011) Variation in intra-annual radial growth (xylem formation) of *Picea mariana* (Pinaceae) along a latitudinal gradient in western Quebec, Canada. Amer J Botany 98: 792–800.

49. Zhai LH, Bergeron Y, Huang JG, Berninger F (2012) Variation in intra-annual wood formation, and foliage and shoot development of three major Canadian boreal tree species. Amer J Botany 99(5): 827–837.

50. Morin X, Lechowicz MJ, Augspurger C, O'Keefe J, Viner D, et al. (2009) Leaf phenology in 22 North American tree species during the 21st century. Global Change Biol 15: 961–975.

51. Savolainen O, Pyhäjärvi T, Knürr T (2007) Gene flow and local adaptation in trees. Annu Rev Ecol Evol Syst 38: 595–619.

52. Stadt KJ, Huston C, Coates KD, Feng Z, Dale MRT, et al. (2007) Evaluation of competition and light estimation indices for predicting diameter growth in mature boreal mixed forests. Ann For Sci 64: 477–490.

53. Huang JG, Bergeron Y, Denneler B, Berninger F, Tardif J (2007) Response of forest trees to increased atmospheric CO_2. Crit Rev Plant Sci 26: 265–283.

54. Köchy M, Wilson SD (2001) Nitrogen deposition and forest expansion in the northern Great Plains. J Ecol 89: 807–817.

55. Goldblum D, Rigg LS (2005) Tree growth response to climate change at the deciduous boreal forest ecotone, Ontario, Canada. Can J For Res 35: 2709–2718.

20

A Risk-Based Framework for Assessing the Effectiveness of Stratospheric Aerosol Geoengineering

Angus J. Ferraro*, Andrew J. Charlton-Perez, Eleanor J. Highwood

Department of Meteorology, University of Reading, Reading, United Kingdom

Abstract

Geoengineering by stratospheric aerosol injection has been proposed as a policy response to warming from human emissions of greenhouse gases, but it may produce unequal regional impacts. We present a simple, intuitive risk-based framework for classifying these impacts according to whether geoengineering increases or decreases the risk of substantial climate change, with further classification by the level of existing risk from climate change from increasing carbon dioxide concentrations. This framework is applied to two climate model simulations of geoengineering counterbalancing the surface warming produced by a quadrupling of carbon dioxide concentrations, with one using a layer of sulphate aerosol in the lower stratosphere, and the other a reduction in total solar irradiance. The solar dimming model simulation shows less regional inequality of impacts compared with the aerosol geoengineering simulation. In the solar dimming simulation, 10% of the Earth's surface area, containing 10% of its population and 11% of its gross domestic product, experiences greater risk of substantial precipitation changes under geoengineering than under enhanced carbon dioxide concentrations. In the aerosol geoengineering simulation the increased risk of substantial precipitation change is experienced by 42% of Earth's surface area, containing 36% of its population and 60% of its gross domestic product.

Editor: Vanesa Magar, Plymouth University, United Kingdom

Funding: This work is funded by the UK Natural Environment Research Council grant number NE/I528569/1. The funders had no role in study design, data collection and analysis, decision to publish, or preparation of the manuscript.

Competing Interests: The authors have declared that no competing interests exist.

* E-mail: a.j.ferraro@exeter.ac.uk

Introduction

Geoengineering by injection of aerosol into the stratosphere has been proposed as a possible countermeasure to climate warming driven by human emissions of greenhouse gases [1]. Sulphate aerosol is most commonly proposed, though other aerosol types could also be used [2–5]. Climate model simulations have suggested that stratospheric aerosol geoengineering can be used to effectively reduce Earth's global mean surface temperature [6], [7], but that it is not possible to simultaneously minimise changes in both surface temperature and precipitation [8–10].

The impacts of geoengineering are also unlikely to be regionally uniform [9–12]. Therefore some regions may benefit more from geoengineering than others, and there may potentially be some regions for which the impacts of geoengineering are more undesirable than those of unabated CO_2-driven climate change. In addition, individual regions may have different preferences on the amount of cooling required [9], [10]. Therefore, even if there is a universal global benefit associated with geoengineering, inequality of benefits could still lead to conflict [13]. There may also be diverse views on the appropriate goal for geoengineering: for example, should geoengineering be optimised to protect the most people from climate changes, or to protect key global economic regions? [10] The level of inequality in impacts will also depend on the chosen goal [11].

Another possible geoengineering scheme is the placement of reflectors in space to reduce the incoming solar radiation. This technique can be simulated in climate models by reducing the amount of solar radiation reaching top of Earth's atmosphere,

termed the total solar irradiance. For practical reasons some climate model simulations adopt this approach to represent stratospheric aerosol geoengineering. Such simulations include the 'G1' scenario of the Geoengineering Model Intercomparison Project (GeoMIP) [14]. However, the spatial distribution of a reduction in solar irradiance on the global radiation balance may not be the same as a geoengineering aerosol layer. In addition, solar dimming does not represent the radiative effects of aerosol on the stratosphere [2], which may lead to different impacts [15]. Therefore it is important to assess the extent to which solar dimming experiments are useful in quantifying the regional impacts of solar radiation.

Geoengineering can be thought of as an approach to managing climate risk [16]. The success of a geoengineering scheme could be described by the extent to which it reduces the risk of significant climate changes (though the definition of 'significant' is subjective).

In the field of epidemiology a risk-based approach is often adopted to disease treatment trials [17]. A particular view of geoengineering could be analogous to this approach. In this analogy, geoengineering is a treatment for the symptoms of a disease (elevated atmospheric CO_2 concentrations). Note that sunlight reflection geoengineering treats only the symptoms of the disease, which are in this case climatic changes including global-mean surface warming, rather than the disease itself. The success of the treatment is judged by the extent to which it reduces the risk of the planet experiencing the symptoms. This analogy has clear relevance for geoengineering policy, in that the level of risk reduction could be used to justify or prohibit the deployment of the 'treatment'.

In this paper we present a simple risk-based framework for the assessment of the regional impacts of stratospheric aerosol geoengineering. We test the framework using climate model simulations of geoengineering represented by sulphate aerosol and by solar dimming. We find that when geoengineering is represented by solar dimming, the risks associated with geoengineering are underestimated compared to the sulphate aerosol simulations.

Methods

Climate model simulations

The University of Reading Intermediate General Circulation Model (IGCM) [18] is used to simulate high-CO_2 and geoengineered climates. The model is coupled to a mixed-layer 'slab' ocean 100 m in depth. Using a 'slab' ocean allows the model to equilibrate rapidly to perturbations. Each simulation is 80 years in length and the final 65 years are analysed (thus allowing 15 years for the climate to equilibrate to radiative forcings). A 'slab' ocean, being static, needs calibration to represent the effects of the ocean circulation on heat transport. Ocean heat fluxes are calculated from the surface energy imbalance when the IGCM is run with sea surface temperatures fixed with a monthly climatology from the ERA-40 reanalysis [19]. The model is run with a spectral resolution of T42 (triangular truncation of wavenumbers greater than 42 – a horizontal resolution of approximately 2.7 degrees) with 35 vertical layers up to 0.1 hPa.

The climate model simulations are shown in Table 1. The 'Control' simulation represents a 20th Century climate. The '4CO₂' simulation has quadrupled CO_2 concentrations, representing an undesirable climate state in which substantial greenhouse gas emissions have produced global-mean surface warming (4.20 K – see Table 1). Two geoengineering simulations are used. In '4CO₂ + Sulphate' the quadrupling of CO_2 is counterbalanced by prescribing a time-invariant zonally uniform layer of sulphate aerosol in the lower stratosphere (described in Text S1 and illustrated in Figure S1). In '4CO₂ + Solar' the quadrupling of CO_2 is counterbalanced using a reduction in total solar irradiance, after the GeoMIP protocol [14].

The sulphate aerosol in '4CO₂ + Sulphate' interacts with both shortwave and longwave radiation. Representation of the effects of aerosol on the full spectrum of electromagnetic radiation is important because, though sulphate is primarily scattering at visible wavelengths, it produces non-negligible absorption at longer wavelengths [2]. It is assumed to have a lognormal size distribution with a median radius of 0.1 μm and a geometric standard deviation of 2.0, based on previous studies of stratospheric aerosol geoengineering using aerosol microphysical models [20].

Risk analysis framework

We present a novel framework for the assessment of regional climate risk in high-CO_2 and geoengineered climates, based on the probability of exceedance of a target climate threshold at a particular location in any given year. A threshold could be chosen for any climate variable of interest. In this paper we simply consider annual-mean temperature and precipitation; in the Discussion we address the possibility of including other variables.

The risk ratio is defined as:

$$RR = \frac{p_{GE}}{p_{4CO2}}$$

Where p_{GE} is the probability of exceedance of the threshold in the geoengineered climate ('4CO₂ + Sulphate' or '4CO₂ + Solar') and p_{4CO2} is the probability of exceedance of the threshold in the '4CO₂' climate. Therefore if $RR > 1$, geoengineering increases the risk of exceeding the given climate change threshold relative to the un-geoengineered '4CO₂' climate.

For illustrative purposes, in this paper the threshold is taken to be when the annual-mean climate state in a perturbed climate differs from the control climatology by greater than ± 1 standard deviation (σ) of the interannual variability. We assume that climate damages associated with this change do not depend on the sign of this change, i.e. that a negative change in a given climate variable is as undesirable as a positive change [10]. This is not necessarily the most appropriate approach for considering the impacts of climate change, since some regions may be more sensitive to climate changes of a certain sign. The possibility of incorporating sign-sensitivity into the framework is addressed in the Discussion.

The annual probability of exceedance is calculated at each spatial point as the fraction of years exceeding the ± 1 standard deviation threshold.

A risk ratio greater than 1 has two possible implications: either that geoengineering enhances the magnitude of climate change caused by a quadrupling of CO_2, or that the geoengineering produces substantial climate changes where there were none under a quadrupling of CO_2. In the former case geoengineering has exacerbated the existing climate risk caused by CO_2, whereas in the latter geoengineering has introduced climate risk in a region where there was none under a quadrupling of CO_2.

A risk ratio of less than 1 also has two possible implications: that geoengineering has reduced the existing climate risk of CO_2, or that geoengineering has reduced climate risk in a region which was not at risk of substantial climate change under a quadrupling of CO_2 anyway.

Taken together there are four possible outcomes, expressed in Figure 1 as regions on a scatter plot of the probability of exceedance for 4CO₂ against the probability of exceedance for

Table 1. Climate model simulations.

Simulation name	CO₂ concentration (ppmv)	Geoengineering	Global-mean surface temperature change (K)	Global-mean precipitation change (mm/day)
Control	355	-	0	0
4CO₂	1420	-	4.20	0.20
4CO₂ + Sulphate	1420	Prescribed sulphate aerosol layer	−0.28	−0.25
4CO₂ + Solar	1420	3.4% reduction in total solar irradiance	0.10	−0.10

geoengineering. We adopt the following definitions of the outcomes:

- **Damaging**. Risk increased in areas not at risk. In areas where a 1σ change was *less* likely than not (i.e. $p_{4CO2} < 0.5$) under $4CO_2$, geoengineering *increases* the likelihood of a 1σ change.
- **Ineffective**. Risk increased in at-risk areas. In areas where a 1σ change was *more* likely than not (i.e. $p_{4CO2} > 0.5$) under $4CO_2$, geoengineering *increases* the likelihood of a 1σ change.
- **Benign**. Risk reduced in areas not at risk. In areas where a 1σ change was *less* likely than not (i.e. $p_{4CO2} < 0.5$) under $4CO_2$, geoengineering *decreases* the likelihood of a 1σ change.
- **Effective**. Risk reduced in at-risk areas. In areas where a 1σ change was *more* likely than not (i.e. $p_{4CO2} > 0.5$) under $4CO_2$, geoengineering *decreases* the likelihood of a 1σ change.

Thus areas which experience 'damaging' and 'ineffective' changes have a risk ratio of greater than 1, and areas which experience 'benign' and 'effective' changes have a ratio ratio of less than 1. The choice of the change and likelihood thresholds depends on the application (addressed in the Discussion). However, in order to illustrate the framework clearly and generally, we adopt the simple approach outlined above.

Some apparent climate change signals may in fact be due to natural variability. This could lead to misclassification of variability-driven signals as a consequence of geoengineering. This problem occurs when the difference between perturbed (geoengineered or $4CO_2$) and control climates is small. When the probability of exceedance of the threshold is small for *both $4CO_2$ and geoengineering* (i.e. the region close to the origin in Figure 1) no conclusion can be drawn because the distinction between 'damaging' and 'benign' will be dominated by natural variability rather than forced climate changes. We therefore exclude from the analysis those regions where the response (compared to the control simulation) is not statistically significant at the 95% level in *either* the $4CO_2$ or geoengineering simulations. This step ensures that it is likely that the regions admitted to the analysis are comparing real forced signals rather than changes associated with natural variability. Figure S2 shows, within the framework presented in Figure 1, those regions where the response to geoengineering is unclassifiable according to this criterion.

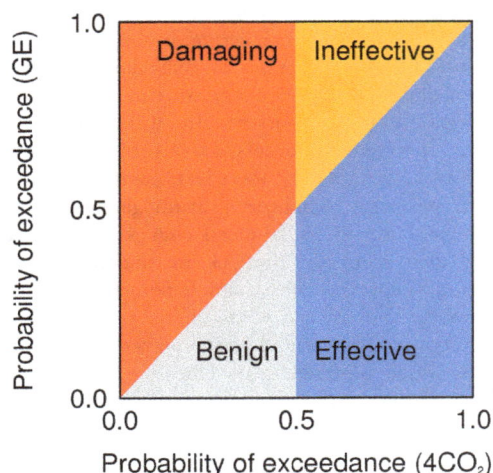

Figure 1. Matrix for classifying impacts of geoengineering (GE) by comparing its effect with a quadrupled-CO_2 scenario.

This framework assigns simple terms to the effects of geoengineering on climate. Each term also gives information on climate risk. If geoengineering is 'damaging', risk has been *introduced* where there was none before. We use the term 'damaging' since this outcome implies climate change from geoengineering in areas which might not be prepared to adapt to climate change produced by greenhouse gases, so resilience may be lower. If geoengineering is 'ineffective', risk has been increased (or sustained) in areas which were at risk from substantial climate change under the high-CO_2 scenario. If geoengineering is 'benign', risk has been decreased, but the response in the high-CO_2 is small enough that there is little risk to begin with. If geoengineering is 'effective', risk has been decreased where there is risk of severe climate change from CO_2. This choice of terms is subjective and applications of this framework to specific climate impacts may be better suited to a different set of terms.

'Ineffective' geoengineering does not necessarily imply that geoengineering has little effect on the climate variable of interest. It simply means that geoengineering has not reduced the risk of severe climate change. The climate response in the $4CO_2$+Sulphate and $4CO_2$+Solar simulations will be a combination of the responses to a quadrupling of CO_2 and geoengineering. If geoengineering is classified as 'ineffective', this implies either:

- The climate response to geoengineering is small and the response in the geoengineering simulations is dominated by the $4CO_2$ component.
- The climate response to geoengineering is large (and potentially of the opposite sign to $4CO_2$), but does not return the local climate to the control baseline.

Thus it is possible for the climate response to geoengineering to be classified as 'ineffective', while at the same time being very different to the response to a quadrupling of CO_2 alone.

Results

Here we present illustrative results of the effects of geoengineering on climate risk using the simple framework described above together with the climate model simulations of geoengineered and quadrupled-CO_2 climates. In this framework, geoengineering is broadly effective at counterbalancing regional changes in annual-mean surface temperature (Figure 2A). This is to be expected since minimisation of global-mean surface temperature change was an explicit goal of the climate model simulation. A greater area is classified as 'ineffective' in the '$4CO_2$ + Sulphate' simulation, indicating more regional inhomogeneity in this simulation than '$4CO_2$ + Solar'. However, nowhere does geoengineering increase the risk of 1σ changes in surface temperature where there was none before (which, under our framework, would be classified as 'damaging'). Since quadrupling CO_2 concentrations produces substantial warming everywhere, none of the spatial points are masked out as statistically insignificant (recall that small changes, when neither the responses to CO_2 or geoengineering are statistically significant, cannot be classified).

Consistent with previous climate modelling studies [8–10], our climate model simulations show that geoengineering to minimise global-mean surface temperature change cannot minimise global-mean precipitation change (Table 1). This is a robust result of the different vertical profiles of the radiative forcings of CO_2 and shortwave flux reductions [8], [12], and is seen in both geoengineering simulations. However, Figure 2B shows that the two geoengineering simulations have different effects on regional climate risk of annual-mean precipitation change. In the Equatorial and subtropical regions sulphate aerosol

A Temperature

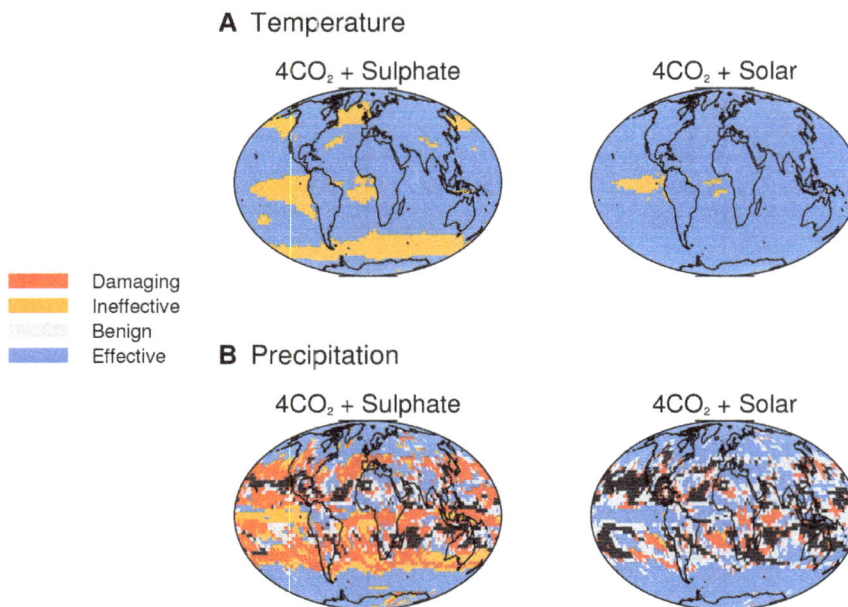

B Precipitation

Figure 2. Maps of outcomes of geoengineering. The risk-based framework (illustrated in Figure 1) is used to classify outcomes for (a) annual-mean climatological surface temperature and (b) annual-mean climatological precipitation. Black shading denotes regions where neither the response to $4CO_2$ or geoengineering are statistically significant at the 95% level (making it impossible to accurately classify the effectiveness of geoengineering).

geoengineering mostly increases climate risk from precipitation change, whereas the area over which solar dimming increases risk is much smaller. A greater area remains unclassified in the $4CO_2$+Solar case than the $4CO_2$+Sulphate case, indicating the magnitude of the regional precipitation response is generally smaller in $4CO_2$+Solar.

Most notably, however, a much larger area of the Earth is 'damaged' by geoengineering in the sulphate case than the solar dimming case (red shading in Figure 2). This indicates that geoengineering introduces risk of substantial climate changes in regions where there was no risk under a quadrupling of CO_2.

In both simulations geoengineering is effective in high-latitude regions, indicating geoengineering has reduced the climate risk of precipitation changes from a quadrupling of CO_2.

We now apply this regional analysis to potential policy-relevant metrics of sensitivity to climate change. We calculate the fraction of the global area affected by dangerous, ineffective, benign and effective geoengineering and compare this with the fraction of global population and the total GDP of the affected regions. Population data for the year 2000 are obtained from the Gridded Population of the World version 3 dataset [21] and GDP data for the year 2005 are obtained from the G-Econ dataset [22].

While some of the Earth experiences ineffective reduction in risk of surface temperature change in '$4CO_2$ + Sulphate' (Figure 2A), all of these regions are oceanic. Consequently, nearly all of the Earth's population and GDP escape this increase in climate risk (Figure 3A), as is the case for the '$4CO_2$ + Solar' simulations.

Sulphate geoengineering, however, approximately doubles the global area experiencing increased risk of substantial precipitation change when compared to the solar dimming simulation (Figure 3B). In '$4CO_2$ + Sulphate' nearly 50% of the Earth's surface area experiences this increase in risk.

A majority of the world's population experiences a reduction in the climate risk from precipitation change in both simulations. This indicates that geoengineering reduces risk for most of the

world's population, but this majority is much smaller in the '$4CO_2$ + Sulphate' case.

This distinction between sulphate and solar dimming geoengineering becomes greater when the contribution of different regions to global GDP is considered. The right-hand panel of Figure 3b shows that around 60% of the world's economic output (as measured by GDP) resides in regions in which sulphate geoengineering increases the risk of precipitation changes, and approximately 40% in regions that are damaged. In the solar dimming simulation the GDP fraction in regions with increased risk of precipitation change is approximately 15%.

Discussion

We have presented a simple, intuitive framework for describing the regional climate impacts of geoengineering. In this framework it is assumed the goal of geoengineering is the reduction of the risk of exceeding a given climate threshold in a given year, and that the sign of the change is not important. In this framework geoengineering may be considered successful if this risk is reduced.

Using this framework, we show that there is substantial regional variation in effectiveness of geoengineering in mitigating precipitation changes (in addition to global-mean changes), and that these impacts and their regional variations are underestimated when geoengineering is represented by a simple reduction in total solar irradiance rather than using a stratospheric aerosol layer. These results suggest caution is required when interpreting climate model experiments which represent geoengineering using solar dimming, and that correct representation of the stratospheric aerosol layer is required to correctly characterise the regional impacts.

Since the risk metrics presented here are based on results from a single climate model of intermediate complexity we do not suggest the specific regional impacts identified in this paper are a good measure of the potential real-world impacts of stratospheric

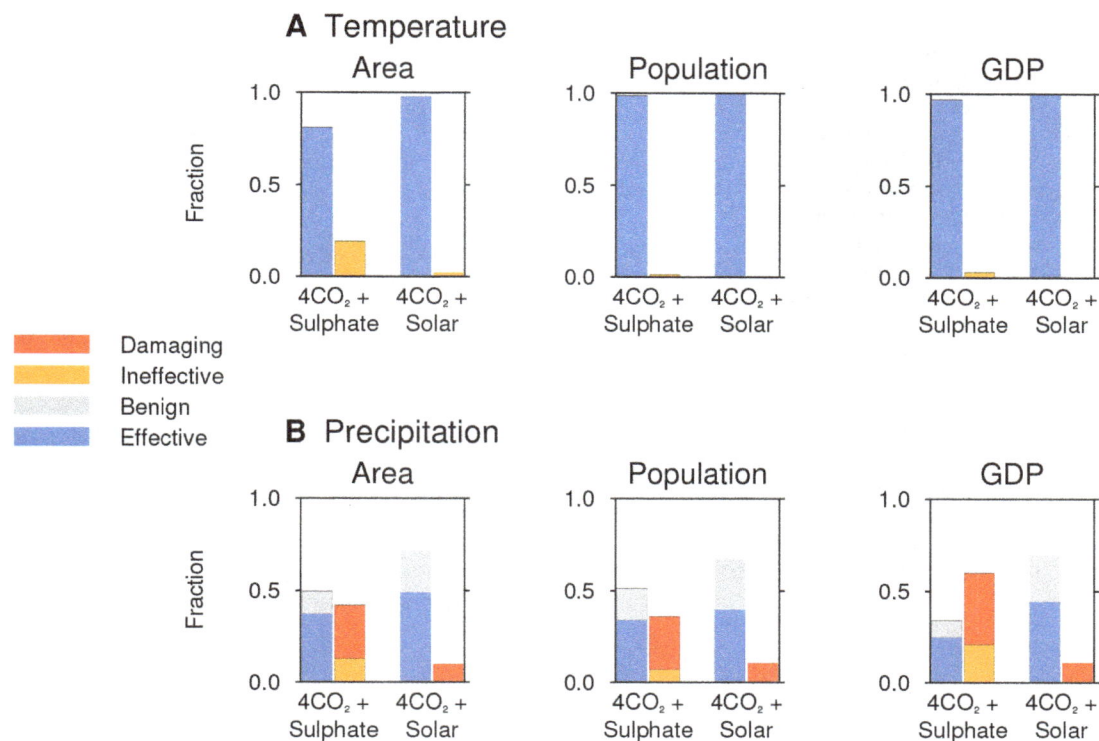

Figure 3. Fraction of global area, population and GDP affected by different outcomes of geoengineering. Each climate model simulation has a pair of bars. The left-hand bar shows the 'benign' and 'effective' outcomes, i.e. where geoengineering reduces risk. The right-hand bar shows the 'damaging' and 'ineffective' outcomes, i.e. where geoengineering increases risk. Regions where neither the response to $4CO_2$ or geoengineering are statistically significant at the 95% level are neglected, so the bars do not sum to 1.0.

aerosol geoengineering. In addition, the representation of the stratospheric aerosol used here is highly idealised, since the aerosol is not allowed to interact with the atmospheric circulation. Other climate model simulations of geoengineering might produce different results, but the simple risk framework presented here, with four clearly-defined outcomes, may be helpful in comparing simulations and assessing the robustness of regional impacts. It may also prove useful in modelling studies attempting to optimise the deployment of geoengineering to minimise negative impacts.

The global area affected by different outcomes of geoengineering in this framework is very sensitive to the chosen definition of 'substantial change'. In this paper, for illustrative purposes, we have assumed the goal of geoengineering to be a reduction in the risk of experiencing a year in which the mean surface temperature or precipitation is outside 1 standard deviation of the current interannual variability. Figures S3 and S4 show results corresponding to Figures 2 and 3 in which the chosen threshold is 2 standard deviations. In this case a large surface area experiences changes in which geoengineering is considered 'benign', but this is because the quadrupled-CO_2 simulation rarely breaches the threshold (due to large interannual variability in precipitation), and so is rarely classified as damaging. Therefore, in this framework the conclusions of the analysis depend strongly on the initial choice of threshold, and this threshold should be carefully selected.

The framework can however be used flexibly, with the goal of geoengineering and the threshold over which climatic changes are damaging chosen according to policy requirements. In addition, depending on the application, a time-resolution of greater than 1 year may also be appropriate. For example, depending on a

particular region's sensitivity, droughts may occur when there is a sustained rainfall deficit over smaller timescales [23]. The framework could also be used to assess the impact of geoengineering on climate risk depending on the season. For example, food production regions would be more sensitive to climatic changes during the growing season.

Since global-mean precipitation is reduced in geoengineering simulations, most of the regional changes are also reductions [8], [10], [12]. The regional precipitation response to carbon dioxide increase is, on the other hand, mixed. Therefore, sometimes geoengineering and carbon dioxide can act to drive similar magnitudes of climate change but of different signs. Depending on the application a weighting could be applied to the risk ratio results to reflect the potential asymmetry in the damage inflicted by increases and decreases in climate variables. Such a weighting could be applied on a regional basis to account for different regional sensitivities to climatic changes.

Multiple variables could also be incorporated into the analysis, potentially by introducing a 'loss function' at each model grid point, representing that region's sensitivity to changes in temperature, precipitation or other climate variables. In the example of food production, changes in soil moisture would be relevant. A loss function would need to describe whether a particular region's agricultural productivity was limited by water supply or by other conditions.

The framework presented in this paper could also be used as part of a cost-benefit study of geoengineering. However, a cost-benefit approach introduces further uncertainties because the conversion of changes in physical variables as simulated by a climate model into meaningful monetary costs and benefits is not straightforward. In addition, there will almost certainly be unknown climate risks associated with any kind of climate change (from greenhouse gases or geoengineering) that cannot be simulated by climate models. We therefore propose this framework primarily as a way to present climate model results in a simple and meaningful fashion, keeping in mind these models' capabilities. The framework allows comparison between results from different climate models as well as the outcomes from different levels of geoengineering. Simple metrics such as the population fraction experiencing increased climate risks could be used to calculate the optimal level of geoengineering when multiple climate variables are taken into account [10] (e.g. simultaneously minimising changes in temperature and precipitation, taking into account different regions' sensitivities).To return to the analogy of disease treatment introduced at the beginning of this paper, we see that geoengineering can be used as a treatment to alleviate the symptoms of elevated atmospheric CO_2 concentrations. However, the treatment itself carries risks, and substantial parts of the world (whether measured by area, population or economic activity) experience greater risk when the geoengineering treatment is applied than when the effects of CO_2 on their climate are unabated.

Supporting Information

Figure S1 Zonal-mean aerosol mass mixing ratio. The aerosol distribution is used in '4CO_2 + Sulphate'. Units are 10^{-6} kg/kg.

Figure S2 Scatter plot of exceedance probabilities in 4CO_2 and geoengineering (GE) scenarios. Each point represents one climate model grid box. The probabilities of exceedence are calculated as

the fraction of years in the climate model simulations exceeding 1 standard deviation of the interannual variability. Shaded regions indicate the classification of the responses according to the framework described in the main text. Black crosses indicate spatial points at which the climatological response is not statistically significant at the 95% level in either the 4CO_2 or GE scenario.

Figure S3 Maps of outcomes of geoengineering using a 2σ threshold for CO_2 changes becoming substantial. The risk-based framework (illustrated in Figure 1) is used to classify outcomes for (a) annual-mean climatological surface temperature and (b) annual-mean climatological precipitation.

Figure S4 Fraction of global area, population and GDP affected by different outcomes of geoengineering, using a 2σ threshold for CO_2 changes becoming substantial. Each climate model simulation has a pair of bars. The left-hand bar shows the 'benign' and 'effective' outcomes, i.e. where geoengineering reduces risk. The right-hand bar shows the 'damaging' and 'ineffective' outcomes, i.e. where geoengineering increases risk.

Author Contributions

Conceived and designed the experiments: AJF AJC-P EJH. Performed the experiments: AJF. Analyzed the data: AJF AJC-P EJH. Contributed reagents/materials/analysis tools: AJF AJC-P EJH. Wrote the paper: AJF.

References

1. Crutzen PJ (2006) Albedo Enhancement by Stratospheric Sulfur Injections: A Contribution to Resolve a Policy Dilemma? Clim Change
2. Ferraro AJ, Highwood EJ, Charlton-Perez AJ (2011) Stratospheric heating by potential geoengineering aerosols. Geophys Res Lett 38: L24706 doi:10.1029/2011GL049761
3. Fujii Y (2011) The role of atmospheric nuclear explosions on the stagnation of global warming in the mid 20th century. J Atmos Solar-Terrestrial Phys 73: 643–652 doi:10.1016/j.jastp.2011.01.005
4. Kravitz BS, Robock A, Shindell DT, Miller MA (2012) Sensitivity of stratospheric geoengineering with black carbon to aerosol size and altitude of injection. J Geophys Res 117: D09203 doi:10.1029/2011JD017341
5. Pope FD, Braesicke P, Grainger RG, Kalberer M, Watson IM, et al. (2012) Stratospheric aerosol particles and solar-radiation management. Nat Clim Chang 2: 713–719 doi:10.1038/nclimate1528
6. Kravitz B, Caldeira K, Boucher O, Robock A, Rasch PJ, et al. (2013) Climate model response from the Geoengineering Model Intercomparison Project (GeoMIP). J Geophys Res Atmos 118: 8320–8332 doi:10.1002/jgrd.50646
7. Rasch PJ, Tilmes S, Turco RP, Robock A, Oman L, et al. (2008) An overview of geoengineering of climate using stratospheric sulphate aerosols. Philos Trans R Soc A 366: 4007–4037 doi:10.1098/rsta.2008.0131
8. Tilmes S, Fasullo J, Lamarque J, Marsh DR, Mills M, et al. (2013) The Hydrological Impact of Geoengineering in the Geoengineering Model Intercomparison Project (GeoMIP). J Geophys Res Atmos (in press) doi:10.1002/jgra.50868
9. Ricke KL, Morgan MG, Allen MR (2010) Regional climate response to solar-radiation management. Nat Geosci 3: 537–541 doi:10.1038/ngeo915
10. Moreno-Cruz JB, Ricke KL, Keith DW (2011) A simple model to account for regional inequalities in the effectiveness of solar radiation management. Clim Change 110: 649–668 doi:10.1007/s10584-011-0103-z
11. Irvine PJ, Ridgwell A, Lunt DJ (2010) Assessing the regional disparities in geoengineering impacts. Geophys Res Lett 37: L18702 doi:10.1029/2010GL044447
12. Robock A, Oman L, Stenchikov GL (2008) Regional climate responses to geoengineering with tropical and Arctic SO2 injections. J Geophys Res 113: D16101 doi:10.1029/2008JD010050

13. Ricke KL, Moreno-Cruz JB, Caldeira K (2013) Strategic incentives for climate geoengineering coalitions to exclude broad participation. Environ Res Lett 8: 014021 doi:10.1088/1748-9326/8/1/014021
14. Kravitz BS, Robock A, Boucher O, Schmidt H, Taylor KE, et al. (2011) The Geoengineering Model Intercomparison Project (GeoMIP). Atmos Sci Lett 12: 162–167 doi:10.1002/asl.316
15. Niemeier U, Schmidt H, Alterskjaer K, Kristjánsson JE (2013) Solar irradiance reduction via climate engineering - Impact of different techniques on the energy balance and the hydrological cycle. J Geophys Res Atmos doi: 10.1002/2013JD020445
16. Keith DW (2000) Geoengineering the climate: history and prospect. Annu Rev Energy Environ 25: 245–284.
17. Viera AJ (2008) Odds ratios and risk ratios: what's the difference and why does it matter? South Med J 101: 730–734 doi:10.1097/SMJ.0b013e31817a7ee4
18. Forster PMF, Blackburn M, Glover R, Shine KP (2000) An examination of climate sensitivity for idealised climate change experiments in an intermediate general circulation model. Clim Dyn 16: 833–849.
19. Uppala SM, Kallberg PW, Simmons AJ, Andrae U, Bechtold VDC, et al. (2005) The ERA-40 re-analysis. Q J R Meteorol Soc 131: 2961–3012 doi:10.1256/qj.04.176
20. Hommel R, Graf H-F (2010) Modelling the size distribution of geoengineered stratospheric aerosols. Atmos Sci Lett. doi:10.1002/asl.285
21. Center for International Earth Science Information Network (CIESIN)/Columbia University, United Nations Food and Agriculture Programme (FAO), and Centro Internacional de Agricultura Tropical (CIAT) (2005) Gridded Population of the World, Version 3 (GPWv3): Population Count Grid. Palisades, NY: NASA Socioeconomic Data and Applications Center (SEDAC). Available: http://sedac.ciesin.columbia.edu/data/set/gpw-v3-population-count. Accessed 2013 Oct 10.
22. Nordhaus WD (2006) Geography and macroeconomics: new data and new findings. Proc Natl Acad Sci U S A 103: 3510–3517 doi:10.1073/pnas.0509842103
23. Marsh T, Cole G, Wilby R (2007) Major droughts in England and Wales, 1800–2006. Weather 62: 87–93 doi:1002/wea.67

21

Beyond a Climate-Centric View of Plant Distribution: Edaphic Variables Add Value to Distribution Models

Frieda Beauregard[1], Sylvie de Blois[2]*

1 Department of Plant Science, McGill University, Sainte Anne-de-Bellevue, Quebec, Canada, **2** Department of Plant Science and McGill School of Environment, McGill University, Sainte Anne-de-Bellevue, Quebec, Canada

Abstract

Both climatic and edaphic conditions determine plant distribution, however many species distribution models do not include edaphic variables especially over large geographical extent. Using an exceptional database of vegetation plots (n = 4839) covering an extent of ~55000 km², we tested whether the inclusion of fine scale edaphic variables would improve model predictions of plant distribution compared to models using only climate predictors. We also tested how well these edaphic variables could predict distribution on their own, to evaluate the assumption that at large extents, distribution is governed largely by climate. We also hypothesized that the relative contribution of edaphic and climatic data would vary among species depending on their growth forms and biogeographical attributes within the study area. We modelled 128 native plant species from diverse taxa using four statistical model types and three sets of abiotic predictors: climate, edaphic, and edaphic-climate. Model predictive accuracy and variable importance were compared among these models and for species' characteristics describing growth form, range boundaries within the study area, and prevalence. For many species both the climate-only and edaphic-only models performed well, however the edaphic-climate models generally performed best. The three sets of predictors differed in the spatial information provided about habitat suitability, with climate models able to distinguish range edges, but edaphic models able to better distinguish within-range variation. Model predictive accuracy was generally lower for species without a range boundary within the study area and for common species, but these effects were buffered by including both edaphic and climatic predictors. The relative importance of edaphic and climatic variables varied with growth forms, with trees being more related to climate whereas lower growth forms were more related to edaphic conditions. Our study identifies the potential for non-climate aspects of the environment to pose a constraint to range expansion under climate change.

Editor: Bruno Hérault, Cirad, France

Funding: The authors acknowledge funding from the Fonds de Recherche du Québec – Natures et Technologies (www.fqrnt.gouv.qc.ca) to F. Beauregard and from the Natural Sciences and Engineering Research Council of Canada (http://www.nserc-crsng.gc.ca) to S. de Blois. The funders had no role in study design, data collection and analysis, decision to publish, or preparation of the manuscript.

Competing Interests: The authors have declared that no competing interests exist.

* E-mail: sylvie.deblois@mcgill.ca

Introduction

Climate is a strong predictor of plant species distribution at regional and continental scales, and therefore climate change is expected to lead to range shifts [1]. Models that predict the impact of climate change on plant distribution, however, often ignore the relative contribution of other potentially important environmental predictors that could limit plant species' ability to establish in areas newly within their climatic niches. When abiotic predictors other than climate are included in species distribution models, they are often those that can be interpreted across large expanses/grain sizes, such as the ones derived from digital elevation models, generalized geological characteristics, or satellite imagery [2]–[4]. Edaphic variables that are typically measured at point locations in the field and that vary at fine spatial scales, such as pH or humus characteristics, are rarely considered in regional or continental assessments, and so their contribution to distribution models relative to that of climate variables remains largely untested.

Commonly, the climatic signal when measured over broad climatic gradients is expected to override the influence of edaphic variables in distribution models [5], with only marginal gain to model fit obtained from adding edaphic data [6]. However, recent studies have also shown that including edaphic variables, along with climate variables, can greatly influence predicted species distribution even at large regional extents, with important consequences for predictions of range expansion or contraction under climate change [7]–[11]. As these studies have focused on a few woody species and grasses, it is recognized that this work needs to be extended to a larger suite of species, growth-forms, and regions since it cannot be assumed that all species would respond to climatic or edaphic gradients uniformly [10], [12], [13].

A common limitation in incorporating edaphic variables in distribution models is the availability of data over large spatial extents. Several edaphic variables are categorical in nature (e.g., drainage class or soil type) and cannot meaningfully be averaged within the large grid cells commonly used in distribution models. Ideally, species presence or absence must be recorded at the location where the edaphic variables are measured, which is often not the case when species are recorded in grid cells. As well, species occurrence records are often obtained from compiled sources such as the Global Biodiversity Information Facility (www.gbif.org) or herbaria which do not provide edaphic information.

Consequently, even if the choice of variables in predicting species distribution should ideally be based on the known ecological requirements of species [6], [7], [10], [14], [15], so that projections are more robust for new areas or time frames [16], [17], few modelling frameworks have incorporated potentially ecologically relevant sets of predictors beyond climate variables. Regions for which information about species occurrence patterns precisely matches the location of edaphic data therefore provide invaluable model systems to assess the relative contribution of climatic and edaphic conditions to plant species distribution and can improve our knowledge of factors limiting species' ranges in a climate change context. Transitional areas between major ecosystems such as that between the northern temperate forest and the boreal forest are ideally suited to test this, since they show important gradients in both climatic and edaphic conditions.

We used an ecological dataset of global significance (one of the 20 largest datasets listed in the Global index of vegetation plot databases, www.givd.info [18]) to test a series of hypotheses about the relative importance of climatic and edaphic conditions on the distribution of plant species. Here we consider edaphic variables broadly to be those describing the nature of the site's substrate, including the topographic position. This forest database covers a large geographic area in eastern North America with a diverse climate, yet with a grain size appropriate to capture the variability of edaphic features. First, we asked whether climate-only species distribution models (cSDM) have better predictive accuracy than edaphic-only models (eSDM); we also tested whether combined edaphic-climatic models (ecSDM) had improved predictions over models with only climate predictors. Second, since growth-forms are associated and adapted to climate regimes [19], we considered whether the value of including climate vs. edaphic variables varied with growth-forms (trees, shrubs and sub-shrubs, seed bearing herbaceous plants, non-seed bearing plants and lichens). The buds of trees and shrubs are exposed to harsh climatic conditions, whereas low-lying species may benefit from more sheltered conditions and micro-climates, or their perennating buds may escape unfavourable climatic conditions underground. Finally, we verified commonly-held assumptions about the relationship between biogeographic attributes of a species within the study area and model outcomes [20], [21]. We expected prevalent species in the study area to have models with lower predictive accuracy than less prevalent species because the latter would be restricted to more specific abiotic conditions. As well, since climate has a strong latitudinal gradient in the study area, we expected it to be a better predictor for species with a range boundary in the study area than for species without a range boundary.

This study shows the potential for fine scale edaphic variables to improve species distribution models even over a large regional extent and climate gradient, while also confirming the greater importance of climate at this scale in controlling distribution. It also highlights previously unrecognized relationships between climatic and edaphic predictors and growth-forms.

Materials and Methods

Study area

The study was carried out in southern Quebec, Canada (south of 52° N, Figure 1). The study area covers more than 55 million hectares. Two major vegetation zones are present: the northern temperate zone in the south, and the boreal zone in the north, with a gradient in importance of broadleaved trees in the south to evergreen trees in the north [22], [23]. Temperature follows a north/south gradient with average annual temperatures ranging from 6.5°C in the south to −4.5°C in the north. The precipitation

pattern is more longitudinal, with the western side of the province having lower annual precipitation than the east; total annual precipitation ranges from 730 mm to 1500 mm [24]. Edaphic characteristics are also diverse. The main soil types are glacial tills, clay and sandy soils from lacustrine and fluvial origins, and peat bogs and marshland with organic soils. There are several mountain ranges in the study area, which have different geologic histories and add to the diversity of soil conditions. The Canadian Shield underlies much of the study area, thus many sites have acidic bedrock. The Appalachian Mountains contain pockets of carbonate rocks (calcareous rocks, dolomites, and marbles). In the Saint Lawrence Lowlands, there are rich lacustrine deposits [22].

Data sources

Species and edaphic data were obtained from the Quebec Ministry of Natural Resources (*Ministère des Ressources naturelles du Québec*) Vegetation Database of Quebec, the original name for this database is the *Point d'Observation Écologique* (POE) [25]. This dataset is available through the Ministry upon their approval of its intended use. At 4,839 uninhabited locations within the study area, 400 m^2 circular quadrats were sampled for the presence of a set list of 314 plant species along with detailed edaphic characteristics. Some variables were obtained via photo-interpretation. The sampling was made between 1987 and 2000. A detailed description of the methods used for data collection can be found in Saucier *et al.* [26] and are briefly outlined below. Edaphic variables that we used for modelling and their units or categories are given in Tables 1 and 2. They were measured as follows: elevation (taken from a 1:20 000 topographic map), relative height (height of the site relative to the surrounding areas taken from a 1:250 000 topographic map, or from a 1:15 000 or 1:40 000 aerial photo if available, or in the field), slope position (assessed in the field), slope angle (measured with a clinometer at the centre of the plot), microtopography (assessed visually in the field with a pictogram key), drainage (assessed in the field based on natural drainage class keys following the Canadian system of soil classification), speckling (indicative of soil saturation, assessed visually within the soil profile), parent material (assessed visually in the field with keys), texture of the B horizon (using tactile tests in the field to match textural classes from the Canadian system of soil classification), humus type (assessed in the field using keys), humus depth (assessed in the field visually and by touch, up to a depth of 1 m), and soil profile depth (assessed in the field visually to the depth of the start of the BC horizon, or the C horizon if present). pH was measured in the field using a Hellige-Truog test kit for several horizon depths; we grouped and averaged these into: pH of surface horizons (all humus layers and A horizons for which pH was measured, except for the eluviated A horizon, which was not consistently taken), pH of B horizons (all B horizons measured above transitional BC horizons), and the pH of the BC or C horizon (the pH taken at the deepest level). Some potentially relevant variables in the database (e.g., depth of the water table, exposure) were not included because of a lack in consistency in their measurements or because the measure included combinations of continuous and categorical values.

All the species modelled are native to the study area and include bryophytes and lichens. Out of the 314 surveyed species, 128 met our requirements of having at least 100 occurrences, and so were retained for modelling (Table S1). This minimum sample size level was chosen because species distribution model accuracy has been shown to decrease with less than 100 occurrences [27], [28].

Thirteen climate variables (Table S2) for the period 1961–1990 were obtained from the USDA Forest Service Rocky Mountain Research station website (http://forest.moscowfsl.wsu.edu). They

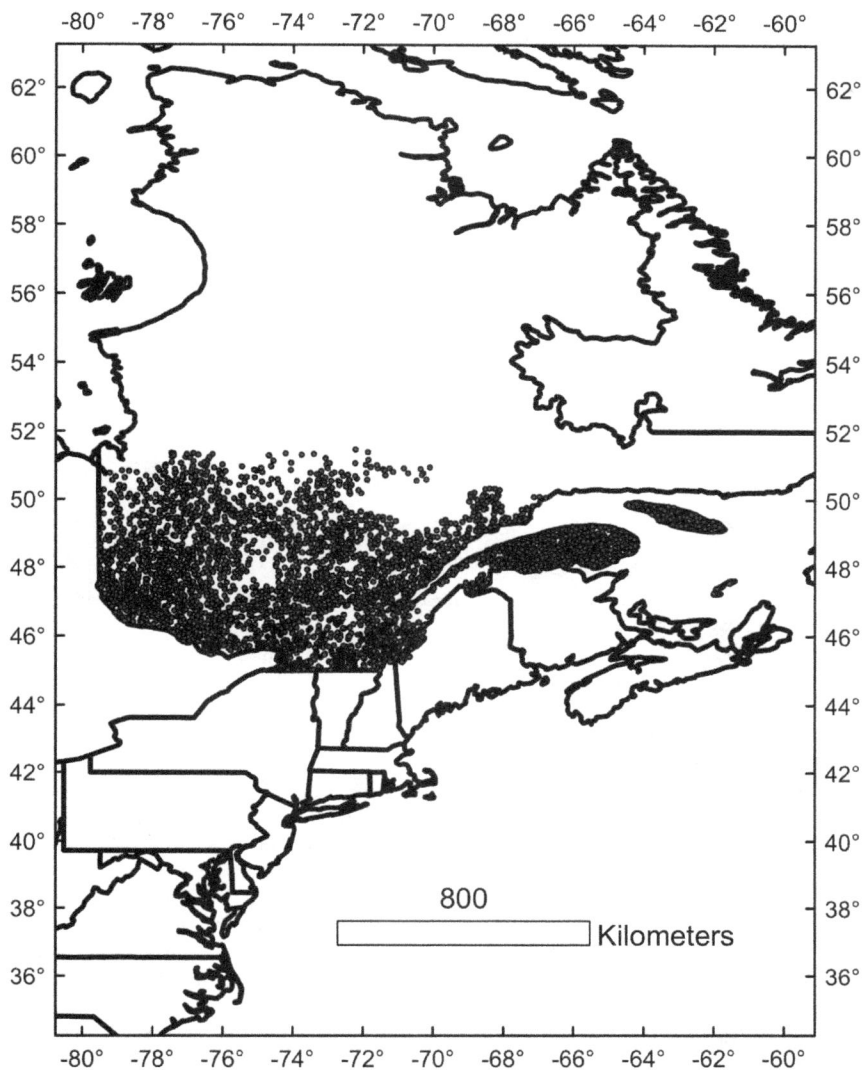

Figure 1. Map of study area showing sampling point locations.

Table 1. Continuous variables used in modelling.

Variable name	Form	Mean	SD
Humus depth (cm)	Edaphic	9.9	6.9
Soil depth (cm)	Edaphic	32	15.7
pH surface horizons	Edaphic	4.4	0.7
pH B horizons	Edaphic	5.8	0.8
pH BC or C horizon	Edaphic	6.7	0.7
Elevation (m)	Edaphic	338	151
Slope angle (%)	Edaphic	11	11.2
Degree days (accumulated first to last frost, 5°C base)	Climatic	1341	254
Minimum temperature (°C)	Climatic	−21	3.6
Precipitation (accumulated April to September, mm)	Climatic	560	58

Table 2. Categorical variables used in modelling.

Variable name[†]	Categories (count)
Relative height	Higher (1000), level (1940), lower (1898)
Humus type	Peat (310), mull (184), moder (1180), mor (3164)
Slope position	Flat (892), depression (364), plateau (848), mid-slope (2075), crest (122), upper slope (457), lower slope (80)
Microtopography	Even (1590), uneven (2140), very uneven (895)
Speckling	Colouration from good aeration (2148), colouration from continuous saturation (2148), colouration from alternating flooded and dry conditions (691)
Texture B horizon	Sand (328), sandy loam (733), loam (1325), loamy sand (1062), clay (154), clay loam (263), loamy clay (890), loamy sandy clay (83)
Parent material origin	Glacial (3015), glacio-fluvial (458), fluvial (55), marine (259), estuarine (42), laucustrine (528), colluvial (262), bed rock close to the surface (102)
Drainage	Excessively drained or somewhat excessively drained (62), well drained (1362), moderately well drained (2254), somewhat poorly drained (939), poorly drained or very poorly drained (221)

[†]All are edaphic variables

were produced using Hutchinson's ANUSPLIN software, which uses a digital elevation model and creates thin-plate spline interpolations of weather station data [29]. These surfaces were produced for North America from data from 11,757 weather stations, 471 of these were within our study area. Full details on the methods used for creation of these surfaces are available in Rehfeldt [30]. Climate data have a resolution of 0.0083 decimal degrees ($\approx$1 km). The sampling points of the POE data were matched to this grid, and the climate data was joined to the other environmental variables for each data point.

Data preparation

Preliminary analysis revealed that climate variables relating to temperature were highly correlated with each other, as were indices derived from precipitation and moisture (such as precipitation-potential evapotranspiration). A pre-selection of climate variables was made to reduce this multicollinearity using the VARCLUS procedure [31] with SAS 9.2 software (SAS Institute Inc., 2008). VARCLUS is an algorithm that produces clusters of variables that have similar patterns of variation. It is an iterative process that works by splitting variables into groups based on which of the first two principal components within a group has the highest correlation, as well as by iteratively reassigning cluster membership to test the effect on this classification. We entered the 13 climatic variables (Table S2) into the clustering analysis, and selected for further analysis one variable from each cluster based on their having a high R^2 with their own cluster, a low R^2 with the other clusters, and on their biological and physiological significance for plants. Climate variables retained were growing degree days (base of 5°C), growing season precipitation (total from April to September), and minimum temperature of the coldest month.

Tests for multicollinearity were further done between all the selected variables (edaphic and climatic). Pearson correlation coefficients and R^2 statistics were calculated between all of the continuous variables. For ordinal and categorical variables, contingency tables were made and the significance of a chi square test was used to assess multicollinearity. One-way ANOVAs were performed between each categorical or ordinal variable and each continuous variable. For all tests, although there were several significant relationships at a $P<0.05$ threshold, none had a R^2 higher than 0.3, and so no further variables were removed. These analyses were made in Statistica 10.0 (Statsoft, Inc., 2010).

Modelling

Model construction. Models were constructed for each species using either climate variables, edaphic variables, or both sets of variables. We chose our modelling approach to be comparable to commonly used SDM techniques, including both model-driven and data-driven approaches to model fitting [15]. Because the choice of a statistical model can influence the result, we tested four statistical model types: generalized additive model (GAM), generalized boosted model (GBM), generalized linear model (GLMs), and Random Forest model (RF) within the BIOMOD platform [32] implemented in R 2.12.1 (R Development Core Team, 2010). For all models a version of a step-wise or iterative approach to model fitting was used, so that the final model may not have included all provided variables. For GLM and GAM, both forward and backward selections were made. GBM and RF both work by producing many models; these are weighted, such that some variables will have greater importance in predicting the outcome than others. We used Akaike's Information Criteria to compare competing model types within this fitting process. For the GLMs, quadratic terms were allowed, but interactions between model terms were not, since this would have created many possible predictors, especially since there were already many categorical variables in the analysis. For GAMs, interaction terms were also not included and degrees of freedom for smoothing were set at three, which is comparable to a quadratic response [15]. GBM and RF include interaction terms between variables because of their tree structure. Ten iterations for each species were made with 70% of the species occurrence data used to fit the models, and 30% to test the models (referred to from now on as cross-validation models); also a full data model was made with 100% of the available data.

Model predictive accuracy. Model predictive accuracy was assessed with the true skills statistic (TSS) and area under the receiver operating characteristic curve (AUC). We calculated these for both the test data for the cross-validation models and for the full data model. The TSS is similar to the Kappa statistic, however, unlike Kappa, the TSS is not as sensitive to prevalence. TSS is calculated as the sensitivity plus the specificity minus one [33], [34]. The threshold chosen for ranking a point as present or absent in order to calculate the specificity and sensitivity was based on the value that would maximise the TSS score. We consider values of TSS greater than 0.6 to be good, 0.4 to 0.6 moderate, and less than 0.4 poor ([35] adapted from [36]). The AUC is taken from the receiver operating characteristic curve which is the curve

Table 3. Summary statistics of predictive accuracy for the different models.

	Full model		Cross validation models	
Climate SDM	AUC	TSS	AUC	TSS
GAM	0.79±0.09 (50%)	0.48±0.17 (27%)	0.78±0.10 (50%)	0.48±0.18 (23%)
GBM	0.85±0.07 (75%)	0.56±0.15 (41%)	0.80±0.09 (55%)	0.50±0.16 (27%)
GLM	0.79±0.10 (49%)	0.48±0.12 (23%)	0.78±0.10 (50%)	0.45±0.12 (24%)
RF	1±0 (100%)	0.97±0.02 (100%)	0.78±0.09 (50%)	0.46±0.17 (22%)
Edaphic SDM				
GAM	0.81±0.07 (56%)	0.49±0.12 (22%)	0.78±0.07 (43%)	0.45±0.12 (13%)
GBM	0.81±0.06 (62%)	0.49±0.11 (18%)	0.77±0.07 (34%)	0.43±0.12 (8%)
GLM	0.81±0.07 (56%)	0.48±0.12 (23%)	0.78±0.07 (42%)	0.45±0.12 (12%)
RF	1±0 (100%)	1±0 (100%)	0.78±0.07 (43%)	0.44±0.12 (12%)
Edaphic-climate SDM				
GAM	0.85±0.07 (79%)	0.57±0.15 (43%)	0.83±0.08 (62%)	0.54±0.15 (37%)
GBM	0.86±0.07 (83%)	0.59±0.13 (44%)	0.83±0.07 (63%)	0.53±0.14 (34%)
GLM	0.85±0.07 (76%)	0.57±0.15 (41%)	0.82±0.08 (63%)	0.53±0.15 (36%)
RF	1±0 (100%)	1±0 (100%)	0.84±0.07 (71%)	0.55±0.14 (38%)

TSS-true skill statistic and AUC-the area under the receiver operating characteristic curve; *climate SDM*-species-distribution model using only climate variables; *edaphic SDM*-species distribution model using only edaphic predictors; *edaphic-climate SDM*-species distribution model using both edaphic and climate predictors. Reported are means for all species for each statistical model type, ± one standard deviation, and percentage of species with a AUC greater than 0.80 or a TSS greater than 0.60; for cross validation models, the mean of each species was first calculated

of sensitivity versus 1 minus the specificity for a range of probability threshold values for ranking points as present or absent. The area under the curve (or AUC) is then used as a measure of model predictive accuracy [34]. We consider AUC values greater than 0.8 to be good, between 0.6 and 0.8 to be moderate, and less than 0.6 to be poor ([37] adapted from [38]). To compare eSDMs, cSDMs and ecSDMs we made a t-test of the ten cross-validation scores of the TSS and AUC of the competing model types for each species. We also made linear regressions between the average cross-validation scores for each predictor set. An investigation of the spatial distribution of this prediction success was also made by mapping the values from the confusion matrix [34] for the different SDMs.

Variable importance estimation. To evaluate the importance of variables with the ecSDMs, the full data models for all statistical model types were used and a measure of variable importance was calculated as one minus the correlation between the model output and the model output with the variable of interest permutated [32]. This metric was chosen because it is comparable across all the statistical model types. High values indicate greater importance.

Species characteristics

We compared species based on their characteristics to test if there were general patterns in model predictive accuracy and variable importance. The characteristics considered were: (1) the presence of a range boundary within the study zone; species ubiquitous in the study area were given a status of no range boundary. Species with no observations in the north, south, east or west of the study site (or combinations) were given the status of having a range boundary. In order to determine this, histograms of species presence with latitude and longitude were made to identify clear breaks in species prevalence, so that species that had no observations within regions of the study zone could be identified and distinguished from species that were merely rare on the

landscape. In case any species had a more complex distribution that was not a north-south or east-west split, maps of each species were also examined. All of the categorisation was done prior to analysis/modelling (results not shown); (2) species prevalence in the sampled sites (a count of occurrences), and (3) plant growth forms of trees, shrubs and sub-shrubs, herbaceous seed bearing plants, or herbaceous non-seed bearing plants and lichens. The significance of these characteristics on model predictive accuracy and variable importance were tested with either one-way ANOVAs, t-tests (Welch's), or Pearson's correlations where appropriate, as well as counts of the number of species in each group with at least one important edaphic predictor.

Results

Predictive accuracies across models

When comparing all models, many had TSS and AUC values in the good range (Table 3). The ecSDM had consistently better predictive accuracy than the cSDM or eSDM for the majority of species regardless of the statistical models used. Considering both the full models and the cross-validation scores, the different statistical model types performed similarly. Of note, RF was the most sensitive to data input/over-fitting, producing full models with perfect classification of presence and absence points but having cross-validation scores that were much lower than these full model scores. Despite this, the RF models can be considered as robust as the other statistical model types, since their cross-validation scores were comparable. There was a large difference when using the TSS or AUC as an assessment of model accuracy, although the two were highly correlated. Based on the threshold levels chosen, the TSS metric produced a more conservative estimate of the number of good models than the AUC metric. Because there was no great difference between statistical models, the rest of the results are presented for the average of the four models.

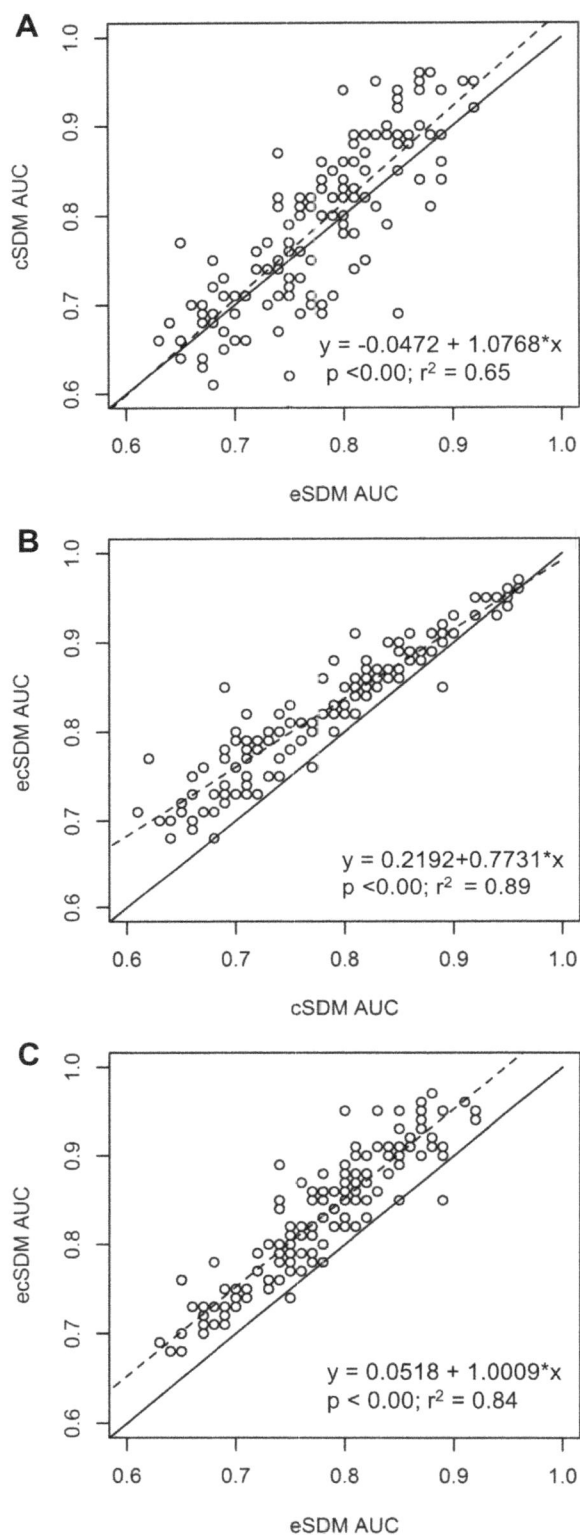

Figure 2. Comparisons of predictive accuracy of different abiotic model types through scatterplots and regressions. Average area under the curve of the receiver operator characteristic (AUC) for each species (mean of the 10 x cross-validation models of each statistical model type); solid lines have a slope of one with no intercept; dashed lines are the linear regression produced from either A) the AUC of the climate species distribution models (SDM) vs. AUC of the

edaphic SDM; B) the AUC of the edaphic-climate SDM vs. the AUC of the climate SDM; or C) the AUC of the edaphic-climate SDM vs. the AUC of the edaphic SDM. Panel A: there is a small improvement to model fit if models are constructed from climate vs. edaphic predictors; panel B: improvement is greatest for models with lower predictive accuracy; panel C: improvement is consistent across strong and weak models, and is greater than that from adding edaphic predictors to climate models.

The improvement to model predictive accuracy obtained by including the other set of abiotic predictors (Table 3, Figure 2) was greater for the eSDMs, and more species had models with higher predictive accuracy if only including climate variables than if only including edaphic variables. The pattern of improvement to model predictive accuracy is illustrated in Figure 2 for the AUC metric, the TSS metric showed similar patterns (but with lower values which are characteristic of the TSS metric).

Model predictive accuracy was generally higher for species with a range boundary within the study zone, but was not significant for the ecSDM (Table S3). As well, less prevalent species had cSDM and eSDM with higher predictive accuracy, reflected in the significant negative correlation between model predictive accuracy and prevalence, but there was no effect on the full ecSDMs (Table S3). There was a clear difference, however, in the general spatial patterns of omission and commission errors for species with a range boundary. The spatial distribution of the predictive success based on the confusion matrix is illustrated in Figures S1–S19 for those species with a north-south range boundary in the study area and an average AUC for all three sets of environmental predictor models of at least 0.90 (N = 19). The cSDMs predicted most sites as positive within the area where a species was prevalent, but few sites outside this area as positive. In contrast, the eSDM predicted sites as positive both within and outside the areas where the species was most prevalent.

We also considered the effect of plant growth form on the predictive accuracy within each set of abiotic predictors. For the ecSDM, ANOVAs of plant growth form and model predictive accuracy had significant differences (P<0.05) for both the TSS and AUC metrics, with trees having a higher mean (TSS = 0.69, AUC = 0.89), than shrubs (TSS = 0.62, AUC = 0.86), herbaceous seed bearing species (TSS = 0.59, AUC = 0.84) and non-seed bearing species and lichens (TSS = 0.58, AUC = 0.84). Tukey's honest significant difference test identified the ecSDM for trees as having a significantly higher predictive accuracy than those for herbaceous seed bearing species. Growth form did not have significant differences on cSDM and eSDM predictive accuracy based on ANOVAs.

To highlight species which had the greatest improvement by adding the edaphic data, we considered those species which had an improvement in model predictive accuracy (AUC scores) in the 90th percentile when comparing the cSDM to the ecSDM (Table 4). There was no discernible pattern to the characteristics of these species. They came from all the plant growth form types; they included species with and without range boundaries, and varied in terms of their prevalence in the study area. Of note is that three of the four *Sphagnum* species were in this group.

Variable importance within ecSDM

If considering the mean importance of predictor variables in the full models of the ecSDM, degree days and minimum temperature were by far the most important variables (Table 5). There were a few species which had high importance of specific edaphic variables (or precipitation). To estimate objectively the number of times a variable had a "high" importance, we considered the

Table 4. Species with the greatest improvement to model predictive accuracy (top 90th percentile) when adding edaphic predictors to climate only models.

Scientific name	Edge	Count	cSDM AUC	ecSDM AUC
Actaea rubra (Aiton) Willd.	+	1194	0.74	0.82
Alnus incana (L.) Moench. ssp. rugosa (Du Roi) Clausen	-	905	0.71	0.82
Athyrium filix-femina (L.) Roth	+	784	0.69	0.77
Coptis trifolia (L.) Salisb.	-	1804	0.61	0.71
Epigaea repens L.	+	213	0.70	0.79
Ilex mucronata (L.) Powell, Savolainen & Andrews	-	983	0.71	0.79
Larix laricina (Du Roi) Koch	-	222	0.62	0.77
Lycopodium obscurum L.	+	1565	0.67	0.76
Maianthemum trifolium (L.) Sloboda	+	222	0.81	0.91
Mitella nuda L.	+	198	0.79	0.88
Populus grandidentata Michx.	+	170	0.75	0.83
Prunus pensylvanica L.	+	1347	0.78	0.86
Rubus pubescens Ruf.	+	1248	0.69	0.78
Sphagnum girgensohnii Russow	+	580	0.70	0.80
Sphagnum magellanicum Brid.	+	221	0.69	0.85
Sphagnum squarrosum Crome	+	123	0.58	0.74
Thalictrum pubescens Pursh	+	116	0.66	0.75

Edge-whether or not there was a range edge within the study area; *Count*-The number of occurrence points within the dataset; *cSDM AUC*-The area under the curve of the receiver operator characteristic for the climate species distribution model; *ecSDM AUC*- The area under the curve of the receiver operator characteristic for the edaphic-climate species distribution model

distribution of variable importance across all species and all variables, and we found it to fit closely a negative exponential distribution, with 85% of variable importance values being below 0.1, and 95% of variable importance values being below 0.3. Based on the criteria of mean importance, count of importance between 0.1 and 0.3, count of variable importance greater than 0.3 and count of times a variable was one of the top two most predictive variables within the model, a clear pattern emerges (Table 5). Degree days is most important, with minimum temperature a close second and precipitation often important but less so than the temperature variables. Drainage, texture, humus type, humus depth, and surface pH are important for several species. Soil depth, pH of B horizons, pH of BC or C horizon, slope angle, relative height, micropotography, parent material origin and elevation are important for a few species. Slope position and speckling are never important.

When considering species growth form effect on variable importance, out of the 18 variables, only soil texture, soil depth and degree days had significant differences based on ANOVAs (with P<0.05, see Table S4). Trees had a marked greater importance of degree days compared to the other groups. Although it was not usually a variable with high importance, soil depth importance was significantly higher for herbaceous plants than trees or shrubs. In least significance difference tests, but not in Tukey's honest significant difference test, soil texture was significantly more important for herbaceous plants and shrubs than for trees.

Out of the 128 species, 125 had models with at least one climate variable which had a high importance (based on a variable importance of 0.1 or greater); whereas 82 species had models with at least one edaphic variable of high importance. Considering growth form, trees had the fewest species with at least one edaphic

predictor with a high importance (12 out of 30 species) and herbaceous seed bearing plants had the most (27 out of 33 species). Shrubs and seedless plants and lichens had about two thirds of species with at least one edaphic predictor having a high variable importance (21 out of 33 and 22 out of 32, respectively).

Variable importance based on t-tests (with P<0.05) was not different between the group of species with a range boundary and those without, except for degree days, which was a more important variable for species with a range boundary (means of 0.34 and 0.26, respectively), and humus depth, which was less important (means of 0.04 and 0.08, respectively). Prevalence, if significant (based on a Pearson correlation with P<0.05) was generally negatively correlated with variable importance, except for humus depth, which had a significant positive correlation (R = 0.19, P< 0.05).

Discussion

For the majority of species, temperature variables are most predictive of distribution over large geographic extents, even when grain size is suitable to capture the variation in edaphic variables. However, for some species, edaphic variables can be important predictors as well, even more so than climate predictors. Surprisingly, models made with only edaphic predictors performed almost as well as those with only climate predictors. which underlines the potential for edaphic variables to provide useful information about species distribution, even over large extents. Whereas cSDMs are definitely valuable on their own when projecting species distribution in future climate, eSDMs, or even better ecSDMs, provide useful information to help reduce the level of uncertainty of cSDMs projected into areas outside of the normal range of edaphic conditions used to train the model.

Table 5. Variable importance across all species for the full edaphic-climatic model.

	Importance (mean ± SD)	Count >0.3	Count 0.1–0.3	Count top two variables
Degree days	0.32±0.25	60	35	89
Minimum temperature	0.21±0.15	28	67	71
Precipitation	0.07±0.07	3	21	17
Humus type	0.06±0.07	2	20	16
pH surface	0.05±0.09	4	16	15
Humus depth	0.05±0.07	2	16	13
Drainage	0.04±0.07	2	17	7
Texture	0.04±0.05	0	14	9
Elevation	0.04±0.05	0	14	4
Parent material origins	0.04±0.05	1	8	5
Slope angle	0.03±0.04	0	7	0
Microtopography	0.02±0.05	1	6	3
Relative height	0.02±0.03	0	2	3
pH BC or C horizons	0.02±0.03	1	0	2
pH B horizons	0.02±0.03	0	2	1
Soil depth	0.01±0.02	0	2	1
Slope position	0.01±0.01	0	0	0
Speckling	0.01±0.01	0	0	0

Importance-variable importance calculated as one minus the correlation between the model output and the model output with the variable of interest randomized, for the means of the final climate-edaphic model for each statistical model; *Count*- the count of the number of species with a variable importance value greater than 0.3, which relates to the top 95% of variables importance values, or with a variable importance value between 0.1 and 0.3, which relates to the 85% to 95% margin of variable importance values; *Count top two variables*-count of the number of species with the variable among the top two most important in the model.

Ignoring edaphic characteristics could lead to significant overestimates of suitable conditions within a given climate or when projecting over new geographic areas (e.g., from temperate forest to boreal forest) where edaphic conditions are not equivalent to those found within the species current range. This is illustrated by the spatial distribution of errors which, despite similar predictive accuracies, was very different for the eSDMs and cSDMs (Figures S1–S19). We interpret the error pattern of the cSDM as the model being able to pick up the climatic constraints on distribution, but making many false positive predictions within the climatically suitable area. Projecting beyond the current range would lead to the same overestimate of suitable conditions. The eSDMs were able to estimate locations with suitable edaphic conditions; these fell both inside and outside the range boundaries, although with more frequency within the range boundaries. This meant that within the range boundaries, where the climate was also suitable, the model was able to accurately assign presence or absence. Outside the range boundaries, the eSDM also predicted species presences where there were suitable edaphic conditions outside of the current range/climate niche. The information garnered from the false positive locations of eSDMs could be useful to identify edaphic homologs to southern areas beyond the current range, for instance to assist migration of southern species in climate change adaptation strategies in conservation or forestry [38]. Overall, the ecSDMs were most accurate in their predictions, with many models fitting the observed distribution closely, underscoring that both edaphic and climatic aspects of the environment are important at this scale in determining species distribution.

Although all growth forms had some species with high importance of edaphic variables, distinct patterns, which are rarely emphasized, also emerged from our analysis. The distribution of trees is more constrained by climate than the distribution of other low growth forms; the latter is more related to edaphic conditions. At the landscape scale (approximately 600 ha in this case), herbaceous and shrub species were also found to be more constrained by edaphic conditions than trees [39]. Whether this means that low growth forms can escape changing climate conditions better than trees is uncertain, but their range expansion in response to climate change is expected to be more restricted by the availability of suitable edaphic conditions. On the other hand, trees, assuming they have longer life cycles than those of lower growth forms, may be more restricted in their capacity to adapt in terms of range expansion as climate warms [40]. The interactions between life history, dispersal strategy, and edaphic requirements, especially at the establishment phase, warrant further investigation in the context of climate change.

Species in the genus *Sphagnum* stood out as being better modelled if edaphic characteristics were included, drainage having a particularly high variable importance score. However, humus characteristics were also important, which brings up the issue of cause and effect between plants and their substrate. *Sphagnum* are bog species, and so are found in areas with poor drainage and thick humus layers, however, they are also the main producers of peat in the bogs in which they grow. In other words, the physical characteristics of humus may limit or promote species presence, and/or the species presence may alter humus characteristics. These complex feedback relationships may not be a problem when relating humus type or depth to the contemporary distribution of

Sphagnum, but they would need to be carefully considered when projecting future distribution of suitable conditions in time and space. There are many other examples of the interconnected nature of vegetation and substrates. For example in our study region, tree species which grow on nutrient rich soils tend to have leaves which decompose easily, thus enhancing nutrient cycling and nutrient richness [41]. These feed-back loops are numerous and complex in ecosystems, and also not fully understood. The difficulty in pulling these relationships apart adds to the uncertainty of using SDMs to make predictions about range shifts under novel climates or geographic areas. As well, climate has an influence on soil biota and chemistry [42]. It may be best to think of these model outputs as being able to highlight areas where environmental conditions would be most favourable to a species given what is known of the current edaphic state of the landscape, but that this state will be dynamic as well, at least for some edaphic variables.

Edaphic conditions may be indicative of other physical or historical processes at the stand level [39]. Within the edaphic variables, the characteristics of the humus layer were the most predictive. The humus could be an indirect measure of several other site conditions; humus characteristics are very indicative of a site's geochemistry and the organisms it supports [43], [44]. The depth of the humus layer, for instance, is often related to disturbance regime or stand age and is expected to be greater for sites that have not been burned recently. It can also be a proxy for nutrient availability, as would humus type. Slow nutrient cycling will result in humus accumulation and these conditions could limit which species would occupy a site, thereby reducing competition. In our study, humus depth was an important variable mostly for common understory species in the boreal forest, which could explain why it was also positively correlated with prevalence.

A potential bias in favour of edaphic variables exists in our study design since we compared fifteen edaphic variables to only three climate variables, thereby increasing the chances that one of the edaphic variables would be picked up in the eSDM and ecSDMs. Whereas many climate variables are available, they are generally all derived from temperature and precipitation and therefore likely to be correlated to each other. There was, however, no justification to exclude *a priori* edaphic variables based on our preliminary analysis for collinearity. Each was fairly unique in the type of information it provided about the site, and all were plausible explanatory variables. Even with the possibility for bias towards edaphic variables, our results support climate variables as being universally important, whereas which edaphic variable relates most to species distribution tended to vary with species. We were nevertheless able to identify a few edaphic variables with consistently low contribution in this region (e.g., slope position, speckling) and future modelling could benefit from this knowledge.

We have used 'edaphic' variables in a broad sense to encompass all aspects of a site's physical nature, i.e. those variables relating to topography and the soil substrate and which are most often considered in ecological studies. We could have divided our predictors into further categories, for instance based on their direct or indirect effects, although these may be hard to evaluate in absolute term. Slope angle, for instance, will determine the amount of insolation, windiness, or erosion patterns and therefore probably also integrates a range of climatic and non-climatic conditions. We have not measured microclimatic conditions in this study given the grain size we used for climatic data and so some of these conditions are probably captured in some of the edaphic variables. Another possibly relevant classification could have been to distinguish between permanent site conditions (e.g., parent material, elevation) vs. dynamic ones (e.g., surface pH, humus

type), assuming the latter will change with species and climate, and therefore could be less relevant when projecting in a future climate. Our results, however, suggest that both types of variables determine species distribution. More importantly, given the rapid rate of climate change, species will have to migrate and establish under current edaphic conditions.

Many SDMs are built within specific geo-political boundaries, often with no consideration for species' range boundaries [45], [46]. Our results support that model predictive accuracy is usually reduced if range boundaries are not included [20] (or if the species is common), but we observed that these effects were buffered if edaphic as well as climate variables were included when model predictive accuracy was assessed with the TSS, but not the AUC metric. As well, the importance of variables could be underestimated if an inadequate study zone is used, such as the lower importance of degree days for species without a range boundary observed in this study.

Conclusion

In this study, we have shown that for a large suite of species native to this area, climate variables are most important in predicting distribution at regional scale, particularly for trees. Despite this, eSDMs produced models almost equal to cSDMs in predictive performance, indicating that edaphic variables also pose important constraints on distribution patterns. The inclusion of edaphic variables in SDMs significantly improved model accuracy for the majority of species, whereas the relative importance of edaphic and climatic variables varied with growth forms. In northern ecosystems such as this one, many species reach their northern edge of distribution and northern range expansion under a future warmer climate is expected [47], [48]. Our study identifies the potential for non-climate aspects of the environment, particularly variables relating to characteristics of the humus layer, to pose a constraint to this expansion. Although some edaphic characteristics are also dynamic and both species and climate are expected to modify the substrate over time, these changes are expected to happen at a slower rate than those predicted by climate models [42] and species will have to migrate under current edaphic conditions. This could result in a decoupling between edaphic and climate conditions. Edaphic SDMs could be valuable tools to locate sites edaphically-analogous to a species' current habitat in areas that are expected to become suitable under rapid climate change. This could aid in the identification of suitable refuges for conservation and management, especially for edaphically sensitive species.

Supporting Information

Figure S1 *Acer saccharum* **Marsh. mapped distributions.** Comparison of omission and commission errors in the different forms of species distribution models (SDM). The statistical model used in these maps is the full data generalized linear model.

Figure S2 *Betula populifolia* **Marsh. mapped distributions.** Comparison of omission and commission errors in the different forms of species distribution models (SDM). The statistical model used in these maps is the full data generalized linear model.

Figure S3 *Chimaphila umbellata* **(L.) Bartram ssp.** *umbellata* **mapped distributions.** Comparison of omission and commission errors in the different forms of species distribution models (SDM). The statistical model used in these maps is the full data generalized linear model.

Figure S4 *Fagus grandifolia* Ehrh. mapped distributions. Comparison of omission and commission errors in the different forms of species distribution models (SDM). The statistical model used in these maps is the full data generalized linear model.

Figure S5 *Fraxinus americana* L. mapped distributions. Comparison of omission and commission errors in the different forms of species distribution models (SDM). The statistical model used in these maps is the full data generalized linear model.

Figure S6 *Mitchella repens* L. mapped distributions. Comparison of omission and commission errors in the different forms of species distribution models (SDM). The statistical model used in these maps is the full data generalized linear model.

Figure S7 *Onoclea sensibilis* L. mapped distributions. Comparison of omission and commission errors in the different forms of species distribution models (SDM). The statistical model used in these maps is the full data generalized linear model.

Figure S8 *Ostrya virginiana* (Mill.) Koch mapped distributions. Comparison of omission and commission errors in the different forms of species distribution models (SDM). The statistical model used in these maps is the full data generalized linear model.

Figure S9 *Polypodium virginianum* L. mapped distributions. Comparison of omission and commission errors in the different forms of species distribution models (SDM). The statistical model used in these maps is the full data generalized linear model.

Figure S10 *Populus balsamifera* L. mapped distributions. Comparison of omission and commission errors in the different forms of species distribution models (SDM). The statistical model used in these maps is the full data generalized linear model.

Figure S11 *Prunus serotina* Ehrh. mapped distributions. Comparison of omission and commission errors in the different forms of species distribution models (SDM). The statistical model used in these maps is the full data generalized linear model.

Figure S12 *Quercus rubra* L. var. *ambigua* (Gray) Fernald mapped distributions. Comparison of omission and commission errors in the different forms of species distribution models (SDM). The statistical model used in these maps is the full data generalized linear model.

Figure S13 *Solidago rugosa* Mill. mapped distributions. Comparison of omission and commission errors in the different forms of species distribution models (SDM). The statistical model used in these maps is the full data generalized linear model.

Figure S14 *Spiraea alba* du Roi mapped distributions. Comparison of omission and commission errors in the different forms of species distribution models (SDM). The statistical model used in these maps is the full data generalized linear model.

Figure S15 *Tiarella cordifolia* L. mapped distribution. Comparison of omission and commission errors in the different forms of species distribution models (SDM). The statistical model used in these maps is the full data generalized linear model.

Figure S16 *Tilia americana* L. mapped distributions. Comparison of omission and commission errors in the different forms of species distribution models (SDM). The statistical model used in these maps is the full data generalized linear model.

Figure S17 *Tsuga canadensis* (L.) Carriere mapped distributions. Comparison of omission and commission errors in the different forms of species distribution models (SDM). The statistical model used in these maps is the full data generalized linear model.

Figure S18 *Ulmus americana* L. mapped distributions. Comparison of omission and commission errors in the different forms of species distribution models (SDM). The statistical model used in these maps is the full data generalized linear model.

Figure S19 *Viburnum lantanoides* Michx. mapped distributions. Comparison of omission and commission errors in the different forms of species distribution models (SDM). The statistical model used in these maps is the full data generalized linear model.

Table S1 List of modelled species, their characteristics and model predictive accuracy. *Cross-validation scores*-are the means of four different statistical model types: generalized boosted models, generalized linear regression models, generalized additive models, and random forest models, each with ten iterations of data-splitting for model building and evaluation.; *SDM*- species distribution model; *AUC*-the area under the curve of the receiver operating characteristic; *TSS*-true skill statistic; *Edge*- indicates if the species had an observable range boundary within the study points; *Count*-the count of the number of occurrences.

Table S2 List of climate variables included in the VARCLUS analysis.

Table S3 Model predictive accuracy and the influence of species biogeographic characteristics on these metrics. *AUC*-the area under the curve of the receiver operating characteristic; *TSS*-true skill statistic; *SDM*-species distribution model. Reported are the means of the final climate-edaphic model for each statistical model type; comparisons of the presence of a range boundary in the study area (t-test) and the effect of number of occurrences (Pearson correlation) significant results are in bold, *P<0.1; **P<0.05.

Table S4 Differing importance of variables between plant growth form groups. Reported are significant ANOVA results of variable importance (calculated as one minus the correlation between the model output and the model output with the variable of interest randomized) for the means of the final climate-edaphic model for each statistical model type between the plant form groups: tree, shrub, herbaceous seed bearing, and seedless plants (including lichens).

Acknowledgments

The authors thank the *Ministère des Ressources naturelles du Québec* and the Ouranos consortium for providing this project with data, specifically the help of Catherine Périé and Travis Logan; Nicolas Casajus and Marie-Claude Lambert for statistical assistance; and Chantal Gagnon in the developmental stages of the project. The fruitful exchanges we had within the CC-BIO (climate change and biodiversity in Quebec) project also helped shape this work. The comments of four reviewers have greatly contributed to improve the original manuscript.

Author Contributions

Conceived and designed the experiments: FB SdB. Performed the experiments: FB SdB. Analyzed the data: FB. Wrote the paper: FB SdB.

References

1. Thuiller W, Lavorel S, Araujo MB, Sykes MT, Prentice IC (2005) Climate change threats to plant diversity in Europe. Proc Natl Acad Sci USA 102: 8245–8250.
2. Young N, Stohlgren T, Evangelista P, Kumar S, Graham J, et al. (2012) Regional data refine local predictions: modeling the distribution of plant species abundance on a portion of the central plains. Environ Monit Assess 184: 5439–5451.
3. Rupprecht F, Oldeland J, Finckh M (2011) Modelling potential distribution of the threatened tree species *Juniperus oxycedrus*: How to evaluate the predictions of different modelling approaches? J Veg Sci 22: 647–659.
4. Aranda SC, Lobo JM (2011) How well does presence-only-based species distribution modelling predict assemblage diversity? A case study of the Tenerife flora. Ecography 34: 31–38.
5. Heikkinen RK, Luoto M, Araújo MB, Virkkala R, Thuiller W, et al. (2006) Methods and uncertainties in bioclimatic envelope modelling under climate change. Prog Phys Geogr 30: 751–777.
6. Syphard AD, Franklin J (2009) Differences in spatial predictions among species distribution modeling methods vary with species traits and environmental predictors. Ecography 32: 907–918.
7. Coudun C, Gegout JC, Piedallu C, Rameau JC (2006) Soil nutritional factors improve models of plant species distribution: an illustration with *Acer campestre* (L.) in France. J Biogeogr 33: 1750–1763.
8. Austin MP, Van Niel KP (2011) Impact of landscape predictors on climate change modelling of species distributions: A case study with *Eucalyptus fastigata* in southern New South Wales, Australia. J Biogeogr 38: 9–19.
9. Sormunen H, Virtanen R, Luoto M (2011) Inclusion of local environmental conditions alters high-latitude vegetation change predictions based on bioclimatic models. Polar Biol 34: 883–897.
10. Bertrand R, Perez V, Gegout JC (2012) Disregarding the edaphic dimension in species distribution models leads to the omission of crucial spatial information under climate change: the case of *Quercus pubescens* in France. Glob Chang Biol 18: 2648–2660.
11. Dubuis A, Giovanettina S, Pellissier L, Pottier J, Vittoz P, et al. (2013) Improving the prediction of plant species distribution and community composition by adding edaphic to topo-climatic variables. J Veg Sci 24: 593–606.
12. Hanspach J, Kühn I, Pompe S, Klotz S (2010) Predictive performance of plant species distribution models depends on species traits. Perspect Plant Ecol Evol Syst 12: 219–225.
13. Thuiller W (2013) On the importance of edaphic variables to predict plant species distributions - limits and prospects. J Veg Sci 24: 591–592.
14. Marage D, Gegout JC (2009) Importance of soil nutrients in the distribution of forest communities on a large geographical scale. Glob Ecol Biogeogr 18: 88–97.
15. Franklin J (2009) Mapping species distributions: spatial inference and prediction. Cambridge: Cambridge University Press. 338 p.
16. Elith J, Leathwick JR (2009) Species distribution models: ecological explanation and prediction across space and time. Annu Rev Ecol Evol Syst 40: 677–697.
17. Austin MP, Van Niel KP (2011) Improving species distribution models for climate change studies: variable selection and scale. J Biogeogr 38: 1–8.
18. Dengler J, Jansen F, Glöckler F, Peet RK, De Cáceres M, et al. (2011) The Global Index of Vegetation-Plot Databases (GIVD): a new resource for vegetation science. J Veg Sci 22: 582–597.
19. Olson DM, Dinerstein E, Wikramanayake ED, Burgess ND, Powell GVN, et al. (2001) Terrestrial Ecoregions of the World: A New Map of Life on Earth. Bioscience 51: 933–938.
20. Lobo JM, Jiménez-Valverde A, Real R (2008) AUC: a misleading measure of the performance of predictive distribution models. Glob Chang Biol 17: 145–151.
21. Chambers D, Périé C, Casajus N, de Blois S (2013) Challenges in modelling the abundance of 105 tree species in eastern North America using climate, edaphic, and topographic variables. For Ecol Manage 291: 20–29.
22. Rousseau C (1974) Géographie floristique du Québec-Labrador: distribution des principales espèces vasculaires. Quebec City: Presses de l'Université Laval. 798 p.
23. Saucier JP, Robitaille A, Bergeron JF (2003) Vegetation zones and bioclimatic domains in Quebec. Quebec City: *Ministère des Ressources naturelles du Québec*. 2 p.
24. Philips DW, Sanderson M (1991) Canada precipitation. Ottawa: National Atlas Information Service, Geographical Services Division, Centre for Mapping, Energy, Mines and Resources. Map document.
25. Gosselin J, Major M (2012) Vegetation Database of Québec (MRNF). In: Dengler J, Oldeland J, Jansen F, Chytrý M, Ewald J, et al., editors. Vegetation databases for the 21st century. Biodivers Ecol 4: 432–432.
26. Saucier JP, Berger JP, D'Avignon H, Racine P, Robitaille A, et al. (1994) *Le point d'observation écologique*. Quebec City: *Ministère des Ressources naturelles du Québec*. 124 p.
27. Stockwell DRB, Peterson AT (2002) Effects of sample size on accuracy of species distribution models. Ecol Modell 148: 1–13.
28. Wisz MS, Hijmans RJ, Li J, Peterson AT, Graham CH, et al. (2008) Effects of sample size on the performance of species distribution models. Divers Distrib 14: 763–773.
29. Hutchinson MF (2002) ANUSPLIN Version 4.2 User's Guide. Canberra, Australia: Australian National University, Centre for Resource and Environmental Studies. 50 p.
30. Rehfeldt GE (2006) A spline model of climate for the Western United States. Fort Collins, Colorado: United States Department of Agriculture, Forest Service.
31. Nelson BD (2001) Variable reduction for modeling using PROC VARCLUS. Cary, North Carolina: Proceddings of the twenty-sixth annual SAS Users Group International Conference.
32. Thuiller W, Lafourcade B, Engler R, Araújo MB (2009) BIOMOD—a platform for ensemble forecasting of species distributions. Ecography 32: 369–373.
33. Landis JR, Koch GG (1977) The measurement of observer agreement for categorical data. Biometrics 33: 159–174.
34. Fielding AH, Bell JF (1997) A review of methods for the assessment of prediction errors in conservation presence/absence models. Environ Conserv 24: 38–49.
35. Jones CC, Acker SA, Halpern CB (2010) Combining local- and large-scale models to predict the distributions of invasive plant species. Ecol Appl 20: 311–326.
36. Araujo MB, Pearson RG, Thuiller W, Erhard M (2005) Validation of species-climate impact models under climate change. Glob Chang Biol 11: 1504–1513.
37. Swets J (1988) Measuring the accuracy of diagnostic systems. Science 240: 1285–1293.
38. Wang T, Campbell EM, O'Neill GA, Aitken SN (2012) Projecting future distributions of ecosystem climate niches: Uncertainties and management applications. For Ecol Manage 279: 128–140.
39. de Blois S, Domon G, Bouchard A (2001) Environmental, historical, and contextual determinants of vegetation cover: a landscape perspective. Landsc Ecol 16: 421–436.
40. Lenoir J, Gegout JC, Marquet PA, de Ruffray P, Brisse H (2008) A significant upward shift in plant species optimum elevation during the 20th century. Science 320: 1768–1771.
41. Cote B, Fyles J (1994) Nutrient concentration and acid–base status of leaf litter of tree species characteristic of the hardwood forest of southern Quebec. Can J For Res 24: 192–196.
42. Lafleur B, Paré D, Munson AD, Bergeron Y (2010) Response of northeastern North American forests to climate change: Will soil conditions constrain tree species migration? Environ Rev 18: 279–289.
43. Ponge JF (2003) Humus forms in terrestrial ecosystems: a framework to biodiversity. Soil Biol Biochem 35: 935–945.
44. Ascher J, Sartori G, Graefe U, Thornton B, Ceccherini M, et al. (2012) Are humus forms, mesofauna and microflora in subalpine forest soils sensitive to thermal conditions? Biol Fertil Soils 48: 709–725.
45. Mateo RG, Croat TB, Felicísimo AM, Muñoz J (2010) Profile or group discriminative techniques? Generating reliable species distribution models using pseudo-absences and target-group absences from natural history collections. Divers Distrib 16: 84–94.
46. Attorre F, Alfò M, De Sanctis M, Francesconi F, Valenti R, et al. (2011) Evaluating the effects of climate change on tree species abundance and distribution in the Italian peninsula. Appl Veg Sci 14: 242–255.
47. Berteaux D, de Blois S, Angers JF, Bonin J, Casajus N, et al. (2010) The CC-Bio Project: Studying the effects of climate change on Quebec biodiversity. Divers 2: 1181–1204.
48. McKenney DW, Pedlar JH, Rood RB, Price D (2011) Revisiting projected shifts in the climate envelopes of North American trees using updated general circulation models. Global Change Biol. 17: 2720–2730.

Permissions

List of Contributors

Susan K. Skagen
United States Geological Survey, Fort Collins Science Center, Fort Collins, Colorado, United States of America

Valerie Steen
United States Geological Survey, Fort Collins Science Center, Fort Collins, Colorado, United States of America
Department of Fish, Wildlife, and Conservation Biology, Colorado State University, Fort Collins, Colorado, United States of America
Graduate Degree Program in Ecology, Colorado State University, Fort Collins, Colorado, United States of America

Barry R. Noon
Department of Fish, Wildlife, and Conservation Biology, Colorado State University, Fort Collins, Colorado, United States of America
Graduate Degree Program in Ecology, Colorado State University, Fort Collins, Colorado, United States of America

Alicia Torregrosa
Western Geographic Science Center, United States Geological Survey, Menlo Park, California, United States of America

Maxwell D. Taylor
Contractor, Western Geographic Science Center, United States Geological Survey, Menlo Park, California, United States of America

Lorraine E. Flint and Alan L. Flint
California Water Science Center, United States Geological Survey, Sacramento, California, United States of America

Carlos Gay-García and Benjamín Martínez-López
Department of Atmospheric Sciences, Centro de Ciencias de la Atmó sfera, Universidad Nacional Autónoma de México, Mexico City, Mexico

Francisco Estrada
Department of Atmospheric Sciences, Centro de Ciencias de la Atmósfera, Universidad Nacional Autónoma de México, Mexico City, Mexico
Department of Environmental Economics, Institute for Environmental Studies, Vrije Universiteit Amsterdam, The Netherlands

Pierre Perron
Department of Economics, Boston University, Boston, Massachusetts, United States of America

Louise Morin and Agathe Leriche
Commonwealth Scientific and Industrial Research Organisation, Ecosystem Sciences, Canberra, Australian Capital Territory, Australia

Darren J. Kriticos
Commonwealth Scientific and Industrial Research Organisation, Ecosystem Sciences, Canberra, Australian Capital Territory, Australia
International Science & Technology Policy & Practice, Department of Applied Economics, University of Minnesota, St. Paul, Minnesota, United States of America

Robert C. Anderson
University of Hawai'i, College of Tropical Agriculture and Human Resources, Honolulu, Hawaii, United States of America

Peter Caley
Commonwealth Scientific and Industrial Research Organisation, Mathematics Informatics and Statistics, Canberra, Australian Capital Territory, Australia

Sergio A. Estay, Roger D. Sepulveda and Leonardo D. Bacigalupe
Instituto de Ciencias Ambientales y Evolutivas, Facultad de Ciencias, Universidad Austral de Chile, Valdivia, Chile

Fabio A. Labra
Centro de Investigación e Innovación para el Cambio Climático, Facultad de Ciencias, Universidad Santo Tomas, Santiago, Chile

Heather A. Hager, Sarah E. Sinasac and Jonathan A. Newman
School of Environmental Sciences, University of Guelph, Guelph, Ontario, Canada

Zéev Gedalof
Department of Geography, University of Guelph, Guelph, Ontario, Canada

Brian Stone Jr and Dana Habeeb
School of City and Regional Planning, Georgia Institute of Technology, Atlanta, Georgia, United States of America

Jason Vargo
Center for Sustainability and the Global Environment, University of Wisconsin-Madison, Madison, Wisconsin, United States of America

Peng Liu, Armistead Russell, Marcus Trail and Yongtao Hu
School of Civil and Environmental Engineering, Georgia Institute of Technology, Atlanta, Georgia, United States of America

Anthony DeLucia
Quillen College of Medicine, East Tennessee State University, Johnson City, Tennessee, United States of America

Kylie B. Ireland and Giles E. St. J. Hardy
Cooperative Research Centre for National Plant Biosecurity, Canberra, Australian Capital Territory, Australia
Centre for Phytophthora Science and Management, School of Veterinary and Life Sciences, Murdoch University, Perth, Western Australia, Australia

Darren J. Kriticos
Cooperative Research Centre for National Plant Biosecurity, Canberra, Australian Capital Territory, Australia
Commonwealth Scientific and Industrial Research Organisation (CSIRO)Ecosystem Sciences, Canberra, Australian Capital Territory, Australia

Barry Sinervo
Department of Ecology and Evolutionary Biology, University of California Santa Cruz, Santa Cruz, California, United States of America

Alison R. Davis Rabosky
Department of Ecology and Evolutionary Biology, University of California Santa Cruz, Santa Cruz, California, United States of America
Department of Integrative Biology and Museum of Vertebrate Zoology, University of California, Berkeley, California, United States of America
Department of Ecology and Evolutionary Biology, University of Michigan, Ann Arbor, Michigan, United States of America

Ammon Corl
Department of Ecology and Evolutionary Biology, University of California Santa Cruz, Santa Cruz, California, United States of America
Department of Evolutionary Biology, Evolutionary Biology Centre, Uppsala University, Uppsala, Sweden

Heather E. M. Liwanag
Department of Ecology and Evolutionary Biology, University of California Santa Cruz, Santa Cruz, California, United States of America
Department of Biology, Adelphi University, Garden City, New York, United States of America

Yann Surget-Groba
Department of Ecology and Evolutionary Biology, University of California Santa Cruz, Santa Cruz, California, United States of America
Ecological Evolution Group, Xishuangbanna Tropical Botanical Garden, Menglun, Mengla, Yunnan, P. R. China

Guangcai Duan, Rongguang Zhang and Weidong Zhang
Department of Epidemiology, College of Public Health, Zhengzhou University, Zhengzhou, Henan, China

Huifen Feng
Department of Epidemiology, College of Public Health, Zhengzhou University, Zhengzhou, Henan, China
Department of Infectious Diseases, the Fifth Affiliated Hospital of Zhengzhou University, Zhengzhou, Henan, China

James I. Watling, David N. Bucklin, Carolina Speroterra and Frank J. Mazzotti
Ft Lauderdale Research and Education Center, University of Florida, Ft Lauderdale, Florida, United States of America

Laura A. Brandt
U.S. Fish and Wildlife Service, Ft Lauderdale, Florida, United States of America

Stephanie S. Romañ ach
Southeast Ecological Science Center, U.S. Geological Survey, Ft Lauderdale, Florida, United States of America

Rafael D. Loyola
Department of Ecology, Universidade Federal de Goiás, Goiânia, Goiás, Brazil

Priscila Lemes, Frederico V. Faleiro and Joaquim Trindade-Filho
Graduate Program in Ecology and Evolution, Universidade Federal de Goiás, Goiânia, Goiá´s, Brazil

Ricardo B. Machado
Departament of Zoology, Universidade de Brasília, Brasília, Distrito Federal, Brazil

Farzin Shabani, Lalit Kumar and Subhashni Taylor
Ecosystem Management, School of Environmental and Rural Science, University of New England, Armidale, Australia

Rebecca Chaplin-Kramer
Natural Capital Project, Stanford University, Stanford, California, United States of America

Melvin R. George
Plant Sciences Department, University of California Davis, Davis, California, United States of America

Francisco Amorim and Sílvia B. Carvalho
CIBIO/InBIO, Research Center in Biodiversity and Genetic Resources, University of Porto, Vairão, Portugal

João Honrado
CIBIO/InBIO, Research Center in Biodiversity and Genetic Resources, University of Porto, Vairaão, Portugal
Department of Biology, Faculty of Sciences of the University of Porto, Porto, Portugal

Hugo Rebelo
CIBIO/InBIO, Research Center in Biodiversity and Genetic Resources, University of Porto, Vairão, Portugal
School of Biological Sciences, University of Bristol, Bristol, United Kingdom

Raimundo Real, David Romero, Jesús Olivero and Ana L. Márquez
Biogeography, Diversity, and Conservation Research Team, Department of Animal Biology, Faculty of Sciences, University of Malaga, Malaga, Spain,

Alba Estrada
Instituto de Investigaciónen Recursos Cinegéticos IREC, (CSIC-UCLM), Ciudad Real, Spain

Kevin R. Ford, Ailene K. Ettinger and Janneke Hille Ris Lambers
Department of Biology, University of Washington, Seattle, Washington, United States of America

Jessica D. Lundquist and Mark S. Raleigh
Department of Civil and Environmental Engineering, University of Washington, Seattle, Washington, United States of America

Chiyuan Miao and Qingyun Duan
State Key Laboratory of Earth Surface Processes and Resource Ecology, College of Global Change and Earth System Science, Beijing Normal University, Beijing, People's Republic of China

Lin Yang
State Key Laboratory of Resources and Environmental Information System, Institute of Geographical Sciences and Natural Resources Research, Chinese Academy of Sciences, Beijing, People's Republic of China

Alistair G. L. Borthwick
Department of Civil & Environmental Engineering, University College Cork, Cork, Ireland

Jian-Guo Huang, Yves Bergeron and Bernhard Denneler
Chaire industrielle CRSNG-UQAT-UQAM en Aménagement Forestier Durable, Universitédu Québec en Abitibi-Témiscamingue, Rouyn-Noranda, Québec, Canada

Frank Berninger and Lihong Zhai
Département des sciences biologiques, Université du Québecá Montréal, Montréal, Québec, Canada

Jacques C. Tardif
Centre for Forest Interdisciplinary Research (C-FIR), University of Winnipeg, Winnipeg, Manitoba, Canada

Angus J. Ferraro, Andrew J. Charlton-Perez, Eleanor J. Highwood
Department of Meteorology, University of Reading, Reading, United Kingdom

Frieda Beauregard
Department of Plant Science, McGill University, Sainte Anne-de-Bellevue, Quebec, Canada

Sylvie de Blois
Department of Plant Science and McGill School of Environment, McGill University, Sainte Anne-de-Bellevue, Quebec, Canada

Index